C0-AWW-140
3 3073 00260695 0

Biological Reactive Intermediates

Formation, Toxicity, and Inactivation

Edited by

David J. Jollow
Medical University of South Carolina, Charleston

James J. Kocsis
Thomas Jefferson University, Philadelphia

Robert Snyder
Thomas Jefferson University, Philadelphia

and

Harri Vainio
Institute of Occupational Health, Helsinki

Associate Editors

Jussi Saukkonen and Charlotte Witmer
Thomas Jefferson University, Philadelphia

and

Thomas Walle
Medical University of South Carolina, Charleston

PLENUM PRESS · NEW YORK AND LONDON

Library of Congress Cataloging in Publication Data

Main entry under title:

Biological reactive intermediates.

Includes bibliographical references and index.

1. Toxicology—Congresses. 2. Biotransformation (Metabolism)—Congresses. 3. Carcinogenesis—Congresses. I. Jollow, David J. II. Turku, Finland Yliopisto. [DNLM: 1. Biotransformation—Congresses. QV38 I603b 1975]

RA1191.B56 615'.7'04 76-55724

ISBN 0-306-30970-X

Proceedings of an International Conference on Active Intermediates: Formation, Toxicity, and Inactivation, held at the University of Turku, Turku, Finland, July 26–27, 1975

A Division of Plenum Publishing Corporation
227 West 17th Street, New York, N.Y. 10011

Printed in the United States of America

Preface

The concept that detoxication is the inevitable result of biotransformation of xenobiotic compounds by mammalian systems has undergone modification since it was first described. Indeed, despite the fact that R. T. Williams popularized the notion, he was among the first to caution that it was not possible to predict the biological activities of the resulting metabolites. It has become increasingly apparent in recent years that not only do many metabolites of drugs and other chemicals display biological activity but also in many instances these metabolites play an important role in initiating several forms of cancer and are the cause of a variety of types of toxicity. Thus it seems appropriate to collect in one volume a series of reports outlining advances in the study of the formation of chemically active intermediary metabolic products of chemicals, mechanisms of toxicity or carcinogenesis, and pathways for true detoxication of these chemicals.

The work of R. T. Williams, beginning in the late 1920s and early 1930s, marked the first concerted effort to understand the biotransformation of foreign chemicals in animals. He investigated the metabolic pathway of numerous compounds in a wide variety of animal species while training large numbers of our colleagues who have been responsible for further advances in biochemistry, pharmacology, and toxicology. Another pioneer in the study of drug metabolism, B. B. Brodie, was unable to take part in this presentation, but his significant contributions included not only laying the groundwork for the study of human drug metabolism and pharmacokinetics but also teaching many of the participants in this work. The fundamental contributions of H. Remmer to our understanding of enzyme induction and cytochrome P450 as well as his vigorous attempts to promote international scientific interchanges are also worthy of our acclaim. Certainly no one has contributed more to our understanding of metabolic activation of xenobiotics and their covalent bonding to critical cellular elements than J. A. and E. C. Miller. These workers were among the first to demonstrate the metabolism of foreign compounds *in vitro* using liver microsomal preparations, they were the first to demonstrate enzyme induction by polycyclic aromatic hydrocarbons, and they have shown that many carcinogenic and mutagenic chemicals undergo metabolism to reactive electrophilic intermediates that exert toxic effects by binding to critical sites on DNA, proteins, and other macromolecules. We are fortunate in having all of these scientists represented in this volume either directly or through their selected representatives and it is to them that we dedicate this book.

The papers published in this volume represent the proceedings of an international conference entitled *Active Intermediates: Formation, Toxicity, and Inactivation* which convened at the University of Turku, Turku, Finland, July 26–27, 1975. The conference was organized by a committee composed of Drs. J. F. Borzelleca, J. R. Gillette, O. Hänninen, D. J. Jollow (Cochairman), J. J. Kocsis, H. Remmer, R. Snyder (Cochairman), and H. Vainio (Cochairman). The chairmen wish to express their deep appreciation to Dr. Gillette for his productive efforts during the organization of the program.

The conference would not have been possible without the generous support, encouragement, and participation of a number of individuals and organizations. Among these were the National Institute of Environmental Health Sciences of the National Institutes of Health, Professor K. Hartiala and the University of Turku, the Turku University Foundation, the Ministry of Education of Finland, Thomas Jefferson University, Philadelphia, Pennsylvania, the Rohm and Haas Company, the Hoffman-LaRoche Company, the Haskell Laboratories of E. I. DuPont de Nemours and Company, and Lääke Oy, Turku, Finland.

The editors decided not to present a verbatim reproduction of the discussion that accompanied each group of papers presented at the conference. Instead, Drs. J. Saukkonen and C. M. Witmer of Thomas Jefferson University and Dr. T. Walle of the Medical University of South Carolina prepared summaries of the discussions with the intent of presenting only their salient features in a concise manner. We also wish to express our gratitude to Drs. R. T. Williams, K. J. Netter, J. A. Miller, S. Orrenius, R. Sato, and H. Gelboin for their efforts as session chairmen.

We wish to acknowledge the hospitality, the facilities, and the stimulating scientific atmosphere that we enjoyed in Turku. International cooperation and exchange of ideas should never be considered extraordinary. Communication is the lifeblood of science. That our fundamental purpose can be achieved while enjoying cordial relationships with our colleagues around the world is a fringe benefit of untold value.

D. J. J.
J. J. K.
R. S.
H. V.

Contents

I. Reactive Intermediates in Toxicology and Carcinogenesis: Role of Covalent Binding

II. Formation of Reactive Intermediates

*Poster sessions.

IV. Specific Reactive Intermediates

*Poster sessions.

VI. Reactive Intermediates in Carcinogenesis

I

Reactive Intermediates in Toxicology and Carcinogenesis: Role of Covalent Binding

1

Introduction to the Concept of Reactive Intermediates

R. T. Williams
Department of Biochemistry
St. Mary's Hospital Medical School
London W2 1PG, United Kingdom

The concept of reactive intermediates being formed during the metabolism of foreign compounds and being responsible for toxic effects and therapeutic responses has been developing since the beginning of this century. In the last 10 years or so, it has become evident that chemically reactive metabolites may be the cause of serious forms of toxicity including cellular necrosis, mutagenesis, carcinogenesis, teratogenesis, hypersensitivity, and blood disorders such as hemolytic anemia, blood dyscrasias, and methemoglobinemia.

In the early part of the century, hydroxylamines were postulated as being involved in the toxic effects of aromatic nitro and amino compounds. In 1905 Meyer (1) postulated that phenylhydroxylamine was an intermediate in the conversion of nitrobenzene to *p*-aminophenol in the body, and in 1920 Ellinger (2) claimed to have found *N*-hydroxyacetanilide in the blood of cats poisoned with acetanilide. In 1944, Channon *et al.* (3) actually isolated a hydroxylamine (4-hydroxylamino-2,6-dinitrotoluene) from the urine of rabbits receiving TNT (2,4,6-trinitrotoluene). The importance of hydroxylamine intermediates as toxic agents will be actively pursued in several papers in this symposium.

Paul Ehrlich (4) in 1909 also postulated an active intermediate during his study of arsenical drugs. He found that atoxyl (Na salt of *p*-aminophenylarsonic acid) had no trypanocidal action *in vitro* but was active *in vivo*. He therefore suggested that atoxyl, which contains pentavalent arsenic, was reduced in the body to the trivalent *p*-aminophenylarsenoxide which was active *in vivo* and *in vitro* but highly toxic to both host and parasite.

One of the most important therapeutic discoveries of this century was that of the antibacterial dyes prontosil and neoprontosil by Domagk (5) in 1935. This led to the well-known group of antibacterial drugs, the sulfonamide or "sulfa" drugs, when it was found by Tréfouel *et al.* (6) in 1935 that the activity of prontosil was due to a metabolite, sulfanilamide.

Further steps in the active intermediate story were made in 1950 by Boyland (7), who postulated arene oxides or epoxides as intermediates in the metabolism of polycyclic hydrocarbons, by the Millers (8) of Wisconsin (1947), who showed that carcinogenic azo dyes were bound to tissue proteins, and by Heidelberger and his co-workers (9), who showed the binding of carcinogenic hydrocarbons to tissues.

These earlier observations led to the concept of the covalent binding of active intermediates of drugs and toxic agents with vital macromolecules such as DNA, RNA, and various proteins in the body, thus leading to the adverse responses such as carcinogenesis, mutagenesis, and necrosis mentioned earlier. Developments in this field are now indicating, as we shall see from the various articles in this symposium, that in some cases conjugates, often regarded as detoxication products, may be proximate carcinogens. These are reactive electrophilic conjugates such as acetamidofluorene *N*-sulfate, *N*-acetoxy-2-aminofluorene, safrole-1′-sulfate, and the *N*-sulfate of *N*-methyl-4-aminoazobenzene (10). Reactive metabolites even of simple molecules such as CS_2 and CCl_4 have now been found and these may be responsible for their toxic effects.

REFERENCES

1. E. Meyer, Ueber das Verhalten des Nitrobenzols und einiger anderer aromatischer Nitrokörper im Organismus, *Hoppe-Seyler's Z. Physiol. Chem.* **46**, 497–509 (1905).
2. A. Ellinger, Ueber den Mechanismus der Methaemoglobinbildung durch Acetanilid und seine Abkömmlinge, *Hoppe-Seyler's Z. Physiol. Chem.* **111**, 86–125 (1920).
3. H. J. Channon, E. T. Mills, and R. T. Williams, Metabolism of 2:4:6:-trinitrotoluene, *Biochem. J.* **38**, 70–85 (1944).
4. P. Ehrlich, Ueber den jetzigen Stand der Chemotherapie, *Ber. Dtsch. Chem. Ges.* **42**, 17–47 (1909).
5. G. Domagk, Ein Beitrag zur Chemotherapie der backteriellen Infektionen, *Dtsch. Med. Wochenschr.* **61**, 250–253 (1935).
6. J. Trefouel, (Mme) J. Tréfouel, F. Nitti, and D. Bouvet, Activité du *p*-aminophenylsulfamide sur les infections streptococciques expérimentales de la souris et du lapin, *C. R. Seances Soc. Biol. Paris* **120**, 756–758 (1935).
7. E. Boyland, The biological significance of metabolism of polycyclic compounds, *Biochem. Soc. Symp.* **5**, 40–54 (1950).
8. E. C. Miller and J. A. Miller, Presence and significance of bound aminoazo dyes in livers of rats fed *p*-dimethylaminoazobenzene, *Cancer Res.* **7**, 468–480 (1947).
9. P. M. Bhargava and C. Heidelberger, Studies on the structure of the skin protein-bound compounds following topical application of 1,2,5,6-dibenzanthracene-9,10-C^{14}. II.

Nature of the 2-phenylphenanthracene-3,2′-dicarboxylic acid-protein bond, *J. Am. Chem. Soc.* **78**, 3671 (1956).

10. J. H. Weisburger and E. K. Weisburger, Biochemical formation and pharmacological, toxicological and pathological properties of hydroxylamines and hydroxamic acids, *Pharmacol. Rev.* **25**, 1–66 (1973).

2

The Concept of Reactive Electrophilic Metabolites in Chemical Carcinogenesis: Recent Results with Aromatic Amines, Safrole, and Aflatoxin B_1

James A. Miller and Elizabeth C. Miller
McArdle Laboratory for Cancer Research
University of Wisconsin Medical Center
Madison, Wisconsin 53706, U.S.A.

Chemical carcinogens appear to be the first class of foreign compounds demonstrated to be converted *in vivo* to reactive metabolites which bind covalently with tissue macromolecules (1,2), and a large literature now exists on this subject and its relevance to carcinogenesis by these agents (3–5). Considering the great structural heterogeneity of chemical carcinogens (5–8) and of drugs in general, it is not surprising that at high dosage levels a few drugs are known to be similarly activated *in vivo* to reactive metabolites which can exert acute toxic reactions through covalent binding with tissue components, especially proteins (9,10). Thus, although most drugs and their metabolites probably interact only noncovalently and thus reversibly with cellular molecules in their pharmacological actions, closer inspection of the metabolism of various drugs will doubtless reveal various degrees of covalent interactions *in vivo*. Such interactions may pose carcinogenic, mutagenic, teratogenic, allergenic, necrogenic, and possibly other hazards that must be weighed against the benefits provided by each drug in this category.

METABOLISM OF CHEMICAL CARCINOGENS AND MECHANISMS OF CHEMICAL CARCINOGENESIS

Carcinogenesis appears to involve at least a quasi-permanent change of phenotype of normal cells involving the control of growth. Hence it is axiomatic that carcinogens must interact directly or indirectly with critical informational macromolecules that control cell replication. The identity of neither the critical macromolecule(s) nor their critical modification(s) has yet been achieved unequivocally in studies on the mechanism of action of any chemical, viral, or physical carcinogen. Approaches to these goals in chemical carcinogenesis have centered primarily about the metabolism of chemical carcinogens, especially to reactive forms, the natures of the interactions of chemical carcinogens and their metabolites with macromolecules, and the relations of these interactions to tumor formation.

In 1947 the authors (1) reported the covalent binding of the hepatocarcinogenic dye 4-dimethylaminazobenzene in the livers of rats fed this agent. Several observations were made that pointed to a causal role of this binding in the formation of liver tumors by the dye, and subsequent work with related dyes amplified these findings (11). In 1951 (2), fluorescent protein-bound derivatives of 3,4-benzpyrene were noted in the skin epithelium of mice treated topically with this carcinogen. Neither of these carcinogens nor any of their known metabolites reacted with protein *in vitro* in this manner, so it seemed likely that other metabolites of these carcinogens reacted directly or in enzyme-mediated reactions to yield the protein-bound forms. Hydrolysis of these protein adducts yielded highly polar derivatives of the carcinogens, presumably carcinogen–amino acid adducts. In the early work with 4-dimethylaminoazobenzene, no binding to the hepatic nucleic acids was detected with the colorimetric method employed; the structures of the recently characterized major nucleic acid-bound derivatives of this carcinogen in rat liver now provide an explanation of our failure to detect them. Subsequently, in 1956, the first observation of nucleic acid binding of a carcinogen *in vivo* was made with a ^{14}C-labeled nitrogen mustard (12). By 1966 numerous observations of the binding *in vivo* of chemical carcinogens to tissue protein and nucleic acids had been made (3), and today a wide variety of carcinogens are known to bind covalently with macromolecules *in vivo* (4,5).

The amounts of the bound residues of chemical carcinogens found in proteins and nucleic acids *in vivo* are of the order of one carcinogen residue per 10^4 to 10^7 monomer residues. In some cases, fair to good correlations have been found between the total binding to the total protein and/or nucleic acids of tissues and the carcinogenic activities of a series of similar structures (11,13–15). In other cases the correlations between carcinogenic activities and the levels of bound forms are poor (16–19). Noncarcinogenic compounds closely related to chemical carcinogens are frequently bound at inappreciable or only low levels, but some weakly carcinogenic or apparently noncarcinogenic compounds bind appreciably and some binding of chemical carcinogens occurs in apparently nontarget tissues (11,14,20). However, at present no case is known in which an adequately studied chemical carcinogen has failed to exhibit covalent binding to macromolecules in its target tissue *in vivo*. Attempts to correlate the gross amount of binding of a chemical carcinogen with carcinogenicity are no longer profitable in view of our present knowledge of the numerous sites of binding of chemical carcinogens in tissue macromolecules and the multiple reactive forms of some chemical carcinogens.

The early studies on the natures of the interactions of chemical carcinogens with tissue macromolecules *in vivo* were hampered by the small amounts of the bound forms in tissues, by the lack of adequate separation and characterization techniques, and by the lack of reactive forms of the chemical carcinogens for studies *in vitro*. These hindrances have been largely overcome, and it is now evident that the majority of the classes of chemical carcinogens are *precarcinogens* which must be converted *in vivo* to *ultimate reactive carcinogens.* These conversions are usually mediated by enzymes and may involve the formation of intermediate or *proximate carcinogens* (Fig. 1). Furthermore, despite the great structural heterogeneity of chemical carcinogens, it now appears that the ultimate carcinogens are all strong electrophilic reactants (4,21). Thus ultimate carcinogens contain relatively electron-deficient carbon or nitrogen atoms which can react covalently with electron-rich or nucleophilic nitrogen, oxygen, or sulfur atoms in cellular components such as the nucleic acids and proteins.

The ultimate reactive forms of chemical carcinogens are presumably identical with the ultimate carcinogenic forms of these agents. This is especially evident with the carcinogenic alkylating agents. These structures are electrophils *per se* and evidently, in most cases at least, need no metabolic activation. As with the precarcinogens, the alkylating agents are structurally very diverse, but they have in common the abilities to act as electrophils and carcinogens. Data on the identities of the ultimate reactive and carcinogenic forms of precarcinogens are accruing. Multiple metabolic pathways to electrophilic metabolites are being noted for some of the more complex carcinogenic structures such as 2-acetyl-

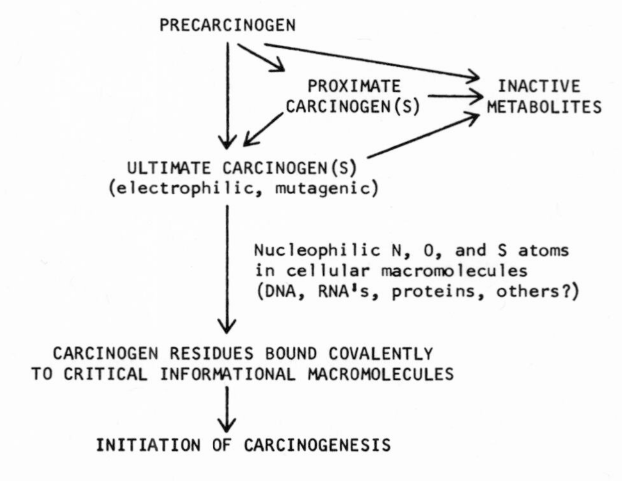

Fig. 1. General scheme for the metabolic activation and inactivation of chemical carcinogens and the role of ultimate carcinogens in chemical carcinogenesis.

aminofluorene and the polycyclic aromatic hydrocarbons (5). It seems likely that these multiple electrophilic forms will have various tissue and even species selectivities in the induction of tumors.

The electrophilic nature of ultimate chemical carcinogens appears to be a consequence of the structures of the major components of cells such as the proteins and nucleic acids. These and many other molecules in cells contain primarily nucleophilic centers (at certain N, O, and S atoms). Hence, among the metabolites of foreign molecules the electrophilic reactants are the ones with a potential for binding to cellular molecules. A different situation exists for the multitudinous physiological covalent bond-making and bond-breaking events that occur in normal cell metabolism. The physiological electrophils and nucleophils formed and joined in these reactions are under tight and ordered control at enzyme surfaces and do not enter into reactions in random fashion, while the strong electrophilic forms of chemical carcinogens appear to attack cellular molecules with little discrimination and generally without the aid of enzymes.

As noted above, the critical molecular targets of ultimate carcinogens and the molecular processes that ensue in chemical carcinogenesis have not been elucidated unequivocally. At present it appears possible that either genetic or epigenetic mechanisms, or both, may occur in the carcinogenic processes initiated by ultimate chemical carcinogens (5). The genetic mechanisms assume that somatic mutations, i.e., alterations in nucleotide sequence in the DNA genome, are primary events in carcinogenesis. Such alterations may result not only from direct covalent attacks of ultimate carcinogens on DNA, but conceivably also from indirect mechanisms involving covalent binding to either RNA or specific proteins. Similarly, epigenetic mechanisms involving stable and heritable changes in the expression of a normal genome, akin to those occurring in normal cell differentiation, may be induced by protein- or RNA-bound forms of chemical carcinogens. A mechanistic implication is sometimes drawn from the fact that the ultimate electrophilic forms of chemical carcinogens are mutagenic (22). This derives from the related fact that most, but not all, chemical mutagens are active via electrophilic forms. However, while the primary critical target of chemical mutagens is DNA, the existence of protein- and RNA-bound forms of chemical carcinogens demands that epigenetic mechanisms of chemical carcinogenesis also receive consideration. Thus, at present, the data relating chemical carcinogenesis and chemical mutagenesis constitute a strong formal relationship between these two properties of chemicals, but they do not necessarily imply a causal relationship. However, this strong formal relationship has great potential practical value in the detection of carcinogenic activity of chemicals. Thus mammalian tissue-mediated mutagenicity assays (23), in which metabolic activation of precarcinogens to mutagenic electrophils occurs, appear promising for the detection of potential carcinogenicity among the many man-made and naturally occurring compounds in the environment. With this prescreen the limited facilities for conventional carcinogenicity testing by lifetime administration of chemicals to rodent species may be employed to best advantage.

One approach to the identification of critical molecular targets in chemical carcinogenesis is to correlate the extent to which a given electrophil attacks specific nucleophilic targets in cells undergoing carcinogenesis with the tumor-inducing activity of different carcinogens which yield that electrophil. This is possible with alkylating agents and potential alkylating agents such as the alkylnitrosamides and dialkylnitrosamines, since these agents alkylate nucleic acids *in vivo* at a variety of sites (24,25). For example, a growing body of evidence suggests that the extent and persistence of O-6 alkylation of guanine in DNA *in vivo* may be of special importance in the induction of mutations and tumors (24–29). This minor site of alkylation is apparently of much greater importance in these biological processes than N-7 alkylation of guanine, the major site of alkylation of DNA *in vivo*. So far there are few data from which to assess the importance of alkylations at several other sites in the nucleic acids *in vivo*. The major site of reaction of the electrophilic metabolites of aromatic arylamines and arylamides with hepatic DNA and RNA *in vivo* is the 8-position of guanine (4,5,30,31). Alkylation of this site has not been observed *in vivo* or demonstrated in reactions *in vitro*. On the other hand, reactions of the tertiary phosphate groups of nucleic acids appear to occur with both alkylating agents and electrophils derived from arylamines (32–37). Similarly, covalent reactions of chemical carcinogens with proteins *in vitro* occur at a variety of sites, including methionyl, cysteinyl, histidyl, and tyrosyl residues (3,5,38–43).

As noted in Fig. 1, chemical carcinogens are subject to both activation and inactivation pathways of metabolism *in vivo*. These reactions involve a wide variety of enzymatic pathways, and many of these reactions have been observed with drugs as well. Many of the enzymatic pathways of conversion of chemical carcinogens to reactive electrophils are oxidative in nature and are catalyzed by mixed-function oxidases, especially the NADPH-NADH:flavoprotein:cytochrome P450:O_2 systems in the endoplasmic reticulum. Similar oxidative systems also occur in the nuclear membrane and might be of special importance in the covalent binding of carcinogens to DNA (44–46). The balance between the activation and inactivation pathways for metabolism of chemical carcinogens is surely an important factor in determining the rate of tumor formation; in many cases, this balance must involve the concentration of the metabolites, the K_m's of the enzymes involved, and the pharmacodynamics of the persistence and removal of these metabolites from cells.

RECENT RESULTS WITH AROMATIC AMINES, SAFROLE, AND AFLATOXIN B_1

Aromatic Amines and Amides

The carcinogenic aromatic amines and amides that have been adequately studied are activated by two metabolic steps. *N*-Hydroxy proximate carcinogenic metabolites are first formed, and these derivatives are then converted to electrophilic metabolites (4,5,30). The versatile carcinogen 2-acetylaminofluorene (AAF) and its proximate carcinogenic metabolite *N*-hydroxy-AAF have been studied in considerable detail. The *N*-hydroxylation of AAF is catalyzed by a P450 system in the hepatic endoplasmic reticulum, and the *N*-hydroxy metabolite may be conjugated to form the *O*-glucuronide which is then transported to other tissues and excreted in the urine. In rat livers susceptible to the carcinogenic action of AAF, *N*-hydroxy-AAF is converted to a very reactive and mutagenic major ultimate carcinogen, AAF-*N*-sulfate, by 3′-phosphoadenosine-5′-phosphosulfate (PAPS) and one or more soluble sulfotransferases (43,47–50). Enzymatic formation of other electrophilic metabolites of *N*-hydroxy-AAF has also been observed, and these metabolites are candidate ultimate carcinogens for carcinogenesis by AAF in extrahepatic tissues where *N*-hydroxy-AAF sulfotransferase activity has not been detected. These metabolites include the O-glucuronide of *N*-hydroxy-AAF; this derivative is a major metabolite and a weakly reactive electrophil (5). The *O*-glucuronide of *N*-hydroxy-2-aminofluorene is a strong electrophil (5), but the metabolic formation of this glucuronide has not yet been demonstrated. *N*-Hydroxy-AAF and several other arylhydroxamic acids are subject to the action of an acyltransferase in the soluble fraction of rat liver, in the cytosols of a variety of other rat tissues, and in tissues of other species (51–54). This enzyme, which transfers the *N*-acetyl group of *N*-hydroxy-AAF to its oxygen atom to form the very reactive ester, *N*-acetoxy-2-aminofluorene, may be responsible for the covalently bound 2-aminofluorene residues found, together with covalently bound AAF residues, in hepatic macromolecules. Lastly, peroxidases and H_2O_2 catalyze the conversion of *N*-hydroxy-AAF to a nitroxide free radical that readily dismutates to yield two mutagenic and carcinogenic electrophils, *N*-acetoxy-AAF and 2-nitrosofluorene (55–58). This latter reaction has not yet been demonstrated to occur in mammalian tissues. Most of the above activation reactions have also been demonstrated for the analogous carcinogen 4-acetylaminobiphenyl. Major quantitative differences in the extents of the reactions and in the reactivities of the esters of *N*-hydroxy-4-acetylaminobiphenyl and of *N*-hydroxy-AAF have been observed. These differences are probably important in determining the greater carcinogenicity of *N*-hydroxy-AAF relative to that of the biphenyl derivative (43,59–61).

Recent work in our laboratory has shown that the hepatocarcinogenic arylamine *N*-methyl-4-aminoazobenzene (MAB) undergoes metabolic activation in the rat liver by an *N*-hydroxylation reaction in essentially the same manner as noted for AAF. Initial efforts to synthesize *N*-hydroxy-MAB for studies *in vivo* and *in vitro* did not succeed and the synthetic ester *N*-benzoyloxy-MAB was prepared instead (38). Unlike MAB, this ester was electrophilic and was carcinogenic at sites of injection in the rat. Evidence suggesting that an ester of *N*-hydroxy-MAB was formed when MAB was fed to rats was obtained when 3-methylmercapto-MAB, 3-(homocystein-*S*-yl)-MAB, and 3- and *N*-(tyrosin-3-yl)-MAB were obtained on degradation of the liver protein from rats fed MAB

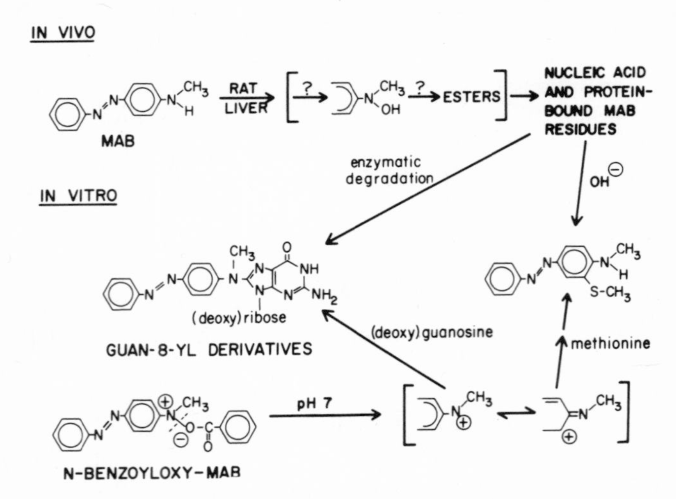

Fig. 2. Metabolic binding of N-methyl-4-aminoazobenzene (MAB) in vivo *to guanine and methionine residues in nucleic acids and proteins, respectively, and comparable reactions of the synthetic ester, N-benzoyloxy-MAB,* in vitro.

and also from reactions *in vitro* of the corresponding amino acids with *N*-benzoyloxy-MAB (Fig. 2) (38,41,42,62). Corresponding evidence for the nucleic acid-bound forms of MAB in the rat liver has been obtained recently. Nucleoside-dye adducts were isolated from the hepatic DNA and RNA of rats administered MAB and characterized as *N*-(deoxyguanosin-8-yl)-MAB and *N*-(guanosin-8-yl)-MAB, respectively (Fig. 2) (31). The same products were isolated from reactions *in vitro* of *N*-benzoyloxy-MAB with nucleic acids and guanine nucleosides or nucleotides.

Further studies *in vitro* of the metabolic activation of MAB were made possible by the recent synthesis (63) of *N*-hydroxy-MAB* in our laboratory by the thermal decomposition of *N*-methyl-*N*-ethyl-4-aminoazobenzene *N*-oxide to *N*-hydroxy-MAB and ethylene. *N*-Hydroxy-MAB is relatively unstable in the presence of oxygen; it was not detected as a product of the incubation of MAB with rat liver microsomes and a NADPH-generating system, and it could not be recovered after it was added to this incubation system. However, incubation of MAB in this system yielded both 4-aminoazobenzene (AB) and its *N*-hydroxy derivative (Fig. 3). The demethylation of MAB to AB by a hepatic mixed function oxidase is a well-established reaction (66), and the *N*-hydroxylation of AB was consistent with the metabolism of MAB and AB in the rat to the urinary metabolite *N*-acetyl-*N*-hydroxy-AB (67).

The formation of *N*-hydroxy-MAB in the microsomal metabolism of MAB was demonstrated by the following reactions (63). The incubation of MAB in

*A synthesis of *N*-hydroxy-MAB by another route has recently been reported (65).

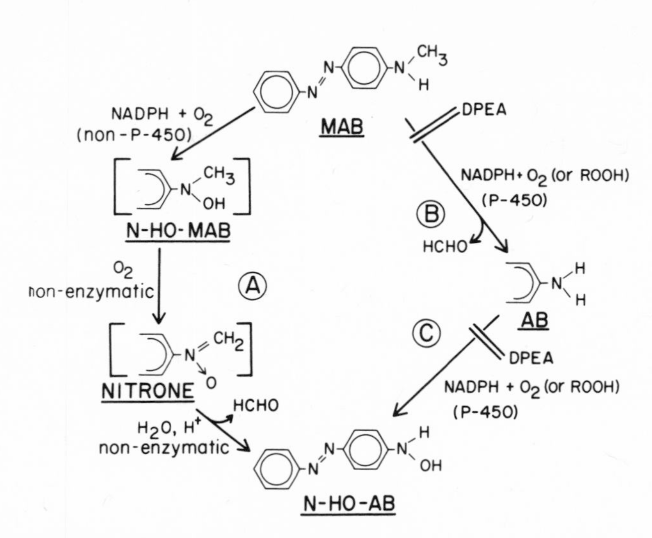

Fig. 3. Metabolism of N-methyl-4-aminoazobenzene (MAB) by rat liver microsomes to N-hydroxy-4-aminoazobenzene (N-HO-AB).

the microsomal system or of *N*-hydroxy-MAB in the presence *or* absence of microsomes resulted in the formation of an extractable intermediate, presumably the nitrone, which decomposed in acid to yield *N*-hydroxy-AB. The *N*-hydroxy-AB formed by the oxidation of AB and that formed by decomposition of the nitrone from *N*-hydroxy-MAB were distinguished by the fact that an acid treatment was required for the generation of *N*-hydroxy-AB from *N*-hydroxy-MAB (Fig. 3). Furthermore, the oxidation of AB to *N*-hydroxy-AB and the preceding oxidative demethylation of MAB to AB were both inhibited by DPEA–2-[(2,4-dichloro-6-phenyl)phenoxy] ethylamine–and could be supported by cumene hydroperoxide in the absence of a NADPH-generating system; these results are expected for a cytochrome P450-dependent system. The formation of the acid-labile nitrone precursor of *N*-hydroxy-AB from MAB was not inhibited significantly by DPEA and was not supported by cumene hydroperoxide in the absence of NADPH.

Thus the rat liver microsomal mixed-function oxidase that *N*-hydroxylates MAB does not appear to contain cytochrome P450 and appears to be very similar to the amine oxidase flavoprotein isolated from porcine liver microsomes (68). The purified porcine liver amine oxidase, in the presence of a NADPH-generating system, oxidized MAB and, with short incubation times, the major product was *N*-hydroxy-MAB; with longer times, the major product recovered was *N*-hydroxy-AB.

The soluble fraction of rat liver contains sulfotransferase activity, which converts *N*-hydroxy-MAB to its reactive sulfuric acid ester in the presence of PAPS (64). In this system there is a PAPS-dependent loss of *N*-hydroxy-MAB

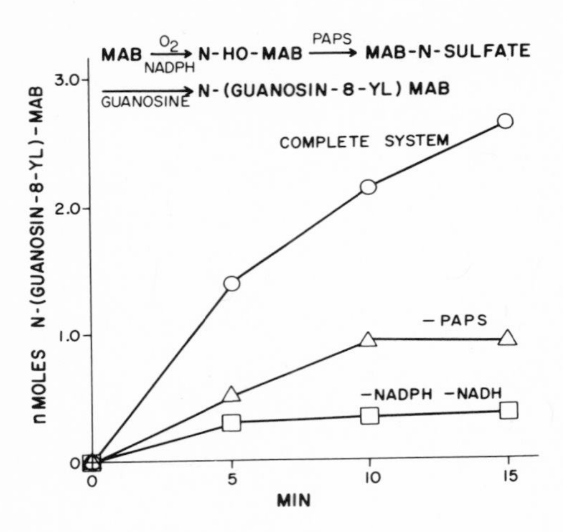

Fig. 4. NADPH- and PAPS-dependent formation of N-(guanosin-8-yl)-N-methyl-4-aminoazobenzene (MAB) in vitro. The complete incubation mixture contained 100 mM tris-(hydroxymethyl)aminomethane-HCl buffer (ph 7.1), 1 mM NaN_3, 5 mM $MgCl_2$, 0.5 mM EDTA, 1 mM PAPS, 1 mM guanosine, 5 mM glucose-6-phosphate, 0.4 mM $NADP^+$, 0.3 mM NAD^+, 1 unit Zwischenferment, 0.5 mM [^{3}H] MAB, 5 mg of protein from a 10,000*g* liver supernatant (64).

and a PAPS-dependent formation of 3-methylmercapto-MAB, *N*-(guanosin-8-yl)-MAB, or RNA-bound MAB when the incubation mixture is supplemented with methionine, guanosine, or RNA, respectively. The rat liver sulfotransferases for *N*-hydroxy-MAB and *N*-hydroxy-AAF are apparently not identical. In an incubation mixture which contained both liver microsomes and cytosol, a NADPH- and PAPS-dependent formation of *N*-(guanosin-8-yl)-MAB from MAB was demonstrated (Fig. 4). As noted for AAF, the *N*-sulfate of MAB appears to be a major carcinogenic metabolite since the incidence of liver tumors in rats fed the closely related amine 3′-methyl-4-dimethylaminoazobenzene was increased by supplementation of the diet with sodium sulfate (69).

Safrole

Safrole or 4-allyl-1,2-methylenedioxybenzene is a major component of oil of sassafras and some oils of *Heterotropa* species and a minor constituent of various spice oils. It has been known for many years to be hepatotoxic. The demonstration in 1960 of its weak hepatocarcinogenicity in the rat upon oral administration led to its being banned as an additive for flavoring soft drinks in the United States. Somewhat later safrole was found to be a relatively strong hepatocarcinogen upon parenteral administration to preweaning mice. These data [see (70)] led to an examination of safrole in our laboratory as an exercise in the prediction of its mode of activation as a carcinogen. Safrole is not electrophilic *per se,* but the fact that its 1′-carbon is a benzylic and allylic center suggested metabolic 1′-hydroxylation and subsequent esterification of the allylic and benzylic alcohol as one means to form an electrophilic ultimate carcinogen. 1′-Hydroxysafrole, as the *O*-glucuronide, was indeed found as a urinary metabolite of safrole in the rat and mouse (Fig. 5) (70). Tests of this new metabolite in these species showed that it was more hepatocarcinogenic than safrole and thus

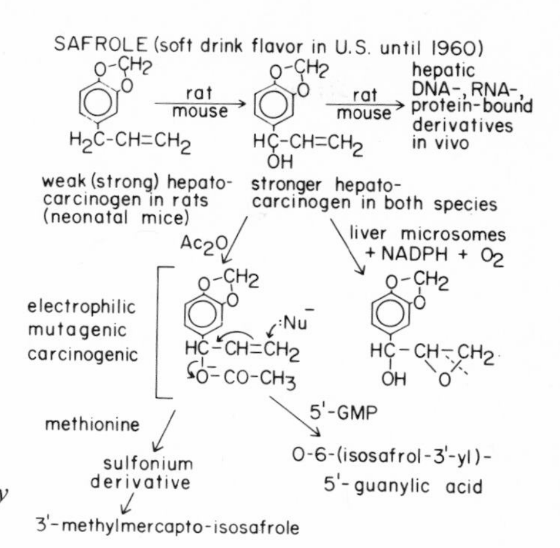

Fig. 5. Metabolic activation of safrole by rat and mouse liver in vivo *and* in vitro.

established it as a proximate carcinogenic metabolite (71). Further studies showed that the synthetic ester, 1′-acetoxysafrole, was more active as a carcinogen at sites of application than 1′-hydroxysafrole. The ester is a moderately strong electrophilic reactant (70,71); it reacts at neutral pH with 5′-guanylic acid and methionine *in vitro* to yield ^{16}O-(isosafrol-3′-yl)-5′-guanylic acid and 3′methylmercaptoisosafrole, respectively.

More recent studies (72, 73) have shown that [2′,3′-^{3}H] 1′-hydroxysafrole binds to added RNA in mouse and rat liver cytosols in a PAPS-dependent reaction. This tritiated 1′-hydroxysafrole also forms hepatic DNA-, RNA-, and protein-bound derivatives upon administration to rats and mice. Degradation of the hepatic protein released a small amount of a tritiated derivative which cochromatographed with 3′-methylmercaptoisosafrole. The amounts of PAPS-dependent binding of tritiated 1′-hydroxysafrole *in vitro* to RNA in liver cytosols and the yield of 3′-methylmercaptoisosafrole from hepatic protein *in vivo* were very low as compared to the yields of the analogous products from *N*-hydroxy-AAF and MAB. These findings are consistent with the much lower carcinogenicity of 1′-hydroxysafrole as compared to that of the latter two carcinogens. These findings also suggest that other electrophilic metabolites of 1′-hydroxysafrole may be quantitatively more important in the protein binding of this proximate carcinogen than are its esters. No degradation products of the DNA- and RNA-bound forms of tritiated 1′-hydroxysafrole from the livers of rats or mice have yet been identified.

Other possible ultimate carcinogenic and electrophilic metabolites of safrole and 1′-hydroxysafrole such as 1′-oxosafrole, and the 2′,3′-oxides of safrole, 1′-hydroxysafrole, and 1′-acetoxysafrole were each examined for their reactivity and carcinogenicity (73). Each of these derivatives reacted with guanosine to different degrees (Fig. 6). 1′-Oxosafrole also reacted to a smaller extent with each of the other common nucleosides. A considerable fraction of these

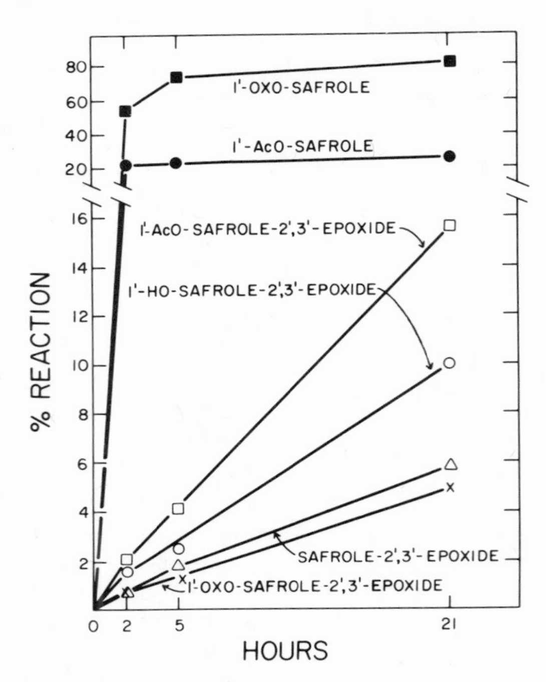

Fig. 6. Reactions of electrophilic derivatives of safrole with [8-^{14}C]guanosine as a function of time. Each safrole derivative (45 μmol) was incubated with 0.9 μmol of guanosine at neutrality in 50% acetone. The products were separated by thin-layer chromatography on silica (72).

products probably represented Michael condensation products between the activated 2′,3′-double bond and amino or imino groups in the nucleosides since hot alkali released the parent nucleosides in considerable yields. No evidence is available on the direct formation of 1′-oxosafrole from safrole or 1′-hydroxysafrole either *in vivo* or *in vitro,* although the urinary excretion of tertiary aminomethylenedioxypropiophenones by rats given safrole provides indirect evidence that 1′-oxosafrole was formed either in the gastrointestinal tract or in tissues, or both, and then reacted with secondary amines (74). Despite its high electrophilicity and close relationship to 1′-hydroxysafrole, 1′-oxosafrole did not display any carcinogenic activity in rats and mice under a variety of conditions (73). Its uptake by extracellular nucleophils may have prevented effective concentrations of this electrophil from entering cells after its administration.

Each of the three epoxides also reacted with guanosine (Fig. 6). Evidence for the formation of safrole-2′,3′-oxide *in vivo* has been obtained in the rat (75) and 1′-hydroxysafrole-2′,3′-oxide was isolated as a product of NADPH-dependent metabolism of 1′-hydroxysafrole by rat and mouse liver microsomes (73). 1′-Hydroxysafrole-2′,3′-oxide initiated the formation of papillomas in mouse skin when a total of 90 μmol was applied to the skin over a 6-week period and these treatments were followed by repeated applications of croton oil (73). The other two epoxides did not show appreciable activity in this test system.

Safrole is representative of the large group of naturally occurring allyl and propenyl benzenes which are found in a wide range of essential oils and spices. They thus occur, usually at low levels, in many human diets. Further studies on the metabolism and carcinogenicity of these compounds are needed.

Aflatoxin B_1

Aflatoxin B_1, a strongly hepatotoxic mycotoxin, is the most potent hepatocarcinogen known and induces tumors in the livers of rats and rainbow trout after long-term ingestion of as little as 1 ppb in the diet with total intakes of only a few micrograms (76,77). Despite this high potency it has been evident for some time that these biological activities of aflatoxin B_1 are dependent on its metabolic activation [see (78)], and recent data (78–81) strongly indicate aflatoxin B_1-2,3-oxide is a major ultimate electrophilic and carcinogenic metabolite (Fig. 7). Aflatoxin B_1 interacts only weakly and noncovalently with macromolecules such as nucleic acids *in vitro,* but *in vivo* in the rat liver it forms covalent adducts with the nucleic acids and protein (81,82). Similarly, with certain bacteria aflatoxin B_1 became toxic and mutagenic only in the presence of liver microsomes and a NADPH-generating system (23,79,83,84). Covalent adducts resulted when this mycotoxin was incubated with nucleic acids, homoribopolynucleotides, or protein in the presence of fortified liver microsomes (78,80,81,85,86). Hydrolysis in weak acid of either the nucleic acid adducts obtained with rat or human liver microsomes or the nucleic acid adducts formed

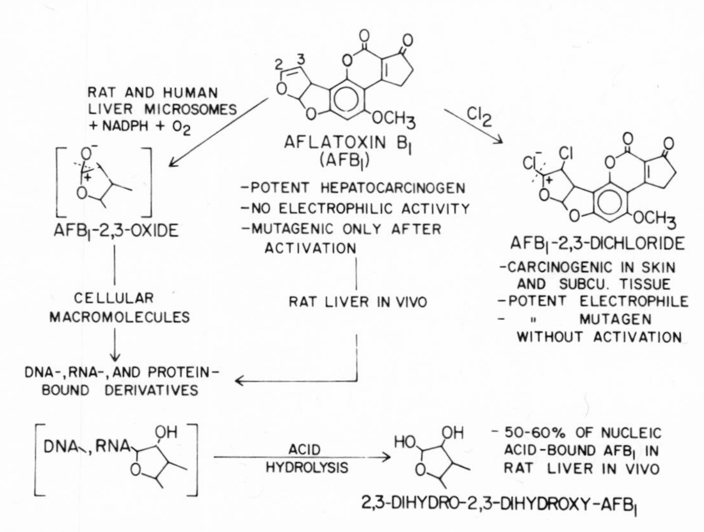

Fig. 7. Metabolism of aflatoxin B_1 to macromolecule-bound forms in rat liver in vivo *and* in vitro *and the properties of its 2,3-dichloride.*

in vivo in the rat liver produced 2,3-dihydro-2,3-dihydroxy-aflatoxin B_1 as the major product (80,81). These observations and the fact that liver microsomes did not bind aflatoxin B_2 (2,3-dihydro-aflatoxin B_1) to added nucleic acid strongly implicate aflatoxin B_1-2,3-oxide as a reactive metabolite. Adduct formation appears to occur at carbon-2 of aflatoxin B_1 since the spectra of the nucleic acid adducts did not exhibit the bathochromic shift at alkaline pH characteristic of 2-hydroxy aflatoxins (80). Carbon-2 is the expected site of reaction of the 2,3-oxide since a positive charge on this carbon would be stabilized by the adjacent furan ring oxygen atom. Furthermore, reaction at carbon-2 with a nucleophilic atom in the nucleic acids would form a glycoside-like bond that would be acid-labile (Fig. 7).

The probable importance of 2,3-epoxidation in the carcinogenicity of aflatoxin B_1 as well as in its electrophilic reactivity is shown by the much lower carcinogenicity of aflatoxin B_2 in the rat liver as compared to that of aflatoxin B_1 (87). Partial conversion of aflatoxin B_2 to aflatoxin B_1 *in vivo* in the rat may occur. Further study of aflatoxin B_1-2,3-oxide has been hindered by our inability to synthesize this apparently highly reactive compound or to isolate it from enzymatic reaction mixtures. These difficulties led us to synthesize aflatoxin B_1-2,3-dichloride (Fig. 7) (88) as a model for the putative 2,3-oxide metabolite. As expected, the dichloride is highly electrophilic at carbon-2 and reacts as such with nucleic acids to form products which, except for the chlorine atom at carbon-3, have the same properties as the adducts from the metabolite of aflatoxin B_1. Aflatoxin B_1-2,3-dichloride is highly toxic, mutagenic, and carcinogenic and is apparently active as such. It is a strong mutagen for both *Salmonella typhimurium* TA 98 and TA 100 in the absence of liver microsome activation. Aflatoxin B_1 is mutagenic for these bacterial strains only in the presence of liver microsomes (84). Subcutaneous injection of the dichloride in rats produced a much higher incidence of sarcomas than did comparable amounts of aflatoxin B_1 (88). Likewise, following topical administration to the skin of mice, aflatoxin B_1 was much less carcinogenic than the dichloride in the induction of skin papillomas and in the formation of lung adenomas (88).

A PERSPECTIVE

The recent results on the metabolic activation of 4-monomethylaminoazobenzene, safrole, and aflatoxin B_1 further document the generalization that the ultimate reactive and carcinogenic forms of chemical carcinogens are strong electrophils (4,21). Similarly, these new findings further emphasize that, except for the carcinogenic alkylating and acylating agents, chemical carcinogens generally require metabolic activation for their activity. Frequently these reactions are catalyzed in the endoplasmic reticulum of cells by mixed-function oxidases. When these observations are considered together with the great structural diversity of chemical carcinogens, it appears inevitable that some drugs will be

acyltransferase- and sulfotransferase-catalyzed activations of *N*-hydroxy-*N*-2-fluorenylacetamide, *Cancer Res.* **35,** 906–912 (1975).
55. H. Bartsch and E. Hecker, On the metabolic activation of the carcinogen *N*-hydroxy-*N*-2-acetylaminofluorene. III. Oxidation with horseradish peroxidase to yield 2-nitrosofluorene and *N*-acetoxy-*N*-2-acetylaminofluorene, *Biochim. Biophys. Acta* **237,** 567–578 (1971).
56. H. Bartsch, M. Traut, and E. Hecker, On the metabolic activation of *N*-hydroxy-*N*-2-acetylaminofluorene. II. Simultaneous formation of 2-nitrosofluorene and *N*-acetoxy-*N*-2-acetylaminofluorene from *N*-hydroxy-*N*-2-acetylaminofluorene via a free radical intermediate, *Biochim. Biophys. Acta* **237,** 556–566 (1971).
57. H. Bartsch, J. A. Miller, and E. C. Miller, *N*-Acetoxy-*N*-acetylaminoarenes and nitrosoarenes: One-electron non-enzymatic and enzymatic products of various carcinogenic aromatic acethydroxamic acids, *Biochim. Biophys. Acta* **273,** 40–51 (1972).
58. C. M. King, T. W. Bednar, and E. M. Linsmaier-Bednar, Activation of the carcinogen *N*-hydroxy-2-fluorenylacetamide: Insensitivity to cyanide and sulfide of the peroxidase-H_2O_2 induced formation of nucleic acid adducts, *Chem.-Biol. Interactions* **7,** 185–188 (1973).
59. J. A. Miller, C. S. Wyatt, E. C. Miller, and H. A. Hartmann, The *N*-hydroxylation of 4-acetylaminobiphenyl by the rat and dog and the strong carcinogenicity of *N*-hydroxy-4-acetylaminobiphenyl in the rat, *Cancer Res.* **21,** 1465–1473 (1961).
60. A. R. Forrester, M. M. Ogilvy, and R. H. Thompson, Mode of action of carcinogenic amines. Part I. Oxidation of *N*-arylhydroxyamic acids, *J. Chem. Soc.* 1081–1083 (1970).
61. E. Kriek, On the mechanism of action of carcinogenic aromatic amines. II. Binding of *N*-hydroxy-*N*-acetylaminobiphenyl to rat-liver nucleic acids *in vivo, Chem.-Biol. Interactions* **3,** 19–28 (1971).
62. J. D. Scribner, J. A. Miller, and E. C. Miller, 3-Methylmercapto-*N*-methyl-4-aminoazobenzene: An alkaline-degradation product of a labile protein-bound dye in the livers of rats fed *N,N*-dimethyl-4-aminoazobenzene, *Biochem. Biophys. Res. Commun.* **20,** 560–565 (1965).
63. Kadlubar, F. F., J. A. Miller, and E. C. Miller, Microsomal *N*-oxidation of the hepatocarcinogen *N*-methyl-4-aminoazobenzene and the reactivity of *N*-hydroxy-*N*-methyl-4-aminoazobenzene, *Cancer Res.* **36,** 1196–1206 (1976).
64. Kadlubar, F. F., J. A. Miller, and E. C. Miller, Hepatic metabolism of *N*-hydroxy-*N*-methyl-4-aminoazobenzene and other *N*-hydroxy arylamines to reactive sulfuric acid esters, *Cancer Res.* **36,** 2350–2359 (1976).
65. Y. Hashimoto and M. Degawa, Synthesis of *N*-hydroxy-4-(methylamino)-azobenzene and its acetate, and their reactivity to amino acids, *Gann* **66,** 215–216 (1975).
66. G. C. Mueller and J. A. Miller, The metabolism of methylated aminoazo dyes. II. Oxidative demethylation by rat liver homogenates. *J. Biol. Chem.* **202,** 579–587 (1953).
67. K. Sato, L. A. Poirier, J. A. Miller, and E. C. Miller, Studies on the *N*-hydroxylation and carcinogenicity of 4-aminoazobenzene and related compounds, *Cancer Res.* **26,** 1678–1687 (1966).
68. D. M. Ziegler and C. H. Mitchell, Microsomal oxidase. IV. Properties of a mixed-function amine oxidase isolated from pig liver microsomes, *Arch. Biochem. Biophys.* **150,** 116–125 (1972).
69. J. M. Blunck and C. E. Crowther, Enhancement of azo dye carcinogenesis by dietary sodium sulphate, *Eur. J. Cancer* **11,** 23–32 (1975).
70. P. Borchert, P. G. Wislocki, J. A. Miller, and E. C. Miller, The metabolism of the naturally-occurring hepatocarcinogen safrole to 1′-hydroxysafrole and the electrophilic reactivity of 1′-acetoxysafrole, *Cancer Res.* **33,** 575–589 (1973).
71. P. Borchert, J. A. Miller, E. C. Miller, and T. K. Shires, 1′-Hydroxysafrole: A proximate

carcinogenic metabolite of safrole in the rat and mouse, *Cancer Res.* **33**, 590–600 (1973).
72. P. G. Wislocki, P. Borchert, J. A. Miller, and E. C. Miller, The metabolic activation of the carcinogen 1′-hydroxysafrole *in vivo* and *in vitro* and the electrophilic reactivities of possible ultimate carcinogens, *Cancer Res.* **36**, 1686–1695 (1976).
73. P. G. Wislocki, On the proximate and ultimate carcinogenic metabolites of precarcinogens: Safrole and certain *N*-alkylaminoazo dyes, Ph.D. thesis, University of Wisconsin, Madison, Wis. (1974).
74. E. O. Oswald, L. Fishbein, B. J. Corbett, and M. P. Walker, Identification of tertiary aminomethylenedioxypropiophenones as urinary metabolites of safrole in the rat and guinea pig, *Biochim. Biophys. Acta* **230**, 237–247 (1971).
75. W. G. Stillwell, M. J. Carman, L. Bell, and M. G. Horning, The metabolism of safrole and 2′,3′-epoxysafrole in the rat and guinea pig, *Drug Metab. Dispos.* **2**, 489–498 (1974).
76. G. N. Wogan, S. Paglialunga, and P. M. Newberne, Carcinogenic effects of low dietary levels of aflatoxin B_1 in rats, *Food Cosmet. Toxicol.* **12**, 681–685 (1974).
77. J. E. Halver, Aflatoxicosis and trout hepatoma, in: *Aflatoxin: Scientific Background, Control, and Implications* (L. A. Goldblatt, ed.), pp. 265–306, Academic Press, New York (1969).
78. R. C. Garner, E. C. Miller, and J. A. Miller, Liver microsomal metabolism of aflatoxin B_1 to a reactive derivative toxic to *Salmonella typhimurium* TA 1530, *Cancer Res.* **32**, 2058–2066 (1972).
79. R. C. Garner, Chemical evidence for the formation of a reactive aflatoxin B_1 metabolite by hamster liver microsomes, *FEBS Lett.* **36**, 261–264 (1973).
80. D. H. Swenson, J. A. Miller, and E. C. Miller, 2,3-Dihydro-2,3-dihydroxyaflatoxin B_1: An acid hydrolysis product of an RNA-aflatoxin B_1 adduct formed by hamster and rat liver microsomes *in vitro*, *Biochem. Biophys. Res. Commun.* **53**, 1260–1267 (1973).
81. D. H. Swenson, E. C. Miller, and J. A. Miller, Aflatoxin B_1-2,3-oxide: Evidence for its formation in rat liver *in vivo* and by human liver microsomes *in vitro*, *Biochem. Biophys. Res. Commun.* **60**, 1036–1043 (1974).
82. W. Lijinsky, K. Y. Lee, and C. H. Gallagher, Interaction of aflatoxin B_1 and G_1 with tissues of the rat, *Cancer Res.* **30**, 2280–2283 (1970).
83. R. C. Garner and C. M. Wright, Induction of mutations in DNA-repair deficient bacteria by a liver microsomal metabolite of aflatoxin B_1, *Br. J. Cancer* **28**, 544–551 (1973).
84. J. McCann, N. E. Spingarn, J. Kobori, and B. N. Ames, Detection of carcinogens as mutagens: Bacterial tester strains with R factor plasmids, *Proc. Natl. Acad. Sci. USA* **72**, 979–983 (1975).
85. R. C. Garner, Microsome-dependent binding of aflatoxin B_1 to DNA, RNA, polynucleotides and protein *in vivo*, *Chem.-Biol. Interactions* **6**, 125–129 (1973).
86. H. L. Gurtoo and T. C. Campbell, Metabolism of aflatoxin B_1 and its metabolism-dependent and independent binding to rat hepatic microsomes, *Mol. Pharmacol.* **10**, 776–789 (1974).
87. G. N. Wogan, G. S. Edwards, and P. N. Newberne, Structure-activity relationships in toxicity and carcinogenicity of aflatoxins and analogs, *Cancer Res.* **31**, 1936–1942 (1971).
88. D. H. Swenson, J. A. Miller, and E. C. Miller, The reactivity and carcinogenicity of aflatoxin B_1-2,3-dichloride, a model for the putative 2,3-oxide metabolite of aflatoxin B_1, *Cancer Res.* **35**, 3811–3823 (1975).

3

Kinetics of Reactive Metabolites and Covalent Binding *in Vivo* and *in Vitro*

James R. Gillette
Laboratory of Chemical Pharmacology
National Heart and Lung Institute
National Institutes of Health
Bethesda, Maryland 20014, U.S.A.

It has become increasingly evident that many chemically inert foreign compounds are converted in the body to chemically reactive metabolites that react with a number of substances in tissues and thereby cause changes which result in various toxicites including cancer, mutagenesis, cellular necrosis, immunological reactions, blood dyscrasias, and fetal damage (1–11). Because of the seriousness of these toxicities, my laboratory has been engaged in developing a general approach by which we can determine whether a given toxicity is mediated by chemically reactive metabolites.

In previous studies of mechanisms of toxicities mediated by chemically reactive metabolites, various laboratories have assumed that the toxicity results from the reaction of the metabolite with a single kind of target substance. They have, therefore, expended considerable effort in attempting to identify the target substance that mediates the toxicity and even the macromolecule that may be involved in the toxicity. Such studies have provided valuable information in elucidating the mechanism of various toxicities and especially the mechanisms of carcinogenic compounds. Nevertheless, in the development of a general test system that would determine whether a given toxicity is mediated by a reactive metabolite without necessarily elucidating the mechanism of the toxicity, this approach seemed fruitless for a number of reasons. It seemed obvious to us that the target substance would depend on the toxicity being studied and the reactive metabolite; some toxicities may be mediated by covalent binding to nuclear DNA, others may be mediated by the covalent binding to certain vitally important enzymes, and others may be mediated by lipid peroxidation. Moreover, in some kinds of toxicity such as tissue necrosis and hypersensitivity reactions, the target substance may be any one of a number of different intracellular components and indeed may differ with the reactive metabolite. For example, there is considerable evidence that the chemically reactive metabolite of carbon tetrachloride causes liver damage by reacting with phospholipids in the endoplasmic reticulum, thereby promoting lipid peroxidation (10,11). But many other substances that cause centrilobular necrosis, such as bromobenzene (12,13), dimethylnitrosamine (12,13), acetaminophen, and furosemide (13) do not promote lipid peroxidation and therefore cannot evoke their toxic effects in this way. Moreover, we realized that the specificity of the reaction between reactive metabolites and various cellular substances can vary markedly with the reactive metabolite being studied. At one extreme, some reactive metabolites, particularly those having relatively low chemical reactivities, may become preferentially bound to certain macromolecules in tissues by first combining reversibly with active centers on a specific macromolecule to form a complex that rearranges to form a covalently bound conjugate. Indeed, this mechanism is the basis of affinity labeling of receptor sites by chemically reactive analogues of endogenous chemical mediators (14) and of the preferential inhibition of choline esterases by organophosphate insecticides and their precursors (15). In these situations where relatively few macromolecules are covalently bound to the metabolite or where the toxicity mimics well characterized pharmacological actions, the identification of the target substance is relatively easy. At the other extreme, however, highly reactive metabolites of foreign compounds combine indiscriminately with many different kinds of intracellular components including protein, lipids, glycogen, DNA, RNA, and even many smaller molecules (1–4). However, the relative rates of reaction with different intracellular substances vary with the reactive metabolite and the tissue. Moreover, the reactive metabolite may evoke its toxic action by altering cellular components without actually being covalently bound with them. For example, 6-hydroxydopamine is thought to damage adrenergic neurons by promoting the formation of superoxide during its autoxidation (16). Similarly, some investigators believe that the trichloromethyl free radical formed from carbon tetrachloride initiates lipid peroxidation by reacting with methylene hydrogens in unsaturated fatty acids to form chloroform and a fatty acid free radical (10). Thus the identification of the target substance can be very difficult. Indeed, when a reactive metabolite interacts with a number of biochemical systems simultaneously, it is difficult to determine whether changes in cell function result from a sequence of changes originating from a single biochemical alteration or from the concerted action of a number of different initial biochemical alterations.

Because highly reactive metabolites can react with so many different kinds of substances in tissues, and because we know so little about the diverse mechanisms by which reactive metabolites might lead to different toxicities and much less about how combinations of reactions of the reactive metabolite with different cellular components might evoke toxicities, there seemed little reason for selecting any given product of the reaction between reactive metabolites and cellular components as the basis of a general test system for determining whether a toxicity caused by a given foreign compound is mediated by a reactive metabolite. It also seemed apparent that the finding of radiolabel covalently bound to tissue macromolecules after the administration of a radiolabeled foreign compound would not be sufficient proof that a reactive metabolite mediated the toxicity under investigation or any other toxicity.

Nevertheless, it seemed likely that the incidence and severity of any toxicity mediated by a chemically reactive metabolite would be roughly proportional to the number of target molecules altered by the reactive metabolite in the tissue after the administration of the foreign compound (17,18). It also seemed likely that changes in the effective dose of a highly reactive metabolite would alter not only the number of target molecules altered by the reactive metabolite but also the amount of reactive metabolite covalently bound to tissue macromolecules, including protein. Thus treatments of animals that change the incidence or severity of the toxicity solely by altering the pattern of metabolism of the toxicant and thereby the effective dose of the reactive metabolite should cause parallel changes in the amount of reactive metabolite that is covalently bound to tissue protein. According to this view, it would not be necessary to identify either the reactive metabolite or the target molecule to determine whether the toxicity was mediated by a chemically reactive metabolite, because the changes in the amount of metabolite covalently bound to tissue protein should parallel changes in the incidence or severity of the toxicity even when the target molecule may be a minor tissue protein, when it is not a protein at all, or when it does not form a stable conjugate with the reactive metabolite. Because of the lack of specificity in the reactions of reactive metabolites with tissue components, however, it is evident that by themselves, measurements of covalently bound metabolites in a tissue cannot be used to predict the toxicity of a given series of organic compounds. Indeed, without correlative studies, no more emphasis should be placed on the finding that a reactive metabolite becomes covalently bound to tissue macromolecules than would be placed on the finding that a reversibly acting drug is localized in a given tissue in drug distribution studies. For example, the finding that a reversibly acting drug is localized in a tissue need not mean that it acts in that tissue, and the finding that a reversibly acting drug metabolite is localized in a tissue need not mean that it was formed there. Nevertheless, such measurements of covalent binding of the toxicant to protein may be useful in deciding whether a given treatment of animals that alters toxicity acts by changing the metabolism of the toxicant or by altering events that occur subsequent to the formation of the reactive metabolite.

FORMATION AND ELIMINATION OF COVALENTLY BOUND METABOLITES

The relationship between the magnitude of the covalent binding of reactive metabolites to tissue proteins and the toxicity after the administration of single doses of toxicant should depend not only on the rates at which the reactive metabolite is formed and inactivated in the body but also on the rate at which the altered target molecule is repaired or replaced and on whether repair of the altered target molecule increases or decreases the incidence of toxicity. If the rate of metabolism of the toxicant is rapid compared with the rate of disappearance of the reactive metabolite–macromolecular conjugates from the tissue, then an indirect measurement of the amount of reactive metabolite formed in the tissue plus the amount entering it from the blood may be obtained by measuring the amount of covalently bound metabolite at a time after the foreign compound has been almost completely metabolized but before appreciable amounts of the covalently bound metabolite have disappeared. On the other hand, if the foreign compound is slowly metabolized compared with the rate of disappearance of the covalently bound metabolite, other techniques may be required to determine changes in the amount of reactive metabolite formed in the body.

In one method that might be useful in the latter situation, the amount of covalently bound metabolite is measured at various times after the administration of the toxicant, these amounts of covalently bound metabolite are plotted on rectilinear graph paper, and the area under the curve (AUC) is calculated. This method is based on the assumption that most of the reactive metabolite is covalently bound to intracellular components that have similar turnover times. It also assumes that the elimination of the covalently bound metabolite follows first-order kinetics; that is, the rate of elimination at any given time is proportional to the amount of covalently bound metabolite present at that time. Under these conditions, the AUC should equal the total amount of the foreign compound that is covalently bound (including that which is eliminated with time) divided by the first-order rate constant of elimination of the covalently bound reactive metabolite. Thus, when the values for the size of the dose of foreign compound, the AUC, and the half-life of the covalently bound components are known, one may calculate the fraction of the dose that becomes covalently bound, even though the covalently bound metabolite is rapidly eliminated from the body.

$$\text{Fraction} = \frac{\text{AUC}\,k}{\text{Dose}} = \frac{0.693\ \text{AUC}}{t_{1/2}\ \text{Dose}} \tag{1}$$

Unfortunately, these idealized conditions will seldom be completely valid in living animals. It seems quite likely, for example, that the turnover rates of the various components that serve as nucleophils will differ, and that the area under the curve will equal the sum of a series of areas under the curve, each of which would be the amount of covalently bound metabolite combined to each of these intracellular components divided by its rate constant of elimination.

Thus

$$\text{AUC} = \text{Dose} \times (F_1/k_1 + F_2/k_2 + F_3/k_3 + \cdots) \quad (2)$$

Even in this situation, the effects of various treatments on the AUC may provide clues in the elucidation of the kind of mechanism by which various treatments alter the severity and incidence rates of various toxicities. For example, changes in the AUC may still be used to estimate changes in the fraction of the dose that becomes covalently bound, if the treatment changes only the pattern of metabolism of the toxicant and not the rate constant of elimination of the major covalently bound metabolite.

Many toxicities are manifested only after repetitive administration of the foreign compound. In these situations, the relationships between the formation of reactive metabolites and toxicity are frequently obscure because the pattern of metabolism of the foreign compound may change as it is given repetitively. Such changes, however, may frequently be detected by measuring the AUC of the covalent binding of the reactive metabolite after single doses of radiolabeled foreign compound at various times during the course of repetitive administration of the unlabeled foreign compound. Indeed, this approach may be the simplest way to evaluate whether alterations in the rates of metabolism of the foreign compound along various pathways result in important changes in the formation of reactive metabolites. From such studies it may be possible to estimate the total amount of reactive metabolite to which the animals are exposed during the course of treatment and thereby to determine whether the effects of the reactive metabolite are accumulative.

RATIO CONCEPT OF COVALENT BINDING

The proportion of the dose of a foreign compound that becomes covalently bound to various tissue components may be visualized as a series of ratios, the length of which depends on the number of reactions by which the foreign compound is converted to its chemically reactive metabolite. For example, suppose that a foreign compound is converted first to an intermediate (the "proximate intermediate") which in turn is converted to a reactive metabolite that then reacts with various substances in a tissue and either becomes covalently bound to macromolecules such as DNA, RNA, and proteins or converts tissue components to reactive intermediates (as presumably occurs in lipid peroxidation), or both, as illustrated in the model

$$
\begin{array}{llllll}
\text{I} & \xrightarrow{k_{1\text{i}}} & \text{II} & \xrightarrow{k_{2\text{i}}} & \text{III} & \xrightarrow{k_{3\text{i}}} \text{Covalent binding to macromolecules} \\
\downarrow k_{1\text{iii}} & \searrow k_{1\text{ii}} & \downarrow k_{2\text{iii}} & \searrow k_{2\text{ii}} & \downarrow k_{3\text{iii}} & \searrow k_{3\text{ii}}
\end{array}
$$

In this model,

$$A = \frac{k_{1\text{i}}}{k_{1\text{i}} + k_{1\text{ii}} + k_{1\text{iii}}}$$

$$B = \frac{k_{2\text{i}}}{k_{2\text{i}} + k_{2\text{ii}} + k_{2\text{iii}}}$$

$$C = \frac{k_{3\text{i}}}{k_{3\text{i}} + i_{3\text{ii}} + k_{3\text{iii}}}$$

In this situation, the proportion of the dose that becomes covalently bound to macromolecules (F) may be visualized as

$$F = ABC \tag{3}$$

in which A is the proportion of the dose of the foreign compound that is converted to the proximate intermediate, B is the proportion of this intermediate that is converted to the reactive metabolite, and C is the proportion of the reactive metabolite that becomes covalently bound to the various macromolecules. Changes in the amount of covalent binding of a chemically reactive metabolite to tissue macromolecules after a given dose of the toxicant can occur only by changing one or more of the ratios, A, B, or C. However, each ratio depends not only on the rate at which the toxicant or metabolite is converted to the next compound along the pathway leading to the covalent binding but also on the rate at which it is eliminated from the body by innocuous pathways. When one or more of the enzymes or transport systems that mediate the elimination of the toxicant or metabolite become saturated or when one or more of the cosubstrates required by these enzyme systems become depleted in the tissues, the relative rates may vary with time and thus the ratios may be difficult to visualize. In order to gain an insight into the interrelationships among the various pathways, however, let us assume that all processes are first order, that is, the rates are directly proportional to the concentration of the parent compound or the metabolite. Under these conditions, each ratio equals the rate constant for the formation of the next metabolite along the pathway leading to the covalent binding of the foreign compound to the tissue macromolecules divided by the sum of the rate constants for this reaction and the other pathways by which the foreign compound or the metabolite is eliminated from the body (see model, page 30).

Inspection of the equation for A as a representative of these ratios leads to several principles that frequently have been misunderstood:

1. When nearly all of the foreign compound is converted to the proximate intermediate (that is, ratio A approaches 1.0), changes in the activity of the enzyme that catalyzes the formation of this intermediate may markedly alter the *rate* of covalent binding of the reactive metabolite to tissue macromolecules but will not appreciably alter the *fraction* of the dose that becomes covalently bound to macromolecules, because such changes will alter both the numerator and the denominator of ratio A to about the same extent.
2. On the other hand, when the proportion of the dose that is converted to the proximate intermediate is small (that is, ratio A approaches zero), changes in the activity of the enzyme that catalyzes the formation of the proximate intermediate will markedly change both the *rate* of covalent binding and the *proportion* of the dose that ultimately becomes covalently bound, because the numerator of ratio A will be changed to a greater extent than its denominator. The effect of changes in the rate of elimination of the foreign compound into the urine, bile, or air or of changes in the activity of an enzyme that metabolizes the foreign compound along a pathway of detoxification may markedly alter both the rate of covalent binding of the reactive metabolite and the proportion of the dose that becomes covalently bound, because such changes may alter the denominator of A.
3. When the rates of metabolism of the foreign compound and its metabolites follow first-order kinetics, the denominator of A is the rate constant of elimination, which may be calculated from the biological half-life of the foreign compound in the body. In view of the above discussion in (1) and (2), however, the relationship between changes in the covalent binding of the reactive metabolite and the biological half-life of the foreign compound will depend on the value of A and the pathway that this changed. If ratio A is large, the biological half-life may be markedly changed without appreciably altering the proportion of the dose of the foreign compound that ultimately becomes covalently bound. By contrast, if ratio A is small, various treatments that alter only the activity of the enzyme that catalyzes the formation of the proximate intermediate may markedly alter the proportion of the dose that becomes covalently bound without appreciably altering the biological half-life of the foreign compound. On the other hand, treatments that increase the activity of the enzyme that metabolizes the foreign compound along one of the

pathways that lead to innocuous metabolites will depend on the relative importance of that pathway compared to the other pathways by which the compound is eliminated from the body. Marked changes in the activity of the enzyme that catalyzes the metabolism along a minor pathway will not appreciably change either the rate of covalent binding or the proportion of the dose that becomes covalently bound because such changes would not appreciably alter the numerator or the denominator of A. On the other hand, increases in the activity of the enzyme that catalyzes the metabolism along a major pathway of detoxification will decrease both the biological half-life of the foreign compound and the proportion of the dose that becomes covalently bound, especially when these treatments do not affect the activity of the enzyme that catalyzes the formation of the proximate metabolite. It is evident, therefore, that measurements of the effects of the various treatments on the biological half-life of foreign compounds will be useful only occasionally in predicting their effects on the covalent binding of reactive metabolites of foreign compounds *in vivo*.

Estimation of Ratios from Urinary Metabolites

It is important to determine which ratio or ratios are altered by the various treatments that change the amount of covalently bound metabolite. If a treatment alters ratio A, the concentration of the proximate metabolite as well as the amount of the reactive metabolite will be changed, and thus it would be difficult to decide whether the toxicity was caused by the proximate or the reactive metabolite. In one approach for estimating the ratios, the foreign compound is administered and the relative amounts of the products (unchanged foreign compound and its metabolites) excreted in air, urine, and feces are determined. The relationships between the proportion of the dose that becomes covalently bound and the excretory products become clear when it is realized that both kinds of processes may be viewed as elimination mechanisms. When viewed in this way, it becomes evident that the fraction of the dose of the foreign compound that is eliminated by pathways other than the formation of the proximate intermediate should equal $(1 - A)$. Similarly the fraction of the proximate intermediate that is eliminated by pathways other than that leading to the formation of the reactive metabolite should equal $(1 - B)$ and the fraction of the reactive metabolite that is eliminated by pathways other than covalent binding should equal $(1 - C)$. Since A is the fraction of the dose that is converted to the proximate metabolite and AB is the fraction of the dose that is converted to the reactive metabolite, it follows that the total excretion of the foreign compound into air, urine, and feces may be represented by

$$\text{Excretory products} = \text{Dose} \times [(1 - A) + A(1 - B) + AB(1 - C)] \qquad (4)$$

Note that when B or C approaches zero, the equation degenerates and the total amount of excretory products approaches the dose. By measuring the amounts of various excretory products and grouping them according to the pathways along which they are formed from the foreign compound, investigators may estimate the various ratios. For example, suppose that 10% of a dose of a foreign compound is excreted unchanged, 40% of it is excreted as a nonreactive glucuronide conjugate of the foreign compound, 15% of it is excreted as a proximate metabolite, 25% of it is excreted as a nonreactive glucuronide of the proximate metabolite, and 9% of it is excreted as a nonreactive conjugate of the reactive metabolite. In this idealized case, the investigator would calculate that A equals 0.5, B equals 0.2, and C equals 0.1. In reality, however, the recoveries of the various metabolites are seldom complete and the analytical methods are seldom sufficiently precise to accurately estimate minor metabolites of foreign compounds. Thus the investigator may find it difficult to estimate B and impossible to estimate C from the relative amounts of the urinary metabolites alone.

It is especially noteworthy that urinary changes in the amount of a nonreactive metabolite (M_{3ii}) derived from the reactive metabolite will not always parallel changes in the proportion of the dose that becomes covalently bound. This may be illustrated by the equations for M_{3ii} and covalent binding:

$$\text{Covalent binding} = \text{Dose} \times ABC \tag{5}$$

$$M_{3ii} = \text{Dose} \times [AB(1 - C)] \tag{6}$$

Increases in the proportion of the foreign compound that is converted to the proximate metabolite (ratio A) or increases in the proportion of the proximate metabolite that is converted to the reactive metabolite (ratio B) will increase both the covalent binding and the amount of M_{3ii}. But increases in the proportion of the reactive metabolite that becomes covalently bound (ratio C) will increase the amount of covalently bound metabolite and decrease the amount of M_{3ii}.

The equations for covalent binding of a reactive metabolite and for M_{3ii} may be used to illustrate another point. Suppose that about one-tenth of the reactive metabolite was covalently bound in one animal species ($C = 0.1$) but that only about one-thousandth of the metabolite was covalently bound in another species ($C = 0.001$). Suppose also that ratio A and ratio B were identical in the two animal species. Since the covalently bound metabolite in the first species would be 100 times that in the second species, the foreign compound may produce a toxicity in the first species but not in the second, even though the pattern of urinary metabolites and the biological half-life of the foreign compound would be virtually identical in the two species. These considerations not only illustrate the dangers in attempting to relate species differences in toxicity to the pattern of drug metabolism, but they also suggest the usefulness of covalent binding studies in elucidating species differences in toxicities mediated by reactive metabolites, because only by measuring the amount of covalent binding would the significant species difference be revealed.

Estimation of Ratios from the Concentration of the Foreign Compound and Its Metabolites in Blood Plasma

When all processes are first order, measurement of the area under the plasma concentration curve (AUC) of the proximate metabolite (M_{2i}) after the intravenous administration of the foreign compound may also be useful in estimating ratio A and the denominator or ratio B. The equation for $\mathrm{AUC}_{M_{2i}}$ of the proximate metabolite is

$$\mathrm{AUC}_{M_{2i}} = \frac{A' \text{ (Dose of foreign component)}}{kV_d} \tag{7}$$

in which k is the rate constant of elimination of the proximate metabolite (that is, the denominator of B) and V_d is its volume of distribution. The value of kV_d may be estimated by measuring the area under the plasma curve of the proximate metabolite after the intravenous administration of the proximate metabolite and substitution into the following equation:

$$\mathrm{AUC}_{(M_{2i})} = \frac{\text{Dose of the proximate metabolite}}{kV_d} \tag{8}$$

This value may then be substituted into equation (7) and the equation solved for A'. The value A' calculated in this way may occasionally be smaller than the value of A calculated from the urinary metabolites. This will occur when an appreciable amount of proximate metabolite formed in a given tissue is converted to a secondary metabolite before it leaves the tissue. Thus a combination of the two methods for determining A may be useful in estimating a "metabolite first-pass effect," because the value of $1 - (A'/A)$ will represent the fraction of the proximate metabolite that is converted to secondary metabolites before it leaves the tissue in which it is formed. This method may also be useful in estimating the fraction of a reactive metabolite that is metabolized to other metabolites before it leaves the tissue in which it is formed, provided that the reactive metabolite has a sufficiently long half-life in the body to permit accurate measurements.

Superposition in First-Order Reactions

As long as all of the metabolic and elimination processes that determine the metabolic fate of a foreign compound follow first-order kinetics, the amount of the foreign compound that becomes covalently bound should be directly proportional to the dose. Thus it is important to measure the amounts of covalently bound metabolite after the administration of several different doses. If the proportion of the dose that becomes covalently bound declines as the dose is increased, it may be inferred that one or more of the numerators of the ratios is decreased. This will occur when the enzymes that catalyze the formation of the proximate metabolite or the reactive metabolite become saturated, when a cosubstrate required for the formation of the reactive metabolite becomes depleted, or when any one of these enzymes is either inhibited or destroyed by a metabolic product. Under these conditions, the toxicity may be self-limiting. However, covalent binding and toxicity may frequently be increased by dividing a large dose into smaller doses and administering them at various time intervals. On the other hand, if the proportion of the dose that becomes covalently bound increases as the dose is increased, it may be inferred that one or more terms in the denominators of the ratios is decreased. This will occur when one or more of the processes that eliminate the foreign compound along nonreactive pathways become saturated or when a cosubstrate required for the formation of a nonreactive conjugate becomes depleted. Under these conditions, there will be either a dose threshold for covalent binding and toxicity or a sharp increase in the amount of covalent binding and the severity of toxicity. The finding of such dose thresholds is especially important in drug therapy because they imply that the drug may be safe in therapeutic doses and dangerous only when the use of the drug is abused.

Studies on the pattern of urinary metabolites after low and high doses of the foreign compound frequently reveal which of the ratios is altered as the dose is increased, and thereby provide valuable clues in elucidating which pathway of metabolism of the foreign compound leads to the formation of the chemically reactive metabolite.

DIFFERENCES BETWEEN *IN VITRO* AND *IN VIVO* STUDIES

Studies *in vitro* are almost invariably required to elucidate the mechanisms and to identify the enzymes that catalyze the formation of reactive metabolites. But studies *in vitro* by themselves seldom reflect all of the interrelationships of the various factors that determine the amount of covalent binding of foreign compounds *in vivo*. Clearly, *in vitro* experiments alone cannot predict the proportions of the drug and its metabolites that are going to be excreted unchanged into air, urine, and feces, nor can they predict the conditions under which a cosubstrate in the body will limit the conversion of a foreign compound or metabolite to other metabolites. Thus it is desirable and perhaps essential to carry out both kinds of experiments with the same strain of animal in the same laboratory in order to obtain a clear picture of the relationships between the metabolism of the drug by innocuous pathways and the formation of chemically reactive metabolites.

As an aid in relating the results of *in vitro* experiments to the living animal, it might be useful to show the relationships between enzyme kinetics and pharmacokinetics. The first-order rate constant used for a given reaction in pharmacokinetics equals the clearance of the foreign compound by the enzyme divided by the volume of distribution of the foreign compound. In turn, the clearance of the foreign compound by the enzyme in a given tissue *in vivo* may be defined as the total rate of metabolism of the compound in the tissue divided by the concentration of unbound drug in that tissue. According to the Michaelis–Menten equation,

$$\text{Rate} = V_{\max}\ [S]/(K_m + [S]) \tag{9}$$

and therefore

$$\text{Clearance}_{(\text{tissue})} = \text{Rate}/[S] = V_{\max}/(K_m + [S]) \tag{10}$$

in which $V_{\max}$ is the calculated maximum rate of metabolism at infinite substrate concentration expressed as the amount of substrate metabolized to a given product per unit time. As the substrate concentration is decreased from saturating concentrations, the rate of metabolism decreases, but the clearance increases until the substrate concentration is negligible compared with the value of the Michaelis constant (K_m). At low substrate concentrations, the rate of metabolism and the clearance$_{(\text{tissue})}$ may be expressed as follows:

$$\text{Rate}_{(\text{tissue})} = V_{\max}\ [S]/K_m \tag{11}$$

$$\text{Clearance}_{(\text{tissue})} = \text{Rate}_{(\text{tissue})}/[S] = V_{\max}/K_m \tag{12}$$

The first-order rate constant may be calculated from the equation:

$$k = \text{Clearance}_{(\text{tissue})}/V_{d(\text{tissue})} \tag{13}$$

However, the "volume of distribution," $V_{d(\text{tissue})}$, used in this equation is not the "volume of distribution" as usually defined by pharmacokineticists, which is the total amount of foreign compound in the body at any given time after the distribution phases are complete divided by the concentration of the compound in the blood plasma at that time. Instead, it is defined as the total amount of foreign compound present in the body at any given time after the distribution phases are complete divided by the concentration of *unbound drug in the immediate environment of the enzyme* at that time. Thus the $V_{d(\text{tissue})}$ differs from the usual $V_{d(\text{plasma})}$ by the product of two ratios: (1) the concentration of unbound drug in the tissue divided by the concentration of unbound drug in the plasma (f_1) and (2) the proportion of the drug in the plasma that is unbound (f_2). Thus

$$V_{d(\text{plasma})} = (f_1 f_2) V_{d(\text{tissue})} \tag{14}$$

If the foreign compound is metabolized to a given metabolite in several tissues and if the concentration of unbound foreign compound at any given time is nearly identical in all the tissues in which it is metabolized, the total body clearance of the foreign compound by a given reaction would be the sum of the clearances by each tissue and may be represented by the following equations:

$$\text{Clearance}_{(\text{metabolite})} = \frac{V_{\max 1}}{K_{m1} + [S]} + \frac{V_{\max 2}}{K_{m2} + [S]} + \cdots \tag{15}$$

$$\lim_{S \to 0} \text{clearance}_{(\text{metabolite})} = (V_{\max 1}/K_{m1}) + (V_{\max 2}/K_{m2}) + \cdots \tag{16}$$

in which $V_{\max 1}$ and K_{m1} are the maximum rate of metabolism and the Michaelis constant in one tissue, $V_{\max 2}$ and K_{m2} are these values in another tissue, etc.

From these equations, the first-order constant for the formation of a given metabolite may be calculated as follows:

$$k_1 = \frac{V_{\max 1}/K_{m1}) + V_{\max 2}/K_{m2} + \cdots}{V_{d(\text{plasma})}/f_1 f_2} \tag{17}$$

Moreover, the total body rate constant for elimination of the foreign compound by all of the pathways from the body is the sum of the individual constants for the formation of the various metabolites and the elimination of the unchanged drug.

$$k_{(\text{elimination})} = k_1 + k_2 + k_3 + \cdots \tag{18}$$

Thus the rate constant of elimination cannot be calculated from *in vitro* experiments alone because such experiments cannot be used to determine the volume of distribution of the foreign compound, the proportion of the drug that is excreted unchanged, or the relative activities of the enzymes catalyzing the conversion of a foreign compound to its metabolites in the different tissues. Nevertheless, if a foreign compound is metabolized almost solely in a single tissue, and if virtually none of it is eliminated unchanged, it may be possible to obtain reasonable estimates of the ratios *A, B, C,* etc., from *in vitro* studies under certain conditions. For example, as pointed out above, ratio A is the ratio of the rate constant for the formation of the proximate metabolite divided by the rate constant of elimination. If the foreign compound is eliminated from the body by reactions that are catalyzed almost entirely by enzymes in a single organ, then both the $V_{d(\text{tissue})}$ and the concentration of unbound foreign compound will be the same for all of the reactions represented in ratio A. Under these conditions, the following equations may be used to estimate ratio A:

$$\text{Ratio } A = \frac{k_{1\text{i}}}{k_{1\text{i}} + k_{1\text{ii}} + k_{1\text{iii}}} \tag{19}$$

$$\text{Ratio } A = \frac{\text{Cl}_{1\text{i}}}{\text{Cl}_{1\text{i}} + \text{Cl}_{1\text{ii}} + \text{Cl}_{1\text{iii}}} \tag{20}$$

$$\text{Ratio } A = \frac{\text{Rate}_{1\text{i}}}{\text{Rate}_{1\text{i}} + \text{Rate}_{1\text{ii}} + \text{Rate}_{1\text{iii}}} \tag{21}$$

in which $\text{Cl}_{1\text{i}}$, $\text{Cl}_{1\text{ii}}$, and $\text{Cl}_{1\text{iii}}$ are the clearance values for the various reactions by which the foreign compound and $\text{Rate}_{1\text{i}}$, $\text{Rate}_{1\text{ii}}$, and $\text{Rate}_{1\text{iii}}$ are the rates of formation of the various primary metabolites under steady-state conditions in which the substrate concentration is negligible compared with the various K_m values for the different enzymes that catalyze these reactions. Thus, under these highly restrictive conditions, it is at least plausible to obtain an estimate of the value of ratio A by measuring the rate of formation of the proximate metabolite and the rate of disappearance of the foreign compound. Similarly, estimates of the other ratios theoretically may be obtained by measuring the rates of disappearance of the proximate metabolite and its conversion to the reactive metabolite and by measuring the rates of conversion of reactive metabolite to innocuous metabolites and to covalently bound metabolites. Thus the proportion of the dose that becomes covalently bound in the tissue may be represented by the following equations:

$$(\text{Covalent binding/dose}) = F'' = A'' B'' C'' \tag{22}$$

$$F'' = \frac{\text{Cl}_{1\text{i}}''}{\text{Cl}_{1\text{i}}'' + \text{Cl}_{1\text{ii}}'' + \text{Cl}_{1\text{iii}}''} \times \frac{\text{Cl}_{2\text{i}}''}{\text{Cl}_{2\text{i}}'' + \text{Cl}_{2\text{ii}}'' + \text{Cl}_{2\text{iii}}''} \times \frac{\text{Cl}_{3\text{i}}''}{\text{Cl}_{3\text{i}}'' + \text{Cl}_{3\text{ii}}'' + \text{Cl}_{3\text{iii}}''} \tag{23}$$

$$F'' = \frac{\text{Rate}_{1i}''}{\text{Rate}_{1i}'' + \text{Rate}_{1ii}'' + \text{Rate}_{1iii}''} \times \frac{\text{Rate}_{2i}''}{\text{Rate}_{2i}'' + \text{Rate}_{2ii}'' + \text{Rate}_{2iii}''}$$

$$\times \frac{\text{Rate}_{3i}''}{\text{Rate}_{3i}'' + \text{Rate}_{3ii}'' + \text{Rate}_{3iii}''} \tag{24}$$

Under steady-state conditions,

$$\text{Rate}_{1i}'' = \text{Rate}_{2i}'' + \text{Rate}_{2ii}'' + \text{Rate}_{2iii}'' \tag{25}$$

$$\text{Rate}_{2i}'' = \text{Rate}_{3i}'' + \text{Rate}_{3iii}'' + \text{Rate}_{3iii}'' \tag{26}$$

On substitution of (25) and (26) into (24),

$$F'' = \frac{\text{Rate}_{3i}''}{\text{Rate}_{1i}'' + \text{Rate}_{1ii}'' + \text{Rate}_{1iii}''} \tag{27}$$

Duncan *et al.* (20) have used this approach to estimate the "binding index" of different polycyclic hydrocarbons to protein, RNA, and DNA by liver cells. These workers found that the "binding index" of the reactive metabolites of these substances to DNA was approximately proportional to their carcinogenic activity in animals.

Unfortunately, most studies *in vitro* are carried out under conditions that do not reflect those occurring *in vivo.* For example, in most studies the concentrations of the foreign compound or one or more of its metabolites are so high that the enzymes that catalyze the formation of the proximate metabolite and the innocuous metabolites become saturated; hence the ratios *A, B, C,* etc., estimated under these conditions differ markedly from those that occur *in vivo.* Moreover, in many experiments, homogenates of the tissues of subcellular fractions of the tissue homogenates are used and thus some of the enzymes that catalyze the metabolism of the foreign compound or its proximate metabolites are either absent or their relative concentrations in the assay system differ markedly from those occurring in the intact tissues either because relative K_m's differ from those occurring within the tissue cells or because the concentrations of cofactors differ. For these reasons, it seems likely that results obtained from *in vitro* studies will frequently differ from those obtained *in vivo.* Nevertheless, the ratio concept provides a way of testing when the results of metabolic studies of a given foreign compound *in vitro* will adequately reflect its metabolism *in vivo* and when they will not.

CONCLUSION

In this chapter, I have discussed the approach our laboratory has followed in relating the formation of chemically reactive metabolites with the incidence and severity of toxicity and caused by foreign compounds. This approach emphasizes the need to correlate the findings of several different kinds of studies including measuring the biological half-life of the foreign compound, isolating the various urinary metabolites of the foreign compound, and determining the amount of covalent binding of reactive metabolites both *in vivo* and *in vitro*. Although such studies by themselves can neither predict the toxicity of foreign compounds nor elucidate the mechanism by which it occurs, they nevertheless have served an essential role in determining that many drugs and other foreign compounds cause liver damage through the formation of reactive metabolites. The concepts discussed in this chapter have been especially useful in reconciling seemingly conflicting data obtained in the different kinds of studies. For example, pretreatment of animals with phenobarbital increases the toxicity of acetaminophen in mice (21) but decreases it in hamsters (22). Moreover, phenobarbital pretreatment decreases covalent binding after low doses of bromobenzene but increases it after high doses of the toxicant (23). The concepts also predict possible pitfalls in attempting to relate species differences in drug toxicity with species differences in the pattern of urinary metabolites and in the biological half-lives of foreign compounds. It seems likely that studies of covalent binding will also be useful in determining whether alterations in the incidence and severity of various toxicities are due to differences in the metabolism of the foreign compound or to modifications in the mechanism of toxicity.

REFERENCES

1. E. C. Miller and J. A. Miller, Mechanisms of chemical carcinogenesis: Nature of proximate carcinogens and interactions with macromolecules, *Pharmacol. Rev.* **18,** 805–838 (1966).
2. J. A. Miller, Carcinogenesis by chemicals: An overview. G. H. A. Clowes Memorial Lecture, *Cancer Res.* **30,** 559–576 (1970).
3. P. N. Magee and J. M. Barnes, Carcinogenic nitroso compounds, *Adv. Cancer Res.* **10,** 163–246 (1967).
4. J. H. Weisburger and E. K. Weisburger, Biochemical formation and pharmacological, toxicological and pathological properties of hydroxylamines and hydroxamic acids, *Pharmacol. Rev.* **25,** 1–66 (1973).
5. A. Hollaender (ed.), *Chemical Mutagens: Principles and Methods for Their Detection,* Vols. 1 and 2, Plenum Press, New York (1971).
6. L. Fishbein, W. G. Flam, and H. L. Falk (eds.), *Chemical Mutagens: Environmental Effects on Biological Systems,* Academic Press, New York (1970).
7. B. B. Brodie, Idiosyncrasy and tolerance, in: *Drug Responses in Man,* pp. 188–213, Little, Brown, Boston (1967).

8. T. A. Loomis (ed.), *Toxicological Problems,* Vol. 2 of *Proceedings of the Fifth International Congress on Pharmacology,* pp. 48–208, Karger, Basel (1973).
9. J. D. Judah, A. E. McLean, and E. K. McClean, Biochemical mechanism of liver injury, *Am. J. Med.* **49,** 609–616 (1970).
10. R. O. Recknagel, Carbon tetrachloride hepatotoxicity, *Pharmacol. Rev.* **19,** 145–208 (1967).
11. T. F. Slater, Necrogenic action of carbon tetrachloride in the rat: A speculative mechanism based on activation, *Nature (London)* **209,** 36–40 (1966).
12. E. S. Reynolds, Comparison of early injury to liver endoplasmic reticulum by halomethanes, hexachloroethane, benzene, toluene, bromobenzene, ethionine, thioacetamide and dimethylnitrosamine, *Biochem. Pharmacol.* **21,** 2555–2561 (1972).
13. J. R. Gillette, Mechanisms of hepatic necrosis induced by halogenated aromatic hydrocarbons, in: *Workshop on Experimental Liver Injury,* pp. 239–254, Medical and Technical Publishing Co., Lancaster, England.
14. S. J. Singer, Affinity labelling of protein active sites, in: *Molecular Properties of Drug Receptors* (R. Porter and M. O'Connor, eds.), pp. 229–242, J. and A. Churchill, London (1970).
15. R. D. O'Brien, *Toxic Phosphorus Esters,* pp. 73–113, Academic Press, New York (1960).
16. R. M. Kostrzewa and D. M. Jacobowitz, Pharmacological actions of 6-hydroxydopamine, *Pharmacol. Rev.* **26,** 199–288 (1974).
17. J. R. Gillette, Factors that affect the covalent binding and toxicity of drugs, in: *Toxicological Problems* (T. A. Loomis, ed.), pp. 187–202, Vol. 2 of *Proceedings of the Fifth International Congress on Pharmacology,* Karger, Basel (1973).
18. J. R. Gillette, A perspective on the role of chemically reactive metabolites of foreign compounds in toxicity. I. Correlation of changes in covalent binding of reactive metabolites with changes in the incidence and severity of toxicity, *Biochem. Pharmacol.* **23,** 2785–2794 (1974).
19. J. R. Gillette, A perspective on the role of chemically reactive metabolites of foreign compounds in toxicity. II. Alterations in the kinetics of covalent binding, *Biochem. Pharmacol.* **23,** 2927–2938 (1974).
20. J. Duncan, P. Brookes, and A. Dipple, Metabolism and binding to cellular macromolecules of a series of hydrocarbons by mouse embryo cells in culture, *Int. J. Cancer* **4,** 813–819 (1969).
21. J. R. Mitchell, D. J. Jollow, W. Z. Potter, D. C. Davis, J. R. Gillette, and B. B. Brodie, Acetaminophen induced hepatic necrosis. I. Role of drug metabolism, *J. Pharmacol. Exp. Ther.* **187,** 185–194 (1973).
22. W. Z. Potter, S. S. Thorgeirsson, D. J. Jollow, and J. R. Mitchell, Acetaminophen induced hepatic necrosis. V. Correlation of hepatic necrosis, covalent binding and glutathione depletion in hamsters, *Pharmacology* **12,** 129–143 (1975).
23. W. D. Reid and G. Krishna, Centrolobular hepatic necrosis related to covalent binding of metabolites of halogenated aromatic hydrocarbons, *Exp. Mol. Pathol.* **18,** 80–99 (1973).

4

Biochemical Aspects of Toxic Metabolites: Formation, Detoxication, and Covalent Binding

David J. Jollow and Carole Smith
Department of Pharmacology
Medical University of South Carolina
Charleston, South Carolina 29401, U.S.A.

The concept that the metabolism of drugs and other xenobiotics *in vivo* leads to chemically inert, readily excretable products has changed dramatically in recent years. It is now appreciated that reactive alkylating, arylating, or acylating derivatives may arise during the metabolism of xenobiotics and that the covalent binding of these reactive metabolites to cellular macromolecules may initiate events leading to the destruction of specific cellular components such as cytochrome P450, or to more severe lesions such as malignant transformation, mutagenicity, and cell death (1–9). Similar mechanisms may underlie many other kinds of serious toxicities, including allergic reactions, blood dyscrasias, hemolytic anemia, and teratogenicity.

However, the fact that a drug or other xenobiotic is converted in the cell to a chemically reactive intermediate does not of necessity mean that cellular damage will result. Recent studies from several laboratories have clearly shown that the extent to which the potential toxicity of a drug is expressed depends on at least two sets of biochemical factors: (1) those concerned with the formation or detoxication of the toxic metabolite, and (2) those which influence the biological response of the cell to the initial chemical insult. Evidence for the role of the second set of biochemical factors in cytotoxicity has been presented by Farber and colleagues in their studies on mucosal cell necrosis in rats (10). These workers suggested that cell death after chemical insult may be an active process which requires the synthesis of specific proteins. Although the biochemical factors concerned with the biological response of the cell to the initial insult are clearly important for an understanding of the fundamental mechanism(s) of drug-induced cytotoxicity, the mechanisms involved are obscure and will not be discussed in this presentation.

There is now abundant evidence for the importance of the first set of biochemical factors, those concerned with the formation or detoxification of reactive metabolites. For example, depression of sulfotransferase activity in the liver by depletion of active sulfate stores *in vivo* has been shown to prevent the carcinogenic effects of *N*-hydroxy-2-acetylaminofluorene (2). Further, alteration in toxicity due to alteration in the activity of hepatic drug-metabolizing enzymes has been shown or implicated in the hepatotoxicity of a variety of compounds including 2-acetylaminofluorene, dimethylnitrosamine, pyrrolizidine alkaloids, carbon tetrachloride, bromobenzene, acetaminophen, furosemide, and isoniazid (2,6,7,11–21).

Studies on the role of the hepatic drug-metabolizing enzymes in the hepatotoxicity of bromobenzene and acetaminophen have been particularly revealing. These studies have shown that the chemically reactive hepatotoxic metabolites of bromobenzene and acetaminophen are normal metabolic intermediates in the hepatic metabolism of these compounds, and that the reactive intermediates are formed after nontoxic as well as toxic doses (22–26). Covalent binding and hepatotoxicity occur at doses at which the formation of the reactive metabolite intermediates exceeds the capacity of the glutathione transferase and (for bromobenzene) epoxide hydratase pathways to convert the reactive intermediates to innocuous products (20,23–26). Thus the severity of liver necrosis in animals after bromobenzene or acetaminophen may be increased either by pretreatment of the animals with inducers of the cytochrome P450-dependent toxic pathways or by treatment with diethylmaleate to decrease the capacity of the glutathione transferase detoxication pathway (20–26). Conversely, the animals may be protected from liver damage either by treatments which inhibit the P450-dependent toxic pathways or by treatment with cysteine and cysteamine to supplement the glutathione transferase pathway (20–26).

However, during the course of studies on the effect of pretreatment regimens on the hepatotoxicity of bromobenzene and acetaminophen, several anomalous observations have suggested that the enzymological and pharmacokinetic relationships may be more complex. For example, pretreatment of mice with phenobarbital increases their susceptibility to acetaminophen-induced liver necrosis but has little effect on the metabolic clearance of acetaminophen *in vivo* (21). In contrast, pretreatment of hamsters with phenobarbital enhances the metabolism of acetaminophen *in vivo* but has little or no effect on acetaminophen's hepatotoxicity (25).

These and other anomalous observations may be more readily understood if it is appreciated that xenobiotics are usually metabolized in the liver by more than one pathway and that inducers and inhibitors of metabolism may act simultaneously on more than one pathway of metabolism. As indicated in Fig. 1, the amount of the reactive intermediate available for reaction with cellular macromolecules (and hence the severity of the lesion) should depend on the relative activity of several pathways. Since the drug is a common substrate for toxic and nontoxic pathways, the proportion of the dose metabolized to the

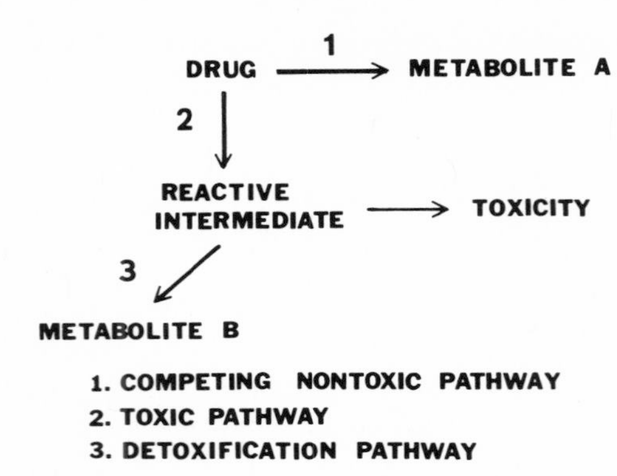

Fig. 1. Relationship between the toxic pathway, the competing nontoxic pathway, and the detoxification pathway.

reactive intermediate would depend on the relative activity of these pathways (ratio of pathway 2 to pathway 1). In turn, the amount of reactive metabolite available for covalent binding to cell macromolecules should depend on the relative activity of the pathway(s) concerned in its formation and with its detoxification (ratio of pathway 2 to pathway 3). Thus an inducer or inhibitor of metabolism acting simultaneously on more than one pathway may cause no change, an increase, or a decrease in tissue injury resulting from the action of a chemically reactive metabolite. This concept will be illustrated in this presentation by the use of two examples: the effect of 3-methylcholanthrene (3-MC) on hepatotoxicity of bromobenzene in rats, and the effect of salicylamide on acetaminophen-induced liver necrosis in hamsters.

EFFECT OF 3-METHYLCHOLANTHRENE ON BROMOBENZENE-INDUCED LIVER NECROSIS

Administration of bromobenzene to rats causes centrilobular liver necrosis (18,27). The liver damage results from the action of a chemically reactive metabolite (18–20). As shown in Table 1, pretreatment of the rats with phenobarbital enhances the metabolism of bromobenzene *in vivo* and potentiates bromobenzene-induced liver damage (18,20). In contrast, SKF 525-A inhibits the metabolism of bromobenzene and protects from liver injury (18–20). The correlation of high prolonged levels of bromobenzene in the livers of SKF 525-A-treated rats with protection from liver damage after bromobenzene clearly indicated that the hepatic necrosis was not caused by the parent compound. The association of enhanced metabolism of bromobenzene in phenobarbital-pretreated rats with enhanced liver damage led Brodie *et al.* (18) to suggest that liver damage resulted from the action of a chemically reactive metabolite of bromobenzene. Subsequent kinetic studies *in vitro* and *in vivo*

Table 1. Effect of Pretreatments on the Metabolic Half-Life of Bromobenzene and the Severity of Hepatic Centrilobular Necrosis in the Rat[a]

Pretreatment	Dose of bromobenzene (mmol/kg)	Severity of centrilobular necrosis	Metabolic half-life, whole body (min)
None	10	Extensive necrosis	9.8 ± 0.80
SKF 525A	10	Little or no effect	15.5 ± 1.80
3-MC	10	Little or no effect	9.2 ± 0.70
Phenobarbital	1.5	Massive necrosis	5.5 ± 0.50

[a]Data from Zampaglione *et al.* (20).

have strongly suggested that the chemically reactive metabolite is 3,4-bromobenzene oxide (23).

However, during the course of these studies there was an anomalous observation. Pretreatment of rats with another inducer of the hepatic drug-metabolizing enzymes, 3-MC, was found to protect the animals from bromobenzene-dependent liver injury even though companion *in vitro* studies had shown that 3-MC pretreatment resulted in a modest induction (two- to three-fold) of bromobenzene metabolism (20). Thus it was apparent that induction of bromobenzene metabolism could be associated with protection as well as potentiation of liver damage.

To examine whether the protective effect of 3-MC was due to an alteration in the metabolic disposition of bromobenzene, the effect of this treatment on the composition of the urinary metabolites of bromobenzene was determined (20). These studies indicated that the 3-MC-pretreated rats excreted appreciably larger amounts of bromocatechol and bromophenyldihydrodiol metabolites of bromobenzene than did the normal animals (Table 2). Since these metabolites arise from the chemically reactive metabolite 3,4-bromobenzene oxide by way

Table 2. Effect of 3-Methylcholanthrene (3-MC) on the Composition of Rat Urinary Metabolites of Bromobenzene (10 mmol/kg, i.p.)[a]

	Percent of total urinary metabolites				
	Bromophenyl-mercapturic acid	4-Bromo-phenol	Bromo-catechol	Bromophenyl-dihydrodiol	2-Bromo-phenol
Normal rats	48 ± 5	37 ± 4	6 ± 2	4 ± 1	4 ± 1
3-MC-induced rats	31 ± 4	20 ± 3	10 ± 1	17 ± 2	21 ± 3

[a]Bromobenzene was administered i.p. in oil. The urinary metabolites were collected for 48 h over dry ice, and fractionated after glucuronidase and sulfatase digestion by thin-layer chromatography. The values given are the mean ± SD for eight rats in each group. Recovery of administered bromobenzene was >90%. Data from Zampaglione *et al.* (20).

of the epoxide hydrase pathway, these data suggested (1) that 3-MC pretreatment enhanced hepatic epoxide hydrase activity *in vivo,* and (2) that the extent of enhancement of epoxide hydrase activity was greater than any enhancement in the rate of formation of 3,4-bromobenzene oxide. These conclusions have been supported by parallel *in vitro* studies (26,28) and by the observations of Oesch, Jerina, and Daly (29,30) that 3-MC pretreatment of rats enhances a variety of epoxide hydrase-catalyzed activities. Thus it seemed reasonable to conclude that the protective effect of 3-methylcholanthrene results from its ability to enhance the capacity of the liver to detoxify 3,4-bromobenzene oxide by conversion of the chemically reactive oxide to innocuous bromophenyldihydrodiol.

However, the protective effect of 3-MC may not be due entirely to this effect. The urinary metabolite studies (Table 2) also indicated that 3-MC pretreatment caused a marked increase in the metabolism of bromobenzene to 2-bromophenol. We speculated that 2-bromophenol might arise from non-enzymatic rearrangement of a 2,3-epoxide in a fashion analogous to that in which 4-bromophenol arises from the 3,4-epoxide (20,31). Further fractionation studies revealed that each of the metabolite fractions, bromophenylmercapturic acid, bromocatechol, and bromophenyldihydrodiol, could be resolved into two components by thin-layer chromatography. A combination of chemical, mass spectrometric, and nuclear magnetic resonance studies led to the identification of the two catechol metabolites as 2,3-bromocatechol and 3,4-bromocatechol, and the two dihydrodiol metabolites as the 2,3- and 3,4-dihydrodiol derivatives (31). Since the formation of these 2,3 metabolites of bromobenzene was inhibited by SKF 525-A to a similar extent as the formation of the 3,4 metabolites, these data indicate that bromobenzene is metabolized in rats by two primary cytochrome P450- (or P448-) dependent oxidative pathways—a 3,4-bromobenzene oxide synthetase and a 2,3-bromobenzene oxide synthetase (Fig. 2).

Quantitative studies on the relative importance of 2,3- and 3,4-bromobenzene metabolites (Table 3) indicated that 2,3-bromobenzene oxide synthetase was of only minor importance in the normal rats and that this enzyme activity was not enhanced by phenobarbital pretreatment. In contrast, in 3-MC-pretreated rats, the 2,3 metabolites and the 3,4 metabolites occurred in approximately equal quantities. Since the induction of bromobenzene metabolism in these rats by 3-MC was approximately two- to threefold (20), this observation suggests that the enhancement of primary oxidation of bromobenzene by 3-MC pretreatment was largely due to the induction of the 2,3-bromobenzene oxide synthetase activity with little or no change in the activity of the hepatic 3,4-bromobenzene oxide synthetase. It follows that, in contrast to normal rats, 3-MC-pretreated rats convert only half of the dose of bromobenzene to the previously identified chemically reactive metabolite, 3,4-bromobenzene oxide. Thus, the 2,3-oxide synthetase activity in 3-MC-pretreated rats may represent a significant competing nontoxic pathway, diverting bromobenzene from its toxic pathway.

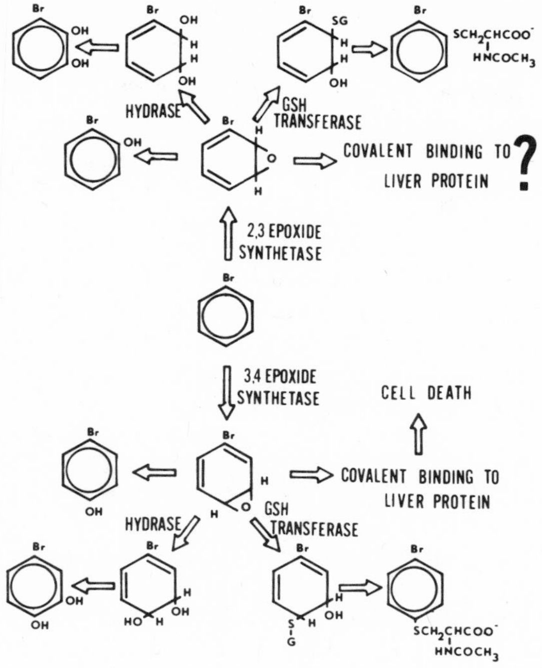

Fig. 2. Pathways of bromobenzene metabolism.

It is apparent that this argument would be valid only if 2,3-bromobenzene oxide is innocuous. Little direct evidence is available on this point. Indirect evidence suggests that although 2,3-bromobenzene oxide is highly electrophilic in nature, it lacks sufficient stability to move from its site of formation into proximity with vital cellular macromolecules (28,31, and unpublished observations). Regardless of this uncertainty, it is clear that the hepatotoxicity of bromobenzene depends on the relative activity of several enzymatic pathways and not merely on the rate of formation of its reactive toxic metabolite.

EFFECT OF SALICYLAMIDE ON ACETAMINOPHEN-INDUCED NECROSIS

Our present understanding of the relationship between the metabolism of acetaminophen and its hepatotoxicity is summarized in Fig. 3. Conjugation reactions at the 4-hydroxy group form the nontoxic glucuronide and ethereal sulfate derivatives, which account for about 80% of the clearance of the drug in all species studied (26,32). In addition, acetaminophen is oxidized by the cytochrome P450-dependent drug-metabolizing enzyme(s) to a reactive intermediate postulated to be the hydroxamic acid and/or its acetimidoquinone

Table 3. Effect of Pretreatments on the Composition of Rat Urinary Metabolites of Bromobenzene[a]

Pretreatment	Percent of total urinary metabolites						
	Bromophenol		Bromocatechol		Bromophenyldihydrodiol		
	4-isomer	2-isomer	3,4-isomer	2,3-isomer	3,4-isomer	2,3-isomer	Ratio 2,3-isomer/3,4-isomer
None	40	4	4	TR[b]	3	TR	0.1
3-MC	20	21	8	5	8	11	1.05
Phenobarbital	38	1	8	N.D.[c]	6	N.D.	0.01

[a] ^{14}C-Bromobenzene was administered i.p. in sesame oil. The urinary metabolites were collected for 48 h over dry ice. The urines were digested with glucuronidase and sulfatase and the liberated metabolites were separated into 2-bromophenol, 4-bromophenol, bromocatechol, and bromophenyldehydrodiol fractions by thin-layer chromatography. The 2,3- and 3,4-isomers of bromocatechol and bromophinyldihydrodiol were resolved by further thin-layer chromatography on silicic acid using the open tank technique (solvent system: chloroform–methanol–acetic acid, 97:3:0.2). Data from Jollow *et al.* (31) and unpublished observations of D. Jollow, J. R. Mitchell, N. Zampaglione, and J. R. Gillette.

[b] TR, trace.

[c] N.D., not detected.

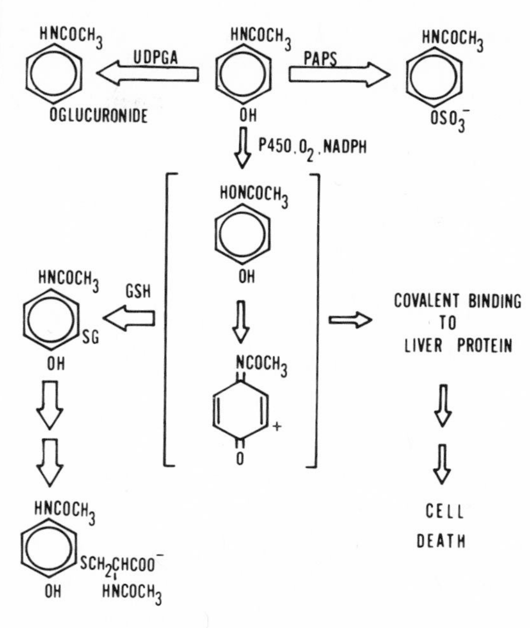

Fig. 3. Pathways of acetaminophen metabolism. *N*-Hydroxyacetaminophen and *N*-acetylbenzoquinoimine are postulated intermediates.

derivative (6). This intermediate is preferentially conjugated with hepatic glutathione, and after further metabolism appears in the urine as the mercapturic acid derivative (24,26). The contribution of this pathway to the total clearance varies from 3–5% in rats and man, to 12–15% in mice and hamsters (26,33). Covalent binding of the reactive intermediate to hepatic macromolecules is insignificant after low doses of the drug (34). However, after high doses of acetaminophen, conjugation of the reactive intermediate with glutathione acts to deplete hepatic glutathione stores. When the depletion of glutathione exceeds 70% of the initial hepatic levels, covalent binding of the reactive metabolite to hepatic protein occurs and, in some unknown fashion, appears to initiate the events leading to cell death (24,26,34).

Since the glucuronidation and sulfation pathways may be regarded as competing nontoxic pathways, the theoretical considerations illustrated in Fig. 1 predict that the inhibition of these pathways would result in enhanced clearance by the toxic pathway, and hence potentiation of acetaminophen's hepatotoxicity. Salicylamide was used to test this postulate since it is known that coadministration of salicylamide delays the metabolic clearance of acetaminophen by these pathways (35).

As shown in Table 4, coadministration of salicylamide with acetaminophen significantly potentiates liver necrosis after acetaminophen. Thus a dose of 200

Table 4. Effect of Salicylamide on Acetaminophen-Induced Liver Necrosis in Normal and 3-Methylcholanthrene (3-MC) Pretreated Hamsters[a]

Acetaminophen (mg/kg)	Salicylamide (mg/kg)	Number of hamsters	Percent of animals: Extent of necrosis[b]					
			0	+	++	+++	++++	Mortality[c]
Normal hamsters								
200	Nil	25	100	0	0	0	0	0
300	Nil	31	19	22	29	26	3	0
400	Nil	23	0	9	13	21	43	13
100	500	20	100	0	0	0	0	0
150	500	32	34	37	19	9	0	0
200	500	30	0	10	27	37	27	0
Nil	500	35	100	0	0	0	0	0
3-MC-pretreated hamsters								
100	Nil	20	100	0	0	0	0	0
150	Nil	20	35	30	25	10	0	0
250	Nil	20	0	10	30	20	30	10
20	500	25	100	0	0	0	0	0
50	500	17	65	35	0	0	0	0
100	500	20	0	10	10	20	55	5

[a] D. Jollow and C. Smith (unpublished observations).
[b] Necrosis was scored 0, absent; +, <5%; ++, 6–25%; +++, 26–50%; ++++, >50% of hepatocytes.
[c] Animals surviving for more than 12 h but less than 24 h after acetaminophen.

mg of acetaminophen/kg, which when given alone did not cause liver necrosis, was highly hepatotoxic when administered with salicylamide. Salicylamide when given alone did not cause liver necrosis in these animals (Table 4), even at doses greater than 500 mg/kg (data not shown). The effect of salicylamide was even more pronounced in hamsters which had been pretreated with 3-MC (Table 4). In these animals, 100 mg acetaminophen/kg caused no liver necrosis when given alone, but elicited severe necrosis in all animals when salicylamide was coadministered. Significant liver damage in animals receiving both drugs was seen even after only 50 mg of acetaminophen/kg.

To determine the effect of salicylamide on acetaminophen metabolism *in vivo*, [^{3}H]acetaminophen was administered to groups of hamsters with and without coadministration of salicylamide. At various time intervals, animals were sacrificed and unmetabolized [^{3}H]acetaminophen in the whole carcass plus excreta was determined. Calculation of the metabolic half-life of acetaminophen from these data revealed that salicylamide dramatically delayed the metabolic clearance of acetaminophen. The metabolic half-life increased from about 75 min in animals receiving acetaminophen alone to over 500 min in animals receiving both drugs.

pretreated rats less 3,4-bromobenzene oxide is formed in hepatocytes which have an increased capacity to detoxify it (increased epoxide hydratase activity plus possibly unchanged glutathione transferase activity). The balance between the toxic pathway and the protective mechanisms within the hepatocytes appears to have altered in favor of the protective mechanism.

It should be noted that inherent in all the above arguments is the assumption that the inhibitors and inducers used in these studies do not affect the biological response of the hepatocytes to the chemical insult (covalent binding). Clearly, interactions on this level may be expected to have profound effects on the severity of drug-induced lesions. Although we have no information about how the various treatment regimens may alter the biological response of hepatocytes, this possibility should be carefully considered since it may form the basis for a broadly applicable treatment regimen for drug overdose and other adverse drug effects in man.

Since salicylamide and acetaminophen are frequently compounded in analgesic preparations, the observation that salicylamide dramatically potentiates acetaminophen-induced liver necrosis in hamsters raises the possibility of a significant drug–drug interaction in man. Hence it must be emphasized that the doses of salicylamide used in this study are appreciably greater than those used therapeutically. Indeed, other studies on the dose of salicylamide needed for potentiation clearly indicate that liver damage is not to be expected after therapeutic doses of these drugs (Jollow, unpublished observations).

However, there are two areas of clinical concern which require special study. The first is the neonate, since it is well known that the neonate may have low glucuronidation capacity. Thus the favorable therapeutic ratio of acetaminophen in adults (greater than 10:1) may not be applicable to the newborn. Further depression of glucuronidation by competition of salicylamide may decrease clearance by this major pathway and hence increase clearance by the toxic pathway, with concomitant enhancement of risk. It should be emphasized, however, that the neonate may have an equally low activity of the toxic pathway (equivalent to equal inhibition of pathways 1 and 2, Fig. 1) and hence not be more at risk than the adult population. Clinical studies are in progress to establish the pattern of metabolism of acetaminophen in the neonate and the capacity of its glutathione threshold.

Acetaminophen causes fulminant hepatic necrosis in man when overdoses are ingested (37,38). Although hepatic damage may be expected after doses greater than 15 g, no clear relationship has been established between dose and severity of liver injury (38). Prescott and colleagues (38) have reported that liver damage is more extensive after large doses of acetaminophen in those individuals who clear the drug slowly (plasma half-life greater than 4 h compared with about 2 h in individuals at less risk). Since normal volunteers given therapeutic doses of acetaminophen showed plasma half-lives of the drug of less than 4 h, Prescott *et al.* suggested that the longer half-life of acetaminophen in overdosed patients at greatest risk of severe liver necrosis might be due to loss of drug-metabolizing activity in an early stage of liver damage. This possibility however, has not been supported by recent animal studies which indicated that drug-metabolizing activity in the livers of hamsters receiving hepatotoxic doses of acetaminophen declined significantly only at a late stage of the development of the liver necrosis (39). The present studies on salicylamide indicate that the enhancement of acetaminophen hepatotoxicity results from suppression of glucuronidation and sulfation activities, with a resultant increase in the metabolic half-life of acetaminophen and an increase in the proportion of the dose metabolized by the toxic pathway. Capacity-limited metabolism in man for the sulfation pathway has been reported in the metabolism of acetaminophen, salicylamide, and salicylic acid after high therapeutic doses of these drugs (35,40,41). Although less certain, the data also suggested that clearance of these drugs by glucuronidation may also be capacity-limited in some individuals. Thus the second area of clinical concern is whether there is a population at greater risk of liver damage after acetaminophen due to a lower capacity of glucuronidation and sulfation pathways. Clinical studies are in progress to determine if individuals who show evidence of saturation of sulfation and glucuronidation pathways of metabolism after high therapeutic doses of acetaminophen metabolize more of the drug by the toxic pathway (excrete more acetaminophen-mercapturic acid) at these doses than at lower therapeutic doses and whether this effect is aggravated by coadministration of salicylamide.

The concept that variation in susceptibility to acetaminophen-induced liver necrosis may be due to variation in the ratio of the activities of the competing nontoxic pathways to the toxic pathway has broad applications to the idiosyncratic nature of drug-induced lesions in man. Idiosyncratic drug-induced liver lesions in man are characterized by low incidence (generally much less than 1 to 100) and an apparent lack of a dose–response relationship (42,43). These characteristics are consistent with the postulate that drug-induced lesions result from the action of a chemically reactive toxic metabolite, produced in all individuals, but exerting toxicity only in the rare individual relatively deficient in the normal protective mechanisms. Thus the idiosyncratic nature of drug-induced tissue lesions in man may be viewed as a consequence of an unfavorable balance between the biochemical pathways responsible for the formation of a toxic metabolite and those pathways responsible for alternate nontoxic metabolism and/or the detoxification of the reactive intermediate.

ACKNOWLEDGMENT

Support by U.S.P.H.S. Grant GM18176 for the studies on the effect of salicylamide on acetaminophen-induced liver necrosis is gratefully acknowledged.

REFERENCES

1. E. C. Miller and J. A. Miller, Mechanisms of chemical carcinogenesis: Nature of proximate carcinogens and interactions with macromolecules, *Pharmacol. Rev.* **18,** 805–838 (1966).
2. J. A. Miller, Carcinogenesis by chemicals: An overview, G. H. A. Clowes Memorial Lecture, *Cancer Res.* **30,** 559–576 (1970).
3. P. N. Magee and J. M. Barnes, Carcinogenic nitroso compounds, *Adv. Cancer Res.* **10,** 163–246 (1967).
4. J. H. Weisburger and E. K. Weisburger, Biochemical formation and pharmacological, toxicological and pathological properties of hydroxylamines and hydroxamic acids, *Pharmacol. Rev.* **25,** 1–66 (1973).
5. E. Farber, Biochemistry of carcinogenesis, *Cancer Res.* **28,** 1859–1869 (1968).
6. J. R. Mitchell and D. J. Jollow, Metabolic activation of drugs to toxic substances, *Gasteroenterology* **68,** 392–410 (1975).
7. J. R. Mitchell, D. J. Jollow, J. R. Gillett, and B. B. Brodie, Drug metabolism as a cause of drug toxicity, *Drug. Metab. Disp.* **1,** 418–23 (1973).
8. B. J. Norman, R. E. Poore, and R. A. Neal, Studies of binding of sulfur released in the mixed-function oxidase catalyzed metabolism of diethyl-*p*-nitrophenylphosphorothionate (parathion) to diethyl *p*-nitrophenol phosphate (paraoxon), *Biochem. Pharmacol.* **23,** 1733–1744 (1974).
9. B. N. Ames, W. E. Durston, E. Yamasaki, and F. D. Lee, Carcinogens are mutagens: A simple test system combining liver homogenates for activiation and bacteria for detection, *Proc. Natl. Acad. Sci. USA* **70,** 2281–2285 (1973).
10. E. Farber, R. S. Verbin, and M. Lieberman, Cell suicide and cell death, in: *Symposium on Mechanisms of Toxicity* (W. Aldridge, ed.), pp. 163–173, MacMillan and Co., London (1971).
11. E. C. Miller, J. A. Miller, R. R. Brown, and J. C. MacDonald, On the protective action of certain polycyclic aromatic hydrocarbons against carcinogenesis by aminoazo dyes and 2-acetylaminofluorene, *Cancer Res.* **18,** 469–477 (1958).
12. L. Fiume, G. Campadelli-Fiume, P. N. Magee, and J. Holsman, Cellular injury and carcinogenesis (Inhibition of metabolism of dimethylnitrosamine by aminoacetonitrile), *Biochem. J.* **120,** 601–605 (1970).
13. P. F. Swamm and A. E. M. McLean, Cellular injury and carcinogenesis (The effect of a protein-free high-carbohydrate diet on the metabolism of dimethylnitrosamine in the rat), *Biochem. J.* **124,** 283–288 (1971).
14. A. R. Mattocks, Acute hepatotoxicity and pyrrolic metabolites in rats dosed with pyrrolizidine alkaloids, *Chem.-Biol. Interactions* **5,** 227–242 (1972).
15. I. N. H. White, A. R. Mattocks, and W. H. Butler, The conversion of the pyrrolizidine alkaloid Petrorsine to pyrrolic derivatives *in vivo* and *in vitro* and its acute toxicity to various animal species, *Chem.-Biol. Interactions* **6,** 207–218 (1973).
16. A. E. M. McLean and E. K. McLean, Diet and toxicity, *Br. Med. Bull.* **25,** 278–281 (1969).
17. G. J. Traiger and G. L. Plaa, Differences in the potentiation of carbon tetrachloride in rats by ethanol and isopropanol pretreatments, *Toxicol. Appl. Pharmacol.* **20,** 105–112 (1971).
18. B. B. Brodie, W. D. Reid, A. K. Cho, G. Sipes, G. Krishna, and J. R. Gillette, Possible mechanism of liver necrosis caused by aromatic organic compounds, *Proc. Natl. Acad. Sci. USA* **68,** 160–164 (1971).
19. J. R. Mitchell, W. D. Reid, B. Christie, J. Moskowitz, G. Krishna, and B. B. Brodie, Bromobenzene-induced hepatic necrosis: Species differences and protection by SKF 525-A, *Res. Commun. Chem. Pathol. Pharmacol.* **2,** 877–888 (1971).

20. N. Zampaglione, D. J. Jollow, J. R. Mitchell, B. Stripp, M. Hamrick, and J. R. Gillette, Role of detoxifying enzymes in bromobenzene-induced liver necrosis, *J. Pharmacol. Exp. Ther.* **187,** 218–227 (1973).
21. J. R. Mitchell, D. J. Jollow, W. Z. Potter, D. C. Davis, J. R. Gillette, and B. B. Brodie, Acetaminophen-induced hepatic necrosis. I. Role of drug metabolism, *J. Pharmacol. Exp. Ther.* **187,** 185 (1973).
22. J. R. Mitchell and D. J. Jollow, Role of metabolic activation in chemical carcinogenesis and in drug-induced hepatic injury, in: *Drugs and the Liver* (W. Gerok and K. Sickinger, eds.), pp. 395–416, Schattaner Verlag, Stuttgart (1975).
23. D. J. Jollow, J. R. Mitchell, N. Zampaglione, and J. R. Gillette, Bromobenzene-induced liver necrosis. Protective role of glutathione and evidence for 3,4-bromobenzene oxide as the hepatotoxic metabolite, *Pharmacology* **11,** 151 (1974).
24. J. R. Mitchell, D. J. Jollow, W. Z. Potter, J. R. Gillette, and B. B. Brodie, Acetaminophen-induced hepatic necrosis. IV. Protective role of glutathione, *J. Pharmacol. Exp. Ther.* **187,** 211–217 (1973).
25. W. Z. Potter, S. S. Thorgeirsson, D. J. Jollow, and J. R. Mitchell, Acetaminophen-induced hepatic necrosis. V. Correlation of hepatic necrosis, covalent binding and glutathione depletion in hamsters, *Pharmacology* **12,** 129–143 (1974).
26. D. J. Jollow, S. S. Thorgeirsson, W. Z. Potter, M. Hashimoto, and J. R. Mitchell, Acetaminophen-induced hepatic necrosis. VI. Metabolic disposition of toxic and non-toxic doses of acetaminophen, *Pharmacology* **12,** 251–271 (1974).
27. D. Koch-Weser, J. De La Huerga, and H. Popper, Hepatic necrosis due to bromobenzene and its dependence upon available sulfur amino acids, *Proc. Soc. Exp. Biol. Med.* **79,** 196–198 (1952).
28. D. J. Jollow, N. Zampaglione, and J. R. Gillette, Mechanism of protection from bromobenzene hepatotoxicity by 3-methylcholanthrene, *Pharmacologist* **13,** 288, abst. (1971).
29. F. Oesch, Mammalian epoxide hydrases: Inducible enzymes catalyzing the inactivation of carcinogenic and cytotoxic metabolites derived from aromatic and oleofinic compounds, *Xenobiotica* **3,** 305–340 (1973).
30. F. Oesch, D. M. Jerina, and J. Daly, A radiometric assay for hepatic epoxide hydrase activity with [7-^{3}H]styrene oxide, *Biochim. Biophys. Acta* **227,** 685 (1971).
31. D. J. Jollow, J. R. Mitchell, N. Zampaglione, and J. R. Gillette, 2,3-Bromophenyl epoxide as a probable metabolite of bromobenzene: Induction by 3-methylcholanthrene, *Fifth International Congress on Pharmacology,* abst. 698 (1972).
32. B. B. Brodie and J. Axelrod, The estimation of acetanilide and its metabolic products, aniline, *N*-acetyl-*p*-aminophenol and *p*-aminophenol (free and total conjugated) in biological fluids and tissues, *J. Pharmacol. Exp. Ther.* **94,** 22–28 (1948).
33. J. R. Mitchell, S. S. Thorgeirsson, W. Z. Potter, and D. J. Jollow, Acetaminophen-induced hepatic injury: Protective role of glutathione in man and rationale for therapy, *Clin. Pharmacol. Ther.* **16,** 676–684 (1974).
34. D. J. Jollow, J. R. Mitchell, W. Z. Potter, D. C. Davis, J. R. Gillette, and B. B. Brodie, Acetaminophen-induced hepatic necrosis. II. Role of covalent binding *in vivo, J. Pharmacol. Exp. Ther.* **187,** 195–202 (1973).
35. G. Levy and H. Yamada, Drug biotransformation in man. III. Acetaminophen and salicylamide, *J. Pharm. Sci.* **60,** 215–221 (1971).
36. W. Z. Potter, D. C. Davis, J. R. Mitchell, D. J. Jollow, J. R. Gillette, and B. B. Brodie, Acetaminophen-induced hepatic necrosis. III. Cytochrome-P-450-mediated covalent binding *in vitro, J. Pharmacol. Exp. Ther.* **187,** 203–210 (1973).
37. A. T. Proudfoot and N. Wright, Acute paracetamol poisoning, *Br. Med. J.* **3,** 557–558 (1970).

38. L. F. Prescott, N. Wright, P. Roscoe, and S. S. Brown, Plasma-paracetamol half-life and hepatic necrosis in patients with paracetamol overdosage, *Lancet* **i**, 519–522 (1971).
39. S. S. Thorgeirsson, H. Sasame, W. Z. Potter, W. L. Nelson, D. J. Jollow, and J. R. Mitchell, Biochemical changes after acetaminophen- and furosemide-induced liver injury, *Pharmacology,* **14**, 205–217 (1976).
40. G. Levy and T. Matsuzawa, Pharmacokinetics of salicylamide elimination in man, *J. Pharmacol. Exp. Ther.* **156**, 285–293 (1967).
41. G. Levy and J. A. Procknal, Drug biotransformation interactions in man. I. Mutual inhibition in glucuronide formation of salicylic acid and salicylamide in man, *J. Pharm. Sci.* **57**, 1330–1335 (1968).
42. G. Klastin, Drug-induced hepatic injury, in: *The Liver and Its Diseases* (F. Schaffner, S. Sherlock, and C. Leevy, eds.), pp. 163–178, Intercontinental Medical Book Corp., New York (1974).
43. H. Zimmerman, Hepatic injury by therapeutic agents, in: *The Liver: Normal and Abnormal Functions,* Part A (F. Becker, ed), pp. 225–284, Dekker, New York (1974).

5

Discussion

Snyder asked Miller whether the non-P450 microsomal mixed-function oxidase which metabolized MAB is inducible and if it is subject to specific inhibitors. Miller replied that the enzyme activity is not very inducible by 3-MC or phenobarbital. He considered the enzyme to be essentially the same as the primary and secondary amine oxidase isolated by Ziegler from pig liver. The purified flavoprotein enzyme, supplied by Ziegler, catalyzes the *N*-hydroxylation of MAB. *N*-Hydroxy-MAB was isolated from the reaction mixture. Miller noted that *N*-hydroxy-MAB undergoes further oxidation in the microsomal fractions used in his study, and is not recoverable. Further discussion concerned the substrate specificity of Ziegler's amine oxidase and the capacity of cytochrome P450-dependent microsomal enzymes to oxidize amides.

Jollow was asked whether salicylamide increased the mortality after acetaminophen. He replied that the salicylamide did not greatly increase the number of hamsters dying within a few hours after the administration of the doses of acetaminophen used in these studies, as might be expected if salicylamide increased acute toxicity (e.g., CNS depression) due to the enhanced levels of acetaminophen in the plasma. However, salicylamide did significantly increase the number of animals dying at later time periods, i.e., 12–24 h after acetaminophen. Since examination of the livers of several hamsters which had died at 14–16 h after acetaminophen administration indicated that the livers were very extensively damaged, Dr. Jollow felt that these deaths were probably a consequence of the liver injury. However, since the livers of such late-dying hamsters were not always suitable for histological examination, these animals were listed separately in the experimental results.

In reply to a request by Oesch for the evidence that acetaminophen is *N*-hydroxylated by a cytochrome P450-dependent pathway, Jollow replied that the *N*-hydroxylation of acetaminophen has not been demonstrated directly due to the instability of this compound. However, using 2-acetylaminofluorene and *p*-chloroacetanilide as substrates, microsomal *N*-hydroxylation was found to be CO sensitive, and to be inhibited by piperonyl butoxide and by a specific antibody prepared against cytochrome *c* reductase. Pretreatment of the animals with cobaltous chloride to reduce microsomal P450 suppressed *N*-hydroxylation. The formation of the chemically reactive covalent-binding metabolite of acetaminophen by microsomes is similarly inhibited by these procedures.

In regard to the suggestion that the protective effect of 3-MC pretreatment for bromobenzene-induced liver necrosis might be due to the induction of epoxide hydratase, Oesch pointed out that phenobarbital is a better inducer of epoxide hydratase in rats than is 3-MC. Oesch agreed that protection could be due to induction by 3-MC of an alternate, less toxic pathway of bromobenzene metabolism but suggested that a closer coupling between the monooxygenase and the epoxide hydratase in 3-MC rats as compared to normals could also explain the protection. In reply, Jollow pointed out that the induction of epoxide hydratase in rats by phenobarbital is offset by an equally great or greater induction of 3,4-bromobenzene oxide synthetase (about eight- to tenfold), whereas in 3-MC rats the induction of the hydratase appears to be greater than the enhancement of the rate of primary oxidation (about twofold). He emphasized that the relative activity of the toxic and detoxication pathways must be considered in the assessment of the effects of treatments on the susceptibility of the liver to injury.

Jerina described some recent microsomal trapping experiments with chlorobenzene as substrate, in which it was possible to trap the 3,4-chlorobenzene oxide but not the 2,3-chlorobenzene oxide. The experiments indicated that 2,3-chlorobenzene oxide may be less toxic since it appears not to leave the endoplasmic reticulum intact, entering the aqueous medium mainly as the phenol. Jerina noted that extrapolation of these *in vitro* data to the intact animal is difficult since the epoxide hydratase-catalyzed metabolites (dihydrodiol and catechol) were not produced in these microsomes but do occur as urinary metabolites. Jollow described similar experiments comparing the metabolism of bromobenzene *in vivo* and *in vitro* (9000*g* supernatant fraction of liver). Agreement between *in vivo* and *in vitro* metabolism was good for normal animals but poor for 3-MC-treated animals. The 9000*g* supernatant fraction from 3-MC animals showed decreased production of the 2,3-diol, the 2,3-catechol, and the 2,3-mercapturic acid, and enhanced formation of 2-bromophenol. These data also suggest that the 2,3-epoxide formed in the membrane may not be stable enough to get to the phase II enzymes. The enhanced conversion to 2-bromophenol suggests that less 2,3-bromobenzene oxide would be available to arylate supernatant protein, and hence presumably would be less toxic.

II

Formation of Reactive Intermediates

6

Mechanism of Microsomal Monooxygenases and Drug Toxicity

Volker Ullrich
Department of Physiological Chemistry
D-665 Homburg-Saar
German Federal Republic

Lipophilic chemicals and drugs are generally converted to more water-soluble derivatives in the body, allowing a more rapid excretion of these frequently harmful compounds. The term "detoxication mechanisms" was introduced for these conversions, although it was recognized very early that a number of compounds were metabolized to more toxic agents (1). Today there is definite proof that the enzymatic reactions involved in the elimination of one drug may cause an activation of another drug which could result in cell damage or even cancer. Therefore, the booming research in drug metabolism during the last two decades has also provided the knowledge for our understanding of drug toxicity.

The enzymes of major importance for the metabolism of foreign compounds are associated with the endoplasmic reticulum in a variety of organs and belong to the group of monooxygenases (2,3). Using molecular oxygen and pyridine nucleotides as cofactors, these enzymes introduce an oxygen atom into a substrate. This occurs by a rather sophisticated multistep mechanism (4), and, as will be shown later, each step may be involved in drug toxicity. Before discussing the details of the monooxygenation mechanism, it is necessary to define the components of the system and the way they interact with each other.

COMPONENTS OF THE MICROSOMAL MONOOXYGENASE SYSTEM

The enzyme that binds a lipophilic chemical or drug is a hemoprotein called cytochrome P450 (5). Electrons for the reduction of this cytochrome are

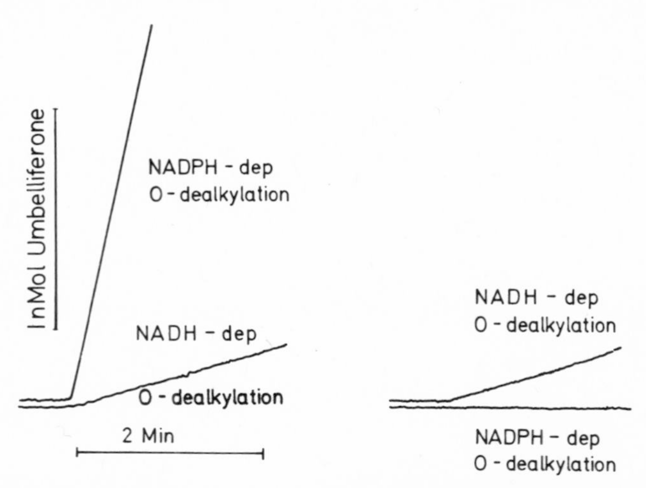

Fig. 1. Effect of trypsin digestion on the NADPH- and NADH-dependent O-dealkylation of 7-ethoxycoumarin. The assays were performed as previously described (12). A suspension of microsomes from rat liver (phenobarbital-induced) containing 5 mg of protein/ml was incubated with 15 μg of trypsin for 10 min at 30°C. Trypsin was omitted in the control (left side).

provided mainly by NADPH, with about 10–30% of the total activity being provided by NADH. Since pyridine nucleotides cannot donate electrons directly to the ferric heme, additional electron transport proteins must be part of the monooxygenase system. An NADPH-specific flavoprotein has been isolated and characterized (7) which mediates the electron transfer from NADPH to cytochrome P450 and which, together with purified cytochrome P450 and a lipid factor, reconstitutes the monooxygenase activity (8). The reductase can be easily digested from the microsomal membrane by trypsin, causing complete loss of the microsomal NADPH-supported monooxygenase activity (9). As can be seen from Fig. 1, the NADH-dependent activity remains after trypsin treatment, indicating that a different electron transport system is involved.

A similar conclusion could be derived from antibody experiments. When immunoglobulins against microsomal NADPH-cytochrome *c* reductase were employed, only the NADPH pathway was blocked but not the NADH-dependent monooxygenase reaction (10). On the other hand, the antibody against cytochrome b_5 was found to inhibit an NADH-dependent *N*-dealkylation (11), and the same antibody as well as the antiserum against NADH-cytochrome b_5 reductase prepared by Omura also inhibited the NADH-supported *O*-dealkylation of ethoxycoumarin but not with NADPH as a cosubstrate, as shown in Figs. 2 and 3.

The high sensitivity of the 7-ethoxycoumarin test (12) allows one to study the activity down to very low NADH concentrations. As can be seen from Fig. 4, the specific activity remains constant even in the micromolar range. This agrees

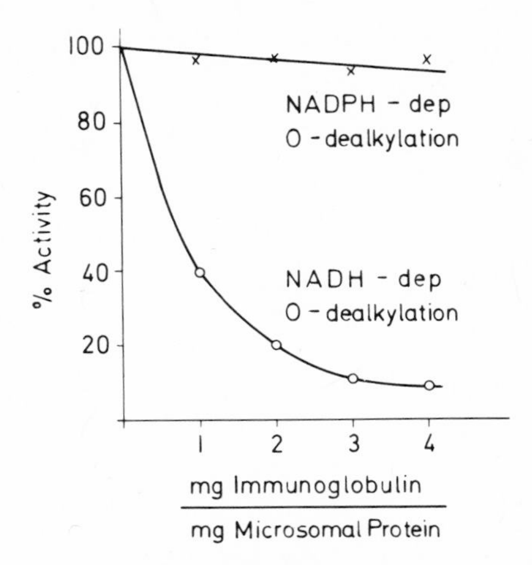

Fig. 2. Effect of rabbit antiserum against rat liver NADH-cytochrome b_5 reductase on the NADH-dependent O-dealkylation of 7-ethoxycoumarin. The fluorescence was recorded for 3 min in the presence of NADPH or NADH and then the immunoglobulin fractions (kindly provided by Omura) were added.

with the high affinity of NADH-cytochrome b_5 reductase for NADH (13). It should be noted, however, that only a small fraction of the NADH electrons is channeled to cytochrome P450; the major part is accepted by other microsomal components. The rate-limiting step of the NADH-supported monooxygenations seems to be the electron transfer from cytochrome b_5 to cytochrome P450 since

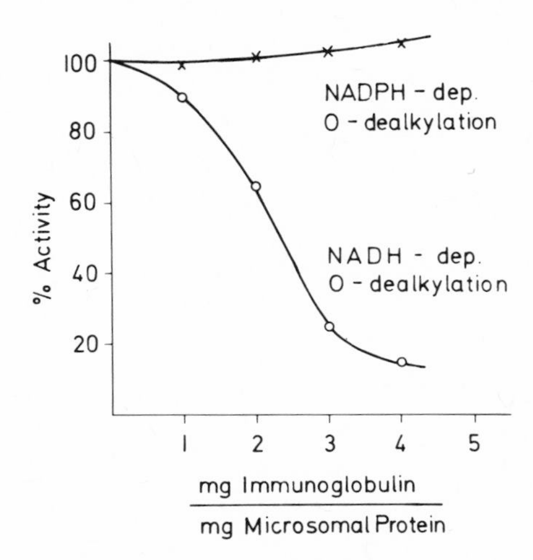

Fig. 3. Effect of rabbit antiserum against rat liver cytochrome b_5 on the NADPH- and NADH-dependent O-dealkylation of 7-ethoxycoumarin. The decrease in fluorescence was recorded for 3 min with NADPH or NADH as cosubstrates and then the immunoglobulin fractions (kindly provided by Omura) were added.

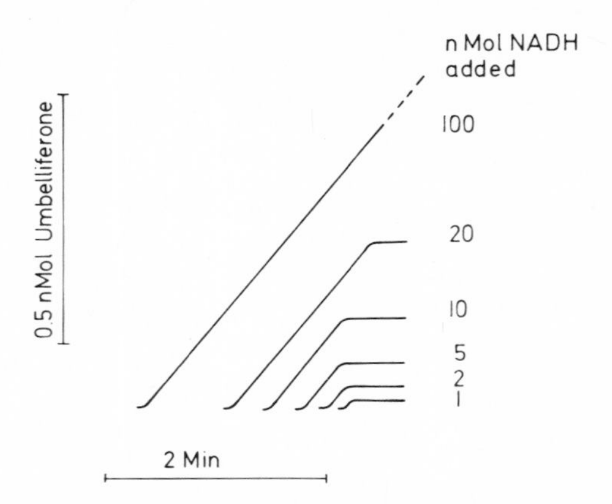

Fig. 4. Effect of various NADH concentrations on the O-dealkylation of 7-ethoxycoumarin. The original recordings of umbelliferone fluorescence of a microsomal suspension from phenobarbital-pretreated rats (2 mg of protein/ml) were redrawn and superimposed.

addition of detergent-solubilized cytochrome b_5 can increase the *O*-dealkylation rate for 7-ethoxycoumarin proportional to its incorporation into the microsomal membrane (Table 1) (14).

Trypsin-solubilized cytochrome b_5, lacking the lipophilic portion of the peptide chain, is ineffective. With NADPH, the rate-limiting step probably is also the reduction of cytochrome P450, since the turnover of cyclohexane hydroxylation is comparable to the first-order rate constant of the reduction of the cytochrome P450–cyclohexane complex (15). This also explains why NADH exerts an additional effect on the cyclohexane hydroxylation (16). An explanation for the more-than-additive effects of NADH is given in another paper (16).

There has been considerable discussion as to whether cytochrome P450 is a

Table 1. Effect of Cytochrome b_5 on the Microsomal *O*-Dealkylation of 7-Ethoxycoumarin[a]

	30 min preincubation in presence of		
Microsomal activity (nmol/mg protein)	–(control)	t-b_5	d-b_5
Cytochrome b_5 (NADH-reducible)	0.66	0.60	1.35
NADH-dependent *O*-dealkylation	0.11	0.10	0.24
NADPH-dependent *O*-dealkylation	4.75	4.70	4.80
NADH + NADPH-dependent *O*-dealkylation	6.40	6.30	7.35

[a] Rat liver microsomes (5 mg of protein/ml) were preincubated for 30 min with either trypsin-solubilized cytochrome b_5 (t-b_5) (30 nmol/ml) or detergent-solubilized cytochrome b_5 (d-b_5) (30 nmol/ml). The microsomal membranes were sedimented and resuspended in tris buffer, pH 7.6. The control contained the same amount of d-b_5 boiled for 3 min before the preincubation period.

homogeneous protein occurring in various conformational states or whether it includes a variety of species with similar spectral properties. This question must be answered today in favor of the latter hypothesis. Two different forms induced with phenobarbital and benzpyrene have been isolated with retention of small differences in the CO spectrum and of differences in the substrate and product specificity (17) which were first seen in intact microsomes (18). Differences in the binding spectra and metabolism of enantiomers of barbiturates were early indications that more than one species exists even in normal and phenobarbital-pretreated rats (19,20). Each of the various species obviously has a characteristic spectrum of substrates; e.g., the benzpyrene-induced form preferentially acts on aromatic compounds, while the phenobarbital-induced form shows highest activities with aliphatic substrates. The substrate specificities, however, are broad and overlapping. Likewise, the *O*-dealkylation of 7-ethoxycoumarin is enhanced by various inducers as shown in Table 2.

The cytochrome P450 species induced by the various pretreatments have different kinetic parameters and different pH optima (21). Looking for a tool to differentiate the various forms, we have employed inhibitors which at a certain concentration rather selectively inhibit only one species (22). Metyrapone inhibits mainly the cytochrome P450 present in microsomes after phenobarbital pretreatment, although higher concentrations also affect other species. 7,8-Benzoflavone is almost specific for the benzpyrene-induced form (23). Ethanol pretreatment has been suggested to lead to a third species (24,25) which is supported by the lack of response to the two inhibitors and the finding of tetrahydrofurane as a specific ligand and inhibitor of this form (22). Microsomes from male and female rats also show a different pattern of inhibition, which

Table 2. Effect of Inhibitors on the *O*-Dealkylation Activity for 7-Ethoxycoumarin in Rat Liver Microsomes after Various Pretreatments (22)

		Percent inhibition by		
Pretreatment	Specific activity[a]	Metyrapone 2×10^{-5} M	7,8-Benzoflavone 2×10^{-5} M	Tetrahydrofurane 10^{-2} M
None (controls) ♂	0.7 ± 0.2	52 ± 10	10 ± 5	15 ± 6
None (controls) ♀	0.4 ± 0.2	5 ± 3	12 ± 5	48 ± 20
Phenobarbital (3 d)	2.2 ± 0.6	72 ± 12	7 ± 5	4 ± 2
3,4-Benzpyrene (2 d)	5.4 ± 2.0	2 ± 5	90 ± 6	2 ± 5
Ethanol (20 d)	0.9 ± 0.2	7 ± 3	10 ± 5	79 ± 10

[a]nmol × mg protein^{-1} × min^{-1}.

Table 3. 7-Ethoxycoumarin *O*-Dealkylation and Inhibition Pattern in Human Liver Biopsy Samples[a]

				Percent inhibition by		
Patient	Sex	Age	Specific activity[b]	Metyrapone 10^{-5} M	7,8-Benzoflavone 10^{-5} M	Tetrahydrofurane 10^{-2} M
S. J.	♂	48	0.13	18	25	92
Sch. T.	♂	31	0.10	27	16	52
Sch. E.	♂	43	0.11	38	12	45
L. E.	♀	57	0.12	37	10	32
A. E.	♂	56	0.12	33	32	42
B. E.	♀	63	0.10	22	33	25
W. A.	♀	51	0.20	26	12	28
B. H.	♂	15	0.83	11	86	49
St. K.	♀	70	0.14	33	84	33
C. P.	♂	46	0.30	30	22	33

[a]To 50 μl of a 10,000*g* supernatant equivalent to about 0.25 mg liver wet weight, 50 μl of a 10^{-3} M solution of 7-ethoxycoumarin in 1 M tris buffer, pH 8.0, was added. The reaction was started by addition of 10^{-4} M NADPH and NADH and the fluorescence was recorded above 430 nm after excitation at 405 nm (interference filter) in an Eppendorf photometer with a special cuvette holder.
[b]nmol × mg protein^{-1} × min^{-1}.

indicates that the well-known sex difference in drug metabolism (26) is based on a different pattern of cytochrome P450 monooxygenases (22).

The 7-ethoxycoumarin test in combination with the selective inhibitors proved also to be useful in studies on human biopsy samples. Table 3 contains the results obtained from ten patients. Not only the specific activity but also the inhibition pattern shows wide individual variations suggesting a different monooxygenase pattern in each individual. Nothing can be concluded, however, about the relative amounts of the cytochrome P450 species present since several observations indicate that the P450 enzymes in humans differ from those in rats, and therefore other inhibitors or other concentrations may be required in human tissues.

As a consequence of the various cytochrome P450 species, the pattern of products may vary considerably, as shown for testosterone (18), biphenyl (27), *n*-hexane (28), and other substrates. Usually steric differences in substrate binding at the active site can cause alterations in the pattern of products, but the following experiments suggest that even changes in the reactivity of the active oxygen must be considered (Fig. 5).

7-Methoxycoumarin, like 7-ethoxycoumarin, is a substrate of the microsomal monooxygenase system. Both compounds show similar binding spectra and kinetic parameters. Microsomes from phenobarbital-pretreated rats catalyze mainly the *O*-dealkylation, but a minor metabolite identified as the corresponding 6-hydroxy derivative is also formed (29). When microsomes from benzpy-

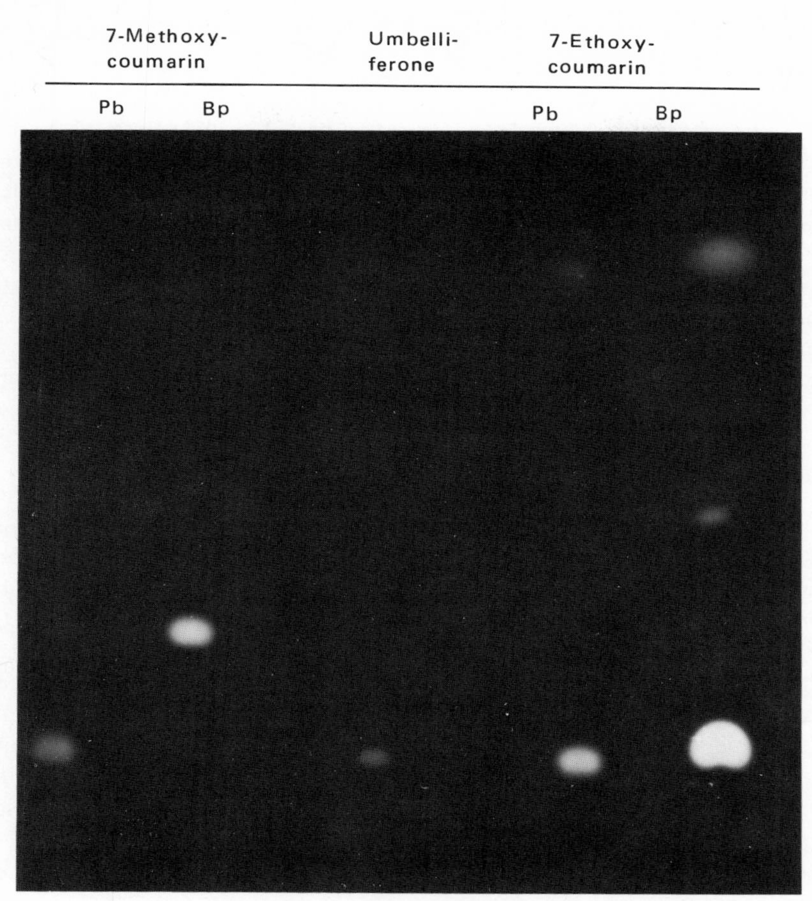

Fig. 5. Chromatogram of phenolic metabolites obtained by incubations of 7-methoxy- and 7-ethoxycoumarin with liver microsomes from phenobarbital- and benzpyrene-pretreated rats (29).

rene-induced rats were used, this metabolite became predominant with 7-methoxycoumarin but not with 7-ethoxycoumarin as a substrate. Comparative studies with model systems also show that ring hydroxylation is preferred when the methoxy ether is used. This is in agreement with the higher activation energy involved in the splitting of a primary CH– bond compared to addition to a π-electron system. Thus, in addition to steric effects, differences in reactivity of the active oxygen induced by slight changes in the ligand field of the various species of cytochrome P450 may also be responsible for changes in the product pattern. Regardless of the mechanisms of these changes, it is evident that the existence of different metabolic patterns has great implications for drug toxicity.

STEPS IN THE MONOOXYGENATION CYCLE RESPONSIBLE FOR DRUG TOXICATION

Toxic effects of chemicals and drugs can be the results of various kinds of interactions with the microsomal monooxygenase system. This is best demonstrated by looking at successive steps in the reaction cycle. Although details of this mechanism are still controversial, the sequence of events in Fig. 6 seems to be generally accepted. Each side of the figure represents one of the four important steps in the monooxygenase mechanism: step 1, substrate binding leading to formation of the enzyme–substrate complex; step 2, reduction and subsequent oxygenation; step 3, oxygen activation; step 4, oxygen atom transfer and release of product. In the following, I shall analyze each step with regard to toxic effects of organic compounds.

Substrate Binding

Since addition of the substrate molecule to the active site of cytochrome P450 is reversible with only minor free energy changes, activation of substrates leading to toxic effects is unlikely to occur at this step. It should be kept in mind, however, that most inhibitors of the monooxygenase system act at this step by competitively blocking the binding of a substrate. In case such an inhibitor blocks the cytochrome irreversibly, this might well be considered a toxic event, since further metabolism of drugs or even endogenous substrates like steroids is prevented. Such a tight binding could be caused if the inhibitor contains an atom with a free electron pair which strongly interacts with the iron (30). Some phosphines (31) and carbenes (32; also Nastainczyk and Ullrich, unpublished) exhibit such strong liganding properties.

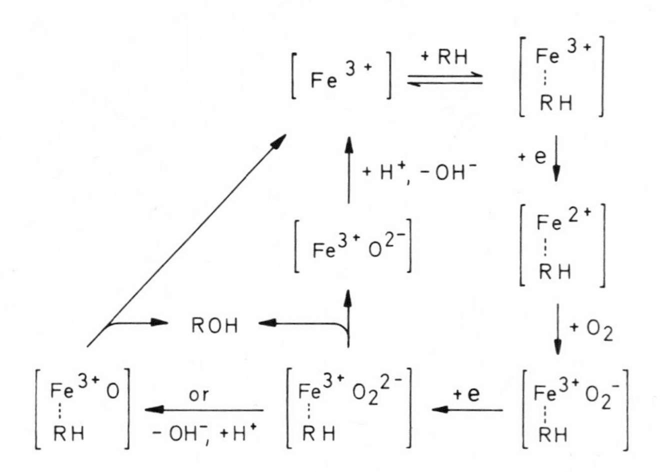

Fig. 6. Catalytic cycle of microsomal cytochrome P450.

Reduction of the Cytochrome P450–Substrate Complex

Cytochrome P450 contains a heme–sulfide linkage to the protein which causes its rather low redox potential (4,33). Hence the reduced cytochrome is a very active reductant and immediately binds oxygen. However, if the bound substrate has oxidizing properties and can accept electrons, it could compete with oxygen reduction. Such substrates are nitro and azo compounds which are converted to amines in the organism by a reaction that is partly inhibited by carbon monoxide. Cytochrome P450, therefore, seems to participate, although NADPH–cytochrome P450 reductase which can also perform similar reductions may be responsible for the CO-insensitive part (34). Similar mechanisms may occur with polyhalogenated hydrocarbons like carbon tetrachloride. Chloroform is a product of the metabolic reduction of CCl_4 (35) but the intermediate $CCl_3\cdot$ radical is also formed (36) and is probably responsible for liver damage. Under anaerobic or hypoxic conditions, dihalogen compounds such as halothane can be reduced by cytochrome P450. We have recently shown that during the reaction of halothane with reduced cytochrome P450 the 1,1,1-trifluoroethane carbene is formed (32), which gives rise to the spectral species absorbing at 471 nm as first described by Uehleke (37). Simultaneous recording of the oxygen concentration and carbene complex formation at 471 nm clearly proves that reduction of halothane and carbene formation starts at low oxygen pressure (Fig. 7A,B).

The spectrophotometric traces also indicate that readmission of oxygen decreases the carbene complex. A free carbene species would be unstable, and covalent binding to cell constituents would be the most likely reaction to occur. This hypothesis would be in accord with the finding of covalently bound halothane metabolites in liver proteins, especially under anaerobic conditions (37) and after phenobarbital pretreatment (38).

Oxygen Activation

The one-electron reduction of the oxy complex of cytochrome P450 yields the oxenoid complex as the active intermediate. Its lifetime is probably very short and is terminated by reaction with a suitable substrate. The oxenoid complex in either one of its possible structures must be considered as a strongly basic species, and protonation should result in inactivation. This would inevitably occur if the substrate could not accept the oxygen atom, as is the case with uncoupling agents (16). The product of such an uncoupled monooxygenation would be hydrogen peroxide, but a microsomal peroxidase (39), catalase (40), or glutathione peroxidase (41) could sufficiently decrease its steady-state concentration so that no toxic effects would occur. Protonation of the active oxygen in rare cases may be mediated also by the substrate itself, if this contains an acidic hydrogen. An example may be fluorene, which has been shown to yield the carbanion as an intermediate (42), which in a nonenzymatic reaction could form

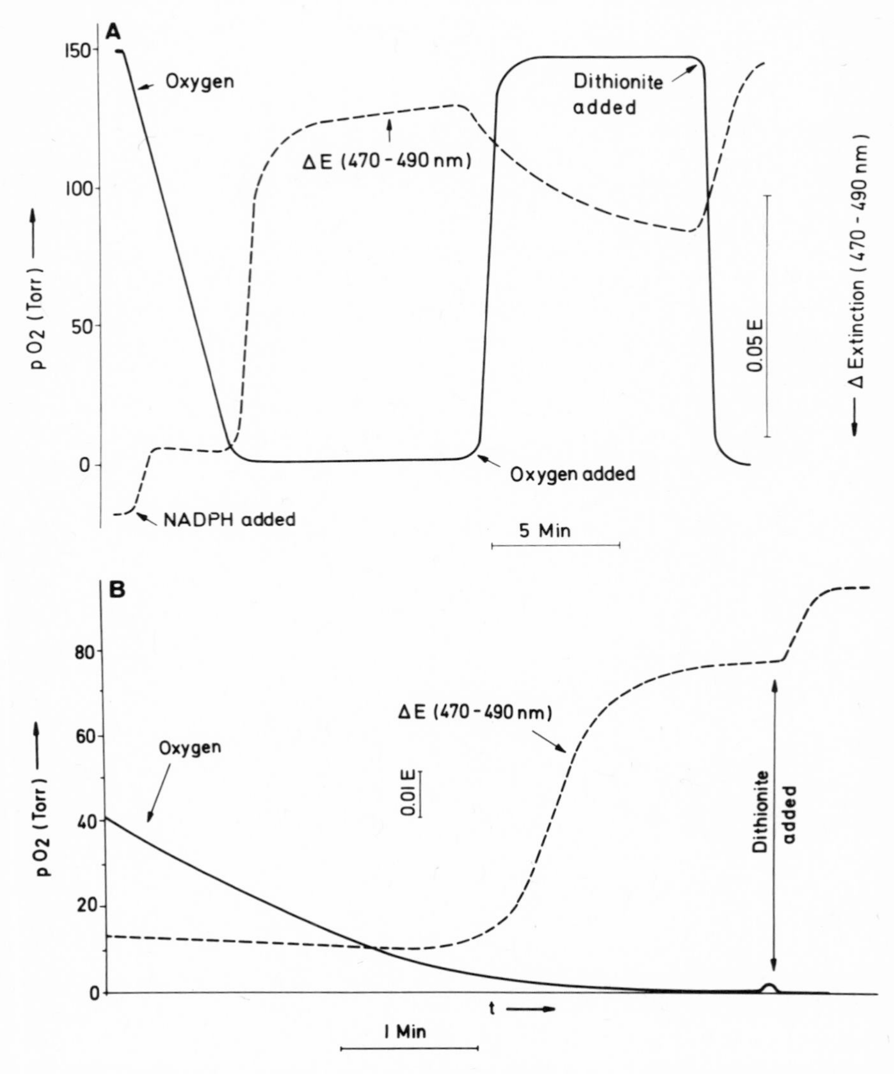

Fig. 7. Simultaneous recording of oxygen uptake and 471 nm peak formation in rat liver microsomes in the presence of halothane. (A) A special cuvette with stirrer and a Clark-type oxygen electrode were placed in an Aminco-Chance DW-2 spectrophotometer. After addition of NADPH to 3 ml of a liver microsomal suspension from phenobarbital-pretreated rats (3 mg of protein/ml), the oxygen concentration and the increase in extinction between 470 and 490 nm were recorded. In (B) the sensitivity of the oxygen recording was increased.

the fluorene-9-peroxide (43). Since this is also a substrate for the microsomal peroxidase, no toxic effects may be expected from the oxygen activation process.

Oxygen Atom Transfer

The most critical step in drug metabolism occurs with the introduction of the oxygen atom into the substrate. Keeping in mind that the binding of the substrate mainly involves hydrophobic interactions and is rather unspecific, it is evident that the attack of the oxenoid complex is determined to a large extent by the chemical reactivity of the substrate molecule. This could be demonstrated by a comparison with model systems creating electrophilic hydroxylating species. Essentially the same pattern of products was obtained by chemical hydroxylations as with the liver microsomal system (4,45,46). From these experiments, some general rules about the orientation of the oxygen atom can be derived which can be helpful in predicting the nature of the metabolites.

1. π-Systems, either as an isolated double bond or as aromatic systems, undergo electrophilic additions by the oxenoid species, leading primarily to epoxides. If rearomatization is favorable, rearrangement to phenols can take place. Migration of substituents during this process has been observed (47).
2. CH– bonds are hydroxylated in the sequence tertiary>secondary> primary. The differences in the rates of hydroxylation among the three bonds are extremely large, thus characterizing the oxenoid species as a rather weakly attacking species.
3. Heteroatoms with a hydrogen atom like NH can also be hydroxylated or can be oxidized at a lone electron pair, yielding $N \rightarrow O$ or $S \rightarrow O$.

From this, it can be derived that drug toxicity is inherent in the substrate molecule and will occur if a product of the monooxygenase reaction is active enough to react in the body before it is excreted.

Examples of these toxic products first described in the literature are *N*-hydroxy compounds (49) and semiquinones (50). Epoxides have long been suggested to be active and carcinogenic intermediates in the monooxygenation of polycyclic hydrocarbons (51). The main interest has been focused on the K-region epoxides, but only recently it became evident that those compounds are not reactive enough and are hydrolyzed or conjugated to a large extent before they can bind to cell constituents (52). Epoxides from monooxygenated products have now been found to be far more active and seem to be mainly responsible for the carcinogenic properties of polycyclic hydrocarbons (53).

Another general pathway of toxication is opened when hydroxylation takes place at a carbon which already has two halogen substituents. Then immediate dehalogenation can follow, leading to acid halides. The formation of carbon monoxide from methylene chloride could be formulated as follows:

$$\mathrm{H{-}\overset{H}{\underset{Cl}{C}}{-}Cl} \xrightarrow{[O]} \mathrm{H{-}\overset{OH}{\underset{Cl}{C}}{-}Cl} \xrightarrow[-2\,HCl]{} CO$$

Halogens are known as good leaving groups but there are indications that the hydroxyl group introduced by cytochrome P450 can also be eliminated:

$$\mathrm{>C(OH)H} \xrightarrow[-H_2O]{} \mathrm{>C\!:}$$

This reaction will proceed only if the resulting carbene is stable or if it is stabilized as a ligand at a metal ion. As already outlined, reduced cytochrome P450 itself can bind ligands and, among others, also carbenes (32). As a first example for this working hypothesis the interaction of the insect synergist piperonyl butoxide with liver microsomes may be considered. Under hydroxylating conditions, the formation of a ligandlike spectrum has been observed (55). The demethylation of aminopyrine is inhibited after the formation of this complex, but the inhibitor can be released in the presence of light (30). The photochemical action spectrum for the dealkylation of aminopyrine is shown in Fig. 8. The spectrum is clear evidence for a ligand binding of a piperonyl butoxide metabolite to reduced cytochrome P450. A series of chemical studies carried out by Mansuy (unpublished) indicates that substitution of the methylene carbon with alkyl groups abolishes the complex formation but introduction of an ethoxy group does not. Since the latter can be dealkylated to yield the unstable 1-hydroxy metabolite, the following mechanism seems to be involved:

$$\text{(methylenedioxy)}\ \mathrm{>CH_2} \quad\text{or}\quad \mathrm{>C(H)OC_2H_5} \xrightarrow[-C_2H_4O]{} \underset{\text{unstable}}{\mathrm{>C(OH)H}} \xrightarrow[-H_2O]{Fe^{2+}} \mathrm{>C{:}Fe^{2+}}$$

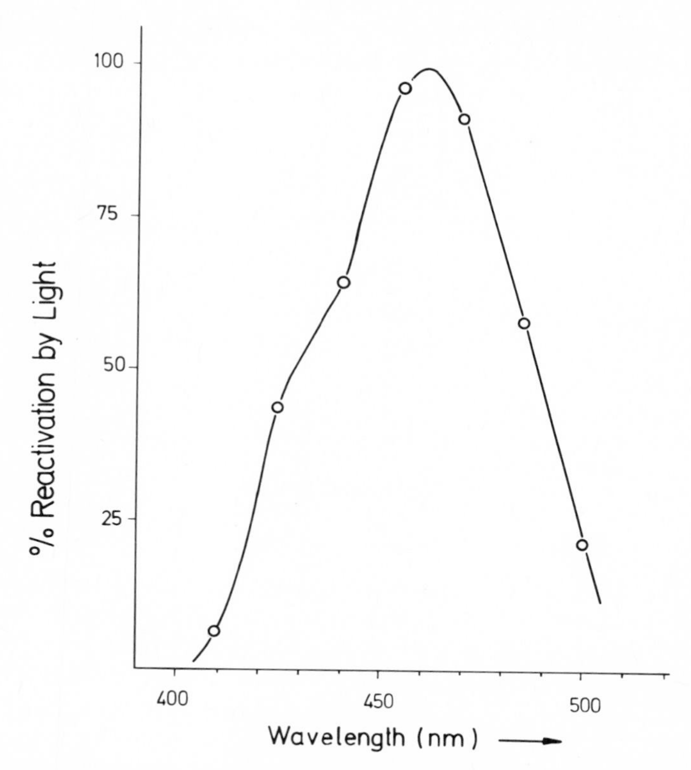

Fig. 8. Photochemical action spectrum of the piperonyl butoxide-inhibited N-dealkylation of aminopyrine. The *N*-dealkylation by liver microsomes of phenobarbital-pretreated rats was inhibited by piperonyl butoxide in the dark and the same assay was incubated under light of the given wavelengths. The reversal of inhibition was corrected for equal light intensities. Details of the method will be published elsewhere.

The earlier hypothesis of a carbanion ligand (30) could be ruled out since no acidity of the methylene hydrogens could be found (V. Ullrich, unpublished).

In the methylenedioxy compounds, the oxygen atoms guaranteed a high stability of the carbene–cytochrome P450 complex, but there may be other compounds for which similar carbene species are less stable. Release of the carbene ligand is most likely to occur after oxidation of the ferrous complex. The high reactivity of such free carbenes has been discussed in connection with the reductive formation of carbenes from polyhalogenated hydrocarbons.

The chemistry and biochemistry of cytochrome P450 and its catalysis are unusual in many respects. Unlike other enzymes, the microsomal monooxygenases have to be nonspecific in order to handle all the chemicals and drugs to which man may be exposed. The price paid for this lack of specificity is the risk of substrate activation instead of inactivation. A better knowledge of the mechanisms of monooxygenase systems may help us to recognize potential hazards of a chemical even before it is synthesized.

ACKNOWLEDGMENT

This work was supported by the Deutsche Forschungsgemeinschaft, SFB 38.

REFERENCES

1. R. T. Williams, *Detoxication Mechanisms,* 2nd ed., Wiley, New York (1959).
2. B. B. Brodie, J. Axelrod, J. R. Cooper, L. Gaudette, B. N. LaDu, C. Mitoma, and S. Udenfriend, Detoxication of drugs and other foreign compounds by liver microsomes, *Science* **121**, 603–604 (1955).
3. H. S. Mason, in: *Advances in Enzymology,* Vol. XIX (F. F. Nord, ed), pp. 79–233, Interscience, New York (1957).
4. V. Ullrich, Enzymatic hydroxylations with molecular oxygen, *Angew. Chem. Int. Ed. Engl.* **11,** *701*–712 (1972).
5. T. Omura and R. Sato, The carbon monoxide-binding pigment of liver microsomes. II. Solubilization, purification and properties, *J. Biol. Chem.* **239,** 2370–2378 (1964).
6. K. Krisch and Hj. Staudinger, Untersuchungen zur enzymatischen Hydroxylierung: Hydroxylierung von Acetanilid und dessen Beziehungen zur mikrosomalen Pyridinucleotidotidoxydation, *Biochem. Z.* **334,** 312 (1961).
7. A. Y. H. Lu, K. W. Junk, and M. J. Coon, Resolution of the cytochrome P-450-containing ω-hydroxylation system of liver microsomes into three components, *J. Biol. Chem.* **244,** 3714–3721 (1969).
8. A. Y. H. Lu, H. W. Strobel, and M. J. Coon, Hydroxylation of benzphetamine and other drugs by a solubilized form of cytochrome P-450 from liver microsomes: Lipid requirement for drug demethylation, *Biochem. Biophys. Res. Commun.* **36,** 545–551 (1969).
9. S. Orrenius, A. Berg, and L. Ernster, Effects of trypsin on the electron transport systems of liver microsomes, *Eur. J. Biochem.* **11,** 193–200 (1969).
10. H. A. Sasame, S. S. Thorgeirsson, J. R. Mitchell, and J. R. Gillette, NADPH and NADH electron transport systems in rat liver microsomes, *Pharmacologist* **15,** 170 (1973).
11. G. J. Mannering, S. Kuwahara, and T. Omura, *Biochem. Biophys. Res. Commun.* **57,** 476–481 (1974).
12. V. Ullrich and P. Weber, The *O*-dealkylation of 7-ethoxycoumarin by liver microsomes: A direct fluorometric test, *Hoppe-Seyler's Z. Physiol. Chem.* **353,** 1171–1177 (1972).
13. P. Strittmatter and S. F. Velick, The purification and properties of microsomal cytochrome reductase, *J. Biol. Chem.* **228,** 785–799 (1957).
14. J. Poensgen and V. Ullrich, The binding of cytochrome b_5 to liposomes, *Hoppe-Seyler's Z. Physiol. Chem.* **355,** 1238 (1974).
15. H. Diehl, J. Schadelin, and V. Ullrich, Studies on the kinetics of cytochrome P-450 reduction in rat liver microsomes, *Hoppe-Seyler's Z. Physiol. Chem.* **351,** 1359–1371 (1970).
16. H. Staudt, F. Lichtenberger, and V. Ullrich, The role of NADH in uncoupled microsomal monoxygenations, *Eur. J. Biochem.* 46, 99–106 (1974).
17. A. Y. H. Lu, R. Kuntzman, S. West, M. Jacobson, and A. H. Conney, Reconstituted

liver microsomal enzyme system that hydroxylates drugs, other foreign compounds, and endogenous substrates. II. Role of the cytochrome P-450 and P-448 fractions in drug and steriod hydroxylations, *J. Biol. Chem.* **247**, 1727–1734 (1972).
18. A. H. Conney, W. Levin, M. Jacobson, R. Kuntzman, D. Y. Cooper, and O. Rosenthal, in: *Microsomes and Drug Oxidations* (J. R. Gillette, A. H. Conney, G. J. Cosmides, R. W. Estabrook, J. R. Fouts, and G. J. Mannering, eds.), pp. 279–295, Academic Press, New York (1969).
19. E. Degkwitz, V. Ullrich, and Hj. Staudinger, Metabolism and cytochrome P-450 binding spectra of (+)- and (–)-hexobarbital in rat liver microsomes, *Hoppe-Seyler's Z. Physiol. Chem.* **350**, 547–553 (1969).
20. W. Bohn, V. Ullrich, and H. Staudinger, Species of cytochrome P-450 in rat liver microsomes with different stereoselectivity for the binding and monooxygenation of (+)- and (–)-methylphenobarbital, *Naunyn-Schmiedebergs Arch. Pharmakol.* **270**, 41–55 (1971).
21. V. Ullrich, U. Frommer, and P. Weber, Characterization of cytochrome P-450 species in rat liver microsomes. I. Differences in the *O*-dealkylation of 7-ethoxycoumarin after pretreatment with phenobarbital and 3-methylcholanthrene, *Hoppe-Seyler's Z. Physiol. Chem.* **354**, 514–520 (1973).
22. V. Ullrich, P. Weber, and P. Wollenberg, Tetrahydrofurane–an inhibitor for ethanol-induced liver microsomal cytochrome P-450, *Biochem. Biophys. Res. Commun.* **64**, 808–813 (1975).
23. L. Diamond and H. V. Gelboin, Alpha-naphthoflavone: An inhibitor of hydrocarbon cytotoxicity and microsomal hydroxylase, *Science* **166**, 1023–1025 (1969).
24. E. Rubin, C. S. Lieber, A. A. Alvares, W. Levin, and R. Kuntzman, Ethanol binding to hepatic microsomes, its increase by ethanol consumption, *Biochem. Pharmacol.* **20**, 229–331 (1971).
25. K. Comai and J. L. Gaylor, Existence and separation of three forms of cytochrome P-450 from rat liver microsomes, *J. Biol. Chem.* **248**, 4947–4955 (1973).
26. R. Kato, Sex related differences in drug metabolism, *Drug Metab. Rev.* **3**, 1–32 (1974).
27. P. J. Creaven and D. V. Parke, The effect of pregnancy on the microsomal metabolism of foreign compounds, *Proceedings of the 2nd Meeting of the Federation of the European Biochemistry Society* **128**, 88–89 (1965).
28. U. Frommer, V. Ullrich, Hj. Staudinger, and S. Orrenius, The monooxygenation of *N*-heptane by rat liver microsomes, *Biochim. Biophys. Acta* **280**, 487–494 (1972).
29. I. Ellin, A. Ellin. W. Nastainczyk, D. Mansuy, and V. Ullrich, in preparation.
30. V. Ullrich and K. H. Schnabel, Formation and binding of carbanions by cytochrome P-450 of liver microsomes, *Drug Metab. Dispos.* **1**, 176–184 (1973).
31. D. Mansuy, W. Duppel, H.-H. Ruf, and V. Ullrich, Phosphines as ligands to microsomal cytochrome P-450, *Hoppe-Seyler's Z. Physiol. Chem.* **355**, 1341–1349 (1974).
32. D. Mansuy, W. Nastainczyk, and V. Ullrich, The mechanism of halothane binding to microsomal cytochrome P-450, *Naunyn-Schmiedebergs Arch. Pharmakol.* **285**, 315–324 (1974).
33. S. Koch, S. C. Tang, R. H. Holm, R. B. Frankel, and J. A. Ibers, Ferric porphyrin thiolates, possible relationship to cytochrome P-450 enzymes and the structure of *p*-nitrobenzene thiolate in iron (III), protoporphyrin IX, dimethylester, *J. Am. Chem. Soc.* **97**, 916–918 (1975).
34. P. H. Hernandez, P. Mazel, and J. R. Gillette, Studies on the mechanism of action of mammalian hepatic azoreductase. II. The effects of phenobarbital and 3-methylcholanthrene on carbon monoxide sensitive and insensitive azoreductase activities, *Biochem. Pharmacol.* **16**, 1877–1888 (1967).
35. J. Schädelin and B. S. Masters, unpublished.
36. M. C. Villarruel and J. A. Castro, Irreversible binding of carbon tetrachloride to

microsomal phospholipids: Free radical nature of the reactive species and alterations in the physico-chemical properties of the target fatty acids, *Res. Commun. Chem. Pathol. Pharmacol.* **10,** 105–116 (1975).
37. O. Reiner und H. Uehleke, Bindung von Tetrachlorkohlenstoff an reduziertes mikrosomales Cytochrom P-450 und an Häm, *Hoppe-Seyler's Z. Physiol. Chem.* **352,** 1048–1052 (1971).
38. R. A. Van Dyke and C. L. Wood, Binding of radioactivity from ^{14}C-labeled halothane in isolated perfused rat livers, *Anesthesiology* **38,** 328–332, (1973).
39. E. G. Hrycay and P. J. O'Brien, Microsomal electron transport. II. Reduced nicotinamide, adenine dinucleotide-cytochrome b_5 reductase and cytochrome P-450 as electron carriers in microsomal NADH-peroxidase activity, *Arch. Biochem. Biophys.* **160,** 230–245 (1974).
40. H. Sies, Biochemistry of the peroxisome in the liver cell, *Angew, Chem. Int. Ed. Engl.* **13,** 706–718 (1974).
41. G. C. Mills, Hemoglobin catabolism. I. Glutathione peroxidase, an erythrocyte enzyme which protects hemoglobin from oxidative breakdown, *J. Biol. Chem.* **229,** 189–197 (1957).
42. V. Ullrich and K. H. Schnabel, Formation and ligand binding of the fluorenyl carbanion by hepatic cytochrome P-450, *Arch. Biochem. Biophys.* **159,** 240–248 (1973).
43. C. Chen and C.-C. Lin, (9-^{14}C)Fluorene hydroperoxide as a possible intermediate in the hydroxylation of (9-^{14})fluorene by rat liver homogenate, *Biochim. Biophys. Acta* **184,** 634–640 (1969).
44. V. Ullrich, J. Wolf, E. Amadori, and H. Staudinger, The mixed function oxygenation of 4-halogenoacetanilides in rat liver microsomes and model systems, *Hoppe-Seyler's Z. Physiol. Chem.* **349,** 85 (1968).
45. U. Frommer and V. Ullrich, Hydroxylation of aliphatic compounds by liver microsomes. 3. Model hydroxylation reactions, *Z. Naturforsch. Teil B* **26,** 322–327 (1971).
46. V. Ullrich, Oxygen activation by the iron (II)-2-mercaptobenzoic acid complex. A model for microsomal mixed function oxygenases, *Z. Naturforsch. Teil B* **24,** 699–704 (1969).
47. G. Guroff, J. W. Daly, D. M. Jerina, J. Renson, B. Witkop, and S. Udenfriend, Hydroxylation-induced migration: The NIH shift, *Science* **157,** 1524–1530 (1967).
48. U. Frommer, V. Ullrich, and H. Staudinger, Hydroxylation of aliphatic compounds by liver microsomes. I. The distribution pattern of isomeric alcohols, *Hoppe-Seyler's Z. Physiol. Chem.* **351,** 903 (1970).
49. E. C. Miller, J. A. Miller, and H. A. Hartmann, *N*-Hydroxy-2-acetylaminofluorene: A metabolite of 2-acetylaminofluorene with increased carcinogenic activity in the rat, *Cancer Res.* **21,** 815–824 (1961).
50. I. L. Riegel and G. C. Mueller, Formation of a protein-bound metabolite of estradiol-16-^{14}C by rat liver homogenates, *J. Biol. Chem.* **210,** 249–257 (1954).
51. P. L. Grover and P. Sims, K-region epoxides of polycyclic hydrocarbons: Reactions with nucleic acids and polyribonucleotides, *Biochem. Pharmacol.* **22,** 661–666 (1973).
52. W. M. Baird, A. Dipple, P. L. Grover, P. Sims, and P. Brookes, Studies on the formation of hydrocarbon-deoxyribonucleoside products by the binding of derivatives of 7-methylbenz(*a*)anthracene to DNA in aqueous solution and in mouse embryo cells in culture, *Cancer Res.* **33,** 2386–2392 (1973).
53. P. Sims, P. L. Grover, A. Swaisland, K. Pal, and A. Hewer, Metabolic activation of benzo(a)pyrene proceeds by a diol-epoxide, *Nature (London)* **252,** 326–327 (1974).
54. V. L. Kubic and M. A. Anders, Metabolism of dihalomethanes to carbon monoxide. II. *In vitro* studies, *Drug Metab. Dispos.* **3,** 104–112 (1975).
55. R. M. Philpot and E. Hodgson, A cytochrome P-450-piperonyl butoxide spectrum similar to that produced by ethyl isocyanide, *Life Sci.* **10,** 503–512 (1971).

7

Discussion

In response to a question by Schulman, Ullrich stated that the microsomal desaturase, peroxidase, and P450 mediated pathways can be differentiated using metabolic inhibitors, but that CN is not a good inhibitor for this purpose. The oxidation of ethanol has not been proven to be mediated by a species of P450 which ethanol induces; however, this P450 does differ from the types of P450 inhibited by metyrapone and 7,8-benzoflavone (see text).

Ullrich agreed with Lenk that the species of P450 which discriminates among primary, secondary, and tertiary carbon atoms as substrates for hydroxylation is also differentially inhibited by CO and differs in its affinity for oxygen.

To Lenk's suggestion that the C–F bond could be split in halothane metabolism, Ullrich commented that he did not think that this splitting is possible, despite the fact that the splitting off of fluoride from aromatic compounds is known. This defluoridation takes place by a two-step mechanism, in which initial epoxide formation is followed by removal of F from the epoxide intermediate. He emphasized that this type of dehalogenation cannot occur with halothane. Zimmerman commented that this concept is supported by the fact that all metabolites of halothane which have been identified are trifluorinated.

Ullrich also commented that the reactivity of the oxygen attached at the sixth ligand to the heme Fe in P450 is modified by the characteristics of the fifth ligand in which the Fe is tied to a sulfhydryl component of the protein. For example, the length of the Fe–S bond has a strong effect on the oxenoid intermediate, i.e., the P450–oxygen–substrate complex; the more stable the complex, the less reactive the oxygen. Therefore, the defluoridation following the addition of oxygen to an aromatic π-electron system to form an epoxide, such as in fluorinated aromatic hydrocarbons, is a preferred pathway, whereas epoxidation of halothane resulting in the release of fluoride does not occur because the oxygen is less reactive.

In response to a request by Schleyer that he comment on the relative kinetics of the reductive and oxidative pathways of halothane, Ullrich said that the reductive pathway takes place only at low oxygen pressures, i.e., 3–4 Torr. It is important to determine whether or not such low oxygen tensions are achieved during narcosis. However, metabolism itself is blocked during narcosis, and therefore the reductive pathway does not produce many metabolites. Under these conditions, the process is not cyclic and if oxygen is introduced the regular cyclic pathway is resumed.

8

Aryl Hydrocarbon Hydroxylase: Induction

John B. Schenkman, Kathy M. Robie, and Ingela Jansson
Department of Pharmacology
Yale University School of Medicine
New Haven, Connecticut 06510, U.S.A.

Knowledge about the enzyme system responsible for the metabolism of aryl hydrocarbons dates back to the early studies of Conney *et al.* (1), who showed the enzyme(s) responsible for the oxidation of benzpyrene to be located in the microsomal fraction of the liver. In that study it was shown that the enzyme system required both NADPH and molecular oxygen to function, and produced a number of mono- and dihydroxy products.

Although in this pioneering study aryl hydrocarbon hydroxylase (AHH) activity (benzpyrene hydroxylation) was reported to be restricted to the liver (1), with the development of more sensitive assays based on appearance of polar metabolites (2), most tissues have been found to contain AHH. Table 1 shows an example of the tissue distribution of AHH and representative activities reported by different authors. With regard to activity of the enzyme system, it is generally very low in most tissues, with the highest activity being located in the liver microsomes. Differences in activity values reported by different laboratories vary over 2 orders of magnitude for the same tissues, and even reports of relative activities between different tissues differ markedly between laboratories. Some of these differences are due to assay conditions, but to a great extent they are due to species and strain differences. For example, as shown by Nebert and Gelboin (3) in a comparison of six mouse strains, control AHH activity ranged from a low of about 1 pmol/min/mg liver to a high of 12 pmol/min/mg liver in different strains of male mice. Activity in lung ranged over a factor of 26 between low and high values.

Induction of AHH activity was first discussed by Conney *et al.* (1), and the highest dose used, 20 mg/kg, gave the highest extent of induction of hepatic AHH, i.e., sevenfold in 24 h. Other investigators have also examined the induction of AHH. For example, Alvares *et al.* (4) have shown that after a single i.p. injection of 3-methylcholanthrene the V_{max} of benzpyrene hydroxylase activity is still rising at 24 h after 25 mg/kg injection, being about threefold higher at this time. The reported extent of hepatic AHH induction by polycyclic hydrocarbons *in vivo* is between 1.5-fold and sevenfold (Table 2). Slightly higher extents of induction are obtained with other organs, but in general their starting activities are considerably lower than in the liver. Much higher levels of induction are obtained with the same inducers *in vitro,* using cell cultures. Table 3 shows a partial list of the more active compounds tested; the extent of induction usually exceeds twentyfold, depending on the duration of induction. Such increases, while generally not obtained *in vivo,* have on occasion been reported. However, since stress conditions like partial hepatectomy will allow a twentyfold induction by benzanthracene (5) in animals where normally only a fourfold induction is obtained, it is probable that the extent of cellular induction, which is under genetic control, is modified by the whole organism. Further evidence of modification of the extent of induction by the body is seen by differences observed in male and female animals (3).

The elevation of AHH by various hydrocarbons and foreign compounds represents a true induction, i.e., increase the level of enzyme in tissues. Inhibitors of protein synthesis (Tables 2 and 3) like puromycin, ethionine, and cycloheximide, as well as the inhibitors of RNA synthesis, actinomycin D and mercapto-(pyridethyl)-benzimidazole, all interfere with the induction of AHH activity. Hydroxyurea, which at 50 mM level inhibited thymidine incorporation into DNA by 99%, did not prevent induction of AHH in fetal liver explants (6). Thus it would appear that active DNA synthesis is not required for induction of the enzyme, while both RNA synthesis and protein synthesis are required for increase in AHH levels. In fetal liver explants, a lag of 12 h in AHH induction was observed with fresh explants incubated with β-naphthoflavone. If the cul-

Table 1. Tissue Distribution of AHH[a,b]

Animal	Liver	Kidney	Skin	Lung	Small intestine	Adrenal
Rat	2.71 (19)			0.49 (19)	0.18 (19)	
	5.8 (20)	0.37 (20)		0.13 (20)		2.3 (21)
	16.6 (21)		0.0066 (23)		7.3 (21)	
	0.027 (1)		0.033 (24)		6.048 (2)	
	0.013 (4)		0.001–0.004 (25)			
	0.03 (10)					
Mouse	11.26 (3)	0.03 (3)		0.02–0.2 (3)	0.06–1.7 (3)	
	12 (22)	0.03 (22)	0.67 (22)	0.9 (22)	0.9 (22)	
Monkey	2.5 (3)	0.38 (3)	0.02 (3)	0.20 (3)	0.1 (3)	

[a]Data converted to common units, pmol phenolic products/min/mg tissue (wet weight).
[b]Parentheses contain reference numbers.

Table 2. Compounds Capable of Induction of AHH[a]

Tissue	Inducer	Fold induction	Blocked by
Liver	3-Methylcholanthrene	2.1 (10)	Actinomycin D, puromycin (10)
		5–7 (13)	Actinomycin D, ethionine (13)
		1.75 (26)	
		3.2 (4)	Puromycin (4)
	Immersion oil	10 (24)	
	Benzpyrene	10 (1)	Ethionine (1)
		1.75 (30)	
	Dimethylbenzanthracene	1.5 (26)	
	Benzanthracene	1.6 (26)	
		4 (5)	Actinomycin D, puromycin (5)
	Dibenz[*a,h*]anthracene	1.79 (26)	
	2,6-Dimethylbenz-anthracene	1.67 (26)	
	Cannabis resin	10 (27)	
	Aroclor 1254	5–10 (28)	Actinomycin D (28)
Skin	Immersion oil	11 (24)	
	7,12-Dimethyl benz[α]anthracene	4 (25)	
	3-Methylcholanthrene	9 (25)	
	Benzanthracene	7 (23)	
Lung	Cannabis resin	3 (27)	
	Benzanthracene	10 (22)	
Intestine	Benzanthracene	6 (2)	
Kidney	Benzpyrene	9 (29)	Cycloheximide (29)
	Benzanthracene	3–40 (22)	
Placenta	Smoking	*ca.* 30 (22)	
Lympho-blasts	Dibenz[*a,h*]anthracene	3–6 (31)	
	3-Methylcholanthrene	2–4 (31)	
	Phenobarbital	1.2–1.8 (31)	
Alveolar macrophages	Smoking	5 (32)	

[a]Reference numbers in parentheses.

tures were preincubated for 44 h before addition of the inducer, no lag was observed. When the inhibitor of RNA synthesis was added 2–3 h after the inducer, AHH elevation was not blocked; cycloheximide, however, was still capable of blocking the induction, at least partially, 4 h after the inducer was added (6). Studies by Nebert and Gelboin (7) have shown similar separation of transcription

Table 3. Compounds Inducing AHH in Cell Culture[a]

Tissue cells	Inducer	Fold induction	Blocked by
Fetal rat liver explant	β-Naphthoflavone (6)	5–10	Mercapto(pyridethyl)benzimidizole, cycloheximide
	4-Bromoflavone (6)	5	Cycloheximide
Hamster embryo	Benzanthracene (33)	8	Actinomycin D
Hamster fetus cells	7,12-Dimethylbenzanthracene (34)	8	
	Benzanthracene (34)	25	
	2,5-Diphenyloxazole (34)	17	
	3-Methylcholanthrene (34)	4	
Hepatocyte culture	3-Methylcholanthrene (35)	10	
	Phenobarbital (35)	8	
	Benzanthracene (35)	11	
	Illumination of culture containing riboflavin (36)	6	
Fetal liver cells	β-Naphthoflavone (37)	24	
	α-Naphthoflavone (37)	20	
	3-Methylcholanthrene (37)	20	
	Metyrapone (37)	15	

[a]Reference numbers in parentheses.

and translation in the induction process with hamster fetus cell subcultures, and have estimated that full transcription (programming for maximal induction) occurs within 8 h and is independent of translation (synthesis of the proteins called for by the messenger RNA). The requirement for RNA synthesis in the induction process indicates an involvement of specific genes for the enzyme, a suggestion borne out by the work of Nebert (see Chapter 12 for a discussion of the genetics of AHH).

Many laboratories have been interested in the enzyme system induced by the polycyclic hydrocarbons. Evidence has accumulated which indicates the AHH is a cytochrome P450-containing enzyme that differs from that normally present in the liver microsomes. The terminal oxidase moiety is a hemoprotein, containing iron protoporphyrin IX, like cytochrome P450 (8). The latter hemoprotein was shown by Omura and Sato (9) to be a *b*-type cytochrome by formation of a reduced pyridine hemochrome. One of the characteristics of the terminal oxidase, from which its name P450 was derived, is an absorption peak at 450 nm when reduced in the presence of carbon monoxide. Alvares *et al.* (10) found that in animals pretreated with polycyclic hydrocarbons the absorption maximum of cytochrome P450 was elevated and shifted to 448 nm, whereupon they named the newly induced hemoprotein "P448." Other investigators also found differences in the newly induced terminal oxidase. When native microsomal P450 interacts with the fragrant ligand ethylisocyanide in the presence of a chemical reductant like dithionite ($Na_2S_2O_4$), two Soret peaks appear, one at 455 nm and the other at 430 nm. The relative magnitudes of these peaks are pH dependent. While at pH 7.6 the two peaks are equal in height for the control microsomes (11), after induction with 3-methylcholanthrene the pH for equal 430 nm and 455 nm peak heights drops to 6.9 (12). Mannering's laboratory subsequently named the newly induced hemoprotein "$P_1$450." Similar findings have been made in other laboratories (13).

Another characteristic of the P450 hemoprotein is its ability to interact with substrates of the mixed-function oxidase, forming changes visible by difference spectroscopy. One of the spectral change types was termed a "type I" spectral change and its dissociation constant (K_s) was shown to be similar to the K_m for substrate metabolism (14); this spectral change was suggested to be the visible indication of an enzyme–substrate complex (14). Of interest was the observation that although enhancing the microsomal content of the P450 hemoprotein, treatment of rats with benzpyrene or 3-methylcholanthrene did not elevate the magnitude of the type I spectral change formed by hexobarbital addition to the microsomal suspension (15). Phenobarbital pretreatment, which also elevates the cytochrome P450 content, did elevate the magnitude of the hexobarbital-induced type I spectral change. In agreement with these observations, hexobarbital oxidation by the microsomes is enhanced by phenobarbital but not by polycyclic hydrocarbons. Of even more importance was the finding that while benzpyrene addition to microsomes does not evoke a type I spectral change in

Table 4. K_m Values Reported for BP Hydroxylase in Control and Polycyclic Hydrocarbon-Induced Rat Liver

Substrate range (μM)	Microsomal protein (mg/ml)[a]	K_m apparent: Control	K_m apparent: Induced	References
5–400	0.1	20.0	10.0	38
2.5–15	0.013	14.1	2.0	10
2.5–15	0.026	11.3	1.7	4
0.4–8	0.1[b]	2.95	1.0	39
0.2–2	0.015	1.7	0.27	41

[a]In assay medium.
[b]H. L. Gurtoo (personal communication).

microsomes of untreated rats, it does so after benzpyrene or 3-methylcholanthrene induction of AHH (15). This latter observation clearly indicates that the newly induced hemoprotein, P_1450, shows a different substrate specificity from the native species. Further indication that the induced enzyme differs from the native hemoprotein has been provided by kinetic measurements of benzpyrene metabolism by liver microsomes. Some representative values are listed in Table 4. Note that in all cases after induction of AHH the apparent K_m values decrease to one-third or one-tenth the value for the untreated animal.

Several studies have been performed on the time course of AHH induction. For example, Conney *et al.* (1) showed that peak levels are reached in about 24 h after i.p. injection of 20 mg of benzpyrene per kilogram of body weight in rats. In that study, 2 mg/kg peaked in 12 h, at about half the hepatic content of AHH activity. In another study (4), induction with 25 mg/kg allowed a fairly linear threefold increase in benzpyrene hydroxylase activity over a 24 h period. Attendant with their increase in V_{max} was a rapid decline in the K_m value for benzpyrene to a low point by 12 h (4). A number of similarities have been observed in our laboratory with the above findings.

In agreement with the results of Conney *et al.* (1), the inductive process is dose dependent in rats (Fig. 1). One milligram of 3-methylcholanthrene per kilogram of body weight was capable of elevating AHH activity in 24 h by 1.8-fold. The activity declined thereafter even when the inducer was applied daily. Five milligrams of 3-methylcholanthrene daily induced AHH synthesis at a rate greater than 1 mg/kg, but similar to induction by 20 mg/kg (Fig. 1). However, with the latter dose, activity levels did not fluctuate as markedly from day to day. With continuous treatment with inducer, activity levels show a tendency to decline after 3 days, even at the higher doses. In agreement with the report of Alvares *et al.* (4), the K_m for benzpyrene reached a minimal value by 12 h. The same K_m value was obtained regardless of the amount of inducer used (Table 5) and was maintained for at least 8 h.

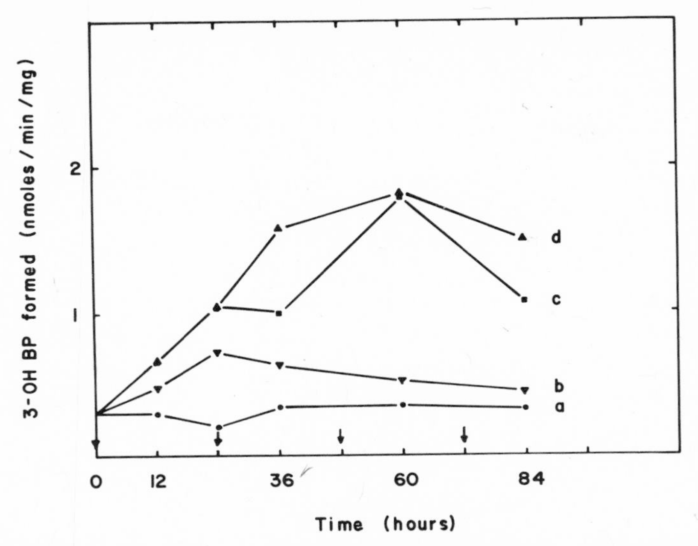

Fig. 1. Time course of AHH induction in rat liver microsomes. Rats were treated with (a) corn oil (●) or (b) 1 mg (▼), (c) 5 mg (■), or (d) 200 mg (▲) of 3-methylcholanthrene/kg body weight at 0, 12, 24, and 48 h. Fifty-gram male Sprague Dawley rats were used. Assay medium consisted of 50 mM tris-HCl–3 mM $MgCl_2$, containing 1 mg bovine serum albumin/ml, 100 μM benzpyrene, and 0.5 mM NADPH. The reaction was initiated by addition of NADPH 1 min after addition of 25 mg microsomes/ml. Aliquots were withdrawn at 1-min intervals over a 10-min period. Temperature was 37°C. Hydroxybenzypyrene formation was measured directly after clarification of 3-ml aliquots with 0.3 ml of 10% Emulgen–1% EDTA–1 N NaOH, using 522 nm emission and 466 nm excitation; standardization was with 3-OH-benzpyrene.

Table 5. Effect of 3-Methylcholanthrene (3-MC) Pretreatment of Rats on Liver Microsomal AHH Kinetic Constants[a]

	Hours after initial MC injection			
	12		84	
Dose of 3-MC (mg/kg)	K_m (μM)	V_{max} (pmol/min/mg)	K_m (μM)	V_{max} (pmol/min/mg)
0	1.59	198		
1	0.25	315	0.26	273
5	0.27	542	0.20	907
20	0.26	584	0.15	1140

[a]Substrate range 0.2–2.4 μM benzpyrene, microsomal concentration 15 μg protein/ml.

The rapid change in K_m for benzpyrene poses somewhat of a puzzle, since it occurs even with the lowest level of inducer, when the extent of induction of AHH is only about 1.5-fold. If differences in the K_m indicate the presence of a different enzyme initially, then one would expect a composite K_m value or a break in the Lineweaver–Burk plot. Yet neither occurred in our study. The only alternative which we can see, which is consistent with our observations here, is that the newly induced oxidase can compete effectively with the native oxidase for some limiting component of the enzyme system (as, for example, cytochrome P450 reductase) and that the observed activity is mainly that of the induced oxidase. Since benzpyrene can form a type I spectral change with the newly induced P_1450 (15) and since type I substrates enhance electron flow to the terminal oxidase (17,18), benzpyrene interaction at the type I site would be the factor causing more effective competition by P448, if this alternative suggestion is correct.

The increase in AHH activity was paralleled by an increase in cytochrome P450 hemoprotein. The percent increase of the hemoprotein was not quite as large as the AHH activity (Fig. 2). Examination of the spectra with respect to absorption maximum showed that in 12 h the peak positions of the CO complex had moved to shorter wavelengths. The extent of the shift was dependent on the amount of inducer used; with 1 mg 3-methylcholanthrene/kg body weight, the

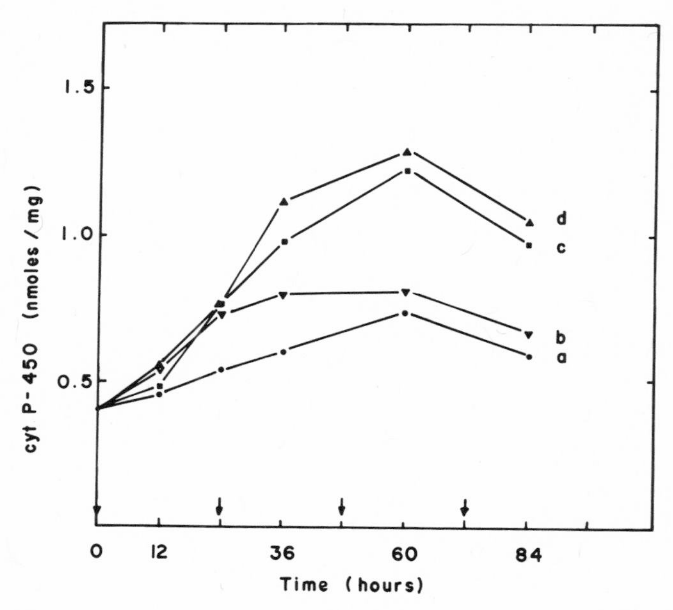

Fig. 2. Time course of induction of P_1450 in rat liver microsomes. Conditions of induction were those of Fig. 1 (see caption). Cytochrome P450 hemoprotein–CO complex was measured by the method of Omura and Sato (9).

absorption maximum was at 449.5 nm in 12 h and 449 nm at 24 h. At 5 mg/kg and 20 mg/kg, the peak had shifted to 449 nm by 12 h and was at 448 nm by 24 h. This continuous shift in wavelength must be considered as a dilution of existing cytochrome P450 with newly synthesized cytochrome P_1450; microsomal aminopyrine *N*-demethylase activity and hexobarbital oxidase activities do not decline, and the type I spectral change elicited by these drugs does not decrease in the first 24 h after 3-methylcholanthrene (unpublished observations). Consequently, the above postulation of more effective competition by P_1450 for reducing equivalents as an explanation for the observed rapid decrease in the K_m for benzpyrene (Table 4) becomes more attractive. Further indication that induction of one species of P450 does not preclude the existence of another species in the microsomes was shown by the additive effect of phenobarbital and 3-methylcholanthrene on microsomal P450 content (16), as well as on total AHH activity (35). It is not known whether in the latter case there is a change in K_m value for benzpyrene.

One last point which should be mentioned concerns the assay for benzpyrene hydroxylation and the K_m values shown in Table 4. Note the reported values span 2 orders of magnitude with microsomes from induced rats and are close to that with control rat microsomes. Part of the reason for the different values resides in the amount of microsomal protein used in the assays, and part is due to differences in the length of time used in the assay. But a good portion of the problem may be the result of the substrate levels employed. Several of these problems were examined by Hansen and Fouts (38). They observed that nonspecific substrate binding to the microsomes causes erroneous effective substrate levels, and that because of high enzyme activity and low substrate levels used a considerable portion of the added substrate was consumed. We have reexamined this problem and found that another potential problem exists, that of substrate aggregation. It is this latter problem, we feel, that is responsible for the differences in reported K_m values. What is generally not considered by most people is that the maximum solubility of benzpyrene in aqueous medium has been reported to be only 9–24 nM (40); consequently, further additions of this hydrophobic hydrocarbon would be expected to cause aggregation or stacking due to π-bonding. If the amount of benzpyrene added is large enough, a suspension of large particles which sediment readily will be obtained. At lower benzpyrene levels, smaller molecular aggregates would be expected. Figure 3 shows the effect of benzpyrene concentration on the extent of aggregation. In the presence of detergent, fluorescence due to benzpyrene increased linearly with the amount added (line 3,4). In the absence of the detergent, the line deviated from linearity (line 1) at about 20 μM benzpyrene. As reported earlier (40), the 410 nm fluorescence shifts to 490 nm when benzpyrene aggregates and comes out of solution. Since benzpyrene at the concentrations shown here did not sediment, it appears that soluble aggregates form in the 20–100 μM range (41).

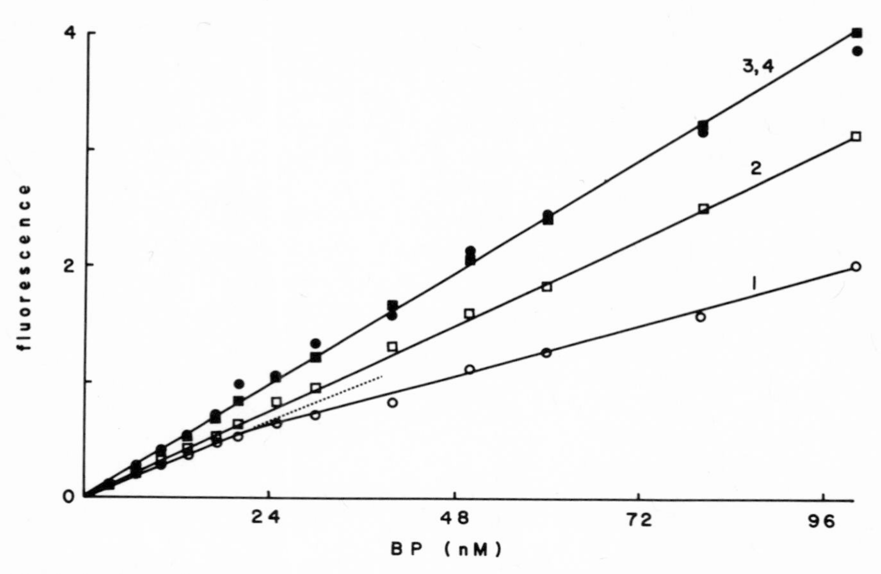

Fig. 3. Evidence for aggregation of benzpyrene in aqueous AHH assay medium. Benzpyrene was added as aliquots of 0.01 μM (in acetone) solution to the assay medium (without NADPH) in Fig. 1, with acetone alone being added to keep volumes constant. Acetone alone was without effect, and 30 μl/3 ml was the amount added. The mixtures were shaken briefly and centrifuged at 140,000*g* for 45 min in the Spinco centrifuge; fluorescence was measured using 365 nm excitation and 410 nm emission. When added, 0.3 ml of 10% Emulgen–1 N NaOH was mixed per 3 ml of sample. 1, No other additions; 2, 1 mg albumin/ml; 3, Emulgen added to No. 1; 4, Emulgen added to No. 2.

These have a lower molar fluorescence. The addition of detergent to the samples in line 1 increased fluorescence to that of line 3. Albumin, which "solubilizes" benzpyrene (line 2), also kept the fluorescence linear, but lower than that in the presence of detergent (line 4); addition of detergent to samples with albumin also increased fluorescence to that of detergent alone. This indicates that benzpyrene, when bound to albumin, is more dispersed and less likely to form large aggregates.

Since benzpyrene aggregates in aqueous solution, the extent of such aggregation will be dependent on the amount of the compound added. It is probable, then, that the extent of aggregation will affect both availability of substrate to the enzyme and the apparent affinity of the enzyme for the aggregated form(s). The data in Table 4 suggest that the K_m one obtains is strongly related to the substrate levels employed.

In future studies, a point which will have to be considered is that of metabolite profiles. If indeed AHH can accept and oxidize molecular aggregates, the position and extent of oxidation may differ depending on the aggregate size.

ACKNOWLEDGMENTS

This work was supported in part by American Cancer Society Grant BC 164 and United States Public Health Service Grant GM 17021. J. B. Schenkman is a Research Career Development Awardee (GM K4-19601) of the NIH.

REFERENCES

1. A. H. Conney, E. C. Miller, and J. A. Miller, Substrate-induced synthesis and other properties of benzpyrene hydroxylase in rat liver, *J. Biol. Chem.* **228,** 753–766 (1957).
2. L. W. Wattenberg, J. L. Leong, and P. J. Strand, Benzpyrene hydroxylase activity in the gastrointestinal tract, *Cancer Res.* **22,** 1120–1125 (1962).
3. D. W. Nebert and H. V. Gelboin, The *in vivo* and *in vitro* induction of aryl hydrocarbon hydroxylase in mammalian cells of different species, tissues, strains, and developmental and hormonal states, *Arch. Biochem. Biophys.* **134,** 76–89 (1969).
4. A. P. Alvares, G. Schilling, and R. Kuntzman, Alteration in the microsomal hemoprotein and the kinetics of 3,4-benzpyrene hydroxylase induced by 3-methylcholanthrene: Time course study and effects of puromycin, *Life Sci.* **10,** 129–136 (1971).
5. T. Spencer and P. W. F. Fischer, The induction of microsomal hydroxylases in regenerating rat liver, *Chem.-Biol. Interactions* **4,** 41–47 (1972).
6. K. R. Cutroneo and E. Bresnick, Induction of benzpyrene hydroxylase in fetal liver explants by flavones and phenobarbital, *Biochem. Pharmacol.* **22,** 675–687 (1973).
7. D. W. Nebert and H. V. Gelboin, The role of ribonucleic acid and protein synthesis in microsomal aryl hydrocarbon hydroxylase induction in cell culture: The independence of transcription and translation, *J. Biol. Chem.* **245,** 160–168 (1970).
8. M. D. Maines and M. W. Anders, Characterization of the heme of cytochrome P-450 using gas chromatography/mass spectrometry, *Arch. Biochem. Biophys.* **159,** 201–204 (1973).
9. T. Omura and R. Sato, The carbon monoxide-binding pigment of liver microsomes. II. Solubilization, purification, and properties, *J. Biol. Chem.* **239,** 2379–2385 (1964).
10. A. P. Alvares, G. R. Schilling, and R. Kuntzman, Differences in the kinetics of benzpyrene hydroxylation by hepatic drug-metabolizing enzymes from phenobarbital and 3-methylcholanthrene-treated rats, *Biochem. Biophys. Res. Commun.* **30,** 588–593 (1968).
11. Y. Imai and R. Sato, Spectral interactions of protoheme with ethyl isocyanide and some other ligands in neutral pH region, *J. Biochem.* **64,** 147–159 (1968).
12. N. E. Sladek and G. J. Mannering, Evidence for a new P-450 hemoprotein in hepatic microsomes from methylcholanthrene-treated rats, *Biochem. Biophys. Res. Commun.* **24,** 668–674 (1966).
13. A. P. Alvares, G. Schilling, W. Levin, and R. Kuntzman, Alteration of the microsomal hemoprotein by 3-methylcholanthrene: Effects of ethionine and actinomycin D, *J. Pharmacol. Exp. Ther.* **163,** 417–424 (1968).
14. J. B. Schenkman, H. Remmer, and R. W. Estabrook, Spectral studies of drug interaction with hepatic microsomal cytochrome, *Mol. Pharmacol.* **3,** 113–123 (1967).
15. J. B. Schenkman, H. Greim, M. Zange, and H. Remmer, On the problem of possible other forms of cytochrome P-450 in liver microsomes, *Biochim. Biophys. Acta* **171,** 23–31 (1969).

16. K. Bidleman and G. J. Mannering, Induction of drug metabolism. V. Independent formation of cytochrome P-450 and P_1-450 in rats treated with phenobarbital and 3-methylcholanthrene simultaneously, *Mol. Pharmacol.* **6**, 697–701 (1970).
17. J. B. Schenkman, Effect of substrates on hepatic microsomal cytochrome P-450, *Hoppe-Seyler's Z. Physiol. Chem.* **349**, 1624–1628 (1968).
18. P. L. Gigon, T. E. Gram, and J. R. Gillette, Effect of drug substrates on the reduction of hepatic microsomal P-450 by NADPH, *Biochem. Biophys. Res. Commun.* **31**, 558–562 (1968).
19. R. P. Weber, J. M. Coon, and A. J. Triolo, Nicotine inhibition of the metabolism of 3,4-benzopyrene, a carcinogen in tobacco smoke, *Science* **184**, 1081–1082 (1974).
20. A. J. Paine and A. E. McLean, The effect of dietary protein and fat on the activity of aryl hydrocarbon hydroxylase in rat liver, kidney and lung, *Biochem. Pharmacol.* **22**, 2875–2880 (1973).
21. N. G. Zampaglione and G. J. Mannering, Properties of benzpyrene hydroxylase in the liver, intestinal mucosa and adrenal of untreated and 3-methylcholanthrene treated rats, *J. Pharmacol. Exp. Ther.* **185**, 676–685 (1973).
22. F. J. Wiebel, J. C. Leutz, and H. V. Gelboin, Aryl hydrocarbon (benzo(α)-pyrene) hydroxylase: Inducible in extrahepatic tissues of mouse strains not inducible in liver, *Arch. Biochem. Biophys.* **154**, 292–294 (1973).
23. W. Levin, A. H. Conney, A. P. Alvares, I. Merkatz, and A. Kappas, Induction of benzo(*a*)pyrene hydroxylase in human skin, *Science* **176**, 419–420 (1972).
24. A. P. Alvares, D. R. Bickers, and A. Kappas, Induction of drug-metabolizing enzymes and aryl hydrocarbon hydroxylase by microscope immersion oil, *Life Sci.* **14**, 853–860 (1974).
25. D. W. Nebert, L. L. Bausserman, and R. R. Bates, Effect on 17-β-estradiol and testosterone on aryl hydrocarbon hydroxylase activity in mouse tissues *in vivo* and in cell culture, *Int. J. Cancer* **6**, 470–480 (1970).
26. P. H. Jellinck and G. Smith. Aryl hydroxylase induction of the chick embryo by polycyclic hydrocarbons, *Biochim. Biophys. Acta* **304**, 520–525 (1973).
27. H. P. Witschi and B. Saint-Francois, Enhanced activity of benzpyrene hydroxylase in rat liver and lung after acute *Cannabis* administration, *Toxicol. Appl. Pharmacol.* **23**, 165–168 (1972).
28. A. P. Alvares, D. R. Bickers, and A. Kappas, Polychlorinated biphenyls: A new type of inducer of cytochrome P-448 in the liver, *Proc. Natl. Acad. Sci. USA* **70**, 1321–1325 (1973).
29. R. Grundin, S. Jakobsson, and D. L. Cinti, Induction of microsomal aryl hydrocarbon (3,4-benzo(α)pyrene) hydroxylase and cytochrome P-450_k in rat kidney cortex. I. Characteristics of the hydroxylase system, *Arch. Biochem. Biophys.* **158**, 544–555 (1973).
30. R. M. Welch, Y. E. Harrison, A. H. Conney, P. J. Poppers, and M. Finster, Cigarette smoking: Stimulatory effect on metabolism of 3,4-benzpyrene by enzymes in human placenta, *Science* **160**, 541–542 (1968).
31. G. Kellermann, M. Luyten-Kellermann, and C. R. Shaw, Metabolism of polycyclic aromatic hydrocarbons in cultured human leukocytes under genetic control, *Humangenetik* **20**, 257–263 (1973).
32. E. T. Cantrell, G. A. Wari, D. L. Busbee, and R. R. Martin, Induction of aryl hydrocarbon hydroxylase in human pulmonary alveolar macrophages by cigarette smoking, *J. Clin. Invest.* **52**, 1881–1884 (1973).
33. H. V. Gelboin and F. J. Wiebel, Studies on the mechanism of aryl hydrocarbon hydroxylase induction and its role in cytotoxicity and tumorigenicity, *Ann. N.Y. Acad. Sci.* **179**, 529–547 (1971).

34. D. W. Nebert and H. V. Gelboin, Substrate-inducible microsomal aryl hydroxylase in mammalian cell culture. II. Cellular responses during enzyme induction, *J. Biol. Chem.* **243**, 6250–6261 (1968).
35. J. E. Gielen and D. W. Nebert, Microsomal hydroxylase induction in liver cell culture by phenobarbital, polycyclic hydrocarbons, and *p,p'*-DDT, *Science* **172**, 167–169 (1971).
36. A. J. Paine and A. E. M. McLean, Induction of aryl hydrocarbon hydroxylase by a light-driven superoxide generating system in liver cell culture, *Biochem. Biophys. Res. Commun.* **58**, 482–486 (1974).
37. I. S. Owens and D. W. Nebert, Aryl hydrocarbon hydroxylase induction in mammalian liver-derived cell cultures. Stimulation of "cytochrome P_1-450-associated" enzyme activity by many inducing compounds, *Mol. Pharmacol.* **11**, 94–104 (1975).
38. A. R. Hansen and J. R. Fouts, Some problems in Michaelis–Menten kinetic analysis of benzpyrene hydroxylase in hepatic microsomes from polycyclic hydrocarbon-pretreated animals, *Chem.-Biol. Interactions* **5**, 167–182 (1972).
39. H. L. Gurtoo and T. C. Campbell, A kinetic approach to a study of the induction of rat liver microsomal hydroxylase after pretreatment with 3,4-benzpyrene and aflatoxin B_1, *Biochem. Pharmacol.* **19**, 1729–1735 (1970).
40. E. Boyland and B. Green, The interaction of polycyclic hydrocarbons and purines, *Br. J. Cancer* **16**, 347–360 (1962).
41. K. M. Robie, Y. N. Cha, R. E. Talcott, and J. B. Schenkman, Kinetic studies of benzpyrene and hydroxybenzpyrene metabolism, *Chem.-Biol. Interactions* **12**, 285–297 (1976).

9

Discussion

In response to Kupfer's comment about the need for caution in the interpretation of assays for benzpyrene hydroxylase activity, Schenkman again emphasized that most benzpyrene hydroxylase assays monitor the appearance of a fluorescent metabolite, despite the lack of stoichiometric relationship between the fluorescent compounds produced and the disappearance of benzpyrene. It has been shown that fluorescent metabolites other than the 3-hydroxy derivative are produced by the AHH-mediated reaction, some of which have not been identified; thus their fluorescence coefficients are not known. Nonfluorescent conjugates are also produced from the hydroxylated intermediates in a microsomal system. Thus he agreed that great care must be exercised in interpreting all results obtained in fluorescent assays of benzpyrene hydroxylase activity.

Gelboin emphasized that the substrate concentration in the *in vitro* assay systems determines which of the several metabolic pathways of benzpyrene predominate. With high substrate concentrations, the primary metabolites are formed almost exclusively, i.e., the monophenols, the quinones, and the dihydrodiols; at low substrate concentrations, there is extensive conversion of these primary metabolites to very highly polar conjugates. Schenkman indicated that he had made similar observations.

Ullrich commented that the existence of many metabolic pathways for benzpyrene and the fact that there are several forms of the heme protein, cytochrome P450, emphasize the importance of accurately determining the extinction coefficient of the CO complex for purposes of quantitation of each type. Conney indicated that during the purification processes there is great loss of the heme proteins and it is not known whether current purification procedures result in preparations truly representative of the microsomal cytochrome complement. Nebert also emphasized that the isolated cytochrome need not retain the absorption characteristics of the microsomal material. However, the value of 91 cm^{-1} mM^{-1} found by Omura and Sato (*J. Biol. Chem* **239**, 2370, 1964) appears to be correct within 10% for all heme proteins which they have isolated from rats, rabbits, and both genetically "responsive" and "nonresponsive" mice. Sato indicated that the absorption characteristics of P450 recently purified in his laboratory from both phenobarbital-treated and uninduced rats did not change during the purification process. The millimolar extinction coefficient for the phenobarbital-induced P450 was 92 cm^{-1}, while that for the uninduced heme protein was 85–90 cm^{-1}. Sato emphasized that the extinction coefficient for any of these forms will not be accurately known until the several proteins are obtained in a more highly purified form.

10

Benzo[*a*]pyrene Metabolism: Enzymatic and Liquid Chromatographic Analysis and Application to Human Liver, Lymphocytes, and Monocytes

Harry V. Gelboin, James Selkirk, Takao Okuda, Nobuo Nemoto, Shen K. Yang, Friedrich J. Wiebel, James P. Whitlock, Jr., Herbert J. Rapp, and Robert C. Bast, Jr.

Chemistry and Biology Branches
National Cancer Institute
Bethesda, Maryland 20014, U.S.A.

Although the molecular basis for the variability in human responsiveness to the action of polycyclic aromatic hydrocarbon (PAH) carcinogens is unknown, the initial biogical receptors for the hydrocarbons have been identified as the microsomal mixed-function oxygenases. The polycyclic aromatic hydrocarbons are initially oxygenated by this enzyme system and the metabolites formed can be further oxygenated by the same mixed-function oxygenases. Other routes of metabolism can be by hydration of the epoxide intermediates to dihydrodiols or by conjugation of the oxygenated intermediates to water-soluble products. The carcinogenic potential of the hydrocarbon may be either diminished or enhanced by these enzyme systems (1–3) and variation in the activity of these enzyme systems may be related to the variability of the carcinogenic effects in each individual. Heterogeneity in these enzymes may thus be an important factor in the heterogeneity of individuals to carcinogen susceptibility. Thus it is important to have methods to determine the profiles of these enzyme activities in individuals so that their relationship to carcinogenesis can be understood. In this chapter we present various methods for the study of PAH metabolism. Each method has its unique merits and unique deficiencies. A combination of these approaches may yield a systematic analysis with which the relationship of metabolism and carcinogen susceptibility of an individual might be assessed.

Polycyclic aromatic hydrocarbons are a large class of chemicals common to the environment (4) and are present in the atmosphere (5), waterways and oceans, soil (6), marine life, and the food chain (7).

Major sources of polycyclic aromatic hydrocarbons include emissions from transportation, heat and power generation, refuse burning, industrial processes, and oil contamination by effluent disposal or oil spills into waterways (4). Epidemiological studies indicate that the environment is a significant determinant (4) in the incidence of human cancer (7), although many of the specific causal agents have yet to be identified. One major causal agent is known: inhalation of smoke by cigarette smokers is clearly a major factor in the high incidence of lung cancer among smokers (8). Many of the polycyclic aromatic hydrocarbons present in smoke are powerful carcinogens in experimental animals and are thus logical suspects as the carcinogens causing lung cancer in humans. Since the polycyclic aromatic hydrocarbons are also common environmental pollutants, they are also suspected of contributing to cancer at other organ sites.

In this study, we have examined benzo[*a*]pyrene as a prototype polycyclic aromatic hydrocarbon. Benzo[*a*]pyrene is a powerful carcinogen (9), and a most common hydrocarbon in the environment (4). Its metabolism has been extensively studied (10–14) and the strong fluorescence of its phenolic metabolites has been the basis for a very sensitive assay of a PAH microsomal cytochrome P450 type mixed-function oxygenase, the aryl hydrocarbon (benzo[*a*]pyrene) hydroxylase (AHH) (15). The oxidative metabolism of polycyclic hydrocarbons is catalyzed by the "mixed-function oxygenases" which are part of the endoplasmic reticulum and consist of a terminal cytochrome (P450) and an electron transport chain. The microsomal oxygenases are present in most tissues of mammals (16). They metabolize and are inducible by a wide spectrum of xenobiotics (17) which include pesticides, drugs, food additives, and polycyclic aromatic hydorcarbons (2), and by some endogenous substrates such as steroids (18).

The detoxication role of the microsomal mixed-function oxygenases and related enzymes has been well established and reviewed (1,2,10). It is clear that oxygenation of the PAH at several sites results in detoxified inactive metabolites and the induction of this enzyme system parallels in several cases a marked inhibition of tumorigenesis. For example, the pretreatment of rats with small amounts of polycyclic hydrocarbons decreases tumor formation in the mammary gland induced by subsequently administered polycyclic hydrocarbons (19). Another report showed that the intraperitoneal injection of the enzyme inducer 5,6-benzoflavone inhibited 7,12-dimethylbenz[a]anthracene (DMBA) induced tumorigenesis in the lung and mammary gland of rodents (1). One possible explanation for these results is that prior treatment of the animals with these enzyme inducers causes an increase in mixed-function oxygenase activity in the liver, the major site of metabolism, and this in turn may lower the concentration of the carcinogen in the target tissue, in the latter cases, the lung and mammary gland. Another possibility is that the induced enzyme catalyzes the reactions of the detoxication pathway of metabolism to a greater extent than those of the carcinogen activation pathway (see below).

Although the mixed-function oxygenases clearly serve a detoxication role, the evidence is also clear that some of the polycyclic hydrocarbons are metabolically activated to cytotoxic and carcinogenic intermediates by these enzymes. The evidence for the role of the mixed-function oxygenases in carcinogen activation is as follows: (1) Microsomal NADPH-requiring oxygenases catalyze the formation of hydrocarbon–DNA complexes (20–22). (2) The cellular level of AHH activity is positively correlated with the cytotoxic effects of polycyclic hydrocarbons in a variety of different cells (23,24). (3) Inhibition of enzyme activity by 7,8-benzoflavone (7,8-BF) inhibits polycyclic hydrocarbon toxicity, metabolism, and macromolecule binding in cell culture (25). (4) Inhibition of AHH reduces the covalent binding of DMBA to nucleic acids and protein of mouse skin (26). (5) Inhibition of AHH by 7,8-benzoflavone inhibits tumorigenesis by 7,12-dimethylbenz[*a*]anthracene (DMBA) in mouse skin (26). In the last study, the 7,8-benzoflavone had either no effect or a stimulatory effect on benzo[*a*]pyrene tumorigenesis, indicating that the role of the microsomal hydroxylases may be unique for different hydrocarbons. All of the latter effects clearly relate to the activity of the NADPH-dependent mixed-function oxygenases, but may also be influenced by the action of the related enzymes in hydrocarbon metabolism (see below).

Microsomal AHH is present and inducible in a variety of tissues of different species. It also varies in different mouse strains (16,27) and is affected by age, developmental stage, sex, and hormonal balance (16). The microsomal oxygenases are also present, as measured by the aryl hydrocarbon benzo[*a*]pyrene hydroxylase (AHH) in human tissues. Thus AHH has been found in human liver (28), placenta (29,30), lymphocytes (31,32), monocytes (33,34), and lung macrophages (35).

BENZO[*a*]PYRENE METABOLISM—ANALYSIS BY HIGH-PRESSURE LIQUID CHROMATOGRAPHY

We have found that high-pressure liquid chromatography is a new and most promising technique for metabolite analysis (14,36). This technique has since been used by others (37) for the same purpose and is far superior to thin-layer chromatography. It permits a clean and quantitative separation of at least eight benzo[*a*]pyrene (BP) metabolites. Our current research indicates that further development of the appropriate conditions will enable a comprehensive and quantitative separation of most if not all hydrocarbon metabolites. Figure 1 shows a typical separation of a number of BP metabolites and shows the distinct peaks formed by each compound. There are three distinct diol peaks, two phenol peaks, and a quinone region.

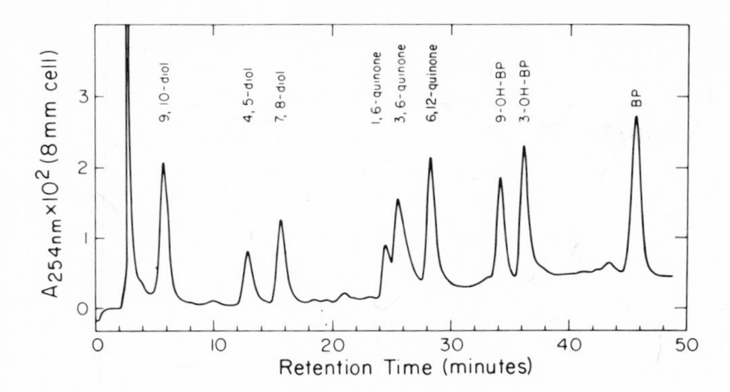

Fig. 1. High-pressure liquid chromatographic separation of benzo[a]pyrene metabolites.

Figure 2 shows some of the known and suspected metabolites of benzo[*a*]-pyrene. Upon incubation of benzo[*a*]pyrene with cells or tissue preparations, various metabolites have been isolated (11,13,38–41). Those separated by high-pressure liquid chromatography (HPLC) include the following: three dihydrodiols–the 7,8-dihydrodiol, 9,10-dihydrodiol, and 4,5-dihydrodiol; two phenols–the 3-OH and the 9-OH; and three quinones–the 1,6-quinone, 3,6-quinone, and 6,12-quinone (14,36). In this HPLC system, the other synthetic

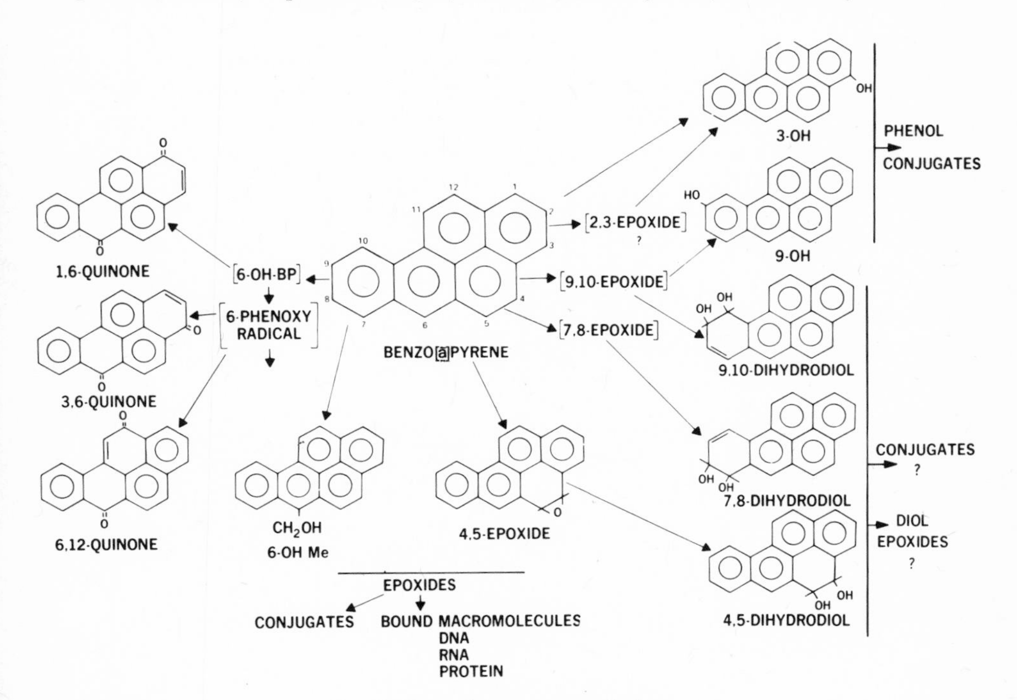

Fig. 2. Benzo[a]pyrene metabolism.

phenols migrate with the 3-OH and 9-OH peaks (37). Under special conditions, the BP 4,5-oxide can be isolated (42). There has been one report that 6-(hydroxymethyl) benzo[*a*] pyrene is also formed during BP metabolism (43). With preparations containing the soluble fraction of cell proteins or with whole cells, unidentified water-soluble metabolites are also produced (44). In several recently reviewed reports (40), the K-region epoxide was suggested to be the active intermediate of BP. In an earlier report (45), evidence was presented that the 7,8-diol was the metabolite most active in the binding of DNA catalyzed by the mixed-function oxygenases. Thus it appeared that a further metabolite of the 7,8-diol may be a very active species with respect to DNA binding. A recent report has suggested that the BP-7,8-diol is converted to a 7,8-dihydrodiol–9,10-epoxide-BP and this compound is very active in the binding to DNA (12).

Figure 3 shows the enzyme systems and Table 1 shows the several assays for the enzymes that are involved in the formation of the various metabolites shown in Fig. 2. The sequence of enzyme action is as follows: the initial oxygenation is catalyzed by the microsomal mixed-function oxygenases, which contain multiple forms of cytochrome P450. This results in the formation of oxides in at least the 4–5, 7–8, and 9–10 positions. The most stable of these, the 4–5-oxide, has been isolated under special conditions, and the 7,8- and 9,10-oxides as well as the 4,5-oxide are likely intermediates (47). Their formation can be deduced from the presence of all three of the corresponding dihydrodiols. These are presumably formed by the action of arene oxide hydratase on the oxide intermediates. Further evidence that this mechanism prevails is the finding that a hydratase inhibitor, 1,1,1-trichloropropylene oxide, completely eliminates dihydrodiol formation (14) and the readdition of a partially purified hydrase results in the appearance of the dihydrodiols (37). The phenols and quinones can either be

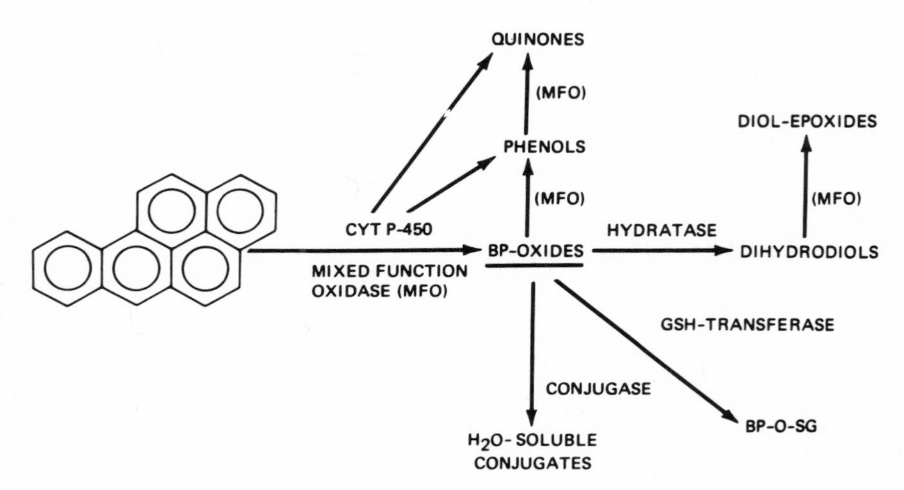

Fig. 3. Enzymatic reactions of benzo[a] pyrene.

Table 1. Assays for Benzo[*a*]pyrene Metabolism

Method	Enzyme measured	Substrate	Products
Aryl hydrocarbon hydroxylase (AHH)	Mixed-function oxidase (P450)	Benzo[*a*]pyrene	Phenols equivalent to 3-OH-BP and 9-OH-BP
Hydratase	BP-4,5-oxide hydratase	[^{3}H] BP-4,5-oxide	[^{3}H] BP-4,5-dihydrodiol
GSH-transferase	Glutathione-*S*-BP-4,5-oxide transferase	[^{3}H] BP-4,5-oxide	[^{3}H] BP-4,5-glutathione
High-pressure liquid chromatography	1. Mixed-function oxidase 2. Hydratase 3. Nonenzymatic reactions	[^{3}H] Benzo[*a*]pyrene	Phenols Dihydrodiols Quinones Hydroxymethyl-BP Diol-epoxides Unknown metabolites
Water-soluble product formation	1. Mixed-function oxidase 2. Hydratase 3. Transferases (GSH), Gluc(?)SO_4 (?)	[^{3}H] Benzo[*a*]pyrene	Water-soluble conjugates
Covalent binding	1. Mixed-function oxidase 2. Hydratase 3. Transferases (GSH), Gluc(?)SO_4 (?)	[^{3}H] Benzo[*a*]pyrene	[^{3}H] BPX-DNA [^{3}H] BPX-RNA [^{3}H] BPX-protein
Mutagen detection	1. Mixed-function oxidase 2. Hydratase 3. Transferase (GSH), Gluc(?)SO_4 (?)	Polycyclic aromatic hydrocarbon	Mutagenic product

formed nonenzymatically from the oxide intermediate or be the result of a direct oxygenation by the mixed-function oxidase. That a mechanism of oxygenation independent of arene oxide intermediacy is operative is suggested by the finding that activation of the benzo[*a*] pyrene and oxygenation occurs at the 6 position (39), with quinone formation at the 1,6, 3,6, and 6,12 positions. The 6-position oxygenation does not sterically permit an oxide intermediate, and thus we may conclude that not all of the oxygenated products are formed solely through oxide intermediates.

MIXED-FUNCTION OXYGENASE–ARYL HYDROCARBON BENZO[*a*] PYRENE HYDROXYLASE (AHH)

The primary catalytic attack on the PAH is by the microsomal mixed-function oxygenase P450-containing enzyme systems. Enzyme purification studies have shown that this system is composed of at least four forms of the enzyme (46). With different steroid and biphenyl substrates, the various forms of the cytochrome P450s show preferential hydroxylation at specific positions. In collaboration with M. Coon and his colleagues (47), we have found that the various forms of highly purified cytochrome P450 have different catalytic activity with respect to the formation of the different benzo[*a*] pyrene metabolites. Figure 4 shows the high-pressure liquid chromatographic separation of benzo[*a*] pyrene metabolites formed by rabbit liver microsomes and by two highly purified cytochrome P450s, cytochrome $P450_{LM2}$ and cytochrome $P450_{LM4}$. The phenolic products were eluted in two peaks, A and B, that contained primarily 9-hydroxy and 3-hydroxy benzo[*a*] pyrene, respectively. The ratio of peak A to peak B was 0.11 for $P450_{LM2}$ and 0.45 for $P450_{LM4}$. Thus the relative amounts of the phenols formed by these two cytochromes differ markedly. The positional specificities of different forms of cytochrome P450 may direct polycyclic aromatic hydrocarbon metabolism into the various activation and detoxication pathways and thereby help determine the cytotoxic and carcinogenic activity of these compounds. The different forms of the cytochrome P450 are also preferentially induced by different inducers of the mixed-function oxygenases such as methylcholanthrene or phenobarbital, and the different forms of the enzyme can also be preferentially inhibited. Thus 7,8-benzoflavone inhibits the MC-induced enzyme in rat liver but does not inhibit the uninduced control rat liver enzyme (48). Induction or inhibition may alter the balance of metabolic pathways (49). An example of altered metabolism after induction (50) is shown in Table 2. The relative amounts of the 9,10-diol and 7,8-diol to the other metabolites are greatly increased after enzyme induction by methylcholanthrene pretreatment. This may be due to the induction of a specific form of cytochrome P450 or to an altered ratio of mixed-function

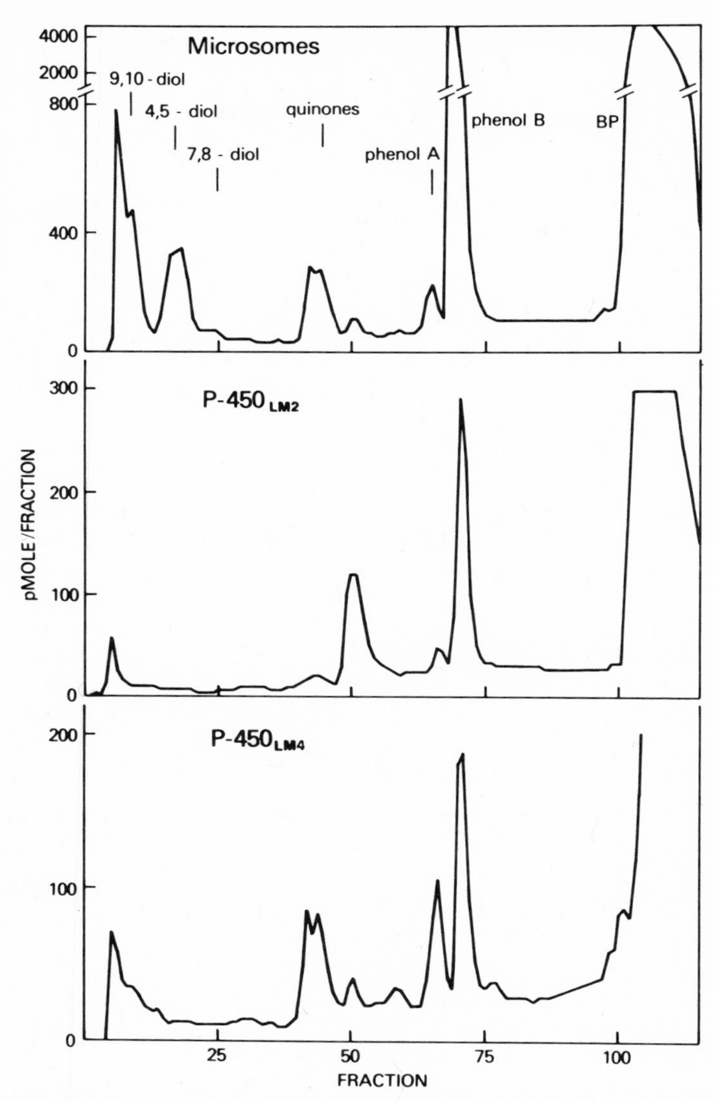

Fig. 4. High-pressure liquid chromatography of benzo[a]pyrene metabolism by rabbit liver microsomes and highly purified cytochrome $P450_{LM2}$ and cytochrome $P450_{LM4}$ (65).

oxygenases and the arene oxide hydratase which results in an altered profile of metabolites. The ratios of mixed-function oxygenase to hydratase can be altered in different ways by pretreatment with either phenobarbital or methylcholanthrene (51), the former having its major effect as an increase in the hydratase and the latter having its major effect as an increase in the oxygenase.

Table 2. Effect of MC Pretreatment on Metabolite Formation[a]

Metabolite	Specific activity[b]		S. A. ratio[c]
	Control	MC-induced	
BP-9,10-diol	16	404	25.2
BP-4,5-diol	37	132	3.5
BP-7,8-diol	10	198	19.8
BP-quinones[d]	117	309	2.6
9-OH-BP	29	140	4.8
3-OH-BP	272	956	3.5
Total	481	2139	4.4

[a]From Yang *et al.* (50).
[b]Specific activities (pmol/min/mg protein) determined at 0.14 microsomal protein/ml of incubation mixture. About 10% errors are estimated to be associated with these determinations.
[c]Specific activity ratios for each metabolite vary with the amount of microsomes used in the *in vitro* incubations (see text for discussion).
[d]Containing 1,6-quinone, 3,6-quinone, and a 6,12-quinone.

BP-4,5-OXIDE HYDRATASE

A second major enzyme system involved in PAH metabolism is arene oxide hydratase. This class of enzymes has recently been reviewed (52). An assay for this enzyme using [^{3}H] BP-4,5-oxide as the substrate has recently been developed (51). The assay measures the formation of the product, BP-4,5-dihydrodiol. Product formation is linear with time and protein concentration (Fig. 5), and stoichiometric with substrate disappearance. This enzyme is very likely similar or identical to epoxide hydrase that has been studied using styrene oxide as the substrate (52). The K-region oxide of BP is a product of benzo[*a*]pyrene metabolism and thus may be more useful in assessing the properties and activity of the hydratase system in polycyclic hydrocarbon metabolism and carcinogenesis. This enzyme is inducible by phenobarbital but is only slightly affected by hydrocarbons, and under certain conditions the AHH can be induced by methylcholanthrene while the hydratase remains unchanged (51). The hydratase can also be inhibited by TCPO which completely eliminates diol formation (14).

GLUTATHIONE-*S*-ARENE OXIDE TRANSFERASE

Other enzyme systems involved in BP metabolism are those engaged in the conversion of the BP-oxygenated intermediates to water-soluble conjugates. One of these is the glutathione-*S*-arene oxide transferase. We have developed an assay

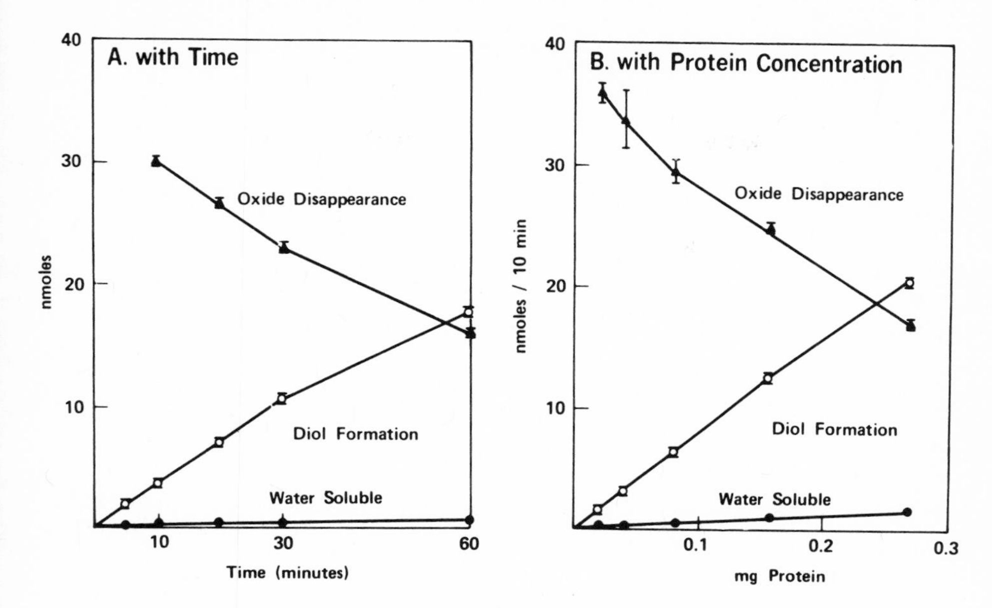

Fig. 5. BP-4,5-oxide hydratase (41).

using BP-4,5-oxide for glutathione-*S*-benzo[*a*]pyrene oxide transferase (53). This enzyme system may be related to that reported for the conjugation of naphthalene oxide (54). This assay can utilize either radioactive glutathione or the ^{3}H-labeled BP-4,5-oxide as the substrate marker for conjugate formation. Figure 6 shows the linearity of conjugate formation with time and protein concentration by this enzyme system. Several highly purified forms of gluta-

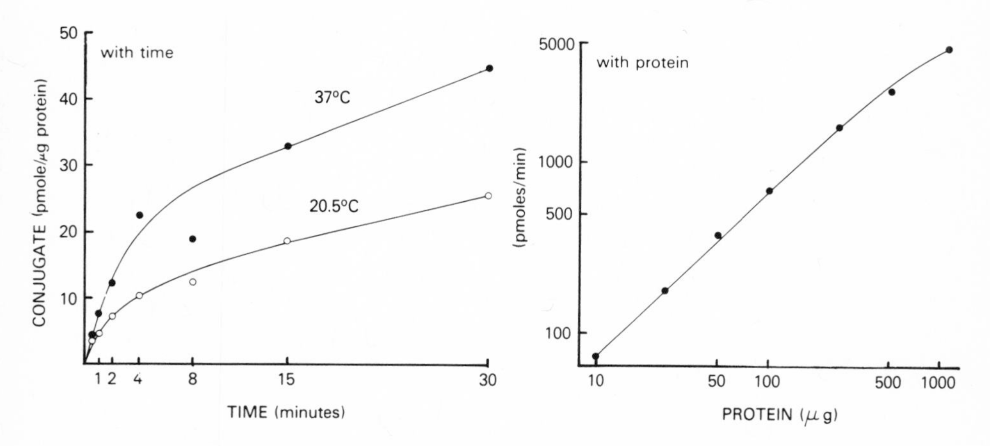

Fig. 6. Glutathione-S-benzo[a]pyrene-4,5-oxide transferase (47).

Table 3. Specific Activity of Homogeneous Rat and Human Glutathione-S-Transferases

Enzyme		Benzo[a]pyrene-4,5-oxide	Alkyl epoxide (nmol min^{-1} mg^{-1})	Bromo-sulfophthalein
Rat liver transferase	A	100	100	530
	B	14	0	6
	C	127	0	18
	E	51	6100	0
	AA	8	0	4
Human liver transferase	β	25	0	10
	δ	33	0	1

thione transferase have been described that conjugate a variety of substrates including halide and oxide compounds (55). In collaboration with Jakoby's laboratory, we have examined these purified homogeneous preparations and have found that there are at least seven forms of the glutathione-*S*-benzo[*a*]pyrene-4,5-oxide transferase, five from rat liver and two from human liver, that have differing activity toward BP-4,5-oxide and two non-PAH substrates, an alkyl epoxide and bromosulfophthalein (56). Table 3 shows that each of these purified forms of the transferase has a unique specificity toward benzo[*a*]pyrene-4,5-oxide. This specificity and the amount of each enzyme present in a given tissue of an individual may affect the efficiency of the detoxication of the arene oxides of polycyclic aromatic hydrocarbons and thus may be important keys in the removal of carcinogenic intermediates.

In addition to the above enzyme systems, there are many reports of the presence of different benzo[*a*]pyrene conjugates in the urine and bile. It is suggested that these may be glucuronides or sulfates, but a final characterization or identification of the enzymatic basis for their formation has not been made.

Table 1 shows some of the assays applicable to the enzyme systems described. The table is largely self-explanatory. Each assay has both advantages and disadvantages.

MIXED-FUNCTION OXYGENASE

The chief virtue of the AHH fluorescence assay is its extraordinary sensitivity (15). The sensitivity is sufficient to detect as little as 1 pmol of phenols equivalent to one of the major phenol products, 3–OH–BP, and thus this assay permits the analysis of minute amounts of tissue. The chief advantage of the HPLC (14) system is its ability to yield a more complete picture of BP metabolism. Thus it accomplishes a separation and quantification of the different known metabolites and has already achieved the separation of new unknown metabolites which have yet to be characterized (see below).

HYDRATASE AND CONJUGASE

The hydratase and GSH-transferase assays relate quite specifically to well-defined reactions in which a single substrate and product are defined. The BP-4,5-oxide is converted to the BP-4,5-dihydrodiol by the action of the hydratase (51) and is conjugated with GSH by the action of the glutathione transferases. These enzymes may also have different specificities toward the various benzo[*a*]pyrene intermediates. The assay for water-soluble product formation gives an indication of the total sequential reactions involved in the metabolism of the BP by the cell. Its disadvantages are largely that it gives no insight into the nature of each product or to the relative levels of each of the enzymes involved in the complete pathway of BP metabolism.

BINDING TO MACROMOLECULES

An additional assay which has potential for use in detecting reactive intermediates of carcinogens is their binding to macromolecules. This can be done *in vitro* and the macromolecule receptor can be either DNA or RNA (20–22, 45) or protein (57). This technique has the potential of detecting reactive species formed by the preparation without assaying for the individual enzymes. Its reliability as a predictive assay, however, depends on the degree of similarity between the *in vitro* system and the *in vivo* reality.

MUTAGEN DETECTION

Another assay which may be useful for the detection of reactive intermediates is the activation system coupled with the bacterial system capable of detecting mutagenic compounds (58). In this assay, very low levels of reactive compounds can be detected by their mutagenic properties. This system may be useful in determining how combinations of the different enzymes, e.g., types of cytochrome P450, hydratases, and transferases, may yield differing levels of reactive compounds. It also may be useful in identifying the "active intermediate" formation potential of different species, tissues, and individuals.

ANALYSES OF BP-METABOLIZING ACTIVITY IN HUMAN TISSUES: MONOCYTES AND LYMPHOCYTES

In assessing the level of enzymes of BP metabolism in human tissues, two conditions must prevail. The tissue must be easily obtainable and the level of the reaction must be of sufficient magnitude to be readily measured. The most sensitive of all the methods described is the AHH assay. We found that this enzyme system is present and inducible in mitogen-stimulated human lymphocytes (31). At the same time, the AHH method was utilized to analyze the inducibility of AHH in lymphocytes obtained from small amounts of blood from human donors (32).

Subsequent reports (59,60) showed a high reproducibility of the AHH assay with human lymphocytes within a given individual and a trimodal distribution of inducibility values among the population surveyed (61). These studies indicated that the AHH in families followed strictly Mendelian inheritance with respect to enzyme inducibility and the values of the children followed the pattern predicted by the AHH levels of their parents. An additional report indicated that lung cancer patients showed an abnormally high distribution of high AHH inducibility (61). Since these exciting reports, the same laboratory has indicated an inability to reproduce some of these results (Shaw, personal communications) and there are conflicting new reports on the reproducibility of the assay as well as on the trimodal distribution of AHH inducibility in the population (62,63). Many laboratories are presently engaged in studies designed to clarify this situation.

None of the solid human tissues in which AHH has been detected can be readily obtained for clinical and epidemiological study. Detection of AHH in macrophages of the human lung, rat liver (9), and guinea pig peritoneum (64) suggested that peripheral blood monocytes might also contain the enzymes. AHH is present and highly inducible in monocytes from human peripheral blood (33) and the enzyme inducibility is much greater in monocytes than in lymphocytes. The advantages of monocytes are that the degree of induction is much greater than in lymphocytes and they do not require prior stimulation with mitogens. Their disadvantages reside mainly in the requirement for greater amounts of blood from the donor. A set of duplicate assays for control and induced levels requires approximately 50–100 ml of blood as compared with only 10 ml for the lymphocyte assay. Table 4 shows typical results on the reproducibility of AHH values in monocytes and lymphocytes from six different individuals who were examined on three different occasions. The mean of the inducibility is approximately eighteenfold with monocytes and only slightly more than twofold with lymphocytes. Furthermore, in our laboratory the reproducibility of the AHH values is considerably greater with monocytes. Thus the variability in lymphocytes is more than twice that observed with monocytes. For monocytes the (standard error/mean) $\times$ 100% was 13.1, 12.0, and 14.2 for control, induced, and induced control, respectively. The corresponding values with lymphocytes were 21.6, 32.7, and 18.1, respectively. Figure 7 shows the distribution of control and induced AHH values in monocytes obtained from different individuals. The control values range in AHH specific activity, in units per million cells, from less than 1 to 3 and induced values range from 2 to 13. The inducibility generally ranges from a low of sixfold to a high of 33-fold. In a few cases, we observed an even higher inducibility. We have observed no distinct

Table 4. Reproducibility and Comparability of AHH in Human Monocytes and Lymphocytes

	Monocytes			Lymphocytes		
Donor No.	Control	BA treated	BA treated/control	Control	BA treated	BA treated/control
1739	0.27 ± 0.09	4.64 ± 0.49	13.4 ± 3.8	1.12 ± 0.26	1.96 ± 0.77	1.64 ± 0.27
2781	0.36 ± 0.04	5.20 ± 0.12	14.6 ± 1.4	1.40 ± 0.37	3.23 ± 1.57	2.09 ± 0.49
2798	0.33 ± 0.04	7.46 ± 1.06	23.2 ± 4.3	0.46 ± 0.12	1.22 ± 0.44	2.42 ± 0.42
1970	0.40 ± 0.05	6.35 ± 0.81	16.9 ± 3.9	0.54 ± 0.17	1.15 ± 0.16	2.35 ± 0.34
2827	0.32 ± 0.01	5.44 ± 0.63	17.2 ± 2.1	0.31 ± 0.05	0.73 ± 0.10	2.35 ± 0.04
3085	0.27 ± 0.05	5.83 ± 1.53	21.7 ± 3.4	0.31 ± 0.02	1.10 ± 0.47	3.72 ± 1.32
Mean	0.32	5.82	17.8	0.69	1.56	2.16
(SE/Mean) × 100%	13.1	12.0	14.2	21.6	32.7	18.1

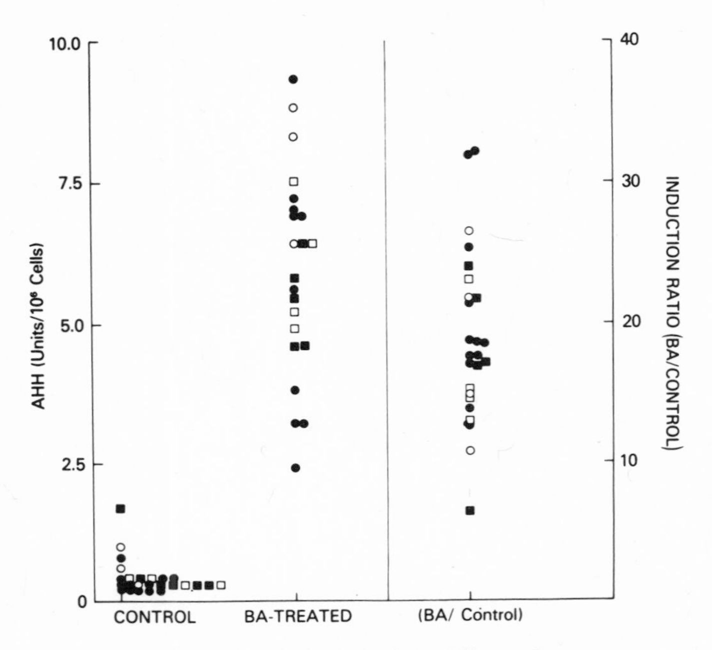

Fig. 7. AHH in monocytes from different donors.

grouping or any evidence of a trimodal distribution in the more than 25 individuals we have examined. Figure 8 suggests that there may be some positive relationship between inducibility in monocytes and lymphocytes. In these individuals examined, the inducibility in monocytes ranged from twelvefold to 24-fold and the range of inducibility in lymphocytes ranged from 1.8- to 3.8-fold with a large clustering around 2.2-fold. Figure 8 shows that the correlation between the activities in the two cell types was positive but weak ($A = 0.39$, $P = 0.05$). It seems clear that there are significant and reproducible differences in the AHH values of the monocytes of different individuals, and although the variability is somewhat greater in lymphocytes the evidence suggests that inducibility in lymphocytes also varies among different individuals. We have not found evidence suggesting a trimodal distribution and we are now examining individuals with lung cancer.

BENZO[*a*]PYRENE METABOLISM IN HUMAN LIVER, LYMPHOCYTES, AND MONOCYTES: ANALYSES BY HIGH-PRESSURE LIQUID CHROMATOGRAPHY

The profile of metabolites of benzo[*a*]pyrene formed by rat liver microsomal metabolism of BP is seen in the dashed line of Fig. 9. A small unidentified

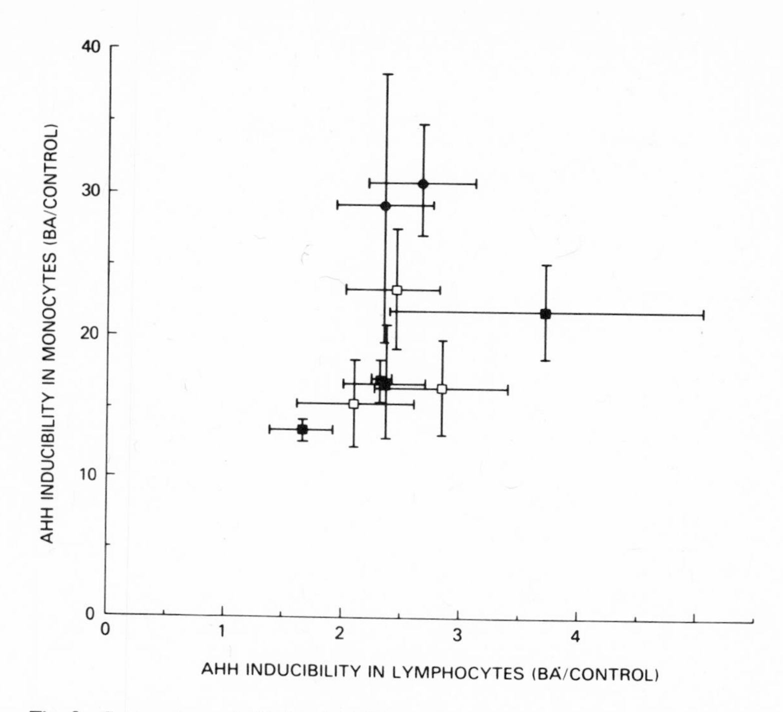

Fig. 8. Comparison of AHH inducibility in human monocytes and lymphocytes.

peak elutes first and is followed by three glycols, the 9,10-dihydrodiol, 4,5-dihydrodiol, and 7,8-dihydrodiol; three quinones, 1,6-quinone, 3,6-quinone, and 6,12-quinone; and two phenol peaks which contain 9–OH–BP and 3–OH–BP, respectively. Although each of these metabolite peaks contains primarily the indicated metabolite, they may possibly also contain small amounts of other unidentified metabolites. A comparison between the metabolism of BP by microsomes from human liver and rat liver is shown in Fig. 9. A large number of metabolites are formed by human liver microsomes which are not formed by rat liver microsomes. In the diol region, a large peak (I) appears just after the 9,10-dihydrodiol (fractions 13–15), and a small peak (II) appears after the 7,8-dihydrodiol (fractions 31–34). The quinone region indicates a major peak (III) at the region of 6,12-quinone. However, we have found that 6-hydroxymethyl-BP cochromatographs with 6,12-quinone in this system and peak III may correspond to the former compound which has been reported as a BP metabolite (43). A smaller peak (IV) follows immediately after (fractions 49–51), with a peak in the region where BP-4,5-epoxide migrates (42). The relative activity of hydratases in human liver may determine the lifetime of the epoxide and the identity of this peak remains to be clarified.

The high background in fractions 31–80 results from small quantities of

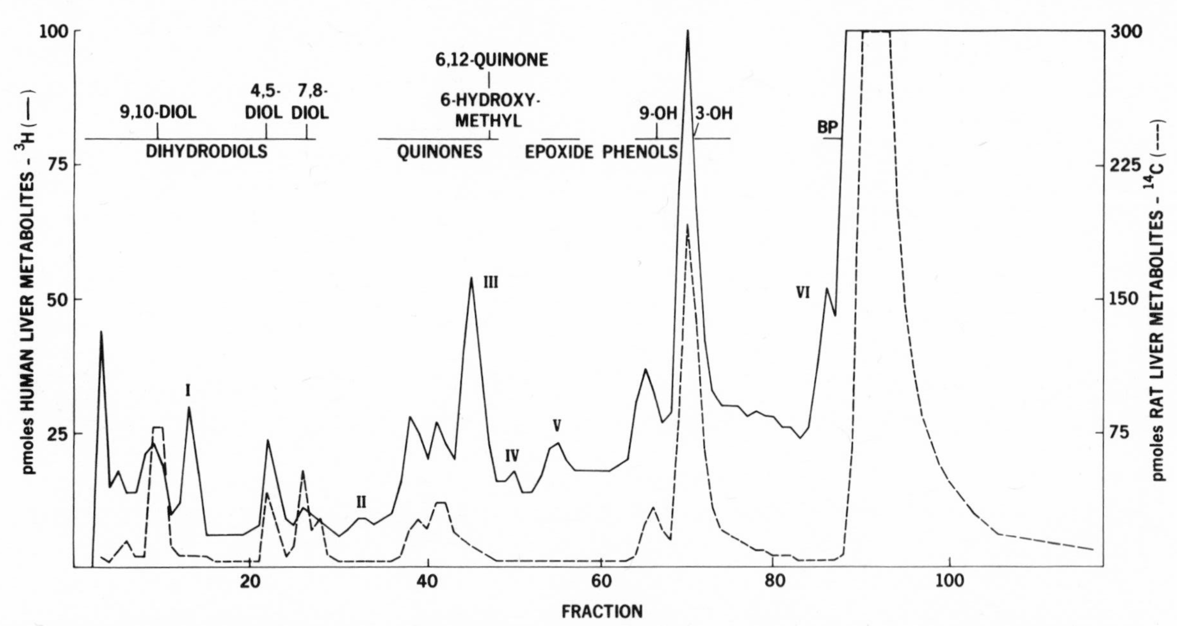

Fig. 9. BP metabolite pattern of the incubation with liver human microsomes. ——, pmol of hydrocarbon metabolites formed by human microsomes with [^{3}H] BP; – – –, pmol of hydrocarbon metabolites formed by rat liver microsomes with [^{14}C] BP. Roman numerals indicate metabolites produced only by human microsomes.

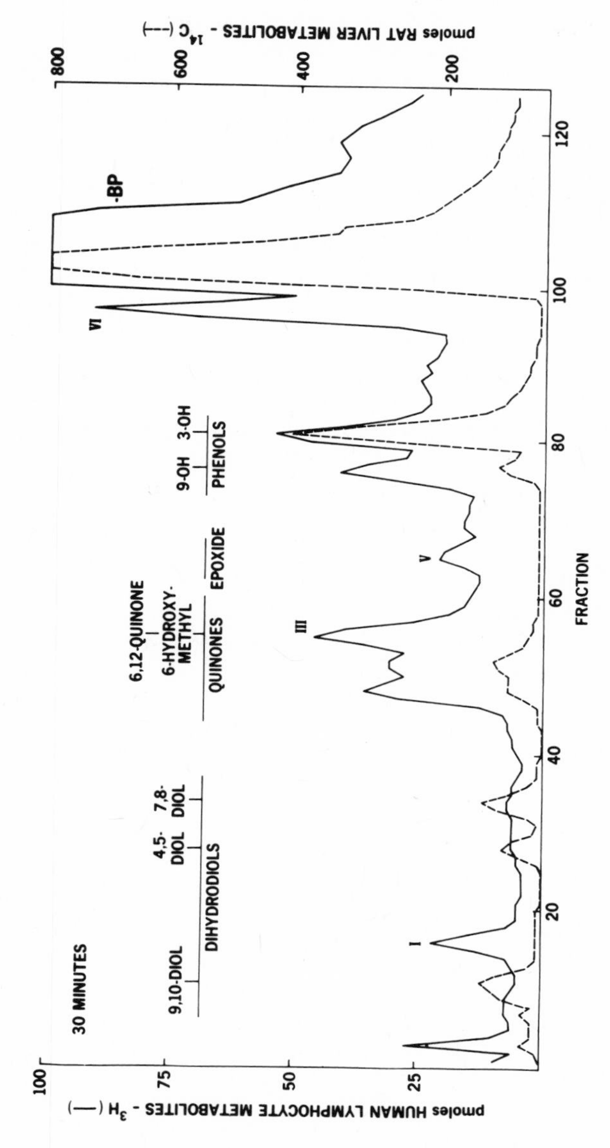

Fig. 10. BP metabolite pattern after incubation for 30 min with human lymphocytes.

[^{3}H] BP leaching from the column and occurs when the ratio of metabolites to [^{3}H] BP is very low. The background of [^{3}H] BP can be reduced by prior removal of the [^{3}H] BP by thin-layer chromatography. This was done prior to the HPLC shown in Fig. 10. The formation of the BP metabolites by human lymphocytes during a 30-min period is seen in Fig. 10. This pattern is quite different from that of rat liver and has characteristics that are quite similar to those of human liver. The patterns of human liver and lymphocytes, however, are not identical and show some distinct differences. None of the three dihydrodiols appeared after a 30-min incubation of [^{3}H] BP with lymphocytes. The metabolites I, III, V, and VI migrate identically as those observed with human liver (Fig. 9) and the small peaks II and IV are absent. In the lymphocytes, the relative amount of peak III in the quinone region was reduced, and the ratio of the 3–OH–BP peak to the 9–OH–BP was considerably reduced. Also, there was a relatively larger amount of peak VI compared to all the other metabolites.

The profile in Fig. 11 shows the metabolites formed during a 24-h incubation of [^{3}H] BP with lymphocytes in culture. In contrast to the absence of diol formation during a 30-min incubation, all three dihydrodiols are formed, with the 7,8-dihydrodiol as the major peak. This result agrees with the report that diols are formed by lymphocytes (65), although the methods used in the report are inadequate for the separation of the individual diols, phenols, or quinones and the new metabolites. The lymphocytes form metabolites which migrate as peaks I, III, and V and correspond to the new metabolites formed by human liver. Peak IV which is observed in human liver (Fig. 9) does not appear but may possibly be hidden between peaks III and V. The bulk of BP was removed prior to TLC by HPLC chromatography in the preparation shown in Fig. 11. Thus peak VI which is seen in the human metabolites in Figs. 9 and 10 may have been removed during the TLC.

In addition to the peaks I–VI, several new peaks are formed during the longer incubation period of BP with lymphocytes. This may reflect a further metabolism of some of the primary metabolites by the mixed-function oxygenase system. There is an additional peak in the dihydrodiol region (VII), and two more peaks in the phenol region labeled VII and IX. Thus our data show that human lymphocytes incubated with BP for 24 h form many metabolites that migrate identically to those formed by human liver. There are, however, additional peaks formed by the lymphocytes, and the ratios of metabolites observed in lymphocytes differ from those in liver.

Figure 12 shows the HPLC pattern of metabolites obtained upon incubation of human monocytes and [^{3}H] BP followed by an HPLC separation. The monocytes form three unidentified new peaks which correspond to those observed with human liver and lymphocytes. The ratios of metabolites are quite different from those observed with human liver. There is no detectable 9,10-diol or 4,5-diol, low levels of the 7,8-diol, and a relatively large amount of the 9–OH phenol peak. The variation in metabolite profile in each tissue examined is likely

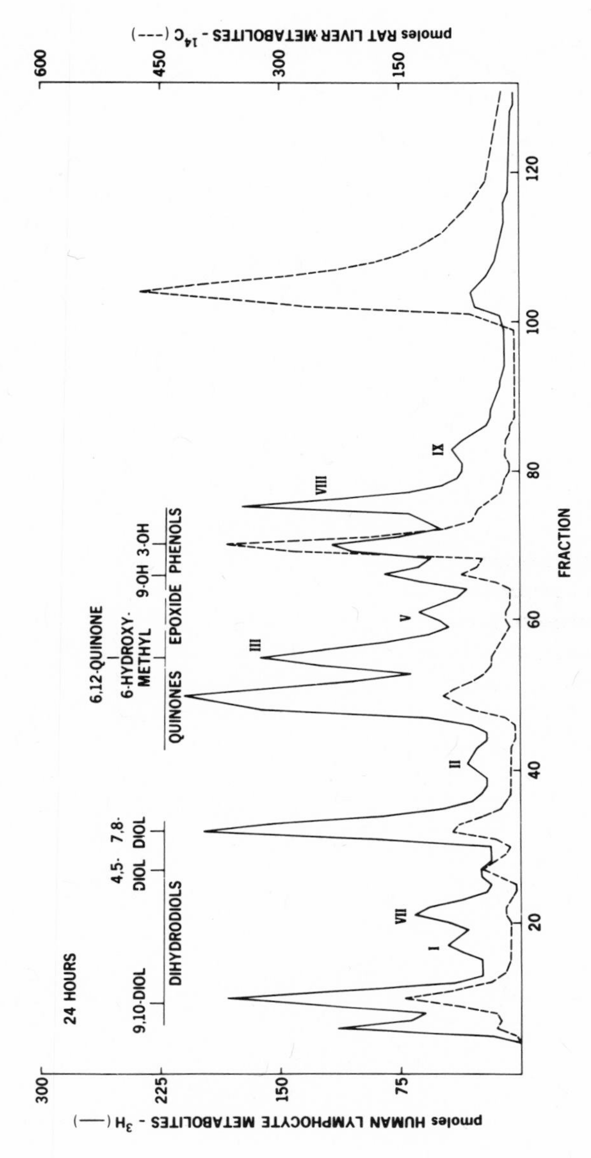

Fig. 11. BP metabolite pattern after incubation for 24 h with human lymphocytes.

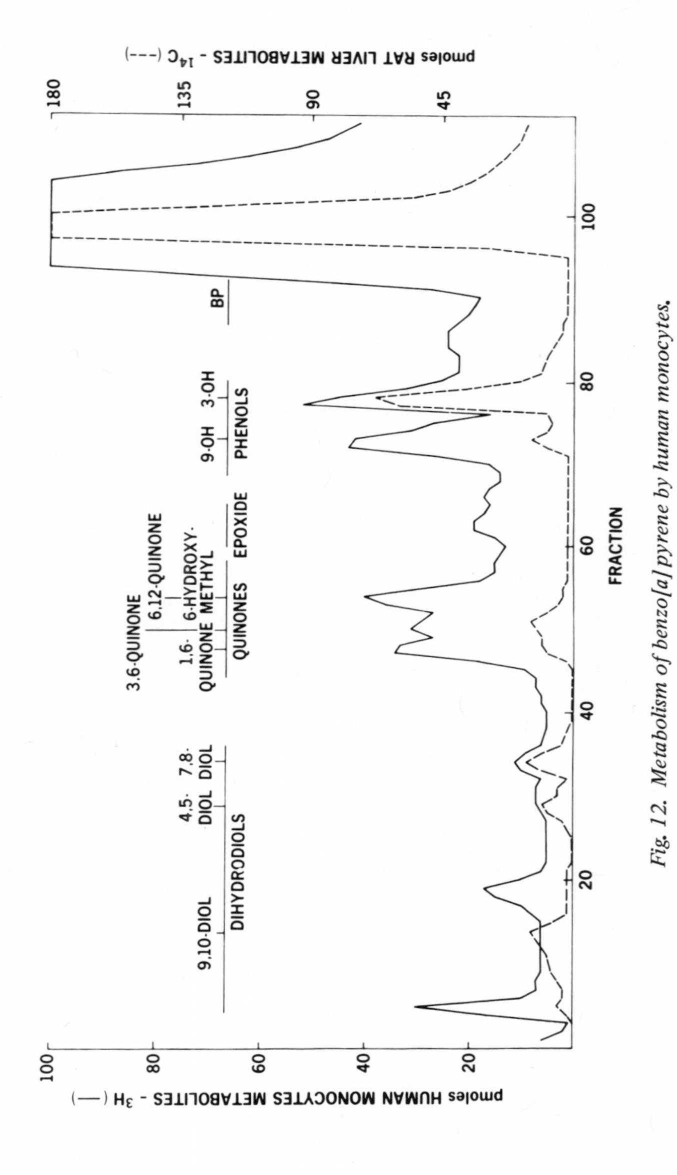

Fig. 12. *Metabolism of benzo[a]pyrene by human monocytes.*

a reflection of the type or ratio of different mixed-function oxygenases present and the relative amounts of the hydratase, GSH transferases, and other conjugases.

SUMMARY

The formation and levels of reactive toxic or carcinogenic intermediates of the polycyclic hydrocarbons depend on the mixed-function oxygenases and subsequent enzymes in the metabolic pathway. We have described various methods and enzyme assays that can be used in examining BP metabolism. These techniques are applicable to human tissues and can be easily modified for application to the metabolism of other polycyclic hydrocarbons. It is possible to assess the activities of mixed-function oxidases by the AHH fluorescence assay and BP-4,5-oxide hydratase and glutathione-*S*-BP-4,5-oxide transferase by radioassays, and to perform an analysis by high-pressure liquid chromatography of metabolites formed by easily obtainable human tissues. With the development of these methods we have obtained more important experimental results:

1. Different highly purified forms of cytochrome P450 convert benzo [*a*] pyrene to a different ratio of phenolic products, indicating that each form of cytochrome P450 exhibits a positional preference for benzo [*a*] pyrene oxygenation.
2. Enzyme induction by methylcholanthrene results in an altered pattern of metabolites.
3. Five homogeneous rat liver and two human liver enzymes show unique glutathione-*S*-BP-4,5-oxide transferase activity.
4. Human monocytes can easily be assayed for AHH. The inducibility ranges from about sixfold to 33-fold in monocytes compared to about two- to fourfold in lymphocytes from different individuals.
5. The metabolic activity of human liver, lymphocytes, and monocytes has been studied by high-pressure liquid chromatography. The human tissues form five new metabolites which are unidentified. Each of the human tissues forms metabolic profiles of benzo [*a*] pyrene that are different from the profiles formed by rat liver and are somewhat different from each other.

Further development of the described methods may permit the analysis of various human populations for the enzyme activities related to polycyclic hydrocarbon metabolism. This may yield information on the relationship between metabolism and the heterogeneity in human susceptibility to carcinogenesis induced by polycyclic aromatic hydrocarbons.

ACKNOWLEDGMENTS

We gratefully acknowledge the valuable technical assistance of E. Plotkin, R. Croy, H. Miller, and J. Leutz, and the secretarial assistance of M. Rexroad.

REFERENCES

1. L. W. Wattenberg and J. L. Leong, Chemoprophylaxis of carcinogenesis: A review, *Cancer Res.* **26,** 1520–1526 (1966).
2. H. V. Gelboin, Carcinogens, enzyme induction, and gene action, *Adv. Cancer Res.* **10,** 1–81 (1967).
3. H. V. Gelbion, N. Kinoshita, and F. J. Wiebel, Microsomal hydroxylases: Induction and role in polycyclic hydrocarbon carcinogenesis and toxicity, *Fed. Proc.* **31,** 1298–1309 (1972).
4. National Academy of Science Reports, USA, Particulate polycyclic organic matter, Committee on Biologic Effects of Atmospheric Pollutants, Division of Medical Sciences, National Research Council, Washington, D.C. (1972).
5. J. B. Andelman and N. J. Suess, Polynuclear aromatic hydrocarbons in the water environment, *Bull. WHO* **43,** 479–508 (1970).
6. L. M. Shabad, Y. L. Cohan, A. P. Illnitsky, A. Y. Khesina, N. P. Sherbak, and G. A. Smirnov, The carcinogenic hydrocarbon benzo[*a*]pyrene in the soil, *J. Natl. Cancer Inst.* **47,** 1179–1191 (1971).
7. W. Haenszel and K. E. Traeuber, Lung cancer mortality as related to residence and smoking histories. II. White females, *J. Natl. Cancer Inst.* **32,** 803–838 (1964).
8. U. S. Department of Health, Education and Welfare, Smoking and Health, Report of the Advisory Committee to the Surgeon General of the Public Health Service, Public Health Service Publication 1103, U.S. Government Printing Office, Washington, D.C. (1964).
9. J. L. Hartwell and P. Shubik, Survey of compounds which have been tested for carcinogenic activity, Public Health Service Publication 149 (2nd ed.), U.S. Government Printing Office, Washington, D.C. (1951).
10. A. H. Conney, E. C. Miller, and J. A. Miller, Substrate-induced synthesis and other properties of benzpyrene hydroxylase in rat liver, *J. Biol. Chem.* **228,** 753–766 (1957).
11. P. Sims, The metabolism of benzo[*a*]pyrene by rat liver homogenates, *Biochem. Pharmacol.* **16,** 613–618 (1968).
12. P. Sims, P. L. Grover, A. Swaisland, K. Pal, and A. Hewer, Metabolic activation of benzo[*a*]pyrene proceeds by a diol epoxide, *Nature (London)* **252,** 326–328 (1974).
13. N. Kinoshita, B. Shears, and H. V. Gelboin, K-region and non-K-region metabolism of benzo[*a*]pyrene by rat liver microsomes, *Cancer Res.* **33,** 1937–1944 (1973).
14. J. K. Selkirk, R. G. Croy, P. P. Roller, and H. V. Gelboin, High-pressure liquid chromatographic analysis of benzo[*a*]pyrene metabolism and covalent binding and the mechanism of action of 7,8-benzoflavone and 1,2-epoxy-3,3,3-trichloropropene, *Cancer Res.* **34,** 3474–3480 (1974).
15. D. W. Nebert and H. V. Gelboin, Substrate-inducible microsomal aryl hydroxylase in mammalian cell culture. I. Assay and properties of induced enzyme, *J. Biol. Chem.* **243,** 6242–6249 (1968).
16. D. W. Nebert and H. V. Gelboin, The *in vivo* and *in vitro* induction of aryl hydrocarbon hydroxylase in mammalian cells of different species, tissues, strains, and developmental and hormonal states, *Arch. Biochem. Biophys.* **134,** 76–89 (1969).

17. A. H. Conney and J. J. Burns, Metabolic interactions among environmental chemicals and drugs, *Science* **178,** 576–586 (1972).
18. A. H. Conney, Pharmacological implications of microsomal enzyme induction, *Pharmacol. Rev.* **19,** 317–366 (1967).
19. C. Huggins, L. Grand, and R. Fukunishi, Aromatic influences on the yields of mammary cancers following administration of 7,12-dimethylbenz[*a*]anthracene, *Proc. Natl. Acad. Sci. USA* **51,** 737–742 (1964).
20. H. V. Gelboin, A microsome-dependent binding of benzo[*a*]pyrene to DNA, *The Jerusalem Symposia on Quantum Chemistry and Biochemistry, Vol. 1: Physico-chemical Mechanisms of Carcinogenesis,* p. 175 (1968).
21. R. L. Grover and P. Sims, Enzyme-catalyzed reactions of polycyclic hydrocarbons with deoxyribonucleic acid and protein *in vitro, Biochem. J.* **110,** 159–160 (1968).
22. H. V. Gelboin, A microsome-dependent binding of benzo[*a*]pyrene to DNA, *Cancer Res.* **29,** 1272–1276 (1969).
23. H. V. Gelboin, E. Huberman, and L. Sachs, Enzymatic hydroxylation of benzpyrene and its relationship to cytotoxicity, *Proc. Natl. Acad. Sci. USA* **64,** 1188–1194 (1969).
24. G. A. Belitskii, Ir. M. Vasiliev, O. I. Ivanova, N. A. Lavrova, E. L. Progozhina, N. L. Samolilina, A. A. Stavrovskaya, A. Y. Khesina, and L. M. Shabad, Metabolism of benzo[*a*]pyrene by cells of different mammals *in vitro* and toxic effect of polycyclic hydrocarbons on the cells, *Vop. Onkol.* **16,** 53–58 (1970).
25. L. Diamond and H. V. Gelboin, Alpha-napthoflavone: An inhibitor of hydrocarbon cytotoxicity and microsomal hydroxylase, *Science* **166,** 1023–1025 (1969).
26. N. Kinoshita and H. V. Gelboin, Aryl hydrocarbon hydroxylase in 7,12-dimethylbenz[*a*]anthracene skin tumorigenesis: On the mechanism of 7,8-benzoflavone inhibition of tumorigenesis, *Proc. Natl. Acad. Sci. USA* **69,** 824–828 (1972).
27. J. E. Gielen, F. M. Goujon, and D. W. Nebert, Genetic regulation of aryl hydrocarbon hydroxylase induction. II. Simple Mendelian expression in mouse tissue *in vivo, J. Biol. Chem.* **24,** 1125–1137 (1972).
28. F. Kuntzman, L. C. Mark, L. Brand, M. Jacobson, W. Levin, and A. H. Conney, Metabolism of drugs and carcinogens by human liver enzymes, *J. Pharmacol. Exp. Ther.* **152,** 151–156 (1966).
29. R. M. Welch, Y. E. Harrison, B. W. Gommi, P. T. Poppers, M. Ernster, and A. H. Conney, Stimulatory effect of cigarette smoking on the hydroxylation of 3,4-benzopyrene and the *N*-demethylation of 3-methyl, 4-mono-methyl amino-azobenzene by enzymes in human placenta, *Clin. Pharmacol. Ther.* **10,** 100–109 (1969).
30. D. W. Nebert, J. Winker, and H. V. Gelboin, Aryl hydrocarbon hydroxylase activity in human placenta from cigarette smoking and non-smoking women, *Cancer Res.* **29,** 1763–1769 (1969).
31. J. P. Whitlock, Jr., H. L. Cooper, and H. V. Gelboin, Aryl hydrocarbon (benzo[*a*]pyrene) hydroxylase is stimulated in human lymphocytes by mitogens and benz[*a*]anthracene, *Science* **177,** 618–619 (1972).
32. D. L. Busbee, C. R. Shaw, and E. T. Cantrell, Aryl hydrocarbon hydroxylase induction in human leukocytes, *Science* **178,** 315–316 (1972).
33. R. C. Bast, Jr., J. P. Whitlock, Jr., H. Miller, H. J. Rapp, and H. V. Gelboin, Aryl hydrocarbon benzo[*a*]pyrene hydroxylase in human peripheral blood monocytes, *Nature (London)* **250,** 664–665 (1974).
34. K. Ptashne, L. Brothers, S. G. Axline, and S. N. Cohen, Aryl hydroxylase induction in mouse peritoneal macrophages and blood-derived human macrophages, *Proc. Soc. Exp. Biol. Med.* **146,** 585–589 (1974).
35. E. T. Cantrell, G. A. Warr, D. Busbee, and R. R. Martin, Induction of aryl hydrocarbon hydroxylase in human pulmonary alveolar macrophages by cigarette smoking, *J. Clin. Invest.* **52,** 1881–1884 (1973).

36. J. K. Selkirk, R. G. Croy, and H. V. Gelboin, Benzo[a]pyrene metabolites: Efficient and rapid separation by high-pressure liquid chromatography, *Science* **184**, 169–171 (1974).
37. G. Holder, H. Yagi, P. Dansette, D. M. Jerina, W. Levin, A. Y. H. Lu, and A. H. Conney, Effects of inducers and epoxide hydrase on the metabolism of benzo[a]pyrene by liver microsomes and a reconstituted system: An analysis by high-pressure liquid chromatography, *Proc. Natl. Acad. Sci. USA* **71**, 4356–4360 (1974).
38. D. M. Jerina and J. W. Daly, Arene oxides: A new aspect of drug metabolism, *Science* **184**, 573–582 (1974).
39. C. Nagata, Y. Tagashira, and M. Kodama, Metabolic activation of benzo[a]pyrene: Significance of the free radical, in: *Chemical Carcinogenesis, Part A* (P. O. P. Ts'o and Joseph A. DiPaolo, eds.), pp. 97–111, Dekker, New York (1974).
40. P. Sims and P. L. Grover, Epoxides in polycyclic aromatic hydrocarbon metabolism and carcinogenesis, *Adv. Cancer Res.* **20**, 165–274 (1974).
41. P. O. P. Ts'o, W. J. Caspary, B. I. Cohen, J. C. Leavitt, S. A. Lesko, Jr., R. J. Lorentzen, and L. M. Schechtman, Basic mechanisms in polycyclic hydrocarbon carcinogenesis, in: *Chemical Carcinogenesis, Part A* (P. O. P. Ts'o and Joseph A. DiPaolo, eds.), pp. 113–147), Dekker, New York (1974).
42. J. K. Selkirk, R. G. Croy, and H. V. Gelboin, Isolation and characterization of benzo[a]pyrene-4,5-epoxide as a metabolite of benzo[a]pyrene, *Arch. Biochem. Biophys.* **168**, 322–326 (1975).
43. J. W. Flesher and K. L. Sydnor, Possible role of 6-hydroxy methyl benzo[a]pyrene as a proximate carcinogen of benzo[a]pyrene and 6-methyl benzo[a]pyrene, *Int. J. Cancer* **11**, 433–437 (1973).
44. L. Diamond, Metabolism of polycylic hydrocarbons in mammalian cell cultures, *Int. J. Cancer* **8**, 451–462 (1971).
45. A. Borgen, H. Darvey, N. Castagnoli, T. T. Crocker, R. E. Rasmussen, and I. Y. Wang, Metabolic conversion of benzo[a]pyrene by Syrian hamster liver microsomes and binding of metabolites to deoxyribonucleic acid, *J. Med. Chem.* **16**, 502–506 (1973).
46. D. A. Haugen, T. A. Van der Hoeven, and M. J. Coon, Purified liver microsomal cytochrome P-450: Separation and characterization of multiple forms, *J. Biol. Chem.* **250**, 3567–3570 (1975).
47. F. J. Wiebel, J. K. Selkirk, H. V. Gelboin, D. A. Haugen, T. A. Van der Hoeven, and M. J. Coon, Position-specific oxygenation of benzo[a]pyrene by different forms of purified cytochrome P-450 from rabbit liver, *Proc. Natl. Acad. Sci. USA* **72**, 3917–3920 (1975).
48. F. J. Wiebel, J. C. Leutz, L. Diamond, and H. V. Gelboin, Aryl hydrocarbon (benzo[a]pyrene) hydroxylase in microsomes from rat tissues: Differential inhibition and stimulation by benzoflavones and organic solvents, *Arch. Biochem. Biophys.* **144**, 78–86 (1971).
49. R. E. Rasmussen and I. Y. Wang, Dependence of specific metabolism of benzo[a]pyrene on the inducer of hydroxylase activity, *Cancer Res.* **34**, 2290–2295 (1974).
50. S. K. Yang, J. K. Selkirk, E. Plotkin, and H. V. Gelboin, Kinetic analysis of the microsomal metabolism of benzo[a]pyrene to phenols, dihydrodiols and quinones by high-pressure liquid chromatography: Effect of enzyme induction, *Cancer Res.* **35**, 3642–3650 (1975).
51. J. C. Leutz and H. V. Gelboin, Benzo[a]pyrene-4,5-oxide hydratase: Assay, properties, and induction, *Arch. Biochem. Biophys.* **168**, 722–725 (1975).
52. F. Oesch, Mammalian epoxide hydrases: Inducible enzymes catalyzing the inactivation of carcinogenic and cytotoxic metabolites derived from aromatic and aliphatic compounds, *Xenobiotica* **3**, 305–340 (1972).

53. N. Nemoto and H. V. Gelboin, Assay and properties of glutathione-*S*-benzo[*a*]pyrene-4,5-oxide transferase, *Arch. Biochem. Biophys.* **170,** 739–742 (1975).
54. T. Hayakawa, R. A. Lemahieu, and S. Udenfriend, Studies on glutathione *S*-arene oxidase [*sic*]: Transferase-A sensitive assay and partial purification of the enzyme from sheep liver, *Arch. Biochem. Biophys.* **162,** 223–230 (1974).
55. W. H. Habig, M. J. Pabst, and W. B. Jakoby, Glutathione *S*-transferases, *J. Biol. Chem.* **249,** 7130–7139 (1974).
56. N. Nemoto, H. V. Gelboin, W. H. Habig, J. N. Ketley, and W. B. Jakoby, K-region benzo[*a*]pyrene-4,5-oxide is conjugated by homogeneous glutathione *S*-transferases, *Nature (London)* **255,** 512 (1975).
57. H. V. Gelboin, J. A. Miller, and E. C. Miller, The *in vitro* formation of protein-bound derivatives of aminoazo dyes by rat liver preparations, *Cancer Res.* **19,** 975–985 (1959).
58. J. McCann, N. E. Spingarn, J. Koborl, and B. N. Ames, Detection of carcinogens as mutagens: Bacterial tester strains with R factor plasmids, *Proc. Natl. Acad. Sci. USA* **72,** 979–983 (1975).
59. G. Kellerman, E. Cantrell, and C. R. Shaw, Variation in extent of aryl hydrocarbon hydroxylase induction in cultured human lymphocytes, *Cancer Res.* **33,** 1654–1656 (1973).
60. G. Kellermann, C. R. Shaw, and M. Luyten-Kellerman, Aryl hydrocarbon hydroxylase inducibility and bronchogenic carcinoma, *N. Engl. J. Med.* **289,** 934–937 (1973).
61. G. Kellerman, M. Luyten-Kellerman, and C. R. Shaw, Genetic variation of aryl hydrocarbon hydroxylase in human lymphocytes, *Am. J. Hum. Genet.* **25,** 327–331, (1973).
62. R. E. Kouri, H. Ratrie, III, S. A. Atlas, A. Niwa, and D. W. Nebert, Aryl hydrocarbon hydroxylase induction in human lymphocyte cultures by 2,3,7,8-tetrachlorodibenzo-*p*-dioxid, *Life Sci.* **15,** 1585–1595 (1974).
63. G. Kellermann, M. Luyten-Kellermann, M. G. Horning, and M. Stafford, Correlation of aryl hydrocarbon hydroxylase activity of human lymphocyte cultures and plasma elimination rates for antipyrine and phenylbutazone, *Drug Metab. Dispos.* **3,** 47–50 (1975).
64. R. C. Bast, Jr., B. W. Shears, H. J. Rapp, and H. V. Gelboin, Aryl hydrocarbon (benzo[*a*]pyrene) hydroxylase in guinea pig peritoneal macrophages: Benzo[*a*]anthracene-induced increase of enzyme activity *in vivo* and in cell culture, *J. Natl. Cancer Inst.* **51,** 675–678 (1973).
65. J. Booth, G. R. Keysall, P. L. Kalyani, and P. Sims, The metabolism of polycyclic hydrocarbons by cultured human lymphocytes, *FEBS Lett.* **43,** 341–344 (1974).

11

Discussion

Schenkman commented that the pH optimum for benzpyrene hydroxylase activity in human lymphocytes is higher than that of the enzyme from rat liver (lymphocytes 8.5, liver 7.5). Gelboin also found different pH optima with these tissues and suggested that the tissues contain different forms of the enzyme.

Gelboin then commented that epoxides may be very important metabolites but that their importance has been overemphasized. In response to a question by Remmer, Gelboin said that it has not yet been established that dihydrodiols are the *in vivo* precursors of these catechols and that he had not studied covalent binding of quinones. He emphasized that the type of metabolite of polycyclic aromatic hydrocarbons which seems to be most important is the diolepoxide, which binds extremely well to both DNA and to protein.

Nebert has found that the reported 6-oxy free radical of benzpyrene is an artifact produced during the solvent extraction of benzpyrene metabolites prior to EPR studies at room temperature. When oxygen is present in the solvent system containing benzene and toluene, there is abstraction of a proton from the benzpyrene and several free radicals are produced, including that found at the 6-oxy position.

Gelboin, commenting on Kupfer's question after the previous paper (Schenkman), said that in most of the cases in which comparisons are made between assays which measure the total metabolism of benzpyrene and those measuring phenol production there is a good correlation between the two methods. The phenol assay is easier than liquid chromatographic methods and is generally acceptable. However, one cannot completely extrapolate from one method to the other and caution should be used in interpreting the results of all benzpyrene hydroxylase assays. The 9-OH metabolite of benzpyrene which he reported has fluorescent properties almost identical to those of the 3-OH derivatives. Also, any dihydrodiols which might be extracted into the alkaline layer do not fluoresce. Therefore, none of these metabolites is a source of error in assays which measure the appearance of a monohydroxylated fluorescent product, previously calculated as the 3-OH derivative.

12

Genetic Differences in Benzo[*a*]pyrene Carcinogenic Index *in Vivo* and in Mouse Cytochrome P_1450-Mediated Benzo[*a*]pyrene Metabolite Binding to DNA *in Vitro*

Daniel W. Nebert and Alan R. Boobis
Section on Pharmacogenetics and Molecular Teratology
Developmental Pharmacology Branch
National Institutes of Health and Human Development
National Institutes of Health
Bethesda, Maryland 20014, U.S.A.

Haruhiko Yagi and Donald M. Jerina
Section on Oxidation Mechanisms
Laboratory of Chemistry
National Institute of Arthritis, Metabolism, and Digestive Diseases
National Institutes of Health
Bethesda, Maryland 20014, U.S.A.

Richard E. Kouri
Department of Biochemical Oncology
Microbiological Associates, Inc.
Bethesda, Maryland 20014, U.S.A.

Carcinogenic polycyclic hydrocarbons, such as the ubiquitous environmental contaminant BP,[1] probably require metabolic activation before initiation of cancer can occur. Boyland first proposed (1) that the initial products of

[1] Abbreviations used include BP, benzo[*a*]pyrene; BP-7,8-diol, *trans*-7,8-dihydroxy-7,8-dihydrobenzo[*a*]pyrene; BP-7,8-diol-9,10-epoxide, (±)-7β,8α-dihydroxy-9β,10β-epoxy-7,8,9,10-tetrahydrobenzo[*a*]pyrene; BP-4,5-diol, *trans*-4,5-dihydroxy-4,5-dihydrobenzo-[*a*]pyrene; BP-4,5-oxide, 4β,5β-epoxy-4,5-dihydrobenzo[*a*]pyrene; [^{3}H]BP and [^{3}H]BP-7,8-diol, the generally tritiated forms of BP and BP-7,8-diol; and MC, 3-methylcholanthrene.

double-bond oxidation are epoxides. When polycyclic hydrocarbon carcinogens are topically applied to mouse skin, covalent binding of the compounds to cellular macromolecules results (2). Subsequently the incubation of BP with microsomes, cofactors, and calf thymus DNA *in vitro* was shown (3) to produce covalent binding of unknown metabolites to the DNA. It was then appreciated that reactive arene oxides are formed by the microsomal cytochrome P450-mediated monooxygenase systems and that these intermediates can rearrange nonenzymatically to form phenols, be hydrated to form *trans*-dihydrodiols, and be conjugated with glutathione (4) [*cf.* refs. 5–8 for reviews].

Although RNA and protein might be the critical intracellular target at which chemical carcinogenesis is initiated, considerable interest has centered on DNA and chemicals which bind covalently to DNA as important early events in the initiation of tumors (9). Because many of the polycyclic hydrocarbons are planar, they possess the capacity to intercalate with DNA base pairs in the helix. The reactive oxides, presumably as alkylating agents, are mutagenic in several systems (10–12) and induce malignant transformation in rodent cell cultures (13).

To understand further the aromatic hydrocarbon–nucleic acid reaction, Baird and Brookes (14) developed a method for the enzymatic degradation of nucleic acid containing bound carcinogens and the fractionation of the resulting mixture by Sephadex LH20 column chromatography. This method has shown great promise in that distinct peaks eluted from the column can be demonstrated to change in elution profile, depending on the carcinogen incubated with microsomes and cofactors, on whether rat liver microsomes or cells in culture are used, and on the use of microsomal enzyme inhibitors *in vitro* (15–19). With this new *in vitro* technique, a correlation between biological activity (i.e., tumorigenesis *in vivo*) and the elution profile remains to be demonstrated. This report provides evidence with BP for a lack of any statistically significant correlation between the quantitative change in the size or shape of any peak formed *in vitro* and tumor formation in the mouse *in vivo.*

AROMATIC HYDROCARBON RESPONSIVENESS IN THE MOUSE

Certain polycyclic aromatic hydrocarbons administered *in vivo* induce the *de novo* synthesis (20) of a new microsomal CO-binding hemoprotein, cytochrome P_1450 (21) [also called P448 (22)]. Mouse strains designated as "aromatic hydrocarbon responsive" (8) have detectable increases of this new cytochrome not only in liver but also in many nonhepatic tissues including skin, following the administration of an aromatic hydrocarbon such as MC; "nonresponsive" strains treated with MC have no detectable increases in hepatic P_1450 and have much less nonhepatic P_1450 (8) compared with MC-treated responsive mice. Aromatic hydrocarbon responsiveness is expressed as a single autosomal dominant trait in offspring from the appropriate genetic crosses between C57BL/6 and DBA/2 strains (8,23–25) and as a simple gene dose (additive inheritance) trait in offspring from the appropriate crosses between C3H/He and DBA/2 strains (8,25,26). Aromatic hydrocarbon responsiveness is associated not only with *de novo* synthesis of P_1450 (20) but also with increases in at least ten microsomal monooxygenase activities (8) including aryl hydrocarbon hydroxylase, the enzyme system known (27) to metabolize BP and several other polycyclic hydrocarbon carcinogens.

DIFFERENCES IN METABOLITE PROFILE CAUSED BY CYTOCHROME P_1450 AND OTHER FORMS OF P450

This trait of aromatic hydrocarbon responsiveness, collectively designated the *Ah* locus (8), has also been shown to be associated with MC-initiated tumorigenesis (28–31), with glutathione depletion and subsequent covalent binding of actaminophen (32), with *in vitro* mutagenicity of MC, 2-acetylaminofluorene, and 6-aminochrysene (but not BP) in combination with the *Salmonella* histidine revertants of Ames (33), and with protection against chemical toxic depression of the bone marrow and resultant shortened survival times in mice exposed to various environmental contaminants (34). These differences between responsive and nonresponsive mice are believed to be caused by varying relative steady-state levels of different reactive intermediates formed by cytochrome P_1450, compared with those formed by other forms of cytochrome P450. There are examples (see Fig. 1) in which predominant metabolites have been identified and shown to differ in quantity between rats treated *in vivo* with MC (resulting in relatively more P_1450) and rats treated *in vivo* with phenobarbital (resulting in relatively more of P450 forms other than P_1450). These examples include biphenyl (35), testosterone (36), 2-acetylaminofluorene (32,37,38), halobenzenes (39–41), *n*-hexane (42), and BP (43,44). The mechanism of acetaminophen-induced toxicity and associated covalent binding is presumably related to glutathione depletion (32,45), a pathway which follows P_1450-mediated *N*-hydroxylation of *N*-acetylarylamines but which does not occur after P450-mediated ring hydroxylation (32). Differences in hepatotoxicity have been demonstrated for bromobenzene (39); however, it is not understood (40) why halobenzene toxicity is much greater in the phenobarbital-treated rat than in the MC-treated rat (*cf.* ref. 46 for most recent review).

It is therefore reasonable to postulate that differences in chemical carcinogenesis may result from differences in the metabolite profiles and in the steady-state levels of metabolites available for reaction with DNA. Ample evidence prior to (18,19,43,44) and during (12,47–49) this symposium indicates that relatively greater amounts of BP-diols at the 7,8 and 9,10 positions are found in animals previously treated with MC *in vivo* than of BP diol at the 4,5 position, and that no such increases occur in phenobarbital-treated animals. This

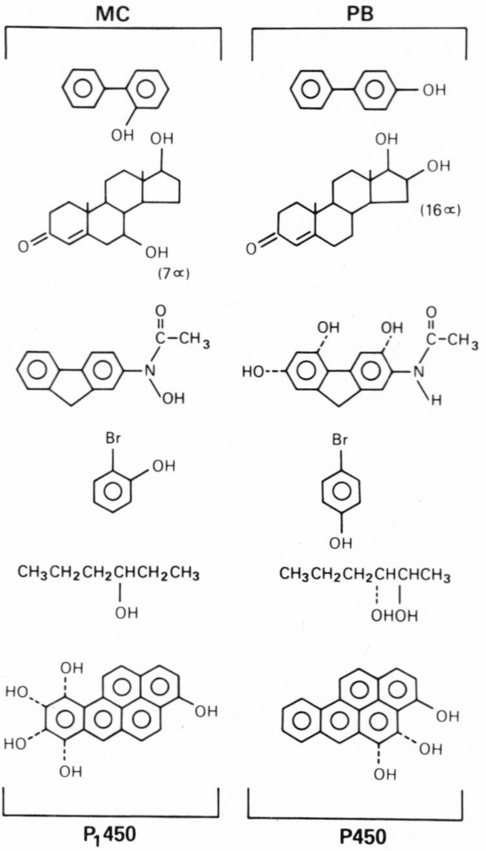

MC | PB

P_1450 | P450

Fig. 1. Chemical structures of known differences in metabolite formation when each of these six substrates is oxygenated in vitro *with liver microsomes from MC- or phenobarbital-treated rats.* PB, phenobarbital.

report will show that, in aromatic hydrocarbon responsive mice having much larger amounts of P_1450 and therefore presumably having the capacity for more epoxide formation in the 7,8 and 9,10 positions than in the 4,5 position, there exists no statistically significant relationship between BP-initiated tumorigenesis *in vivo* and any nucleoside–BP metabolites generated *in vitro.*

Figure 2 shows the dose–response curve for inducible aryl hydrocarbon hydroxylase activity as a function of the microsomal enzyme inducer β-naphthoflavone[2] treatment *in vivo.* At any dose of β-naphthoflavone, the inducible

[2] β-Naphthoflavone functions about the same as 3-methylcholanthrene as an inducer in genetically responsive animals, as compared with a relative lack of inducing effect in genetically nonresponsive mice (8). The flavone appears not to be carcinogenic. For these reasons, β-naphthoflavone instead of MC is often used in our laboratories.

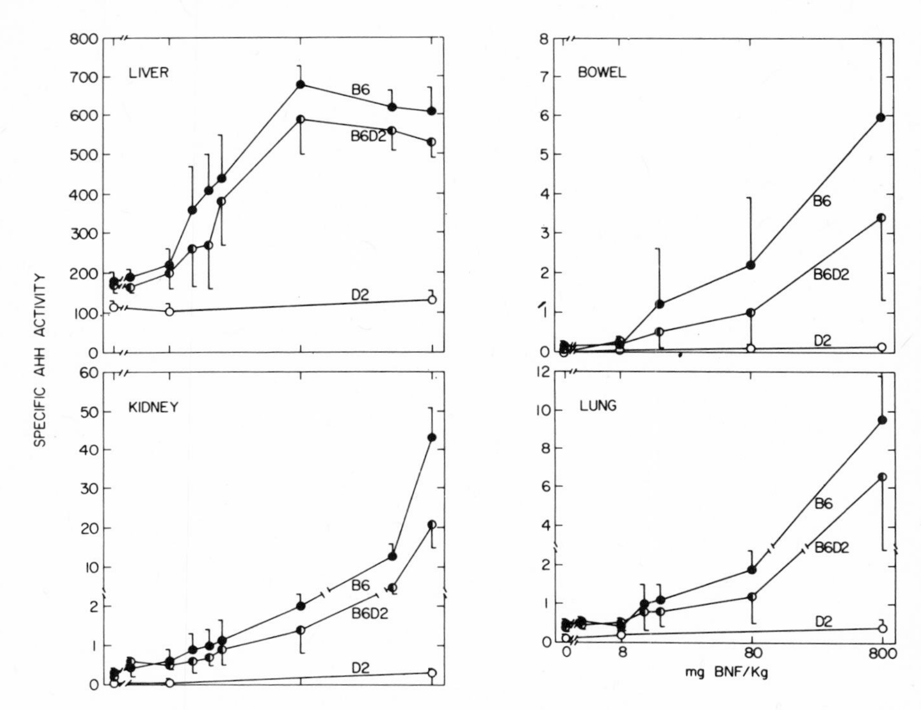

Fig. 2. Specific aryl hydrocarbon hydroxylase (AHH) activity in liver, kidney, bowel, and lung of C57BL/6J (B6), DBA/2J (D2), or (B6D2)F_1 mice as a function of the dosage of β-naphthoflavone (BNF) (50). Brackets denote standard deviation. Each symbol represents the mean of five or six individual determinations per group; all mice used were immature females of identical age. All determinations of enzyme activity in a given tissue were performed in the same assay on the same day.

hydroxylase activity is highest in the inbred C57BL/6 mouse, is somewhat less in the (C57BL/6) (DBA/2)F_1 heterozygote, and is much less in the inbred DBA/2 mouse (8,50). Similar differences in the magnitude of response are shown here for liver, kidney, bowel, and lung but also exist for skin (51) and bone marrow (52). We therefore feel that, in every tissue examined, relative increases in the aromatic hydrocarbon-inducible cytochrome P_1450 occur in the responsive mouse to a much greater extent than in the nonresponsive mouse. Whether or not hepatic cytochrome(s) P_1450 is identical–with regard to electrophoretic, catalytic, and biophysical properties–to cytochrome(s) P_1450 in each of the nonhepatic tissues remains to be determined experimentally.

DIFFERENCES IN THE BP CARCINOGENESIS INDEX *IN VIVO*

In this study different sublines of the inbred strains C3H, C57BL/6, and DBA/2 were used interchangeably. Small but detectable differences in various

Table 1. Carcinogenic Indices and Hepatic Aryl Hydrocarbon Hydroxylase and Total Cytochrome P450 Content in C3H, C57BL/6, and DBA/2 Mice[a]

Inbred strain	Number of tumors per number of mice treated	Tumor incidence (%)	Average latency (days)	Carcinogenic index	Specific hydroxylase activity[b]	Cytochrome P450 content[c]
C3H	33/42	79	142	56	1620	1020
C57BL/6	4/30	13.3	133	10	2260	1330
DBA/2	3/51	5.9	159	3.7	440	690

[a]Four- to six-week-old female mice (C3H/fCum, C57BL/6Cum, and DBA/2Cum) were treated subcutaneously with 150 μg BP per 0.05 ml trioctanoin and palpated weekly for evidence of fibrosarcoma at the site of inoculation (55). Latency was determined when the tumor was approximately 1.0 cm in diameter. The carcinogenic index is defined (56) as percent tumor incidence divided by the average latency in days, multiplied by 100. The carcinogenic indices were calculated 8 months after treatment with BP (*cf.* ref. 28 for details). Groups of 4- to 6-week-old females of the same inbred strains (C3H/HeN, C57BL/6N, and DBA/2N) were also examined for hepatic aryl hydrocarbon hydroxylase activity and total cytochrome P450 content by the procedures described (23,25); the values expressed are the means of groups of 12 mice which had been pretreated *in vivo* with MC 48 h prior to sacrifice. Aryl hydrocarbon hydroxylase induction by MC in the F_1 hybrid derived from the cross between C3H/fCum and DBA/2Cum is expressed as an additive trait (55), a finding previously reported for the F_1 hybrid between C3H/HeJ and DBA/2J (26) and for the F_1 hybrid between C3H/HeN and DBA/2N (25).

[b]Expressed as pmol of phenolic product of BP formed per min per mg microsomal protein.

[c]Expressed as pmol/mg microsomal protein and in the two MC-treated responsive inbred strains is clearly the sum of cytochrome $P_1$450 and other forms of P450 (20,23,25).

enzyme activities and other biochemical markers between sublines of certain inbred strains have recently been reported; these inbred strains include the AKR (25,53), C57BL/6 (25), and BALB/cAn (54) inbred strains. As one will see in this chapter, however, the differences between cancer susceptibility and the elution profile of nucleoside–BP metabolites formed *in vitro* are of such magnitudes that our conclusions remain valid, in spite of the fact that several sublines of the inbred strains were used.

Differences in BP-initiated subcutaneous fibrosarcoma formation have recently been appreciated (28,55) among the C3H, C57BL/6, and DBA/2 inbred strains (Table 1). The BP carcinogenic index for C3H is more than 5 times greater than that for C57BL/6 and about 15 times greater than that for DBA/2 mice. On the other hand, MC-inducible aryl hydrocarbon hydroxylase activity and total cytochrome P450 content in C57BL/6 mice are 30–40% greater than in C3H mice and are about 5 times and 2 times, respectively, more than in DBA/2 mice. Both C3H and C57BL/6 strains are genetically responsive to polycyclic aromatic compounds, whereas the DBA/2 strain is nonresponsive (8). Differences in susceptibility to BP-induced fibrosarcomas between the C57BL/6 and DBA/2 strains might be accounted for by the differences in specific aryl hydrocarbon hydroxylase activity[3] in various tissues (e.g., see Fig. 2) and in cytochrome P_1450 content. The high susceptibility of the C3H strain, however, compared to the C57BL/6 strain, cannot be accounted for by differences in the hydroxylase levels or in total cytochrome P450 content. We therefore wondered if we could find among these three strains any specific peaks representing BP metabolites bound to nucleosides *in vitro* that might account for these differences in the BP carcinogenic index *in vivo.*

ANALYSIS OF NUCLEOSIDE–METABOLITE COMPLEXES

Scheme I shows the method for isolating nucleoside–BP metabolite fractions, as developed by Brookes (14,15,19) and Sims (16–18) and co-workers. Figure 3 illustrates the results obtained with hepatic microsomes from the MC-treated responsive C57BL/6 mouse and the MC-treated nonresponsive DBA/2 mouse. Instead of five peaks designated A through E by Brookes and

[3] With BP as the substrate *in vitro,* "aryl hydrocarbon hydroxylase activity" is equated with the rate of formation of 3-hydroxybenzo[*a*]pyrene and probably other phenols having similar wavelengths of fluorescent activation and emission (8,57). These phenols may be formed either by a direct hydroxylation or in a two-step process via an arene oxide. Following incubation of BP with hepatic microsomes and cofactors, a substantial percentage (35–50%) of the phenolic products of BP formed are not detected (57) by the standard fluorometric assay in alkali (27); however, the proportional amounts of phenolic products from control, MC-, or phenobarbital-treated animals found by the standard fluorometric assay remain proportional, thereby ensuring the continued use of the simple and sensitive standard fluorometric assay (27).

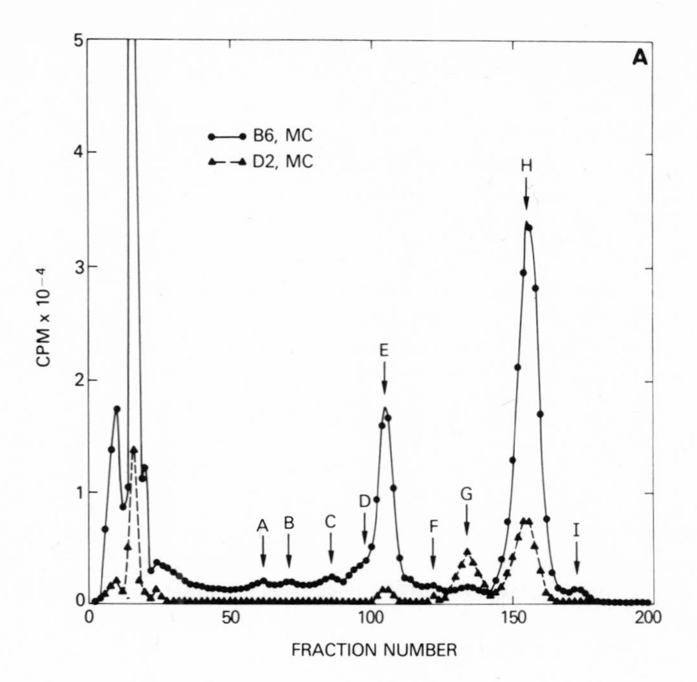
A
B6, MC
D2, MC
CPM x 10⁻⁴
FRACTION NUMBER
A
B
C
D
E
F
G
H
I

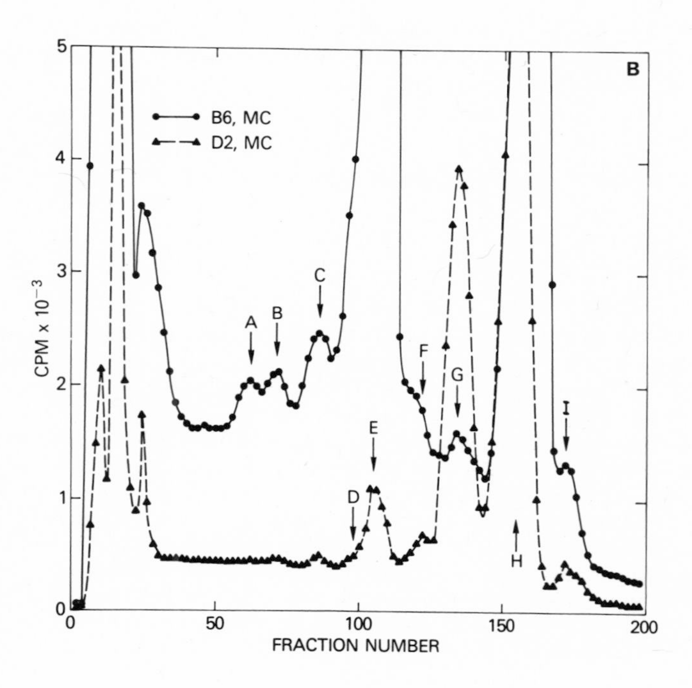
B
B6, MC
D2, MC
CPM x 10⁻³
FRACTION NUMBER
A
B
C
D
E
F
G
H
I

Salmon Sperm DNA (Deproteinized)
↓ *In vitro* Incubation:
  NADPH regenerating system
  [^{3}H]BP or [^{3}H]BP-7,8-diol or BP-7,8-diol-9,10-epoxide
  Hepatic microsomes from MC-treated or control mice of different inbred strains

DNA Reisolated and Purified
↓ Digestion:
  DNase I
  Phosphodiesterase type II
  Alkaline phosphatase type III

Nucleoside-BP Metabolite Complexes
↓ Sephadex LH20 column
  30–100% methanol gradient in water

Peaks Detected Spectrophotometrically or by Radioactivity

SCHEME I

co-workers (19), nine peaks were reproducibly found in this laboratory. Peaks E and H [which correspond to peaks A and D named by Brookes and co-workers (19)] were particularly large with microsomes from the responsive C57BL/6 mouse. Peaks E, G, and H [which correspond to peaks A, C, and D, respectively, named by Brookes and co-workers (19)] were the largest with microsomes from the nonresponsive MC-treated DBA/2 mouse. Whereas peaks E and H were much larger with the C57BL/6 than with the DBA/2 microsomes, peak G was in fact larger with DBA/2 than with C57BL/6 microsomes. Peaks A, B, C, D, F, and I were also larger with microsomes from the responsive strain than with microsomes from the nonresponsive strain. The radioactivity in peak H from microsomes of the C57BL/6 mouse represents 102 pmol of BP metabolite(s) bound to nucleosides; the amount of DNA in this peak cannot be determined in this experiment.

←

Fig. 3. Sephadex LH20 column chromatogram of an enzyme digest of DNA with [^{3}H]BP metabolites bound during an in vitro *incubation with hepatic microsomes from MC-treated C57BL/6N or DBA/2N mice (B6, MC and D2, MC, respectively) (58).* The hepatic microsomal fraction originated from livers combined from four to six mice in each experiment. Deproteinized salmon sperm DNA (20 mg) was incubated with 4 mg of microsomal protein, 25 μmol of $MgCl_2$, 1 μmol of EDTA, 7 μmol of NADPH, 100 μmol of glucose-6-phosphate, 1.4 units of glucose-6-phosphate dehydrogenase, 1 mmol of potassium phosphate buffer, pH 7.5, and 60 nmol of [^{3}H]BP (25 mCi/μmol) added in 200 μl of acetone. The 10-ml reaction mixture was incubated at 37°C for 30 min. The DNA was reisolated, purified, and digested with enzymes, then chromatographed on an 80-cm Sephadex LH20 column eluted with a 30–100% methanol gradient in water (19) at a flow rate of 1 ml/min. Fractions of 5 ml were collected. Radioactivity (in cpm) was determined in 10 ml of Aquasol for 1-ml portions of alternate fractions. The efficiency of tritium counting was found to be 30%. In appropriate experiments, tritium exchange was determined to be negligible. MC treatment of the mice *in vivo* consisted of a single intraperitoneal dose, 80 mg/kg, 48 h prior to sacrifice. The ordinate in (B) is a tenfold expansion of the ordinate (from the same experiment) in (A).

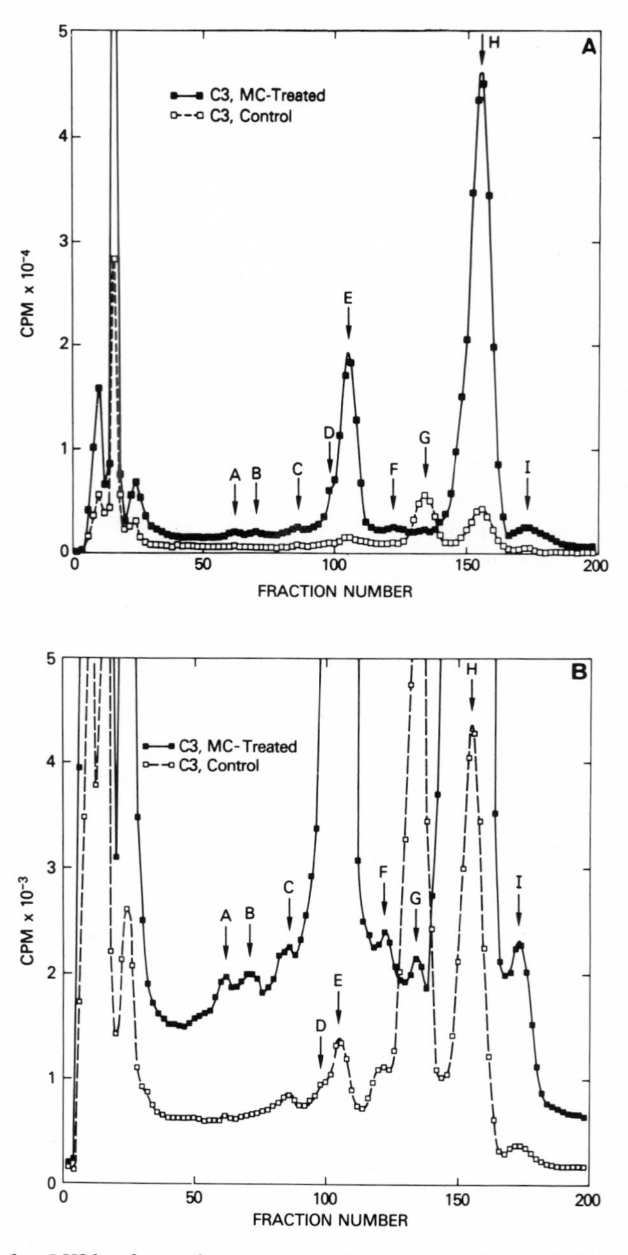

Fig. 4. Sephadex LH20 column chromatogram of an enzyme digest of DNA with [^3H]BP metabolites bound during an in vitro *incubation with hepatic microsomes from control or MC-treated C3H/HeN mice (58).* The experimental protocol was identical in every way to that in Fig. 3. Control C3H/HeN mice received corn oil alone intraperitoneally 48 h prior to sacrifice.

Figure 4 shows a similar experiment with hepatic microsomes from the responsive MC-treated or control C3H mouse. The peaks with microsomes from the MC-treated C3H mouse were relatively similar to those with microsomes from the MC-treated C57BL/6 mouse. The profile obtained with microsomes from the control C3H, C57BL/6, or DBA/2 mouse (Fig. 4 and unpublished data) was very similar to that obtained with microsomes from the MC-treated DBA/2 mouse (Fig. 3). Again, the prominent peak G with microsomes from the control C3H is like that seen with microsomes from the MC-treated DBA/2 mouse and is much greater in intensity than that found with microsomes from either the MC-treated C3H or C57BL/6 responsive mouse. Peak G appears to represent, at least in part, nucleoside–metabolite complexes resulting from BP-4,5-oxide (19). The decrease in peak G which we find in any MC-treated genetically responsive mouse could result from metabolism of BP-4,5-oxide to BP-4,5-diol and perhaps a second epoxidation in another position of the 4,5-diol that is stimulated by the increased P_1450 content of the microsomes.

In Table 2 we have attempted to quantitate the nine peaks by measuring both peak height and the area under each peak. Such quantitation is admittedly very approximate; however, in second and third experiments with similarly treated mice, both the relative peak heights and the relative areas under each peak had coefficients of variance of less than 20%. It appears that eight of the nine peaks are associated, at least in part, with metabolites formed predominantly by cytochrome P_1450. Of particular interest is that no peak from C3H mice was greater than twice that of C57BL/6 mice, yet the carcinogenic index of C3H mice is more than 5 times that of C57BL/6 mice. The carcinogenic index of DBA/2 mice is about 3 times less than that in C57BL/6 mice and about

Table 2. Relative Peak Heights and Estimated Areas Found in Figs. 3 and 4[a]

Inbred Strain	Attempted quantitation of nucleoside–BP metabolite peaks								
	A	B	C	D	E	F	G	H	I
				Peak heights in arbitrary linear chart units					
C3H	43	43	49	(<130)[b]	420	54	49	1000	49
C57BL/6	43	49	54	(<80)	380	(<40)	33	740	27
DBA/2	11	11	11	(<10)	22	16	87	160	11
				Peak areas in arbitrary square units					
C3H	0.5	0.5	0.5	2.7	27	0.5	0.5	100	3.0
C57BL/6	0.8	0.8	0.8	1.4	22	0.8	0.8	70	1.6
DBA/2	<0.3	<0.3	<0.3	<0.3	1.4	0.3	6.5	19	0.5

[a]The areas under the peaks were estimated with a Dietzgen planimeter and are recorded in arbitrary square units, different from the arbitrary linear chart units used to estimate peak heights (58). The largest peak (H in the C3H mice) is arbitrarily normalized to 1000 linear chart units (at top) and 100 square units (at bottom).
[b]Numbers in parentheses denote peak heights which could not be determined accurately, because the peak was a shoulder on a second, larger peak.

15 times less than that in C3H mice, and the relative quantities of all peaks (except peak G) are markedly less in DBA/2 mice. Peaks D, E, H, and I demonstrate relative differences among the three strains that are in the same direction as the carcinogenic index *in vivo;* i.e., the peaks are greater in C3H than in C57BL/6 and both are considerably greater than those in DBA/2 mice. With the use of mouse skin microsomes and cofactors and DNA incubated with $[^3H]$ BP, we find (unpublished data) peaks similar to those formed when hepatic microsomes were incubated; in these experiments, although peaks E, G, and H are prominent following induction of cytochrome P_1450, the size of all three of these peaks was actually greater with C57BL/6 than with C3H skin microsomes.

Differences in the carcinogenic index between the C57BL/6 and DBA/2 strains are therefore correlated not only with MC-inducible aryl hydrocarbon hydroxylase activity and total P450 concentration but also with increases in eight of the nine peaks. Differences in susceptibility to BP-initiated tumorigenesis between C3H and DBA also appear to be highly correlated with increases in eight of the nine peaks. Differences in the carcinogenic index between the C3H and C57BL/6 strains are paralleled by increases in peaks D, E, H, and I; however, the increases in peak size are much smaller than the observed differences in tumor susceptibility. Taking all the data into account for these three inbred strains, we therefore conclude that the carcinogenic index for BP *in vivo* is not necessarily correlated with the size of any of the nine peaks representing nucleoside–BP metabolite complexes.

FACTORS OTHER THAN METABOLIC ACTIVATION THAT MAY PLAY A ROLE IN CHEMICAL CARCINOGENESIS

When tumors are initiated by polycyclic hydrocarbons (Fig. 5), not only metabolic activation with resultant DNA damage but also DNA repair and immunological competence are presumably important factors. The susceptibility or resistance to chemically induced pulmonary tumors is linked to specific genes on six different linkage groups (59), and it has been suggested (30) that the "aromatic hydrocarbon responsiveness" locus is one of these loci. In this chapter we have shown a striking susceptibility to BP-initiated fibrosarcomas in the C3H strain compared with that in the C57BL/6 and DBA/2 strains. The *H-2* alleles are different between the C3H and C57BL/6 strains. C3H has the $H\text{-}2^k$ allele and C57BL/6 has the $H\text{-}2^b$ allele, and differences in these alleles are related to the susceptibility of viral-caused tumors (60,61) and perhaps general immune competence as well (62). The relationship between these alleles and chemically induced cancers remains to be determined.

The possibility exists, of course, that any or all of the nine peaks are mixtures of two or more nucleoside–metabolite complexes. Also, the important quantity of nucleoside–metabolite complexes necessary to initiate cancer may be so minute that it exists as an undetectable amount within any of these nine

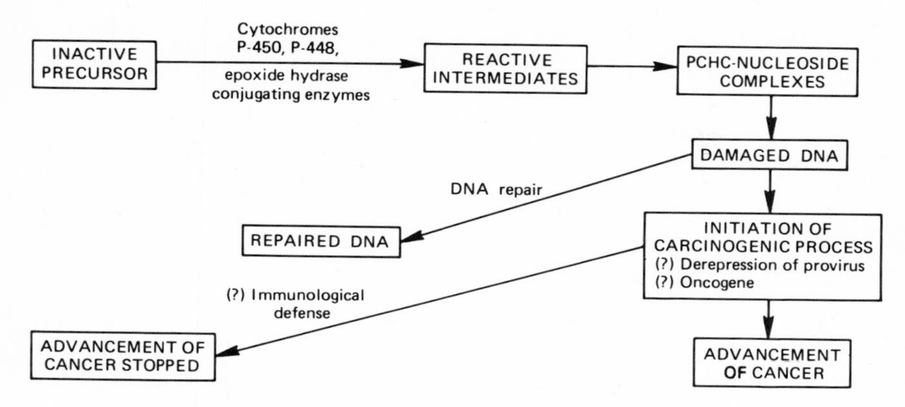

Fig. 5. Hypothetical scheme by which polycyclic hydrocarbons (PCHC) may initiate tumorigenesis. Following metabolic activation by the various drug-metabolizing enzymes (*cf.* refs. 5–8 for reviews), intercalation of reactive intermediates with base pairs may damage DNA. The advancement of cancer, however, most likely can still be prevented by the host's mechanisms of DNA repair and immunological competence.

peaks. The identification of what is included in each of these peaks is therefore of great potential importance.

FURTHER IDENTIFICATION OF PEAK E

The radioactive BP-7,8-diol was reacted with cofactors and hepatic microsomes from various mouse strains in an attempt to identify one or more of these peaks with a further metabolite of the 7,8-diol. Figure 6 demonstrates that peak E is associated with this further metabolite, whether microsomal fractions from MC-treated C57BL/6 or C3H or from control DBA/2 or C3H mice were used. These data confirm the results of Sims and co-workers (18,47) that this particular peak probably represents, at least in part, a complex between a further metabolite of BP-7,8-diol and one or more nucleosides.

COMPARISON OF SPECTROPHOTOMETRIC WITH RADIOMETRIC ANALYSIS OF NUCLEOSIDE–METABOLITE COMPLEXES

BP-7,8-diol-9,10-epoxide,[4] the synthetic isomer which reacts more rapidly with nucleophiles at the epoxide ring because of a neighboring group effect by the 7-hydroxyl group (64,65), was reacted with DNA *in vitro* in the absence of

[4] This stereoisomer has the 7-hydroxyl group and the epoxide ring in the 9,10 positions on the same face of the molecule (64,65).

microsomes and cofactors. Because BP-7,8-diol-9,10-epoxide is not radioactive, it was necessary to analyze it by ultraviolet absorption of the samples eluting from the Sephadex LH20 column. This was done by determining both the absolute absorbance at 254 nm and the peak-to-trough difference in absorption at 344 and 365 nm, respectively. Figure 7A shows the spectrum of the complex

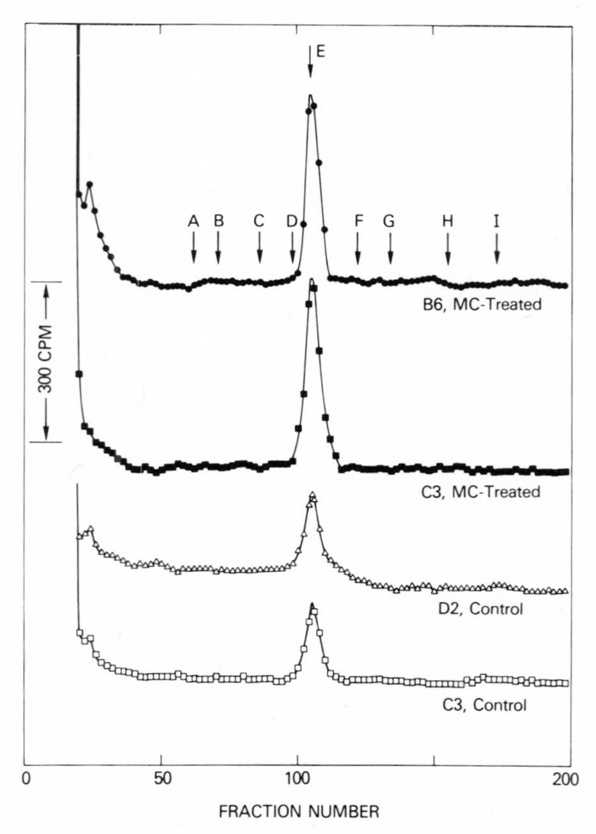

Fig. 6. Sephadex LH20 column chromatogram of an enzyme digest of DNA with bound metabolites of [^{3}H]BP-7,8-diol (52 μCi/μmol) (58). BP-7,8-diol was first made as described (63). [^{3}H]BP-7,8-diol was then synthesized from the tritiated tetrahydrodiol, which had been obtained by reduction of BP-7,8-diol with tritium. The *in vitro* incubation included salmon sperm DNA, the $MgCl_2$-, EDTA-, and NADPH-regenerating system, 6 μM [^{3}H]BP7,8-diol, and hepatic microsomes from MC-treated C57BL/6N (B6) or DBA/2N (D2) or control DBA/2N or C3H/HeN (C3), respectively. The reisolation and digestion of DNA and the chromatography are identical to those described in Fig. 3. The nine arrows, A through I, denote the nine positions where the nucleoside–BP metabolite complexes appear (as shown in Figs. 3 and 4). The elution profiles are placed on the figure at arbitrary positions with regard to the ordinate. If [^{3}H]BP or [^{3}H]BP-7,8-diol was incubated *in vitro* with DNA in the absence of microsomes or any NADPH-regenerating system or incubated *in vitro* with DNA and microsomes and the NADPH-regenerating system at 4°C, we found no peaks such as those illustrated in Figs. 3, 4, and 6.

between BP-7,8-diol-9,10-epoxide and one or more nucleosides following column chromatography; the nucleoside adduct obtained from a reaction between the diol-epoxide and polyguanylic acid (unpublished data) gave maxima at approximately the same five wavelengths: 268, 279, 315, 329, and 345 nm. A broad artifactual peak between 260 and 285 nm can be seen in the spectrum of the blank (30% methanol in water) eluted at the same region of fractionation from a newly prepared column.

Figure 7B illustrates the spectrophotometric analyses of fractions eluted from the column. The absorbance peak at 254 nm, centered around fraction 67, corresponds to the ultraviolet absorption marker *p*-nitrobenzylpyridine routinely used (19). The broad absorption peak at 254 nm, between fractions 111 and 198, is the ultraviolet blank for the gradient and is also found with newly packed columns to which no nucleosides or BP metabolites have been added. Radioactive peak E (and all of the other peak positions) were determined by cochromatographing a mixture of DNA and BP-7,8-diol-9,10-epoxide plus the DNA digest from an incubation containing [^{3}H]BP and hepatic microsomes from MC-treated C57BL/6 mice. There is no detectable ultraviolet absorption in the DNA digest from an incubation containing [^{3}H]BP and hepatic microsomes and cofactors; rather, this was used only as a marker in this experiment for the nine peaks. It is noteworthy in Fig. 7B that radioactive peak E is maximal in fraction number 105, that the 254 nm absorption is maximal in fraction number 107, and that the $\Delta A_{(344-365)}$ is maximal in fraction 109. The radioactive peak therefore does not coincide exactly with the absorption peak at 254 nm, and neither of these maxima corresponds exactly with the maximal $\Delta A_{(344-365)}$ (the nucleoside complex with BP-7,8-diol-9,10-epoxide). These data suggest that the absorption at 254 nm is insufficient as a method for determining nucleoside–BP metabolite complexes and that the radioactive peak E, generated by microsomes, is comprised of more than a single complex between nucleoside and BP-7,8-diol-9,10-epoxide. Hence peak E may contain complexes between nucleoside and other BP metabolites. It is thus likely that peak E, and also any or all of the other peaks, represents more than one metabolite of BP complexing with more than one nucleoside and/or more than one modification of a single nucleoside.

CONCLUSIONS

In summary, the various forms of cytochrome P450 in mouse liver or skin microsomes produce reactive metabolites of BP that bind to DNA *in vitro.* These results confirm the work of other laboratories (14–19,47,48) in which rat liver microsomes or cell cultures were used. Nine peaks, indicating BP metabolites bound to DNA nucleosides, can be reproducibly identified from C3H, C57BL/6, and DBA/2 mice. When either the C3H or C57BL/6 strain is compared with the DBA/2 strain, a positive general correlation is found between the carcinogenic

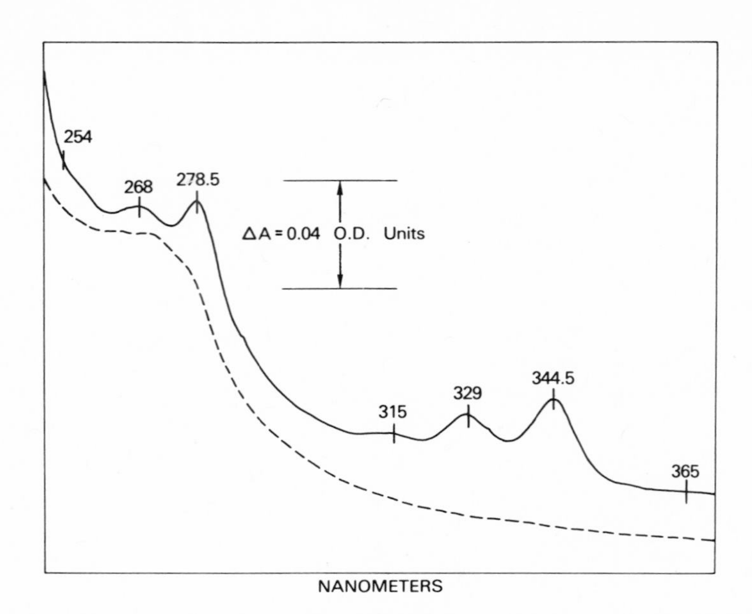

Fig. 7A. Absorption spectrum (solid line), after chromatography, of the digest of DNA which had been reacted in vitro *with BP-7,8-diol-9,10-epoxide in the absence of the NADPH-regenerating system and hepatic microsomes.* Twenty milligrams of DNA was incubated in 20 ml of 0.1 mM potassium phosphate buffer, pH 7.5, at 37°C in the dark. Six additions of 1.65 mg of the diol-epoxide in 0.50 ml of tetrahydrofuran were made over a 6-h period; this amounts to a total of approximately 33 μmol of the diol epoxide. Sufficient acetone was added during the course of the 6 h to prevent precipitation. The DNA was incubated further for 18 h at 37°C and then extracted three times with 20 ml of water-saturated ethyl acetate each time. The DNA was subsequently precipitated and purified as described in the caption of Fig. 3. The control spectrum (dashed line) represents the ultraviolet blank from a fraction eluted from a freshly packed column. Both spectra were obtained from fraction 109, shown in (B), with a blank of 30% methanol in water in the reference cuvette.

index *in vivo* and each of three determinations *in vitro:* the size of eight of the nine peaks, the hepatic aryl hydrocarbon hydroxylase activity, and the total hepatic cytochrome P450 content. Peaks D, E, H, and I demonstrate relative differences among the three strains that are in the same direction as the carcinogenic index *in vivo;* i.e., the peaks are greater in C3H than in C57BL/6, and both are considerably greater than those in DBA/2 mice. When C3H mice are compared with C57BL/6 mice, however, no quantitative correlation is found between the carcinogenic index and the size or shape of any of the nine peaks, the hepatic hydroxylase activity, or the total hepatic cytochrome P450 content. Hence the total quantity of nucleoside–BP metabolites in each of the nine peaks is not necessarily associated with biological activity (i.e., cancer *in vivo*). It is suggested that a defect in the general immune competence of the C3H inbred strain may be an important factor in its markedly increased susceptibility to

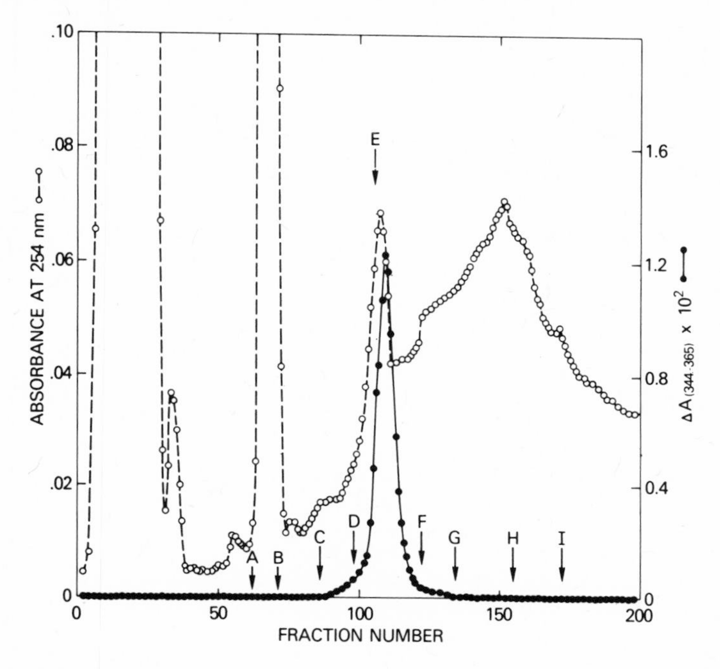

Fig. 7B. Sephadex LH20 column chromatography of the digest of the same material examined spectrophotometrically in (A). The nine arrows, A through I, denote the nine positions where the nucleoside–BP metabolite complexes appear (as shown in Figs. 3 and 4). The closed circles depict $\Delta A_{(344-365)}$, which is the peak-to-trough difference of the absorbance spectrum from samples containing the nucleoside–diol–epoxide complex.

BP-initiated tumorigenesis, compared with the C57BL/6 strain. With hepatic microsomes, peak G is associated, at least in part, with nucleoside–metabolite complexes involving BP-4,5-oxide, whereas the other eight peaks are associated predominantly with metabolites formed by cytochrome P_1450 rather than other forms of P450. With skin microsomes, all peaks including peak G appear to be associated with metabolites formed predominantly by cytochrome P_1450. Peak E is associated predominantly with BP-7,8-diol-9,10-epoxide bound to one or more nucleosides. This reactive intermediate, oxygenated in the non-K region, is associated with genetically mediated increases in aromatic hydrocarbon-inducible cytochrome P_1450.

Note Added in Proof

The *trans* isomer of BP-7,8-diol-9,10-epoxide (in which the 7-hydroxyl group is *trans* to the 9,10-epoxy group) is identical to the *cis* isomer with regard to spectral maxima of the metabolite-nucleoside complex (Fig. 7A) and to its localization in peak E (Fig. 7B).

ACKNOWLEDGMENTS

This study was supported, in part, by contracts from the Council for Tobacco Research and NIH 70-3240 within the Virus Cancer Program, National Cancer Institute, National Institutes of Health, United States Public Health Service.

REFERENCES

1. E. Boyland, The biological significance of metabolism of polycyclic compounds, *Biochem. Soc. Symp.* **5**, 40–54 (1950).
2. P. Brookes and P. D. Lawley, Evidence for the binding of polynuclear aromatic hydrocarbons to the nucleic acids of mouse skin: Relation between carcinogenic power of hydrocarbons and their binding to deoxyribonucleic acid, *Nature (London)* **202**, 781–784 (1964).
3. P. L. Grover and P. Sims, Enzyme-catalyzed reactions of polycyclic hydrocarbons with deoxyribonucleic acid and protein *in vitro*, *Biochem. J.* **110**, 159–160 (1968).
4. D. Jerina, J. Daly, B. Witkop, P. Zaltzman-Nirenberg, and S. Udenfriend, 1,2-Naphthalene, *Biochemistry* **9**, 147–156 (1970).
5. C. Heidelberger, Chemical oncogenesis in culture, *Adv. Cancer Res.* **18**, 317–366 (1973).
6. D. M. Jerina and J. W. Daly, Arene oxides: A new aspect of drug metabolism, *Science* **185**, 573–582 (1974).
7. P. Sims and P. L. Grover, Epoxides in polycyclic aromatic hydrocarbon metabolism and carcinogenesis, *Adv. Cancer Res.* **20**, 165–274 (1974).
8. D. W. Nebert, J. R. Robinson, A. Niwa, K. Kumaki, and A. P. Poland, Genetic expression of aryl hydrocarbon hydroxylase activity in the mouse, *J. Cell. Physiol.* **85**, 393–414 (1975).
9. E. C. Miller and J. A. Miller, Biochemical mechanisms of chemical carcinogenesis, in: *The Molecular Biology of Cancer* (H. Busch, ed.), pp. 377–402, Academic Press, London (1974).
10. B. N. Ames, P. Sims, and P. L. Grover, Epoxides of carcinogenic polycyclic hydrocarbons are frameshift mutagens, *Science* **176**, 47–49 (1972).
11. B. N. Ames, W. E. Durston, E. Yamasaki, and F. D. Lee, Carcinogens are mutagens: A simple test system combining liver homogenates for activation and bacteria for detection, *Proc. Natl. Acad. Sci. USA* **70**, 2281–2285 (1973).
12. A. H. Conney, A. W. Wood, W. Levin, A. Y. H. Lu, R. L. Chang, P. G. Wislocki, R. L. Goode, G. M. Holder, P. M. Dansette, H. Yagi, and D. M. Jerina, Metabolism and biological activity of benzo[*a*]pyrene and its metabolic products, Chap. 37, this volume.
13. E. Huberman, T. Kuroki, H. Marquardt, J. K. Selkirk, C. Heidelberger, P. L. Grover, and P. Sims, Transformation of hamster embryo cells by epoxides and other derivatives of polycyclic hydrocarbons, *Cancer Res.* **32**, 1391–1396 (1972).
14. W. M. Baird and P. Brookes, Isolation of the hydrocarbon-deoxyribonucleoside products from the DNA of mouse embryo cells treated in culture with 7-methylbenz-[*a*]anthracene-^{3}H, *Cancer Res.* **33**, 2378–2385 (1973).

15. W. M. Baird, A. Dipple, P. L. Grover, P. Sims, and P. Brookes, Studies on the formation of hydrocarbon-deoxyribonucleoside products by the binding of derivatives of 7-methylbenz[a]anthracene to DNA in aqueous solution and in mouse embryo cells in culture, *Cancer Res.* **33**, 2386–2392 (1973).
16. J. Booth and P. Sims, 8,9-Dihydro-8,9-dihydroxybenz[a]anthracene 10,11-oxide: A new type of polycyclic aromatic hydrocarbon metabolite, *FEBS Lett.* **47**, 30–33 (1974).
17. A. J. Swaisland, H. Hewer, K. Pal, G. R. Keysell, J. Booth, P. L. Grover, and P. Sims, Polycyclic hydrocarbon epoxides: The involvement of 8,9-dihydro-8,9-dihydroxybenz[a]anthracene 10,11-oxide in reactions with the DNA of benz[a]anthracene-treated hamster embryo cells, *FEBS Lett.* **47**, 34–38 (1974).
18. P. Sims, P. L. Grover, A. Swaisland, K. Pal, and A. Hewer, Metabolic activation of benzo[a]pyrene proceeds by a diol-epoxide, *Nature (London)* **252**, 326–328 (1974).
19. H. W. S. King, M. H. Thompson, and P. Brookes, The benzo[a]pyrene deoxyribonucleoside products isolated from DNA after metabolism of benzo[a]pyrene by rat liver microsomes in the presence of DNA, *Cancer Res.* **34**, 1263–1269 (1975).
20. D. A. Haugen, M. J. Coon, and D. W. Nebert, Induction of multiple forms of mouse liver cytochrome P-450: Evidence for genetic control of *de novo* protein synthesis in response to treatment with β-naphthoflavone or phenobarbital, *J. Biol. Chem.* **251**, 1817–1827 (1976).
21. N. E. Sladek and G. J. Mannering, Evidence for a new P-450 hemoprotein in hepatic microsomes from methylcholanthrene treated rats, *Biochem. Biophys. Res. Commun.* **24**, 668–674 (1966).
22. A. P. Alvares, G. Schilling, W. Levin, and R. Kuntzman, Studies on the induction of CO-binding pigments in liver microsomes by phenobarbital and 3-methylcholanthrene, *Biochem. Biophys. Res. Commun.* **29**, 521–526 (1967).
23. J. E. Gielen, F. M. Goujon, and D. W. Nebert, Genetic regulation of aryl hydrocarbon hydroxylase induction. II. Simple Mendelian expression in mouse tissues *in vivo*, *J. Biol. Chem.* **247**, 1125–1137 (1972).
24. P. E. Thomas, R. E. Kouri, and J. J. Hutton, The genetics of aryl hydrocarbon hydroxylase induction in mice: A single gene difference between C57BL/6J and CBA/2J, *Biochem. Genet.* **6**, 157–168 (1972).
25. J. R. Robinson, N. Considine, and D. W. Nebert, Genetic expression of aryl hydrocarbon hydroxylase induction: Evidence for the involvement of other genetic loci, *J. Biol. Chem.* **249**, 5851–5859 (1974).
26. P. E. Thomas and J. J. Hutton, Genetics of aryl hydrocarbon hydroxylase induction in mice: Additive inheritance in crosses between C3H/HeJ and DBA/2J, *Biochem. Genet.* **8**, 249–257 (1973).
27. D. W. Nebert and H. V. Gelboin, Substrate-inducible microsomal aryl hydroxylase in mammalian cell culture. I. Assay and properties of induced enzyme, *J. Biol. Chem.* **243**, 6242–6249 (1968).
28. R. E. Kouri, R. A. Salerno, and C. E. Whitmire, Relationships between aryl hydrocarbon hydroxylase inducibility and sensitivity to chemically induced subcutaneous sarcomas in various strains of mice, *J. Natl. Cancer Inst.* **50**, 363–368 (1973).
29. R. E. Kouri, H. Ratrie, and C. E. Whitmire, Evidence of a genetic relationship between susceptibility to 3-methylcholanthrene-induced subcutaneous tumors and inducibility of aryl hydrocarbon hydroxylase, *J. Natl. Cancer Inst.* **51**, 197–200 (1973).
30. D. W. Nebert, W. F. Benedict, and R. E. Kouri, Aromatic hydrocarbon-produced tumorigenesis and the genetic differences in aryl hydrocarbon hydroxylase, in: *Chemical Carcinogenesis* (P. O. P. Ts'o and J. A. Dipaolo, eds.), Chapter 12, pp. 271–288, Dekker, New York (1974).

31. R. E. Kouri, H. Ratrie, III, and C. E. Whitmire, Genetic control of susceptibility to 3-methylcholanthrene-induced subcutaneous sarcomas, *Int. J. Cancer* **13**, 714–720 (1974).
32. S. S. Thorgeirsson, J. S. Felton, and D. W. Nebert, Genetic differences in the aromatic hydrocarbon-inducible *N*-hydroxylation of 2-acetylaminofluorene and acetaminophen-produced hepatotoxicity in mice, *Mol. Pharmacol.* **11**, 159–165 (1975).
33. J. S. Felton and D. W. Nebert, Mutagenesis of certain activated carcinogens *in vitro* associated with genetically mediated increases in monooxygenase activity and cytochrome $P_1$450, *J. Biol. Chem.* **250**, 6769–6778 (1975).
34. J. R. Robinson, J. S. Felton, R. C. Levitt, S. S. Thorgeirsson, and D. W. Nebert, Relationship between "aromatic hydrocarbon responsiveness" and the survival times in mice treated with various drugs and environmental compounds, *Mol. Pharmacol.* **11**, 850–865 (1975).
35. P. J. Creaven and D. V. Parke, The stimulation of hydroxylation by carcinogenic and non-carcinogenic compounds, *Biochem. Pharmacol.* **15**, 7–16 (1966).
36. R. Kuntzman, W. Levin, M. Jacobson, and A. H. Conney, Studies on microsomal hydroxylation and the demonstration of a new carbon monoxide binding pigment in liver microsomes, *Life Sci.* **7**, 215–224 (1968).
37. T. Matsushima, P. H. Grantham, E. K. Weisburger, and J. H. Weisburger, Phenobarbital-mediated increase in ring- and *N*-hydroxylation of the carcinogen *N*-2-fluorenylacetamide, and decrease in amounts bound to liver deoxyribonucleic acid, *Biochem. Pharmacol.* **21**, 2043–2051 (1972).
38. S. S. Thorgeirsson, D. J. Jollow, H. A. Sasame, I. Green, and J. R. Mitchell, The role of cytochrome P-450 in *N*-hydroxylation of 2-acetylaminofluorene, *Mol. Pharmacol.* **9**, 398–404 (1973).
39. N. Zampaglione, D. J. Jollow, J. R. Mitchell, B. Stripp, M. Hamrick, and J. R. Gillette, Role of detoxifying enzymes in bromobenzene-induced liver necrosis, *J. Pharmacol. Exp. Ther.* **187**, 218–227 (1973).
40. H. G. Selander, D. M. Jerina, D. E. Piccolo, and G. A. Berchtold, Synthesis of 3- and 4-chlorobenzene oxides. Unexpected trapping results during metabolism of [^{14}C] chlorobenzene by hepatic microsomes, *J. Am. Chem. Soc.* **97**, 4428–4430 (1975).
41. H. G. Selander, D. M. Jerina, and J. W. Daly, Metabolism of chlorobenzene with hepatic microsomes and solubilized cytochrome P-450 systems, *Arch. Biochem. Biophys.* **68**, 309–321 (1975).
42. U. Frommer, V. Ullrich, and S. Orrenius, Influence of inducers and inhibitors on the hydroxylation pattern of *n*-hexane in rat liver microsomes, *FEBS Lett.* **41**, 14–16 (1974).
43. R. E. Rasmussen and I. Y. Wang, Dependence of specific metabolism of benzo[*a*]pyrene on the inducer of hydroxylase activity, *Cancer Res.* **34**, 2290–2295 (1974).
44. G. Holder, H. Yagi, P. Dansette, D. M. Jerina, W. Levin, A. Y. H. Lu, and A. H. Conney, Effects of inducers and epoxide hydrase on the metabolism of benzo[*a*]pyrene by liver microsomes and a reconstituted system: Analysis by high pressure liquid chromatography, *Proc. Natl. Acad. Sci. USA* **71**, 4356–4360 (1974).
45. J. R. Mitchell, D. J. Jollow, W. Z. Potter, J. R. Gillette, and B. B. Brodie, Acetaminophen-induced hepatic necrosis. IV. Protective role of glutathione, *J. Pharmacol. Exp. Ther.* **187**, 211–217 (1973).
46. J. R. Gillette, Kinetics of reactive metabolites and covalent binding *in vivo* and *in vitro*, Chap. 3, this volume.
47. P. Sims, Polycyclic hydrocarbon epoxides as active metabolic intermediates, Chap. 39, this volume.
48. P. Brookes, Role of covalent binding in carcinogenicity, Chap. 54, this volume.
49. H. V. Gelboin, J. Selkirk, T. Okuda, N. Nemoto, S. K. Yang, F. J. Wiebel, J. P.

Whitlock, Jr., H. J. Rapp, and R. C. Bast, Jr., Benzo[*a*]pyrene metabolism: Enzymatic and liquid chromatographic analysis and application to human liver, lymphocytes, and monocytes; Chap. 10, this volume.

50. A. Niwa, K. Kumaki, D. W. Nebert, and A. P. Poland, Genetic expression of aryl hydrocarbon hydroxylase activity in the mouse: Distinction between the "responsive" homozygote and heterozygote at the *Ah* locus, *Arch. Biochem. Biophys.* **166,** 559–564 (1975).
51. W. F. Benedict, N. Considine, and D. W. Nebert, Genetic differences in aryl hydrocarbon hydroxylase induction and benzo[*a*]pyrene-produced tumorigenesis in the mouse, *Mol. Pharmacol.* **9,** 266–277 (1973).
52. R. C. Levitt, J. S. Felton, J. R. Robinson, and D. W. Nebert, A single-gene difference in early death caused by hypoplastic anemia in mice receiving oral benzo[*a*]pyrene daily, *Pharmacologist* **17,** 213 (1975).
53. R. T. Acton, E. P. Blankenhorn, T. C. Douglas, R. D. Owen, J. Hilgers, H. A. Hoffman, and E. A. Boyse, Variations among sublines of inbred AKR mice, *Nature (London) New Biol.* **245,** 8–10 (1973).
54. R. C. Ciaranello and J. Axelrod, Genetically controlled alterations in the rate of degradation of phenylethanolamine *N*-methyltransferase, *J. Biol. Chem.* **248,** 5616–5623 (1973).
55. R. E. Kouri, Relationship between levels of aryl hydrocarbon hydroxylase activity and susceptibility to 3-methylcholanthrene and benzo[*a*]pyrene-induced cancers in inbred strains of mice, in: *Polynuclear Aromatic Hydrocarbons: Chemistry, Metabolism, and Carcinogenesis* (R. I. Freudenthal and P. W. Jones, eds.), pp. 139–151, Raven Press, New York (1976).
56. J. Iball, The relative potency of carcinogenic compounds, *Am. J. Cancer* **35,** 188–190 (1939).
57. G. Holder, H. Yagi, W. Levin, A. Y. H. Lu, and D. M. Jerina, Metabolism of benzo-[*a*]pyrene. III. An evaluation of the fluorescence assay, *Biochem. Biophys. Res. Commun.* **65,** 1363–1370 (1975).
58. A. R. Boobis, O. Pelkonen, R. E. Kouri, H. Yagi, O. Hernandez, D. M. Jerina, and D. W. Nebert, Genetics of benzo[*a*]pyrene-initiated subcutaneous sarcomas and cytochrome P_1-450-mediated benzo[*a*]pyrene metabolite binding to DNA in the mouse, *Mol. Pharmacol.,* in press (1977).
59. W. E. Heston, Genetics of neoplasia, in: *Methodology in Mammalian Genetics* (W. J. Burdett, ed.), pp. 247–268, Holden-Day, San Francisco (1963).
60. F. Lilly, The inheritance of susceptibility to the gross leukemia virus in mice, *Genetics* **53,** 529–539 (1966).
61. F. Lilly, The role of genetics in gross virus leukemogenesis, *Bibl. Haematol.* **36,** 213–220 (1970).
62. H. O. McDevitt and M. L. Tyan, Genetic control of the antibody response in inbred mice, *J. Exp. Med.* **128,** 1–11 (1968).
63. D. T. Gibson, V. Mahadevan, D. M. Jerina, H. Yagi, and H. J. C. Yeh, Oxidation of the carcinogens benzo[*a*]pyrene and benzo[*a*]anthracene to dihydrodiols by a bacterium, *Science* **189,** 295–297 (1975).
64. H. Yagi, O. Hernandez, and D. M. Jerina, Synthesis of (±)-7β,8α-dihydroxy-9β,10β-epoxy-7,8,9,10-tetrahydrobenzo[*a*]pyrene, a potential metabolite of the carcinogen benzo[*a*]pyrene with stereochemistry related to the antileukemic triptolides, *J. Am. Chem. Soc.* **97,** 6881–6883 (1975).
65. D. M. Jerina, H. Yagi, and O. Hernandez, Stereoselective synthesis and reactions of a diol-epoxide derived from benzo[*a*]pyrene, Chap. 40, this volume.

13

Discussion

Brookes confirmed that in high-pressure liquid chromatographic (HPLC) analysis of the products of benzpyrene metabolism in a microsomal system to which DNA had been added, his group had observed the same peaks of metabolite–DNA complexes as those obtained by Nebert's group. However, the first four peaks were not identified by Brookes' group, as they considered these minimal in both size and importance. Peak A of Brookes, therefore, corresponds to peak E of Nebert.

Brookes also commented that his peak D is never seen in an *in vivo* metabolic system. (He defined an *in vivo* system as one in which any total cell system is used as the source of the arene hydrocarbon hydroxylase enzyme system, e.g., mouse skin, embryo cells, or rat muscle.) He emphasized that the relevance of the peaks found using microsomal metabolizing systems to an *in vivo* situation is doubtful, as correlation between these findings and those using a whole cell system varies with the substrate. For example, when 7-methylbenzanthracene is a substrate for a microsomal system (with added DNA), none of the peaks found upon high-pressure liquid chromatography of the products corresponds to those found using an *in vivo* system. However, when benzypyrene is the substrate, peak A is produced in both the *in vivo* and *in vitro* systems. This is one of the few cases in which there is a good correlation between the two systems.

Brookes also emphasized that K-region epoxides are frequently not the important metabolites, e.g., during microsomal metabolism of benzpyrene and 7-methylbenzanthracene in systems to which DNA was added, the K-region epoxides were not seen, and with whole cell systems they are minor metabolites. Peak A of Brookes is the DNA complex of the 7,8-diol, 9,10-epoxide of benzpyrene, which is a non-K-region epoxide.

Gelboin commented that the carcinogenic index is not related to liver carcinogenesis in these mice, but probably to sarcoma production. He also asked whether Nebert had compared peaks from use of target tissues with those from nontarget tissues. Nebert replied that the primary peak seen using skin in an *in vitro* system was peak H (peak D of Brookes' designation). Brookes sees only his peak A in *in vivo* studies with skin.

Nebert also pointed out that after skin is painted with benzpyrene the cytochrome P448 component is roughly 80–95% of the total cytochrome P450 content and suggested that this may explain the almost exclusive appearance of peak A in the experiments using skin enzymes.

Gelboin suggested that the most useful tissue to study in detail would be the one in which the metabolites bound to DNA *in vivo* are identical to those formed by microsomes *in vitro.* A reconstruction of the partially purified enzymes might be useful. Brookes expressed doubt that it would be possible to reconstruct such a system which would correspond to the organization of the enzymes in the intact cells, in which the final product is the result of interaction of all of these enzymes. For example, the relationship between oxygenase and hydrase is very important *in vivo.*

Brookes and Nebert both responded to a question by Snyder that the efficiency of transport of the extranuclear metabolite to the nuclear DNA probably does not determine whether or not a tumor is formed, as the nuclear membrane itself contains the AHH enzyme. Further evidence that the rate of transport across the nuclear membrane is not the critical factor is that the same peaks are obtained using HPLC whether the DNA is supplied to the metabolic system as isolated DNA or by addition of nuclei.

It was suggested by Remmer that conjugation of the metabolite may be a critical process in binding because the conjugated metabolite may have a greatly altered binding capacity for both protein and DNA. The unique role of the conjugating enzymes in the liver is emphasized by the fact that very few primary tumors are found in the liver despite the active metabolic role of this organ. Whether or not the conjugated metabolites form complexes with DNA has not been determined since the analytical studies are carried out on the hydrolyzed complexes. Nebert responded that other enzymes such as epoxide hydrase and glutathione transferase, in addition to the conjugating enzymes, are active in reducing the concentration of the primary metabolites and thus perhaps in influencing tumor formation. Hormonal differences as well as the nutritional state will also cause variation in primary metabolite concentration. The overall correlation of metabolic pattern and tumor formation is not yet clear.

14

Formation of Toxic Intermediates in Fetal Tissues

Olavi Pelkonen
Department of Pharmacology
University of Oulu
SF-90220 Oulu 22, Finland

Pharmacological actions and toxic effects of drugs and other foreign compounds are often mediated by metabolites of the administered compounds rather than by the unchanged compounds themselves. For example, bromobenzene and other halogenated hydrocarbons are converted to reactive epoxides by the action of a drug-oxidizing monooxygenase system in the liver and other tissues, with resultant liver injury (1,2). Most carcinogenic substances, e.g., acetaminofluorene and polycyclic aromatic hydrocarbons, are not carcinogenic *per se,* but have to be transformed into reactive electrophiles, which attack critical target molecules, thus initiating carcinogenesis (3). Many halogenated hydrocarbon pesticides are converted by microsomal enzymes into epoxides, which generally are more toxic than the parent compounds. These examples point to the considerable importance of drug-oxidizing enzyme systems in the toxic effects of xenobiotics.

Components of the microsomal electron transport chain, cytochrome P450 and NADPH–cytochrome P450 (cytochrome *c*) reductase, and related drug-oxidizing enzyme activities have been detected in the human fetal liver, adrenal gland, and placenta (4–8). These findings with human tissues stand in marked contrast to those made in fetuses of common laboratory animals which are essentially devoid of xenobiotic-metabolizing activity. As noted above, the presence of a xenobiotic-oxidizing monooxygenase system is the necessary, although perhaps not sufficient, prerequisite for the production of harmful effects by certain compounds. This chapter will discuss the possible role of the human fetal monooxygenase systems in xenobiotic-induced fetotoxicities. Special attention is given in this chapter to aryl hydrocarbon hydroxylase, an enzyme which metabolizes polycyclic aromatic hydrocarbons, and to the metabolism of benzo[*a*]pyrene, a common environmental carcinogen, in the fetal tissues and in the placenta. A more general review dealing with drug metabolism in human and animal fetuses will be published elsewhere (9).

DRUG-OXIDIZING ENZYMES IN HUMAN FETAL TISSUES

The concentration of cytochrome P450 in the human fetal liver is remarkably high when compared with values in the livers of animal fetuses, i.e., around 0.2–0.3 nmol/mg of microsomal protein and from 3 to 6 nmol/g of tissue (wet weight) (8). Values in the adrenal gland seem to be even higher (8). The activity of NADPH–cytochrome *c* reductase in the human fetal liver is about 30–50% of human adult values (10). Thorgeirsson (11) reported that NADPH–cytochrome P450 reductase in the human fetal liver is about 40% that of the adult human activity. The liver enzyme, however, shows a low substrate specificity, metabolizing a large number of different substrates (Table 1). Also, spectral properties, studies on inhibition of drug-oxidizing enzymes by alternate substrates, and many other aspects suggest that the fetal hepatic enzyme does not differ appreciably from its adult counterpart (12–14). There are, however, some differences: e.g., the low specific metabolism of benzo[*a*]pyrene (10,15) and the apparent change in kinetic properties of different drug-oxidizing activities between fetus and adult (16) reveal that the fetal hepatic enzyme is not totally analogous to the adult enzyme. This may have some important consequences

Table 1. Some Compounds Shown to Be Metabolized by the Human Fetal Liver *in Vitro*[a]

Compound	Metabolic pathway[b]
Ethylmorphine	*N*-Demethylation
Aminopyrine	*N*-Demethylation
N-Methylaniline	*N*-Demethylation
Dimethylnitrosamine	*N*-Demethylation*
Chlorpromazine	*N*-Demethylation, sulfoxidation, aromatic hydroxylation, *N*-oxidation
Diazepam	*N*-Demethylation, aromatic hydroxylation
Desmethylimipramine	Aromatic hydroxylation
Aniline	Aromatic hydroxylation, *N*-oxidation*
Benzo[*a*]pyrene	Aromatic hydroxylation,* epoxide hydration,* quinone formation
N,N-Dimethylaniline	*N*-Oxidation
Hexobarbital	Aliphatic hydroxylation
Naphthalene-1,2-oxide	Epoxide hydration
Aldrin	Epoxidation*
p-Nitrobenzoic acid	Nitro-group reduction

[a]Data are from many sources mentioned in the text proper.
[b]Metabolic pathway marked by an asterisk indicates that an active intermediate or a metabolite is implicated.

with respect to substrate specificity and relative activity toward different substrates.

The fetal hepatic drug-oxidizing enzyme system is detectable already at the age of 6–7 weeks, just at the end of embryogenesis (17), and this finding is in accordance with the histological study of the human fetal liver (18). At the age of 12–14 weeks, the liver system seems to attain a constant level (19).

Aryl hydrocarbon hydroxylase activity using benzo[*a*]pyrene as a substrate has been demonstrated in the human fetal liver, adrenal gland, and placenta (12,15,20–22).

ON THE NATURE OF ARYL HYDROCARBON HYDROXYLASE IN HUMAN FETAL TISSUES AND PLACENTA

At least two types of aryl hydrocarbon hydroxylase activity have been found in animal tissues (23). The ratios of these two forms in a particular tissue may depend on the hormonal status of an animal, its species, strain, pretreatment, or other factors. These two types have been studied most thoroughly in the liver of the rat and the mouse. According to Nebert *et al.* (23), type *b,* the *basal* aryl hydrocarbon hydroxylase activity, is associated with cytochrome P450 and is present in livers of control or phenobarbital-treated animals, whereas type *a,* the *induced* aryl hydrocarbon hydroxylase activity, is associated with cytochrome P448 or P_1450, and is found in livers of polycyclic aromatic hydrocarbon-treated animals. These two types may have different roles in activation and detoxication of polycyclic hydrocarbons, because the metabolite patterns they produce are different (24).

We have tried to differentiate between these forms of aryl hydrocarbon hydroxylase activity in fetal tissues and placenta by using inhibitors showing preference for one or the other form of aryl hydrocarbon hydroxylase (25). Table 2 shows that the fetal hepatic aryl hydrocarbon hydroxylase is inhibited by SKF 525A, aminopyrine, and metyrapone, but not by two naphthoflavones, thus exhibiting properties typical of type *b* aryl hydrocarbon hydroxylase (25). On the other hand, the placental enzyme was strongly inhibited by two naphthoflavones, but not by SKF 525A, aminopyrine, or metyrapone, thus exhibiting inhibitory properties characteristic of the type *a* aryl hydrocarbon hydroxylase (25). The fetal adrenal gland hydroxylase behaved more like the hepatic system. It was also found that hydroxylase activity appearing in the fetal rat and mouse liver after pretreatment of the mother with 3-methylcholanthrene is inhibited by the same compounds as type *a* hydroxylase, quite contrary to the human fetal hepatic type *b* hydroxylase. Also, studies on substrate specificity, drug-induced spectral interactions, and other properties of fetal hepatic, adrenal, and placental monooxygenase systems point to a crucial difference between enzyme systems in the liver and adrenal gland on the one hand and that in the placenta on the other. [For a more complete discussion, see Pelkonen (9).] Thus the evidence

Table 2. Preferential Inhibition of Aryl Hydrocarbon Hydroxylase *in Vitro* by Various Inhibitors Added to Human Fetal Liver and Placental Homogenates[a]

	(Inhibitor)/(benzo[*a*]pyrene) ratio inhibiting aryl hydrocarbon hydroxylase activity by 50%	
Inhibitor	Fetal liver	Placenta
α-Naphthoflavone	Activation	0.3
β-Naphthoflavone	Activation	0.8
Aminopyrine	8	>100[b]
SKF 525A	1	>20
Metyrapone	4	>10
Testosterone	>30	>30
Zoxazolamine	>10	>10

[a]Selection of inhibitors was based on the study of Goujon *et al.* (25).
[b]A 50% inhibition was not attained by the highest inhibitor concentration used.

available at the moment indicates that the monooxygenase systems in the human fetal liver and adrenal gland are "basal," not "induced" like the placental system.

The fetal hepatic aryl hydrocarbon hydroxylase is present at 6–7 weeks of fetal age (17) and attains a rather constant level at the fetal age of 12–14 weeks (19). The activity of the fetal liver is only a few percent that of the adult liver (10,15). It may be proper to say that an effect of a compound (or a metabolite) does not depend only on the actual amount of a compound (or a metabolite) at the receptor site, but also on the sensitivity of the receptor or the target tissue. It is likely that fetal tissues differ from their adult counterparts with respect to sensitivity to such exogenous influences as xenobiotics and their activated metabolites. The low activity of the fetal aryl hydrocarbon hydroxylase does not necessarily mean that harmful effects mediated by the products of its action would be less probable.

INDUCIBILITY OF THE HUMAN FETAL AND PLACENTAL ARYL HYDROCARBON HYDROXYLASE

Aryl hydrocarbon hydroxylase activity in the liver and some other tissues of fetal rat, rabbit, guinea pig, and hamster is inducible by the treatment of the mother during late pregnancy by polycyclic aromatic hydrocarbons, polychlorinated biphenyls, and phenobarbital (9), although to a variable extent, depending on the inducer, species, and tissue. The inducibility of aryl hydrocarbon hydroxylase also depends on the stage of pregnancy, being less inducible the earlier an inducer is administered (26).

It is well demonstrated that maternal cigarette smoking induces aryl hydrocarbon hydroxylase activity in the human placenta (21). Induction first appears during the first trimester of pregnancy and attains its highest level at term (22,23). On the other hand, maternal cigarette smoking does not seem to have any effect on aryl hydrocarbon hydroxylase activity in the human fetal liver (27,28). The apparent noninducibility of hydroxylase of the human fetal liver may be a consequence of a deficient enzyme-forming system or of a low level of inducer reaching the fetal liver cells. Experiments on animals have shown that the dose of polycyclic hydrocarbon inducer required to achieve an effect on the hydroxylase is many times larger for the fetal liver than for the placenta (29,30), thus supporting the former possibility in animal fetuses.

Our recent studies indicate that aryl hydrocarbon hydroxylase activity is present in cell cultures derived from the human fetal liver and other tissues and it is inducible by the exposure of cell cultures to polycyclic aromatic hydrocarbons (31). The basal enzyme in control cultures is not inhibited by naphthoflavone, whereas the induced enzyme is, thus exhibiting the same difference as that seen in the liver microsomal system. The magnitude of induction by benz[*a*]anthracene is dose dependent and varies from 1.5- to 5-fold depending on the individual culture studied (31). Similar variability of induction has been found by several authors in human lymphocyte cultures (32,33) and by Alvares *et al.* (34) in cultures of human newborn foreskin tissue, but we have been unable to find any indication of trimodal variation of induction in fetal liver cell cultures, as shown by Kellermann *et al.* (35) in human lymphocyte cultures.

When considering observations on aryl hydrocarbon hydroxylase activity in human fetal tissues and placenta and in fetal cell cultures with respect to teratogenesis and transplacental carcinogenesis, one must take into consideration that the inducibility of hydroxylase activity by exogenous agents is not necessarily related to the activation of hydrocarbons to toxic or carcinogenic intermediates. Rather, it seems that the induction of aryl hydrocarbon hydroxylase by polycyclic hydrocarbons provides cells with a means of enhanced detoxication of harmful intermediates. In a way, induction of hydroxylase activity seems to be an almost universal defense system of cells against highly lipid-soluble, potentially carcinogenic compounds.

THE SPECIFIC METABOLISM OF BENZO[*a*]PYRENE IN DIFFERENT HUMAN FETAL EXPERIMENTAL SYSTEMS

Benzo[*a*]pyrene was incubated with different human fetal tissues, with homogenates from cultured cells, and in the growth medium of liver cells and

Table 3. Metabolites of Benzo[*a*]pyrene Produced in Different *in Vitro* and *in Vivo* Experimental Systems of Human Fetal Tissues and Placenta

	Tissue homogenates *in vitro*				Cells in culture from		
Metabolites	Liver	Adrenal	Placenta (smokers)	Other tissues[a]	Liver	Lungs	Skin
Ethyl acetate soluble							
Diol I (9,10-)	+	+	+	–	+	+	+
Diol II (7,8-)	+	+	+	–	+	+	+
Diol III (4,5-)	+	+	+	–	+	uncertain	
Hydroxy (3- and/or 9-)	+	+	+	–	+	uncertain	
Quinones (1,6-, 6,12-, 3,6-)	+	+	+	–	+	+	+
Water soluble							
Total	+	+	+	–	+	+	+
β-Glucuronidase susceptible					+		
Sulfatase susceptible					–		
Covalently bound	+	+	+	–	+		

[a]Includes placenta from nonsmokers, and fetal lung, intestine, kidney, and brain.

fibroblasts in culture. Products were resolved into two fractions (36,37): (1) water-soluble products (in some cases identification attempts were made with the aid of β-glucuronidase and sulfatase) and (2) ethyl-acetate-soluble products. The latter fraction was further resolved by thin-layer chromatography into (1) unidentified products located at the origin in the standard TLC system, (2) three dihydrodiol fractions (9,10-, 7,8-, and 4,5-diols), (3) one hydroxy metabolite fraction (3- and/or 9-hydroxybenzo[*a*]pyrene), (4) a quinone fraction (several reference compounds), and (5) unchanged benzo[*a*]pyrene. Qualitative results are shown in Table 3. Benzo[*a*]pyrene was metabolized into oxidized products in homogenates and microsomes from the fetal liver, adrenal gland, and term placenta from smoking mothers. Cell homogenates from liver cell and fibroblast cultures were able to metabolize benzo[*a*]pyrene. When benzo[*a*]pyrene was added to the cultured fetal liver cells and fibroblasts, there was an accumulation of water-soluble radioactivity and a decrease of ethyl-acetate-soluble radioactivity as a function of incubation time (Fig. 1). The decrease of ethyl-acetate-soluble radioactivity was almost totally due to the disappearance of benzo[*a*]-pyrene from the growth medium (Fig. 2). When the production of ethyl-acetate-soluble metabolites was studied as a function of time, an accumulation of diols and hydroxymetabolites was found which reached a maximum at day 1 and then declined. The accumulation of radioactivity into the water phase continued during the whole incubation period of 3 days, thus indicating that ethyl-acetate-soluble products were gradually converted into more water-soluble products, perhaps polyhydroxylated products, as a result of "recycling" and conjugation. Preliminary experiments have indicated that fetal liver cells in culture seem to be

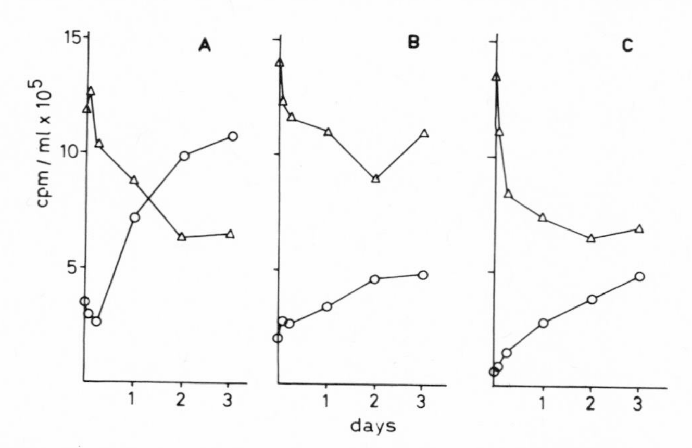

Fig. 1. Ethyl-acetate- (△) and water- (○) soluble products of benzo[a]pyrene in culture medium of fetal liver cells (A), fetal fibroblasts (B), and BS-C cells (C). Abscissa: incubation time in days; ordinate: radioactivity per milliliter of growth medium in cpm × 10^5. Fetal cells were derived from the same fetus at 8 weeks of fetal age. BS-C is an established cell line from monkey kidney tissue. [^{3}H]Benzo[*a*]pyrene was added to the culture medium and samples were taken at 0, 1, 6, 24, 48, and 72 h after the addition.

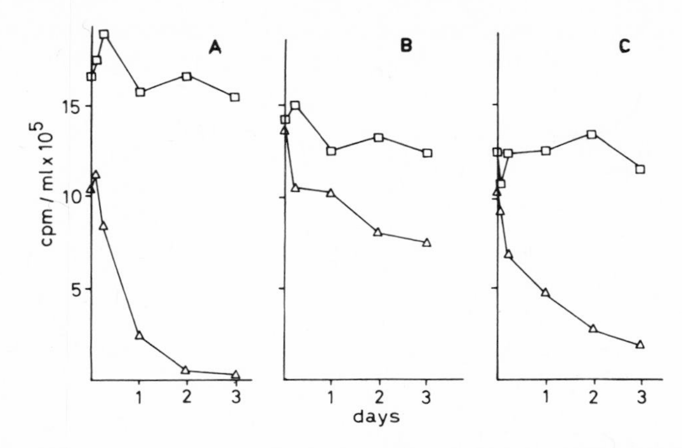

Fig. 2. Disappearance of benzo[a]pyrene from culture medium of fetal liver cells (A), fetal fibroblasts (B), and BS-C cells (C). For explanation, see the caption of Fig. 1. Benzo[*a*]pyrene (△ in experimental and □ in control cultures) was separated from other ethyl-acetate-soluble materials by thin-layer chromatography.

much more active than fibroblasts derived from fetal lungs or skin. Also, the established cell line, BS-C cells from monkey kidney, which were always cultured with fetal cells, was able to metabolize benzo[*a*]pyrene to oxidized metabolites.

Preliminary results indicated that a small part of the water-soluble metabolite fraction was susceptible to hydrolysis by β-glucuronidase, indicating the presence of glucuronic acid conjugates of hydroxybenzo[*a*]pyrene in fetal liver cell cultures. A part of the radioactivity was also covalently bound to cells. Thus, all main classes of benzo[*a*]pyrene metabolites–diols, monohydroxylates, quinones, conjugates, and covalently bound derivatives–were produced by human fetal liver cells in culture. The presence of diols suggests epoxide intermediates (probably 9,10-, 7,8-, and 4,5-oxides) and also an active epoxide hydrase in fetal tissues, fetal cell cultures, and in the placenta from smokers.

POSSIBLE TRANSPLACENTAL TRANSFER OF ACTIVE INTERMEDIATES FROM MOTHER TO FETUS

Although the penetration of the placenta by metabolites of xenobiotics has been thought to occur very slowly or not at all, recent observations suggest that it may be necessary to revise this point of view. For example, the relatively stable epoxides of polycyclic aromatic hydrocarbons described recently by Sims (38) are probably lipid soluble enough to penetrate the placenta readily and to reach the fetal tissues. A recent study by Takahashi and Yasuhira (39) shows that unconjugated metabolites of 3-methylcholanthrene cross the placenta of the pregnant mouse and reach the fetus, whereas conjugated metabolites could not be detected in the fetal side after their administration to the mother.

OTHER POTENTIAL ACTIVATION REACTIONS

Rane and Ackermann (14) demonstrated that when aniline was incubated with human fetal liver microsomes the hemoglobin present was converted to methemoglobin. This was most probably a consequence of the formation of an *N*-oxidized metabolite of aniline. Rane (40) also found considerable *N*-oxidase activity in human fetal liver using *N,N*-dimethylaniline as a substrate. Recently, Pelkonen and Kärki (41) have shown that human fetal liver and adrenal gland preparations (but not placenta) catalyze the epoxidation of aldrin to dieldrin at a rate approximately 30–40% that of rat liver activity.

CONCLUSIONS

Human fetal liver and adrenal gland contain a cytochrome P450-linked monooxygenase system capable of metabolizing benzo[*a*]pyrene to different oxidized metabolites, thus implicating epoxide intermediates. The fetal hepatic system seems to be basically different from the placental system which is present almost exclusively in placentas from smokers. Also, cells in culture derived from fetal liver and other tissues metabolize benzo[*a*]pyrene into a variety of oxidized and conjugated metabolites and into covalently bound derivatives. Other potential activation reactions, *N*-oxidation and pesticide epoxidation, have been shown to occur in human fetal tissues. The significance of the activating systems detected in human fetal tissues is not known, but in the light of our ignorance of the causes of fetal injuries further investigations are needed. The sensitivity of the fetal target tissues themselves to these chemically reactive toxic intermediates may also have to be determined.

ACKNOWLEDGMENTS

This work is dedicated to the memory of Ms. Vuokko Väisänen, the author's laboratory technician, who performed most of the practical work involved in the original research and died accidentally June 1975.

The author wishes to thank Dr. N. T. Kärki, Dr. P. Korhonen, Dr. J. T. Ahokas, Dr. P. Jouppila, Ms. Liisa Tuhkanen, Ms. Leena Pyykkö, and Ms. Ritva Saarikoski for collaboration during different phases of the present work. The original research reported was supported by the Academy of Finland. Reference metabolites of benzo[*a*]pyrene were kindly donated by Dr. H. V. Gelboin, Bethesda, Maryland.

REFERENCES

1. J. R. Gillette, A perspective on the role of chemically reactive metabolites of foreign compounds in toxicity. I. Correlation of changes in covalent binding of reactive metabolites with changes in the incidence and severity of toxicity, *Biochem. Pharmacol.* **23,** 2785–2794 (1974).
2. J. R. Gillette, A perspective on the role of chemically reactive metabolites of foreign compounds in toxicity. II. Alterations in the kinetics of covalent binding, *Biochem. Pharmacol.* **23,** 2927–2938 (1974).
3. J. A. Miller, Carcinogenesis by chemicals: An overview, *Cancer Res.* **30,** 559–576 (1970).
4. A. Rane, C. v. Bahr, S. Orrenius, and F. Sjöqvist, in: *Fetal Pharmacology* (L. Boreus, ed.), pp. 287–301, Raven Press, New York (1973).
5. A. Rane, F. Sjöqvist, and S. Orrenius, Drugs and fetal metabolism, *Clin. Pharmacol. Ther.* **14,** 666–672 (1973).
6. M. R. Juchau, Q. H. Lee, G. L. Louviaux, K. G. Symms, J. Krasner, and S. J. Yaffe, in: *Fetal Pharmacology* (L. Boreus, ed.), pp. 321–333, Raven Press, New York (1973).
7. O. Pelkonen and N. T. Kärki, Drug metabolism in human fetal tissues, *Life Sci.* **13,** 1163–1180 (1973).
8. O. Pelkonen, P. Korhonen, P. Jouppila, and N. T. Kärki, in: *Basic and Therapeutic Aspects of Perinatal Pharmacology* (P. L. Morselli, S. Garattini, and F. Sereni, eds.), pp. 65–74, Raven Press, New York (1975).
9. O. Pelkonen, Transplacental transfer of foreign compounds and their metabolism by the foetus, in: *Progress in Drug Metabolism,* Vol. 2 (L. F. Chasseaud and J. W. Bridges, eds.), in press, Chichester.
10. O. Pelkonen, E. H. Kaltiala, T. K. I. Larmi, and N. T. Kärki, Comparison of activities of drug-metabolizing enzymes in human fetal and adult livers, *Clin. Pharmacol. Ther.* **14,** 840–846 (1973).
11. S. S. Thorgeirsson, Mechanism of hepatic drug oxidation and its relationship to individual differences in rates of oxidation in man, Ph.D. thesis, University of London (1972).
12. O. Pelkonen, P. Arvela, and N. T. Kärki, 3,4-Benzpyrene and *N*-methylaniline metabolizing enzymes in the immature human foetus and placenta, *Acta Pharmacol. Toxicol.* **30,** 385–395 (1971).
13. O. Pelkonen, M. Vorne, P. Jouppila, and N. T. Kärki, Metabolism of chlorpromazine and *p*-nitrobenzoic acid in the liver, intestine and kidney of the human foetus, *Acta Pharmacol. Toxicol.* **29,** 284–294 (1971).
14. A. Rane and E. Ackermann, Metabolism of ethylmorphine and aniline in human fetal liver, *Clin. Pharmacol. Ther.* **13,** 663–670 (1972).
15. A. B. Rifkind, S. Bennett, E. S. Forster, and N. I. New, Components of the heme biosynthetic pathway and mixed function oxidase activity in human fetal tissues, *Biochem. Pharmacol.* **24,** 839–846 (1975).
16. O. Pelkonen, Developmental change in the apparent kinetic properties of drug-oxidizing enzymes in the human liver, *Res. Commun. Chem. Pathol. Pharmacol.* **10,** 293–302 (1975).
17. O. Pelkonen and N. T. Kärki, 3,4-Benzpyrene and aniline are hydroxylated by human fetal liver but not placenta at 6–7 weeks of fetal age, *Biochem. Pharmacol.* **22,** 1538–1540 (1973).
18. L. Zamboni, Electron microscopic studies of blood embryogenesis in humans. I. The ultrastructure of the fetal liver, *J. Ultrastruct. Res.* **12,** 509–524 (1965).

19. O. Pelkonen, Drug metabolism in the human fetal liver: Relationship to fetal age, *Arch. Int. Pharmacodyn. Ther.* **202**, 281–287 (1973).
20. M. R. Juchau, M. G. Pedersen, and K. G. Symms, Hydroxylation of 3,4-benzpyrene in human fetal tissue homogenates, *Biochem. Pharmacol.* **21**, 2269–2272 (1972).
21. R. M. Welch, Y. E. Harrison, A. H. Conney, P. J. Poppers, and M. Finster, Cigarette smoking: Stimulatory effect on metabolism of 3,4-benzpyrene by enzymes in human placenta, *Science* **160**, 541–542 (1968).
22. M. R. Juchau, Human placental hydroxylation of 3,4-benzpyrene during early gestation and at term, *Toxicol. Appl. Pharmacol.* **18**, 665–675 (1971).
23. D. W. Nebert, J. R. Robinson, A. Niwa, K. Kumaki, and A. P. Poland, Genetic expression of aryl hydrocarbon hydroxylase activity in the mouse, *J. Cell. Physiol.* **85**, 393–414 (1975).
24. R. E. Rasmussen and I. Y. Wang, Dependence of specific metabolism of benzo[*a*]pyrene on the inducer of hydroxylase activity, *Cancer Res.* **34**, 2290–2295 (1974).
25. F. M. Goujon, D. W. Nebert, and J. E. Gielen, Genetic expression of aryl hydrocarbon hydroxylase induction. IV. Interaction of various compounds with different forms of cytochrome P-450 and the effect on benzo[*a*]pyrene metabolism *in vitro, Mol. Pharmacol.* **8**, 667–680 (1972).
26. E. Schlede and H.-J. Merker, Effect of benzo[*a*]pyrene treatment on the benzo[*a*]pyrene hydroxylase activity in maternal liver, placenta, and fetus of the rat during day 13 to day 18 of gestation, *Arch. Pharmacol.* **272**, 89–100 (1972).
27. O. Pelkonen, P. Jouppila, and N. T. Kärki, Effect of maternal cigarette smoking on 3,4-benzpyrene and *N*-methylaniline metabolism in human fetal liver and placenta, *Toxicol. Appl. Pharmacol.* **23**, 399–407 (1972).
28. O. Pelkonen, P. Jouppila, E. H. Kaltiala, and N. T. Kärki, in: *Developmental and Genetic Aspects of Drug and Environmental Toxicity* (W. A. M. Duncan, ed.), International Congress Series No. 345, pp. 154–158, Excerpta Medica, Amsterdam (1975).
29. R. M. Welch, B. Gommi, A. P. Alvares, and A. H. Conney, Effect of enzyme induction on the metabolism of benzo[*a*]pyrene and 3′-monomethylaminoazobenzene in the pregnant and fetal rat, *Cancer Res.* **32**, 973–978 (1972).
30. M. R. Juchau and M. G. Pedersen, Drug biotransformation reactions in the human fetal adrenal gland, *Life Sci.* **12 (II)**, 193–204 (1973).
31. O. Pelkonen, P. Korhonen, P. Jouppila, and N. T. Kärki, Induction of aryl hydrocarbon hydroxylase is stimulated in human lymphocytes by mitogens and benz[*a*]anthracene, bons, *Life Sci.* **16**, 1403–1410 (1975).
32. J. P. Whitlock, H. L. Cooper, and H. V. Gelboin, Aryl hydrocarbon (benzopyrene) hydroxylase is stimulated in human lymphocytes by mitogens and benz[*a*]anthracene, *Science* **177**, 618–619 (1972).
33. D. L. Busbee, C. R. Shaw, and E. T. Cantrell, Aryl hydrocarbon hydroxylase induction in human lymphocytes, *Science* **178**, 315–317 (1972).
34. A. P. Alvares, A. Kappas, W. Levin, and A. H. Conney, Inducibility of benzo[*a*]pyrene hydroxylase in human skin by polycyclic hydrocarbons, *Clin. Pharmacol. Ther.* **14**, 30–40 (1973).
35. G. Kellermann, M. Luyten-Kellermann, and C. R. Shaw, Metabolism of polycyclic aromatic hydrocarbons in cultured human leukocytes under genetic control, *Humangenetik* **20**, 257–263 (1973).
36. P. Sims, Qualitative and quantitative studies on the metabolism of a series of aromatic hydrocarbons by rat-liver preparations, *Biochem. Pharmacol.* **19**, 795–818 (1970).
37. A. Borgen, H. Darvey, N. Castagnoli, T. T. Crocker, R. E. Rasmussen, and I. Y. Wang, Metabolic conversion of benzo[*a*]pyrene by Syrian hamster liver microsomes and binding of metabolites to deoxyribonucleic acid, *J. Med. Chem.* **16**, 502–506 (1973).

38. P. Sims and P. L. Grover, Epoxides of polycyclic aromatic hydrocarbons: Metabolism and carcinogenesis, *Adv. Cancer Res.* **20**, 166–274 (1974).
39. G. Takahaski and K. Yasuhira, Chromatographic analyses of 3-methylcholanthrene metabolism in adult and fetal mice and the occurrence of conjugating enzymes in the fetus, *Cancer Res.* **35**, 613–620 (1975).
40. A. Rane, *N*-oxidation of a tertiary amine (*N,N*-dimethylaniline) by human fetal liver microsomes, *Clin. Pharmacol. Ther.* **15**, 32–38 (1974).
41. O. Pelkonen and N. T. Kärki, Epoxidation of xenobiotics in the human fetus and placenta: A possible mechanism of drug-induced injuries, *Biochem. Pharmacol.* **24**, 1445–1448 (1975).

15

Discussion

In response to a question from Conney, Pelkonen replied that he thought that the primary function of human fetal P450 is to metabolize steroids. He went on to suggest that fetal P450 activity was not the result of exposure to environmental chemicals which reach the fetus through the placenta but is a fundamental component of the fetal liver. In contrast, the lack of P450 in the fetus of common laboratory animals suggests that they may not require the ability to metabolize steroids during fetal life.

Schulman reported that lipids extracted from the livers of neonatal rats but not those from adult rats inhibited drug metabolism in adult rat liver microsomes *in vitro.* These results may account for the well-known differences in drug metabolism between neonatal and adult rats.

In response to a question from Jakobsson, Pelkonen commented that (1) all fetuses that he has studied had some detectable aryl hydroxylase activity and (2) the level of P450 in the fetus did not seem to be related to the location in which the mother resided or her exposure to environmental chemicals. He also indicated that subhuman primates exhibited similar fetal enzyme activities.

Netter commented that previous reports that fetal tissues displayed little or no drug-metabolizing-enzyme activity differed from more recent findings such as those of Pelkonen because fetal tissues are now more readily available and can be more accurately studied. He then reported that near-term rabbit fetuses have been found to metabolize ethylmorphine. Furthermore, the low enzyme activities previously reported in many cases may indeed be quite significant in the fetus because they are required for fetal life.

Pelkonen stated that there is, however, a real species difference in fetal drug-metabolizing enzyme levels. For example, fetal liver enzyme activity is found much earlier in human than in laboratory animal fetal livers. He also indicated that the levels of conjugating enzyme in the fetus varied but in some cases were as high as 10% of the adult level.

Hildebrandt questioned whether the cortex or the medulla was used in studies of the fetal adrenal enzymes, and Pelkonen replied that the whole adrenal was used. Orrenius suggested that it would be hard to carry out comparisons between fetal and adult adrenals because during development the adrenals undergo reorganization. He also stated that the only animal known to metabolize drugs in the adrenal is the guinea pig. The human adrenal is not known to metabolize drugs.

Pelkonen, in reply to a question by Remmer, stated that epoxidation had been studied in only a few fetal livers and that no comparison could yet be made between the relative P450 content and the ability of fetal and adult livers to carry out epoxidation.

16

The Possible Role of Trout Liver Aryl Hydrocarbon Hydroxylase in Activating Aromatic Polycyclic Carcinogens

Jorma T. Ahokas, Olavi Pelkonen, and Niilo T. Kärki
Department of Pharmacology
University of Oulu
SF-90220 Oulu 22, Finland

Many compounds induce mutations and cancer only after they are converted to highly reactive intermediates by the organism (1). In recent years, attention has been focused on the cytochrome P450-linked monooxygenase system which hydroxylates polycyclic aromatic hydrocarbons via highly reactive epoxide intermediates. These epoxide intermediates in turn undergo enzymatic reactions mediated by epoxide hydrase or glutathione-*S*-epoxide transferase, or are non-enzymatically converted to various hydroxylated products. Epoxides also react irreversibly with cellular components (2).

Since trout are known to be very susceptible to chemical carcinogenesis (3), trout tissues should be able to form the reactive intermediates that cause cancer. In the present chapter, we describe the various metabolites of benzo[*a*]pyrene formed by trout liver microsomes. The relatively specific inhibitors of aryl hydrocarbon hydroxylase used in this study caused marked changes in the metabolite patterns, which reflect the various pathways for the formation of hydroxylated products.

MATERIALS AND METHODS

Liver microsomes from trout (*Salmo trutta lacustris*) were incubated with the indicated substrates and the metabolites were separated by thin-layer chromatography using the methods described earlier (4). Tritiated benzo[*a*] pyrene (generally labeled, Radiochemical Centre, Amersham, Bucks, England) was used. The inhibitors styrene oxide, 7,8-benzoflavone, and 1,2-epoxy-3,3,3-trichlorpropane were the best commercially available quality.

The authentic reference compounds (9,10-, 7,8-, and 4,5-diols of benzo[*a*]-pyrene, 3- and 9-hydroxybenzo[*a*]pyrene, 1,6-, 6,12-, and 3,6-quinones of benzo[*a*]pyrene) were obtained from H. V. Gelboin (Bethesda, Maryland). The UV-visible spectra of the metabolites were recorded by a Shimadzu MPS-50L multipurpose spectrophotometer.

RESULTS AND DISCUSSION

On TLC at least three dihydrodiols (9,10-, 7,8-, and 4,5-) of benzo[*a*]pyrene were detected (fractions I, II, and III, respectively) (Fig. 1). Fraction IV (Fig. 1) is 3- or 9-hydroxybenzo[*a*]pyrene. Furthermore, a considerable amount of radioactivity was localized in fractions V and VI, these being quinones of benzo[*a*]pyrene. Fraction VII is unchanged benzo[*a*]pyrene.

Fractions I, II, and IV were obtained in sufficient quantities to record their spectra (Figs. 2A,B,C, respectively) and the absorption maxima were compared with literature values (5–7), further supporting the identity of these fractions.

On the basis of the above results, it can be concluded that benzo[*a*]pyrene

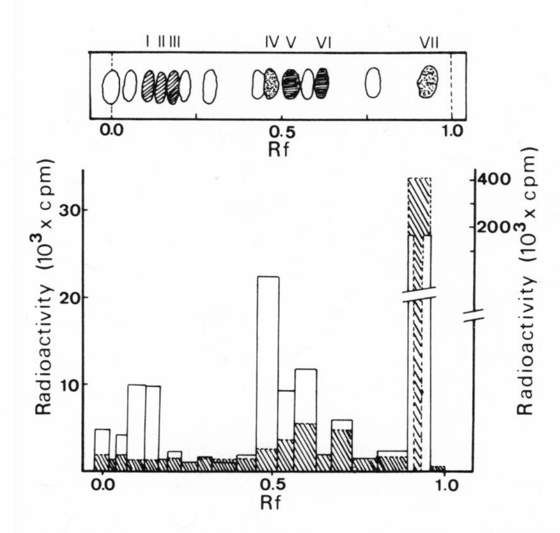

Fig. 1. Pattern of metabolites of benzo[a]pyrene as obtained after incubation with trout liver microsomes. TLC was carried out on silica gel G plates developed with benzene–ethanol (19:1, v/v). The upper trace is that seen under UV light and the lower is the distribution of radioactivity.

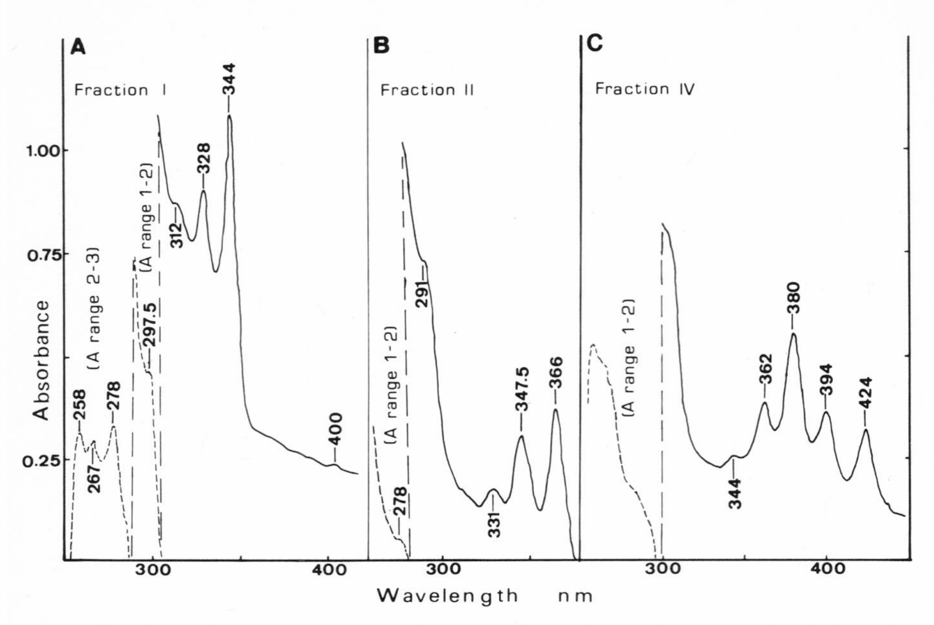

Fig. 2. Ultraviolet-visible spectra of the metabolites of benzo[a]pyrene eluted from the spots I (trace A), II (trace B), and IV (trace C).

metabolism is readily carried out by trout liver microsomes and that dihydrodiols form a major group of metabolites; monohydroxylated and quinone derivatives of benzo[*a*] pyrene are also formed.

To show indirectly that epoxide intermediates and their conversion to dihydrodiols by epoxide hydrase occur, inhibitors of epoxide hydrase were utilized. Styrene oxide, a good substrate of epoxide hydrase (8), very effectively inhibited the production of dihydrodiols (Fig. 3A). This was accompanied by an overall reduction in substrate utilization and a large increase in the production of monohydroxylated and quinone derivatives of benzo[*a*] pyrene.

When comparing the results of styrene oxide and 7,8-benzoflavone inhibition, it can be deduced that styrene oxide mainly inhibits the conversion of intermediates (epoxides) into the final products without greatly affecting the primary step of cytochrome P450-catalyzed epoxidation of the substrate. On the other hand, 7,8-benzoflavone, the inhibitor of the monooxygenase enzyme, caused an overall reduction of all metabolites of benzo[*a*] pyrene (Fig. 3B).

1,2-Epoxy-3,3,3-trichloropropane, although it is an inhibitor of epoxide hydrase (8), probably also inhibits the primary oxygenation step, as judged by its marked reduction of all metabolites (Fig. 3C).

It may be concluded that trout liver has the ability to form the reactive intermediates thought to be responsible for the carcinogenesis caused by polycyclic procarcinogens. Although the mechanisms of chemical carcinogenesis

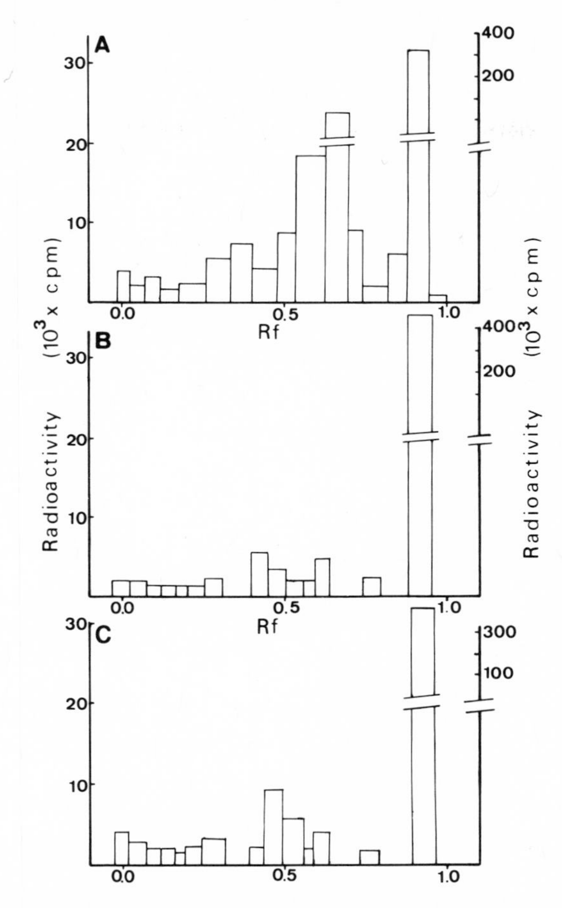

Fig. 3. (A) Pattern of benzo[a]pyrene metabolites as obtained in the presence of styrene oxide (7.5 mM). (B) Effect of 7,8-benzoflavone (0.1 mM) on the metabolite pattern. (C) Effect of 1,2-epoxy-3,3,3-trichloropropane (2.5 mM) on the metabolite pattern.

remain relatively unclear, the high incidence of hepatoma in trout may be the consequence of the highly active aryl hydrocarbon hydroxylase found in trout liver microsomes in these studies.

REFERENCES

1. J. A. Miller, Carcinogenesis by chemicals: An overview, *Cancer Res.* **30,** 559–576 (1970).
2. D. M. Jerina and J. W. Daly, Arene oxides: A new aspect of drug metabolism, *Science* **185,** 573–582 (1974).

3. R. H. Adamson, Drug metabolism in marine vertebrates, *Fed. Proc.* **26,** 1047–1055 (1967).
4. J. T. Ahokas, O. Pelkonen, and N. T. Kärki, Metabolism of polycyclic hydrocarbons by a highly active aryl hydrocarbon hydroxylase system in the liver of a trout species, *Biochem. Biophys. Res. Commun.* **63,** 635–641 (1975).
5. P. Sims, The metabolism of benzo[*a*]pyrene by rat-liver homogenates, *Biochem. Pharmacol.* **16,** 613–618 (1967).
6. P. Sims, Qualitative and quantitative studies on the metabolism of a series of aromatic hydrocarbons by rat-liver preparations, *Biochem. Pharmacol.* **19,** 795–818 (1970).
7. J. F. Waterfall and P. Sims, Epoxy derivatives of aromatic polycyclic hydrocarbons, *Biochem. J.* **128,** 265–278 (1972).
8. F. Oesch, Mammalian epoxide hydrases: Inducible enzymes catalysing the inactivation of carcinogenic and cytotoxic metabolites derived from aromatic and olefinic compounds, *Xenobiotica* **3,** 305–340 (1973).

17

Metabolic Activation of Methyldopa by Cytochrome P450-Generated Superoxide Anion

E. Dybing,* J. R. Mitchell, S. D. Nelson, and J. R. Gillette
Laboratory of Chemical Pharmacology
National Heart and Lung Institute
National Institutes of Health
Bethesda, Maryland 20014, U.S.A.

Renewed interest in the hepatic injury produced by methyldopa (MD) has been stimulated by recent reports that the antihypertensive drug may initiate chronic active liver disease, occasionally with a fatal outcome (1–3). The hepatic damage has been attributed to hypersensitivity rather than to direct toxicity, but careful review reveals that the syndrome is similar to that produced by isoniazid (4). Most individuals fail to show constitutional features indicative of an allergic response and usually demonstrate hepatic injury upon rechallenge only after lengthy reexposure to MD. Moreover, MD produces mild, clinically covert hepatic injury in more than 15% of recipients when liver function tests are monitored (5), and thus the injury is not restricted to rare, idiosyncratic individuals. To determine if the toxicity produced by MD might be due to a reactive intermediate, [^{3}H] MD was incubated with liver microsomes.

RESULTS

A large amount of covalent binding of [^{3}H] MD to protein occurred when rat, mouse, or human microsomes were incubated in the presence of NADPH and O_2 (Fig. 1). This binding was inhibited by a $CO:O_2$ atmosphere (9:1) and

* Present address: Department of Environmental Toxicology, National Institute of Public Health, Oslo 1, Norway.

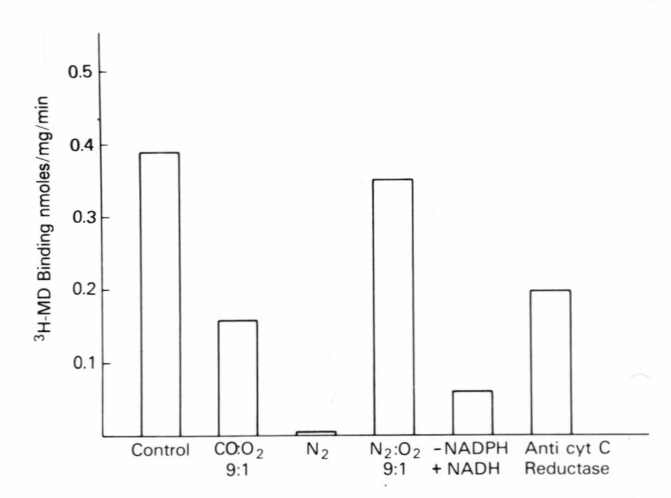

Fig. 1. Cytochrome P450-mediated covalent binding of [³H]MD to rat microsomal protein in vitro.

an antibody against NADPH cytochrome *c* reductase (Fig. 1), whereas SKF 525-A was without effect (Fig. 3). MD did not show cytochrome P450 binding spectra. Pretreatment of animals with cobaltous chloride and piperonyl butoxide decreased MD binding (Fig. 2), but pretreatments with inducers did not increase binding.

Superoxide dismutase, which scavenges superoxide anions (6), totally blocked MD binding (Fig. 3). A xanthine oxidase system (producing superoxide anions) (7) also activated MD. This binding again was blocked by superoxide dismutase. Catalase and benzoate were without effect (Fig. 3). Ascorbic acid, which reduces superoxide (8), semiquinone radicals, and quinones, totally inhibited MD binding (Fig. 3). Ethylenediamine, which scavenges quinones (9), partially blocked binding. Glutathione also inhibited MD binding (Fig. 3), forming a stable conjugate with MD.

MD was also activated by a mushroom tyrosinase system (phenol oxidase). This reaction was not altered by superoxide dismustase, but was inhibited by ascorbic acid, ethylenediamine, and glutathione. A glutathione–MD conjugate formed in this system had the same chromatographic characteristics as that produced by microsomes.

Other catechols were found to inhibit MD binding (Fig. 4), whereas the substituted 3-*O*-methyldopa did not. Radiolabeled catechols all show NADPH-dependent as well as xanthine oxidase-mediated binding to microsomal protein, both reactions being blocked by superoxide dismutase.

The microsomal NADPH-dependent and xanthine oxidase-mediated activation of 2-hydroxyestradiol was also found to be inhibited by superoxide dismutase (Table 1).

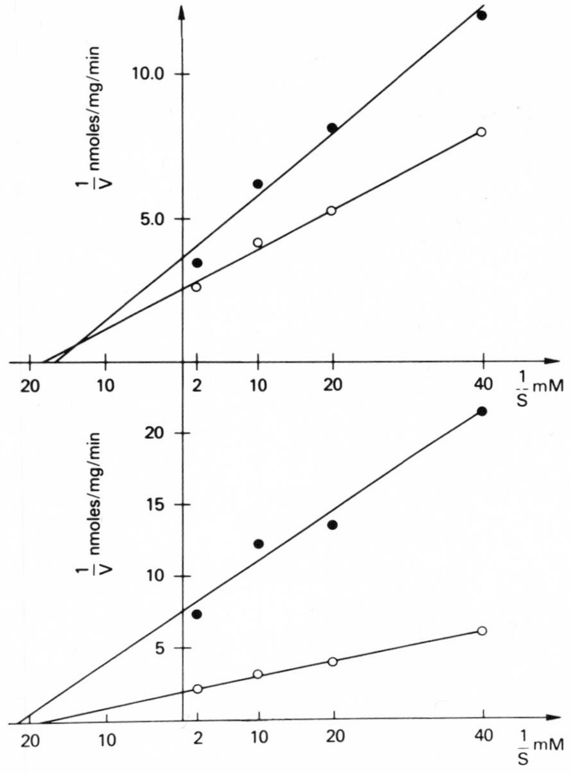

Fig. 2. Effect of piperonyl butoxide pretreatment in mice (upper) and cobaltous chloride pretreatment in rats (lower) on kinetics of [^3H]MD binding to microsomal protein.

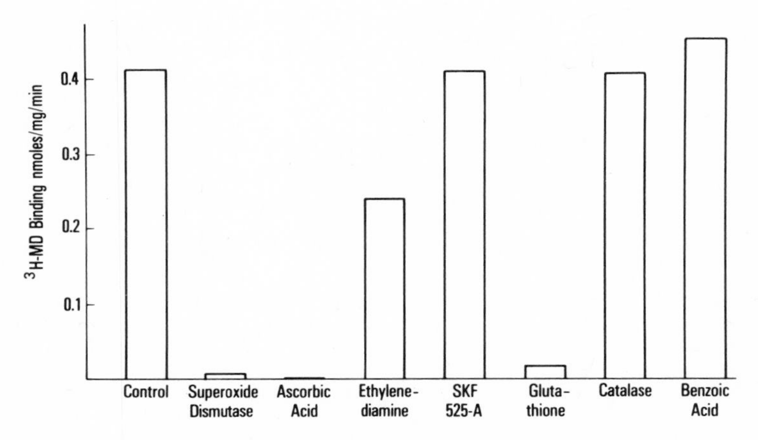

Fig. 3. Conditions for microsomal binding of [^3H]MD.

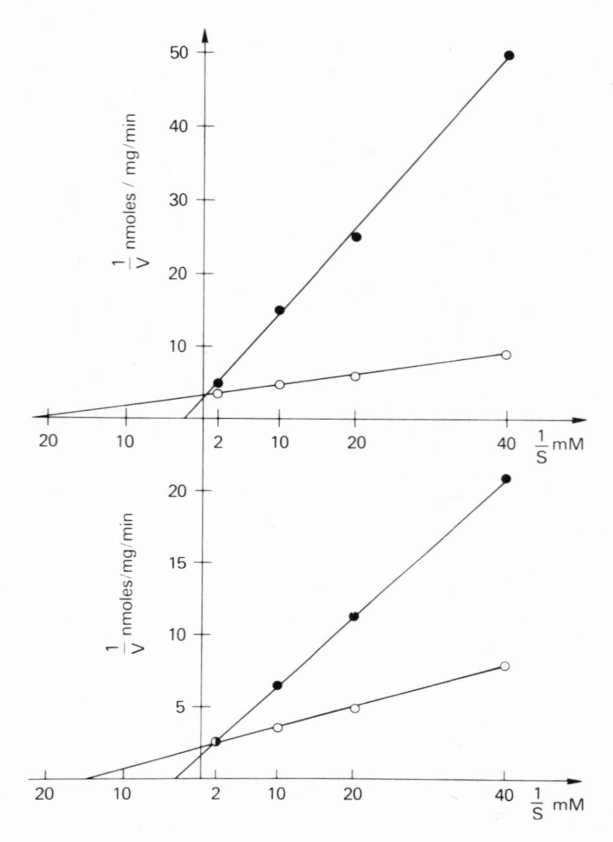

Fig. 4. Effect of dopamine addition (upper) and dopa addition (lower) on kinetics of [^{3}H]MD binding to mouse and rat microsomal protein, respectively.

Table 1. Effect of Superoxide Dismutase on NADPH- and Xanthine Oxidase-Dependent Binding of 2-Hydroxyestradiol and Estradiol in Rat Liver Microsomes

Substrate	NADPH (nmol/mg/min)	NADPH + dismutase (nmol/mg/min)	Xanthine oxidase (nmol/mg/min)	Xanthine oxidase + dismutase nmol/mg/min)
2-OH-Estradiol	0.74	0.54	0.81	0.00
Estradiol	0.65	0.53	0.02	0.01

DISCUSSION

The results presented strongly suggest that superoxide anion radicals can be formed by hepatic microsomal cytochrome P450, leading to the activation of MD. Since MD did not produce binding spectra with cytochrome P450 and SKF 525A was without effect on the activation of MD, this indicates that univalent reduction of oxygen by cytochrome P450 is necessary for irreversible binding of MD, but interaction with the cytochrome P450 system in a manner typical of most substrates is not necessary.

Only suggestive information on the nature of the reactive MD intermediates is available. An explanation consistent with the results presented is the oxidation of the catechol nucleus to its semiquinone free radical by superoxide. The radical can react with tissue macromolecules directly or be further oxidized to an electrophilic quinone which reacts.

If the reaction mechanism presented for MD activation plays a role in the hepatotoxicity observed in MD users, the liver injury could result directly from covalent binding of the reactive MD intermediates to vital macromolecules or might result from peroxidative tissue attack.

The results from experiments using 2-hydroxylated estrogens show that the formation of superoxide is also involved in the *in vitro* activation of these catechols (10, 11).

REFERENCES

1. A. M. Hoyumpa and A. M. Connell, Methyldopa hepatitis: Report of three cases, *Digestive Dis.* **18**, 213–222 (1973).
2. G. B. Goldstein, K. C. Lam, and S. P. Mistilic, Drug-induced active chronic hepatitis, *Digestive Dis.* **18**, 177–184 (1973).
3. I. L. Schweitzer and R. L. Peters, Acute submassive hepatic necrosis due to methyldopa: A case demonstrating possible initiation of chronic liver disease, *Gastroenterology* **66**, 1203–1211 (1974).
4. J. R. Mitchell, S. D. Nelson, S. S. Thorgeirsson, R. J. McMurtry, and E. Dybing, in: *Progress in Liver Disease,* Vol. 4 (H. Popper and F. Schaffner, eds.), pp. 259–279, Grune & Stratton, New York, 1976.
5. S. G. Elkington, W. M. Schreiber, and H. O. Conn, Hepatic injury caused by 1-alpha-methyldopa, *Circulation* **40,** 589–595 (1969).
6. J. M. McCord and I. Fridovich, Superoxide dismutase. An enzymic function for erythrocuprein (hemocuprein), *J. Biol. Chem.* **244,** 6049–6055 (1969).
7. I. Fridovich and P. Handler, Xanthine oxidase. IV. Participation of iron in internal electron transport, *J. Biol. Chem.* **233,** 1581–1585 (1958).
8. M. Nishikimi, Oxidation of ascorbic acid with superoxide anion generated by the xanthine–xanthine oxidase system, *Biochem. Biophys. Res. Commun.* **63,** 463–468 (1975).

9. P. H. Jellinck and L. Irwin, Interaction of oestrogen quinones with ethylenediamine, *Biochim. Biophys. Acta* **78**, 778–780 (1963).
10. F. Marks and E. Hecker, Metabolism and mechanism of action of oestrogens. VII. Structure and mechanism of formation of water-soluble and protein-bound metabolites of oestrone in rat-liver microsomes *in vitro* and *in vivo, Biochim. Biophys. Acta* **187**, 250–265 (1969).
11. H. Kappus, H. M. Bolt, and H. Remmer, Irreversible protein binding of metabolites of ethynylestradiol *in vivo* and *in vitro, Steroids* **22**, 203–225 (1973).

18

Formation of Tertiary Amine *N*-Oxide and Its Reduction by Rat Liver Microsomes

Ryuichi Kato, Hideyo Noguchi, Masahiko Sugiura, and Kazuhide Iwasaki
Research Laboratories
Fujisawa Pharmaceutical Co., Ltd.
Kashima, Yodogawa-ku
Osaka, Japan 532

Tertiary amines are common components of drugs and are readily converted to *N*-oxide by liver microsomes in the presence of NADPH and oxygen (1, 2). The *N*-oxidation of tertiary amines is catalyzed by a flavoprotein enzyme and not by cytochrome P450. On the other hand, as reported in the previous paper, the reduction of tiaramide *N*-oxide in the rat liver microsomes is catalyzed by a reduced form of cytochrome P450 (3).

In the present chapter, we wish to describe the characteristics and mechanism of *N*-oxide reduction by rat liver microsomes using imipramine *N*-oxide (IPNO), tiaramide *N*-oxide (TRNO), and *N,N*-dimethylaniline *N*-oxide (DMANO).

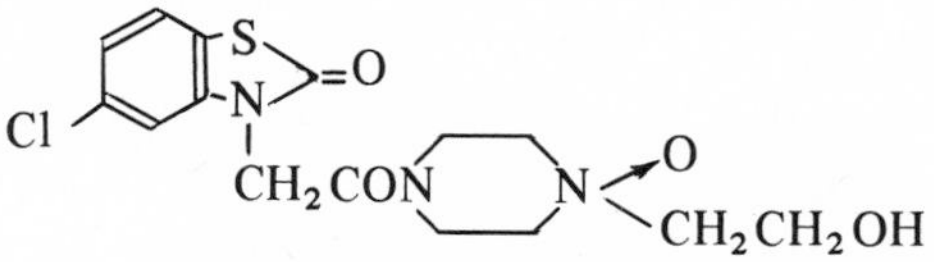

TRNO

Male and female rats (7 weeks old) of the Sprague–Dawley strain were used. The livers were perfused with an isotonic KCl solution to obtain microsomes essentially free from hemoglobin contamination. The standard incubation mixture (5 ml) consisted of 0.1 M phosphate buffer (pH 7.4), 2 mM NADP, 10 mM $MgCl_2$, 10 mM glucose-6-phosphate, 3.0 I.U. glucose-6-phosphate dehydrogenase, the microsomal suspension equivalent to 1.0 g liver, and 1.0 mM tertiary

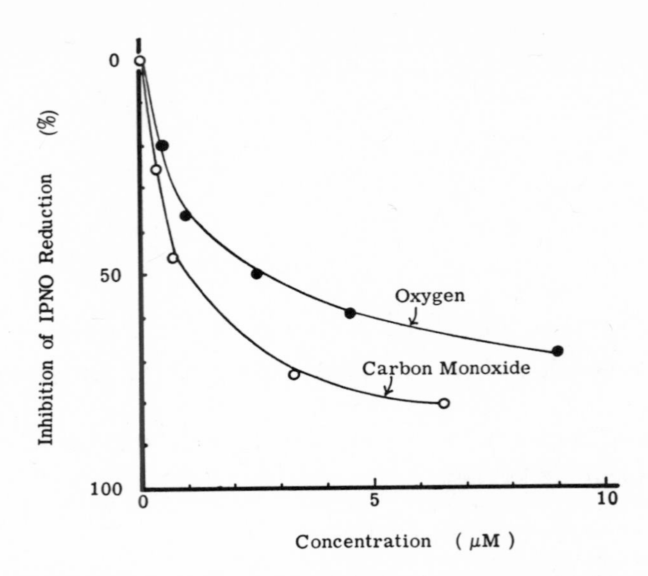

Fig. 1. Inhibition of tertiary amine N-oxide reduction by oxygen and carbon monoxide.

amine *N*-oxide. The incubation was carried out anaerobically at 37°C for 10 min.

The activity of tertiary amine *N*-oxide reductase is located mainly in the microsomal fraction of rat liver, some activity also being found in the mitochondria and negligible activity in the supernatant fraction. The *N*-oxide reductase activity is NADPH dependent, but 50–70% of the total activity is obtained using NADH as cofactor instead of NADPH.

A study of the distribution of *N*-oxide reductase activity among various tissues revealed a close relationship between the *N*-oxide reductase activity and cytochrome P450 levels.

The activity of *N*-oxide reductase is inhibited by both oxygen and carbon monoxide. In an atmosphere of pure oxygen or carbon monoxide, *N*-oxide reductase is inhibited about 85% and 95%, respectively. The concentrations which produce 50% inhibition of imipramine *N*-oxide reductase activity are about 2 μM and 1 μM, respectively, for oxygen and carbon monoxide (Fig. 1). The inhibition of *N*-oxide reductase activity under an atmosphere of 0.5% carbon monoxide was almost completely reversed by exposure to white light (Table 1)(4).

While TRNO, IPNO, and DMANO do not produce any significant spectral change with oxidized cytochrome P450, they do react with reduced cytochrome P450 to produce characteristic spectral changes as reported in the previous paper (3). The intensity of absorption at 450 nm produced by carbon monoxide is decreased by addition of tiaramide *N*-oxide, while absorption at 424 nm pro-

Table 1. Inhibition of Tertiary Amine *N*-Oxide Reductase by Carbon Monoxide and Its Light Reversal

Gas phase	Light exposure	Tertiary amine *N*-oxide reduction IPNO	TRNO (nmol/mg/min)	DMANO
N_2	–	2.08(100%)	2.03(100%)	2.28(100%)
N_2–CO (99.5:0.5)	–	1.05(50%)	1.04(51%)	1.28(56%)
N_2–CO (99.5:0.5)	+	1.87(90%)	1.90(94%)	2.10(92%)

duced by tiaramide *N*-oxide is markedly diminished by carbon monoxide. In addition, type II compounds such as *n*-octylamine and DPEA, which are assumed to interact with cytochrome P450 directly in its heme region, decrease tertiary amine *N*-oxide reductase activity in accord with the decrease in the reduction rate of cytochrome P450. These results support the view that the tertiary amine *N*-oxide combines with the reduced form of cytochrome P450 and accepts electrons directly from ferrous iron throughout the reaction.

Stoichiometric studies clearly show that 1 mol of NADPH is oxidized for each mole of *N*-oxide reduced (4).

While the activity of tertiary amine reductase is very low in newborn rats, it increases progressively with maturation, and the activity in male rats is clearly higher than that in female rats. The administration of phenobarbital markedly increases the reductase activity, while methylcholanthrene produces no significant change. The administration of cobaltous chloride, by decreasing cytochrome P450, decreases the reduction of *N*-oxide but increases the formation of tertiary amine *N*-oxide.

Table 2. Effect of Protease Digestion on Microsomal NADPH-Dependent Tertiary Amine *N*-Oxide Reductase Activity[a]

Treatment		NADPH-cyt *c* reductase (nmol/mg/min)	Cyt b_5 (nmol/mg)	Cyt P450 (nmol/mg)	IPNO reduction rate (nmol/mg/min)
Trypsin	Digested	141	0.32	0.81	0.29
	Supernatant	659	0.13	0.00	0.00
	Pellet	13	0.47	1.36	0.34
Subtilisin	Digested	140	0.36	0.69	0.20
	Supernatant	463	0.81	0.00	0.00
	Pellet	10	0.03	1.02	0.29

[a]Liver microsomes (10 mg protein/ml) were incubated at 0°C with 25 μg trypsin or subtilisin per milligram of protein for 15 h anaerobically in 0.05 M sodium buffer (pH 7.4) containing 25% glycerol.

Table 3. Comparison Between *N*-Oxide Formation and *N*-Oxide Reduction of Tertiary Amine by Rat Liver Microsomes

Item	Formation	Reduction
Substrate	RR′R″N	RR′R″N→O
Oxygen	Requirement	Inhibition
CO inhibition	–	$K_i = 6 \times 10^{-7}$ M Light reversed
Coenzyme requirement	NADPH > NADH	NADPH > NADH
Phenobarbital induction	+	++
Cobalt treatment	Stimulation	Inhibition
Sex	Male > Female	Male > Female
Age	Mature > Newborn	Mature > Newborn
FMN, FAD	–	Stimulation
Inhibition by		
DPEA	–	++
SKF 525A	–	–
1-(1-Naphthyl)-2-thiourea	++	–
Detergent treatment	Resistant	Sensitive
Enzyme	Flavoprotein oxidase	P450-linked reductase

Moreover, the results given in Table 2 show that the *N*-oxide reductase activity is associated with cytochrome P450 whereas NADPH–cytochrome *c* reductase and/or cytochrome b_5 cannot reduce tertiary amine *N*-oxide in the absence of cytochrome P450.

The addition of FMN, FAD, and riboflavin enhances the reduction of *N*-oxide about 5 times and this enhancement is almost completely inhibited by carbon monoxide. Moreover, the addition of methylviologen causes a rapid reduction of tertiary amine *N*-oxide in the presence of cytochrome P450, and carbon monoxide inhibits *N*-oxide reduction by this system.

The comparison between *N*-oxide formation and *N*-oxide reduction by rat liver microsomes is given in Table 3.

After the administration of tiaramide *N*-oxide to rats, a large amount of tiaramide is found in the liver and other tissues. These results indicate that under conditions prevailing *in vivo* tertiary amines may be oxidized to tertiary amine *N*-oxides, reduced again to tertiary amines, and recycled until they are excreted or undergo further metabolic changes. Thus some factors, which only decrease the tertiary amine *N*-oxide reductase activity without affecting the *N*-oxidase activity, will increase the amount of *N*-oxide accumulating *in vitro* and *in vivo*. These results therefore should be taken into consideration in the evaluation of the metabolism and toxicity of tertiary amines.

REFERENCES

1. M. H. Bikel, The pharmacology and biochemistry of *N*-oxides, *Pharmacol. Rev.* **88**, 325–355 (1969).
2. D. M. Ziegler, E. M. McKee, and L. L. Poulsen, Microsomal flavoprotein-catalyzed *N*-oxidation of arylamines, *Drug Metab. Dispos.* **1**, 314–321 (1973).
3. M. Sugiura, K. Iwasaki, H. Noguchi, and R. Kato, Evidence for the involvement of cytochrome P-450 in tiaramide *N*-oxide reduction, *Life Sci.* **15**, 1433–1442 (1974).
4. M. Sugiura, K. Iwasaki, and R. Kato, Reduction of tertiary amine *N*-oxide by liver microsomal cytochrome P-450, *Mol. Pharmacol.* **12**, 322–334 (1975).

III

Inactivation of Reactive Intermediates

19

Epoxide Hydratase: Purification to Apparent Homogeneity as a Specific Probe for the Relative Importance of Epoxides among Other Reactive Metabolites

F. Oesch, P. Bentley, and H. R. Glatt

Section of Biochemical Pharmacology of the University of Mainz
Obere Zahlbacher Strasse 67
D-65 Mainz, German Federal Republic

Aromatic and olefinic compounds can be metabolized by microsomal monooxygenases to epoxides which chemically represent electrophilic species (for reviews, see refs. 1–5). Spontaneous binding of such epoxides to DNA, RNA, and protein has been observed (6–10). Accordingly, such metabolites have been suggested and, in some instances, shown to disturb the normal functions of cells, leading to such effects as mutagenesis (11–14), malignant transformation (15–19), or cell necrosis (20). However, aromatic and olefinic compounds are biotransformed to a vast array of metabolites (*cf.* refs. 21–27), possibly including a considerable number of reactive metabolites other than epoxides. The relative importance of epoxides among other reactive metabolites is at present unknown. With respect to the model compound used in this study, benzo[*a*]pyrene, our previous studies had shown that the 4,5- (K-region-) epoxide metabolite was a potent mutagen for the frameshift-sensitive *Salmonella* strains TA 1537 and TA 1538 (28), that the premutagenic hydrocarbon required a NADPH-supported microsomal monooxygenase system to become mutagenically active, and that the mutagenic response was potentiated by the presence of epoxide hydratase inhibitors at concentrations where no interference with other systems has been observed (28). Yet no conclusion could be reached whether the relative contribution of epoxide metabolites to the overall mutagenic effect of bioactivated benzo[*a*]pyrene was of any significance since the potentiation of the mutagenic effect by epoxide hydratase inhibitors could simply mean that blocking this pathway led to an accumulation of epoxides, making them important in this situation, while in absence of such inhibitors their contribution to the overall mutagenic effect may have been negligible. To probe the relative importance of epoxides among other reactive metabolites, a reagent specifically inactivating epoxides should be used, ideally an epoxide-inactivating enzyme in homogeneous form. We therefore set out to purify epoxide hydratase (2), an enzyme localized in the same subcellular fraction in which epoxides are formed, the microsomes. In this chapter we wish to report the purification of epoxide hydratase to apparent homogeneity, the properties of the apparently homogeneous enzyme, its use in probing the relative importance of epoxides among other reactive metabolites, and its consequences with respect to clinically used drugs.

PURIFICATION OF EPOXIDE HYDRATASE TO APPARENT HOMOGENEITY

Table 1 summarizes the purification procedures. Solubilization of epoxide hydratase from male Sprague-Dawley rat (200–250 g) liver microsomes was effected by resuspension in 10 mM sodium phosphate buffer (pH 7.0) containing 1% Cutscum, a nonionic detergent. This agent had previously been found effective in solubilizing epoxide hydratase from guinea pig (29) and human (30)

Table 1. Purification of Rat Liver Epoxide Hydratase

Purification step	Volume (ml)	Total protein (mg)	Total units	Specific activity[a]	Relative purification	Yield (%)
10,000*g* supernatant (fraction 1)	1660	49,800	82,557	1.67	1	100
Solubilized microsomes (fraction 2)	1660	11,454	102,107	8.91	5.3	124
Ammoniun sulfate precipitate (fraction 3)	360	5,184	99,270	19.15	11.5	120
DEAE-cellulose effluent (fraction 4)	500	740	69,300	93.82	56	84
Concentrated cellulose phosphate effluent (fraction (5)	6	87	19,300	260	156	23
Final preparation	5.4	14	8,400	600	360	10.2

[a]Specific activity expressed as units/mg protein. One unit is defined as that amount of enzyme producing 1 nmol styrene glycol/min under the assay conditions used (30), specifically in absence of Tween 80.

liver microsomes. Ammonium sulfate was then added to a final concentration of 140 g/liter and the precipitate was dissolved in 5 mM sodium phosphate buffer (pH 7.0), dialyzed against the same buffer, and applied to a DEAE-cellulose column that had been equilibrated with the same buffer. In contrast to much of the contaminating material, epoxide hydratase was not retained on the column, which led to a highly purified epoxide hydratase at a relatively high yield. The active fractions were then applied to a cellulose phosphate column which had been equilibrated with the same buffer. The detergent containing a minor portion of epoxide hydratase was not retarded by the column, while the following fractions which were eluted with 5 and 50 mM sodium phosphate buffer (pH 7.0) contained no detectable epoxide hydratase. Epoxide hydratase was then eluted with 50 mM sodium phosphate buffer (pH 7.0) containing 0.5 M NaCl. The fact that a portion of epoxide hydratase was not retarded by this column was not due to overloading since the same portion was not retarded when reapplied to another phosphocellulose column. The observation that the ratio of specific epoxide hydratase activity toward two different substrates, styrene-7,8-oxide and benzo[*a*]pyrene-4,5-oxide, was the same in both fractions did not support the assumption of the presence of two different epoxide hydratases in these two fractions. Indeed, two peaks of enzyme activity eluting from an ion-exchange column do not necessarily imply the presence of two different enzyme proteins since the charged groups of the same enzyme molecule could be masked by nonspecific association with other proteins or lipids, thereby causing different chromatographic properties. On the other hand, the possibility that two different epoxide hydratases are present in the two fractions is not excluded by the observation of identical ratios of specific activities in these two fractions with respect to the investigated substrates only. This point requires further investigation.

Since the hydrophobic character of the enzyme rendered the removal of contaminating hydrophobic proteins difficult, an attempt was made to exploit the relative hydrophobicities of the components of the impure preparation. Hydrophobic arms of different size and geometry were coupled to Sepharose 4B using the method of Er-el *et al.* (31) with some modifications. Selective adsorption of epoxide hydratase followed by elution with low concentrations of detergent (0.05% Cutscum in 5 mM sodium phosphate buffer, pH 7.0) proved most satisfactory when using *n*-butyl residues as hydrophobic arms. The detergent was then removed by chromatography on a small phosphocellulose column. This removal of the detergent at two different stages during purification was necessary since epoxide hydratase was not amenable to hydrophobic chromatography in presence of the detergent. The final preparation was then dialyzed against 50 mM sodium phosphate buffer (pH 7.0) prior to use or storage at 1–5°C.

The final preparations typically had 520–600 units per milligram of protein while the overall yield of epoxide hydratase activity varied between 7 and 10%.

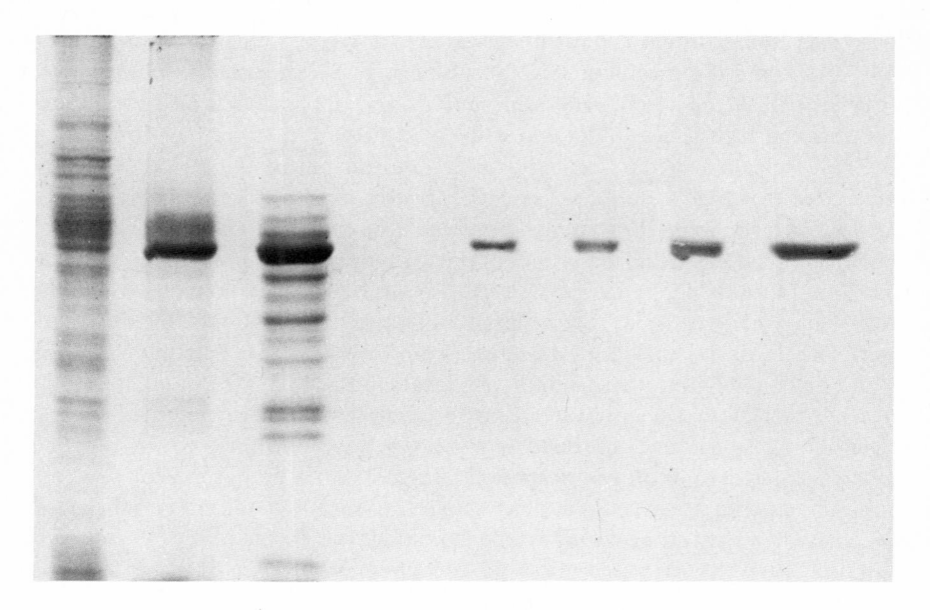

Fig. 1. SDS-gel electrophoresis of various fractions during purification of rat liver epoxide hydratase. Gel slabs (2 mm thick) containing 10% acrylamide and 0.1% SDS were used. From left to right the fractions are solubilized microsomes (20 μg), DEAE-cellulose effluent (20 μg), phosphocellulose effluent (20 μg), and the final preparation (0.5 μg, 1 μg, 2 μg, 5 μg).

During gel electrophoresis in 10% polyacrylamide gels, the enzyme did not migrate into the separating gel in the absence of SDS, while in the presence of SDS one single band was observed (Fig. 1). Similarly, in the analytical ultracentrifuge a single band was observed in the absence of SDS with a high s_{20w} value of 14.5, while in the presence of 0.2% SDS a single symmetrical peak with a s_{20w} value of 3 was observed (protein concentration 2.5 mg/ml). Double-diffusion tests of various concentrations of the final preparation and antiserum raised against the same preparation in New Zealand White rabbits showed one single precipitation line both before (data not shown) and after staining with Coomasie blue (Fig. 2). Thus the preparation appears homogeneous by SDS-electrophoretic, analytical, ultracentrifugal, and immunological criteria.

PROPERTIES OF THE APPARENTLY HOMOGENEOUS EPOXIDE HYDRATASE

The minimum molecular weight of epoxide hydratase was estimated by SDS-gel electrophoresis (32) on 10% polyacrylamide gels in the presence of standard proteins of known molecular weight (Fig. 3). In separate experiments using two different enzyme preparations (with specific activities of 490 and 596

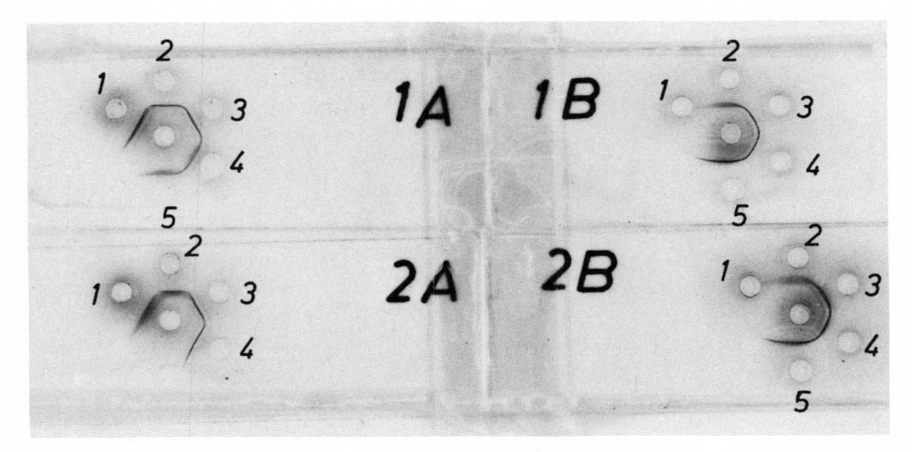

Fig. 2. Pure epoxide hydratase: Ouchterlony double diffusion analysis. 1: Enzyme in the absence of the detergent, Cutscum. (1A) The center well contained 1/2 dilution of antiserum. Wells 1–5 contained pure epoxide hydratase in the following amounts: 7.5, 3.25, 2.5, 1.25, and 0.8 μg. (1B) The center well contained 7.5 μg pure epoxide hydratase. Well 1 contained control serum, wells 2–5 contained antiserum at the following dilutions: 1/8, 1/4, 1/2, undiluted. 2: Enzyme in the presence of the detergent, Cutscum. (2A) The center well contained 1/2 dilution of antiserum. Wells 1–4 contained pure epoxide hydratase in the following amounts: 1.30, 0.65, 0.43, and 0.22 μg. (2B) The center well contained 0.65 μg pure epoxide hydratase. Well 1 contained control serum. Wells 2–5 contained antiserum at the following dilutions: 1/8, 1/4, 1/2, undiluted.

units/mg protein), values for the molecular weight of 49,500 and 48,500 were obtained. This indicates that the minimum molecular weight of the epoxide hydratase is 49,000, which is in agreement with the order of magnitude suggested by the sedimentation coefficient measured in the presence of 0.2% SDS of s_{20w} = 3. In the absence of detergent, the enzyme aggregated, forming a very

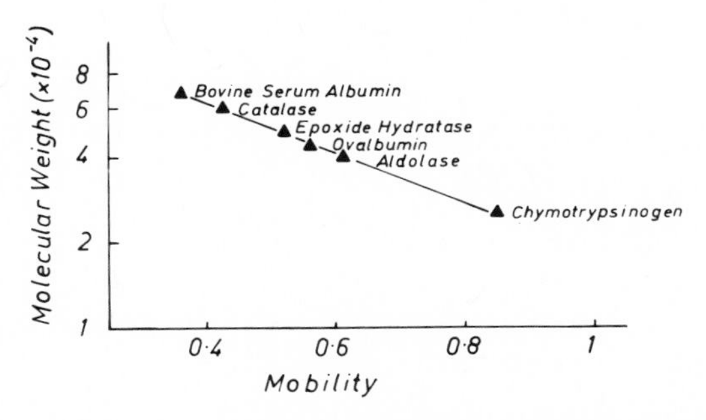

Fig. 3. Molecular weight determination by SDS-polyacrylamide gel electrophoresis. Gel slabs (2 mm thick) containing 10% acrylamide and 0.1% SDS were used. Four micrograms of each protein was applied to the same gel, which was dried after staining. Mobilities were measured on the dried gel. Molecular weights of the standard proteins used are taken from Weber and Osborn (32).

Table 2. Amino Acid Analysis of Epoxide Hydratase

Residue	Residues per monomer					
	Hydrolysis time		Values from			
	24 h	144 h	Extrapolation	Other methods	Integral values	Weight contribution
Lysine	33.7	33.7			34	4352
Histidine	14.1	13.9			14	1932
NH_3	38	46.4	35.6		(<35)	37
Arginine	20	19.4			20	3120
Aspartate (Asx)	30	30			30	3450
Threonine	20.2	19.6	20.3[a]		20	2020
Serine	33.2	31.0	33.6[a]		34	2958
Glutamate (Glx)	45	45			45	5805
Proline	26.8	25.4			26	2522
Glycine	35.1	35.3			35	1995
Alanine	18.9	18.9			19	1349
Cysteine/2	3.33	2	5.0[c]	4.3[a]	5	510
Valine	21.2	24[b]			24	2376
Methionine	12.4	10.9	9[c]		12	1572
Isoleucine	20.5	22.9[b]			23	2599
Leucine	43.7	43.8			44	4972
Tyrosine	23.4	23.6	24[a]	22[d]	23	3912
Phenylalanine	26.8	27.3			27	3969
Tryptophan			8.7[e]	7.5[d]	9	1674
				Total	444	50,950

[a]Subject to time-dependent degradation. Value obtained from extrapolation to zero hydrolysis time.
[b]Time-dependent increase, results based on the value after the longest hydrolysis time.
[c]Determined by performic acid oxidation.
[d]Determined spectrophotometrically.
[e]Determined with 5% thioglycolic acid.

high molecular weight oligomer which sedimented with a sedimentation coefficient of s_{20w} = 14.5 and did not migrate into 10% acrylamide gels during electrophoresis. The functional molecular weight, i.e., the minimum molecular weight at which the enzyme is still active, is unknown.

Table 2 shows the amino acid composition of the pure epoxide hydratase. The hydrophilic amino acids Lys, His, Arg, Asp, Asn, Thr, Ser, Glu, and Gln (33) account for only 44% of the sum of amino acid residues of the protein. This hydrophobic character may account for the aggregation of the protein in the absence of SDS reported above. The partial specific volume calculated from the amino acid analysis is 0.732 ml/g at 25°C.

No *N*-terminal amino acid could be detected as the [^{14}C]dinitrofluoroben-

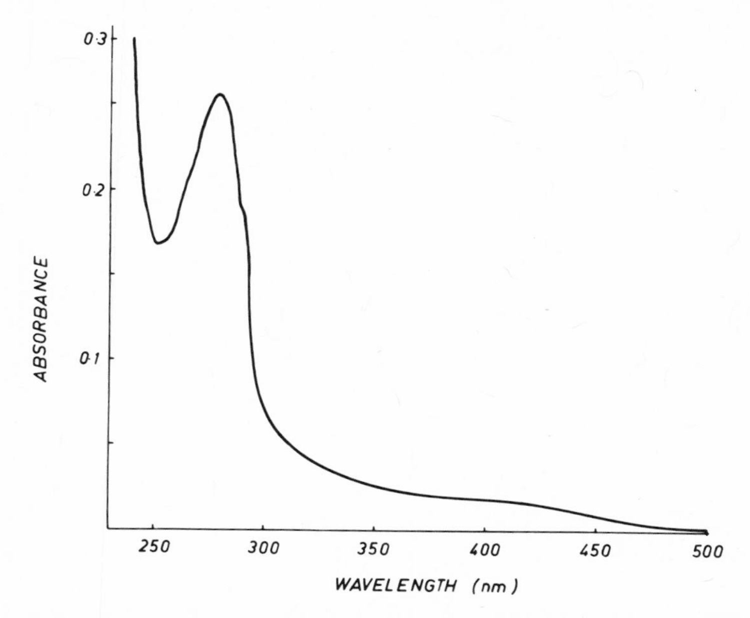

Fig. 4. Absorption spectrum of pure epoxide hydratase. The spectrum was recorded using a Beckman 25 spectrophotometer at 25°C. The enzyme was dissolved in filtered 5 mM sodium phosphate buffer. The protein concentration was 0.14 mg/ml.

zene derivative. Free *C*-terminal amino acids could not be detected by hydrolysis using a mixture of carboxypeptidases A and B or by hydrazinolysis (34). However, since hydrolysis by carboxypeptidases would not release *C*-terminal amino acids adjacent to a proline residue and hydrazinolysis would not detect *C*-terminal basic amino acids or amides, the latter observation does not exclude the presence of free *C*-terminal amino acids.

The absorption spectrum of epoxide hydratase (Fig. 4) showed between 235 and 500 nm one single absorption maximum at 280 nm. A very weak absorption band gradually decreasing toward higher wavelengths was observed between 300 and 500 nm. Accordingly, the preparation appeared very slightly yellow at protein concentrations between 2 and 3 mg/ml. At 290 nm a shoulder was observed in the spectrum, which is indicative of a high tryptophan content, in agreement with the results of the amino acid analysis shown in Table 2. The amounts of tryptophan and tryosine as determined spectrophotometrically were lower than those obtained chemically by 14% and 8%, respectively. This indicates that the preparation is essentially free of the detergent, Cutscum, which absorbs very strongly at the same wavelengths. Moreover, the ratio of the absorption at 280 nm compared to that at 260 nm was 1.46, confirming the detergent-free character of the enzyme, since Cutscum absorbs very strongly between 270 nm and 285 nm (λ_{max} = 278 nm).

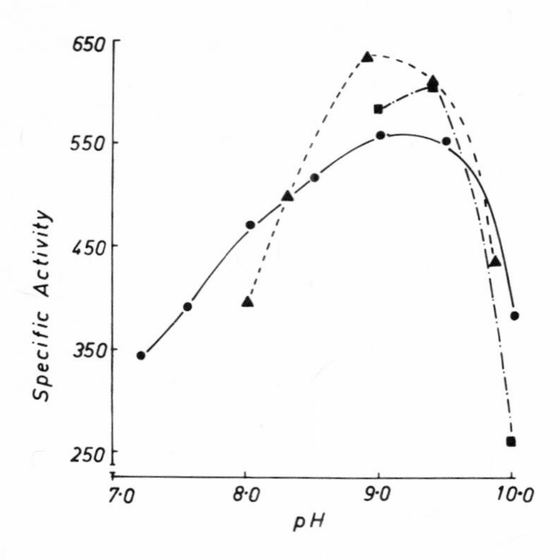

Fig. 5. Activity of the pure epoxide hydratase as a function of the pH. 0.125 M tris (▲), sodium phosphate $I = 0.1$ (●), or 0.05 M glycine (■) was used as buffer. Phosphate solutions above pH 8.5 contained 0.005 M glycine buffer at the appropriate pH and sufficient Na_2HPO_4 solution (pH adjusted with NaOH) to produce the desired ionic strength. Assays were with 2 mM styrene oxide at 37°C for 10 min. Specific activity: units/mg protein.

Figure 5 shows the pH optimum of the pure epoxide hydratase with 2 mM styrene oxide as a substrate. The optimum was at pH 8.9 in tris buffer, 9.0 in phosphate buffer, and somewhat higher in glycine buffer, 9.4. The activity did not appear to depend on the ionic strength of the medium. Thus the activity was not significantly different when the ionic strength of the tris buffer (pH 8.9) was varied between 0.025 M and 0.375 M.

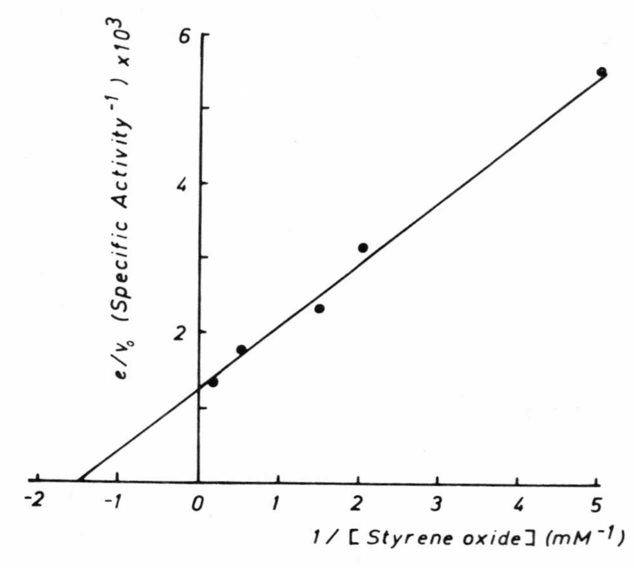

Fig. 6. Lineweaver–Burk analysis. Assays were in 0.125 M tris (pH 8.9) at 37°C at two different protein concentrations for each concentration of substrate. Specific activity: units/mg protein.

A Lineweaver–Burk analysis (Fig. 6) of glycol formation with the pure epoxide hydratase using different concentrations of substrate, styrene oxide, revealed an apparent K_m of 0.67 mM and V_{max} of 800 nmol styrene glycol/min/mg protein, indicating a maximal turnover rate of 40 molecules stryene oxide per molecule of epoxide hydratase per minute when using the minimum molecular weight of 49,000.

The pure epoxide hydratase was active toward a wide spectrum of substrates. Table 3 indicates the specific activities toward three substrates, styrene oxide, benzo[*a*]pyrene-4,5-(K-region)-oxide, and *p*-chlorophenyl-2,3-epoxypropylether of the final preparation as well as those of some earlier fractions during purification. The relative purification factors when comparing the final preparation with the 10,000*g* supernatant are not identical for the three substrates. One single branching-off fraction during purification had measurable epoxide hydratase activity toward the substrates used in this study. However, the ratio of the specific activities toward the three substrates was not different in that fraction from the ratio observed in the corresponding major fraction. All other fractions were inactive. Thus, unless another epoxide hydratase enzyme capable of hydrating one of the three substrates used in this study was very labile, the differences in purification factors may not be due to the presence in liver microsomes of several epoxide hydratases with different but overlapping specificities for the three substrates and sequential removal of some of them during purification. Indeed, a relatively large change in the ratio of specific activity toward styrene oxide as compared to benzo[*a*]pyrene-4,5-oxide occurred between fractions 1 and 2, i.e., possibly due to a change of one single molecular species in conformation and/or environment during solubilization. Moreover, when enough bovine serum albumin was added to make the protein concentration of the final preparation identical to that of the solubilized microsomes (fraction 2), the specific activity toward benzo[*a*]pyrene-4,5-oxide was increased by 45% while that for styrene oxide remained unchanged. "Correction" for these two potentially interfering effects would raise the purification factor with respect to benzo[*a*]pyrene-4,5-oxide as substrate to about 240. Similar effects may be operative with *p*-chlorophenyl-2,3-epoxypropylether as substrate. Moreover, the effects of inhibitors were similar with either styrene oxide or benzo[*a*]pyrene-4,5-oxide as substrate in microsomes, crude fractions during purification, and the pure enzyme (Fig. 7). Thus 1,1,1-trichloropropene-2,3-oxide and cyclohexene oxide were potent inhibitors in all cases, while no significant inhibition was observed with *trans*-stilbene oxide in any case. The apparent somewhat higher potency of cyclohexene oxide against hydration of benzo[*a*]pyrene-4,5-oxide as compared to styrene oxide is probably due to the fact that benzo[*a*]pyrene oxide was used at a lower concentration (150 μM) than styrene oxide (2 mM) since the K_m of the former is lower by more than 2 orders of magnitude as compared to the latter.

To determine whether the enzyme purified to apparent homogeneity was the

Table 3. Changes in Specific Activities of Epoxide Hydratase toward Different Substrates During Purification

Fraction	Substrate					
	Styrene oxide		Benzo[*a*]pyrene-4,5-oxide		*p*-Chlorophenyl-2,3-epoxypropylether	
	Specific activity[a]	Relative purification	Specific activity[a]	Relative purification	Specific activity[a]	Relative purification
1. 10,000*g* supernatant	1.67	1	1.25	1	6.2	1
2. Solubilized microsomes	8.91	5.3	4.49	3.5	–	–
3. Ammonium sulfate precipitate	19.15	11.5	10.61	8.29	–	–
4. DEAE-cellulose effluent	93.8	56	48.2	37.6	303	48.4
5. Phosphocellulose effluent	260	156	73	57.1	711	114
6. Final preparation	600	360	133	104	653	105

[a]Specific activity expressed as nmol product/mg protein/min.

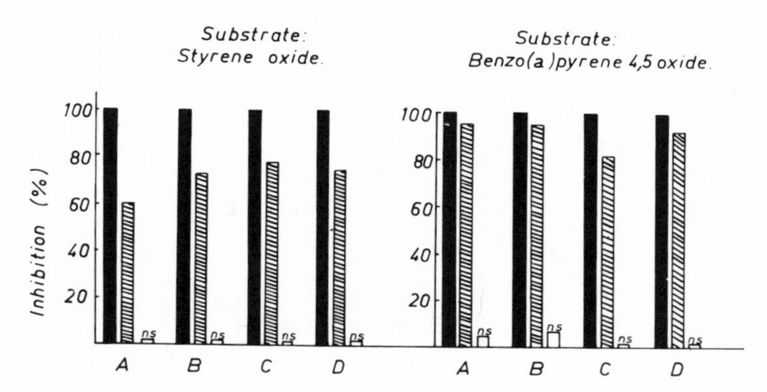

Fig. 7. Effect of inhibitors on epoxide hydratase activity in preparations of different purity. The inhibitors were added in 10 μl of acetonitrile at zero time with no preincubation. The concentrations of 1,1,1-trichloropropene oxide (■), cyclohexene oxide (▨), and *trans*-stilbene oxide (□) and that of styrene oxide were 2 mM; that of benzo[*a*] pyrene 4,5-oxide was 150 μM. Controls had the same amount of acetonitrile. The enzyme preparations used were (A) rat liver microsomes, (B) solubilized microsomes, (C) phosphocellulose effluent, and (D) pure epoxide hydratase. ns, Not significant (level of significance: $P < 0.05$).

only enzyme involved in hydration of styrene oxide (which was used as substrate for following the purification) and in hydration of benzo[*a*]pyrene-4,5-(K-region)-oxide, antibodies against the apparently homogeneous enzyme were raised in New Zealand White rabbits. The inhibitory activity of the antibodies was quite weak (leveling off at 20–30% inhibition when increasing amounts of antibodies were used), indicating that the antibodies were not active-site directed. The maximal inhibition was similar with styrene oxide or benzo[*a*]pyrene-4,5-oxide as a substrate whether the pure enzyme or crude fractions were used. However, immunoprecipitation was complete. When the preparations were centrifuged (8000*g* for 20 min) after exposure to antibodies at 0–5°C for 18 h, epoxide hydratase activity in the resulting supernatant fraction was decreased as a function of the amount of antibody used (Fig. 8). The fact that 100% of epoxide hydratase activity with either styrene oxide or benzo[*a*]pyrene-4,5-oxide as substrate was removed also from the crudest fraction tested, the microsomes immediately after solubilization (Fig. 8), indicates either the presence in liver microsomes of one single enzyme species responsible for the hydration of styrene oxide and benzo[*a*]pyrene-4,5-oxide, immunological cross-reactivity of several epoxide hydratases, or differential and complete loss of some epoxide hydratase species during solubilization. The immunological evidence together with the complete lack of indication of the presence of a different enzyme capable of hydrating styrene oxide and/or benzo[*a*]pyrene-4,5-oxide in any of the branching-off fractions during purification and the similar behavior toward different inhibitors throughout the purification proce-

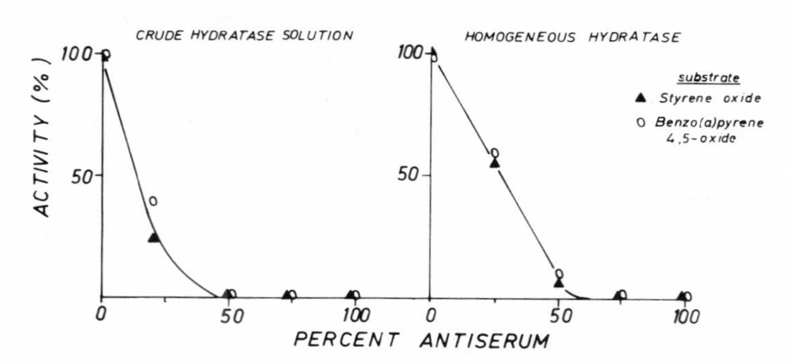

Fig. 8. Effect of antibodies raised against pure epoxide hydratase. Antibodies were raised in New Zealand White rabbits against apparently homogeneous epoxide hydratase purified from rat liver microsomes. 200 μl of crude epoxide hydratase solution (solubilized microsomes, 7 mg protein/ml) or pure epoxide hydratase (46 μg) was kept at 0–5°C for 18 h in the presence of 200 μl serum which contained the indicated percentages of antiserum. Controls were kept for the same amount of time in the presence of control serum. The mixtures were then centrifuged at 8000*g* for 20 min and the enzyme activities were determined in the supernatant fractions.

dure with respect to either of the two substrates strongly suggests that only one microsomal enzyme is responsible for the hydration of styrene oxide and of benzo[*a*]pyrene-4,5-(K-region)-oxide and that this is the enzyme purified to apparent homogeneity in this study.

USE OF THE APPARENTLY HOMOGENEOUS EPOXIDE HYDRATASE AS A PROBE FOR THE RELATIVE IMPORTANCE OF EPOXIDES AMONG OTHER REACTIVE METABOLITES

When benzo[*a*]pyrene was tested for mutagenicity using a set of *Salmonella* strains (TA 1535, TA 1537, TA 1538) constructed by Ames for detecting different (but not necessarily all) types of mutagenic agents (35), it proved inactive in that the number of His^+ revertant colonies was not increased over that occurring spontaneously in plates with no tester compounds added (for TA 1537, see Table 4). Among a great number of derivatives, only the epoxide (benzo[*a*]pyrene-4,5-oxide) was mutagenically active and, moreover, only for the two frameshift-sensitive strains TA 1537 and TA 1538 (for TA 1537, see Table 4; *cf.* also ref. 13).

Similarly, it was observed with other polycyclic hydrocarbons with no side chains, such as benz[*a*]anthracene, that the K-region epoxide was a frameshift mutagen, whereas the parent compound and other derivatives tested were not (data not shown). The methylated hydrocarbon, 7,12-dimethylbenz[*a*]anthracene, behaved quite differently; the frameshift mutational potential of the

Table 4. Mutagenicity of Benzo[*a*]pyrene Derivatives[a]

Derivative	μg per plate	His$^+$ revertant colonies
None (control)	–	8–13
Parent compound	1,5,20,50	7–16
3-Hydroxy	1,5,20	9–13
4- (or 5-) Hydroxy	1,5,20	7–16
6-Hydroxy	1,5,20	9–18
9-Hydroxy	1,5,20	8–10
1,6-Quinone	1,5,20	8–15
3,6-Quinone	1,5,20	8–15
6,12-Quinone	1,5,20	8–14
4,5-*cis*-Dihydrodiol	1,5,20	8–13
4,5-*trans*-Dihydrodiol	0.2,1,5	8–15
7,8-*trans*-Dihydrodiol	0.2,1,5	5–19
9,10-*trans*-Dihydrodiol	0.2,1,5	9–11
4,5-Oxide	0.25	165,198
4,5-Oxide	1	498,523
4,5-Oxide	5	456,573

[a]Test compounds were given in 40 μl dimethylsulfoxide together with 1–2 × 10^8 histidine-dependent *Salmonella typhimurium* TA 1537 bacteria in a histidine-poor top agar onto a petri dish with minimal agar. The number of His$^+$ revertant colonies was counted after incubation at 37°C for two days.

K-region epoxide was exceedingly weak for TA 1537 and fully nonobservable with TA 1538 over a very broad range of concentrations. On the other hand, some alkene oxides such as 4-phenylstyrene-7,8-oxide proved mutagenic with both frameshift-sensitive strains. Differences in penetration are not likely to be responsible for the observed differences in mutagenicity since the tester strains used were constructed by Ames *et al.* (35) in such a way as to allow maximal penetration of test compounds. Moreover, none of the compounds tested had a molecular weight greater than or a relative polarity substantially different from that of compounds known to be active mutagens for the strains used. Thus it is apparent that epoxidation of an aromatic double bond of a polycyclic hydrocarbon is neither a sufficient nor a necessary condition for frameshift mutagenesis to occur. On the other hand, with respect to mutagenic polycyclic hydrocarbons possessing no side chains, the data discussed above tend to suggest that the species responsible for mutagenic effects are epoxide(s).

Although this investigatory approach has the advantage of using known amounts of pure compounds, it has three obvious disadvantages: (1) the relative proportions in which the derivates are metabolically formed is not taken into account; (2) minor metabolites may be overlooked; and (3) since isolation for structure elucidation and testing of metabolites tends to become increasingly difficult with increasing reactivity of the metabolite, the mutagenically most reactive species may not be included. This approach was therefore complemented with some other approaches which have other disadvantages but, when used in conjunction, lead to a more complete picture of the situation.

Benzo[*a*] pyrene was incubated with mouse hepatic microsomes in the presence of a system generating NADPH, the cofactor necessary for monooxygenase activity. The reaction was stopped by extraction of ethyl-acetate-soluble compounds which were fractionated by thin-layer chromatography. Zones of silica gel were scraped off as indicated in Fig. 9 and extracted. In the aqueous phase,

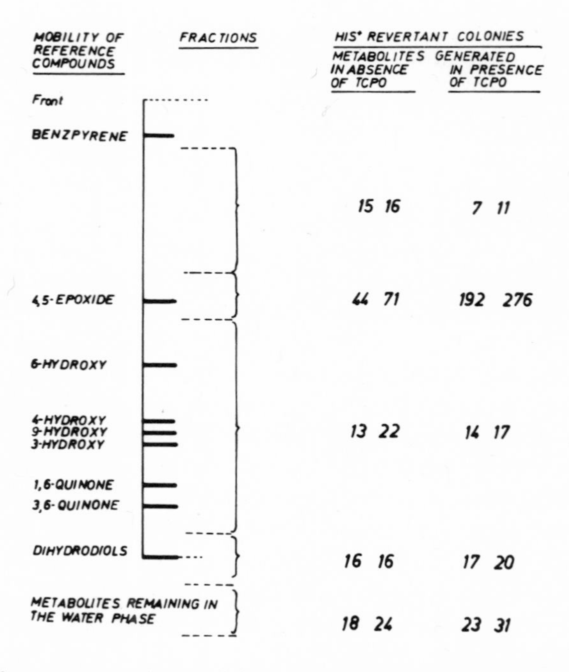

Fig. 9. Crude fractionation of benzo[a] pyrene metabolites: Relationship between mutagenicity and the relative proportions at which metabolites are produced. 100 μM benzo[*a*]-pyrene (BP) was incubated with liver microsomes from female C3H mice (11 mg protein) and NADPH in a total volume of 4.5 ml at 37°C for 45 min in the presence or absence of the epoxide hydratase inhibitor 1,1,1-trichloropropene-2,3-oxide (TCPO, 2 mM). The ethyl acetate extract was chromatographed under argon on silica gel plates with benzene as solvent. Plates were divided in zones as indicated in the figure, which were extracted into ethanol. This extract was concentrated *in vacuo* to about 100 μl and then diluted with dimethyl sulfoxide to 1 ml. At the concentrations used, the two solvents had no effect. Aliquots of 100 μl were used for mutagenicity testing with TA 1537 using the same conditions as indicated in Table 4.

the protein was removed by centrifugation and the clear supernatant was used for the testing. From all fractions, a constant aliquot was tested for mutagenic activity. As a typical example, Fig. 9 shows that for the frameshift-sensitive strain TA 1537 the only fraction leading to a mutational rate higher than that occurring spontaneously was the fraction corresponding to the mobility of epoxide derivatives. Although this indicates that metabolites as reactive as arene oxides may survive the fractionation procedure, less stable mutagenically reactive metabolites are by no means inconceivable, nor does the result imply that all arene oxides have survived. In fact, on grounds of a greater tendency to isomerize to a system with more aromatic character, non-K-region epoxides would be expected to phenolize much more readily than K-region epoxides. On the other hand, these results suggest that the 7,8-dihydrodiol-9,10-epoxide, which has recently been reported to be metabolically produced and to bind very readily to DNA, does not contribute to a major extent in this system to the mutagenicity of benzo[*a*]pyrene metabolites at the proportions in which they are produced since, in contrast to the mutagenically active fraction containing arene oxides, no other fraction including the dihydrodiol fraction, which would also contain mixed dihydrodiol epoxides, lead to an observable increase in the mutational rate above that occurring spontaneously.[1]

Thus testing of classes of metabolites in crude fractions can provide additional information over that obtained with pure compounds, but this method suffers from the fact that highly reactive metabolites may again have escaped the testing. To overcome the need to isolate or fractionate mutagenically active species and still gain information on specific classes of metabolites involved, the precursor compound can be tested with the mutant strains in the presence of activating and inactivating enzyme systems by means of selective or specific modulations of their activities.

When the premutagen benzo[*a*]pyrene was incubated in the presence of the bacterial tester strains with various liver cell fractions and in the presence or

[1] Moreover, when the precursor of the mixed dihydrodiol-epoxide in question, benzo[*a*]-pyrene-7,8-dihydrodiol, was tested in this system in the presence of liver microsomes and a NADPH-generating system, the number of revertant colonies was increased over controls with either TA 1537 or TA 1538 at least ten times less than when benzo[*a*]pyrene was used. Thus benzo[*a*]pyrene-7,8-dihydrodiol-9,10-oxide does not contribute in this system to an observable extent to the overall frameshift mutagenicity of benzo[*a*]pyrene plus its metabolites. With the substitution-mutated strain TA 1535, benzo[*a*]pyrene showed no activity, whether in the presence of microsomes and NADPH or not. However, recently a more sensitive substitution-mutated strain, TA 100, has been developed by Ames and collaborators (51). Benzo[*a*]pyrene was metabolically activated to species mutagenic to this strain. An investigation of whether the same class of metabolites (simple epoxides) is responsible for substitution as observed for frameshift mutation when using the strains TA 1537 and TA 1538 is in progress.

absence of pyridine nucleotide cofactors, it became apparent that the microsomal but not the mitochondrial or the cytoplasmic fraction contained enzyme(s) which, in the presence of NADPH, activate benzo[*a*]pyrene to mutagenic metabolites. Moreover, it was observed (data not shown) that, again typical for microsomal monooxygenase(s), the reaction can also take place in presence of NADH instead of NADPH, but at a much slower rate (approximately 10%), while in the simultaneous presence of both reduced cofactors a synergistic effect was observed. All this is compatible with the assumption of epoxide(s) as the mutagenically reactive species. Moreover, it has been observed by radiotracer trapping technique that benzo[*a*]pyrene is converted to its 4,5-oxide in this system.

When to a system generating *in situ* the mutagenically reactive benzo[*a*]pyrene metabolite(s), epoxide hydratase inhibitors such as 1,1,1-trichloropropene-2,3-oxide (TCPO) or cyclohexene oxide (CHO) were added, a remarkable increase of the mutagenic effect was observed (Fig. 10). This increase was dose

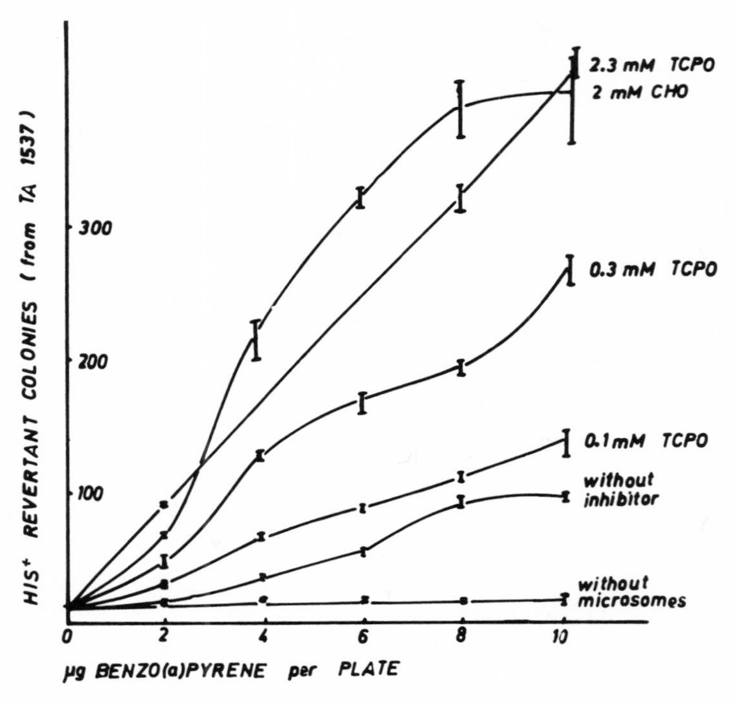

Fig. 10. Metabolic activation of benzo[a]pyrene to a mutagen: potentiation by epoxide hydratase inhibitors. Number of His$^+$ revertant colonies from TA 1537 as a function of benzo[*a*]pyrene concentration in the presence or absence of a microsomal preparation from female C3H mouse liver and epoxide hydratase inhibitors 1,1,1-trichloropropene-2,3-oxide (TCPO) or cyclohexene oxide (CHO). The concentrations indicated are calculated with respect to the top agar. Bars represent the SEM. Conditions were as indicated in Table 4.

dependent and very clearly occurred at concentrations of the inhibitor (0.1–0.3 mM TCPO) where no significant effect on any monooxygenase activity tested was observable (at higher concentrations of the inhibitor a slight apparent activation and at even higher concentrations a slight inhibition of monooxygenase activities were observed). Similar results were obtained with other polycyclic hydrocarbons possessing no side chains, such as benz[*a*]anthracene, while epoxide hydratase inhibitors were fully ineffective with aromatic hydrocarbons bearing methyl (e.g., 7,12-dimethylbenz[*a*]anthracene) or amino (e.g., 4-amino-*trans*-stilbene) groups. With the former (but not with the latter) compounds, these data indicate that metabolically produced epoxides are mutagenically active and that epoxide hydratase is involved in their control. Nevertheless, the data do not allow the exclusion of other metabolites as potentially mutagenically active and perhaps even more important.

The apparently homogeneous epoxide hydratase was therefore used as an epoxide-specific probe. If benzo[*a*]pyrene is activated in the presence of the tester strain, TA 1537, solely to mutagenically reactive epoxides by a microsomal preparation and NADPH, addition of increasing amounts of apparently homogeneous epoxide hydratase would increasingly prevent the mutagenic response until, at high levels of enzyme, the rate of mutation would not be significantly different from that occurring spontaneously in the absence of benzo[*a*]pyrene (Fig. 11, curve representing the experiment without the epoxide hydratase inhibitor cyclohexene oxide, CHO). On the other hand, if metabolites other than epoxides (or derived from epoxides) contributed to the mutagenic effect of the family benzo[*a*]pyrene plus its metabolites to any significant extent, it would be expected that increasing amounts of homogeneous epoxide hydratase would lead to a corresponding decrease in the mutational rate down to a certain level and then to plateau, the remaining activity corresponding to mutagenically active metabolites that are not substrates for the homogeneous epoxide hydratase. As shown in Fig. 11 by the curve representing the experiment without the epoxide hydratase inhibitor cyclohexene oxide (CHO), increments of the homogeneous epoxide hydratase increasingly prevented the mutagenic response to benzo[*a*]pyrene metabolites, and, at 8 units of enzyme, reduced the mutation frequency virtually to background. While the results obtained with benz[*a*]anthracene were similar to the ones obtained with benzo[*a*]pyrene, aromatic compounds bearing methyl groups, such as 7,12-dimethylbenz[*a*]anthracene, or bearing amino groups, such as 4-amino-*trans*-stilbene, behaved very differently in that the pure epoxide hydratase was not effective in lowering the mutagenic activity of their metabolites produced *in situ*, whether a microsomal fraction or a combined fraction of microsomes and cytoplasmic constituents was used for activation. This is in agreement with the results obtained when using epoxide hydratase inhibitors as reported above.

Thus, in contrast to data obtained with epoxide hydratase inhibitors, the results obtained when using the homogeneous enzyme allow the conclusions that

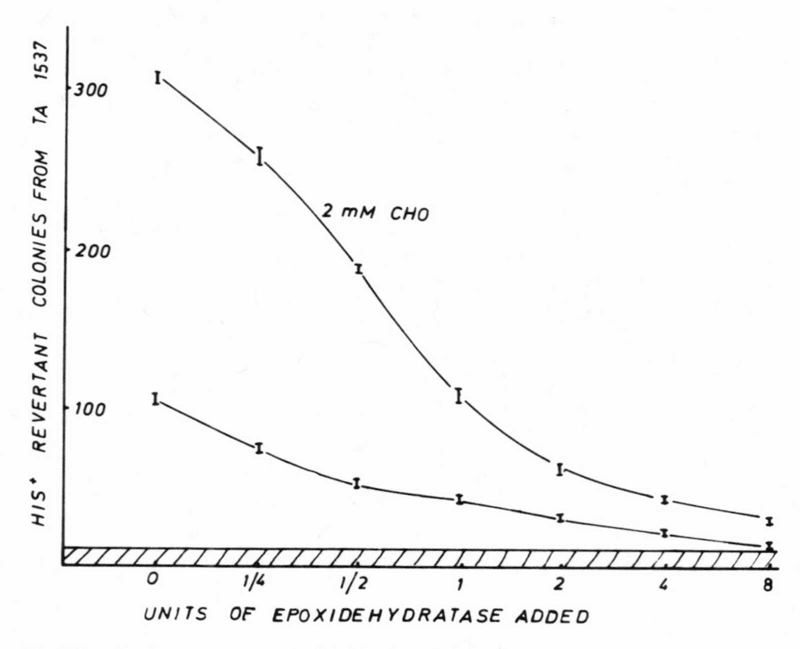

Fig. 11. Titration by pure epoxide hydratase of the mutagenic effect of reactive benzo[a]-pyrene metabolites generated in situ. 10 μg of benzo[*a*]pyrene per plate was incubated with liver microsomes from female C3H mice and in the presence or absence of 2 mM cyclohexene oxide (CHO). Concentrations were calculated with respect to top agar. Units of pure epoxide hydratase were added as indicated. Bars represent the SEM. Conditions were as indicated in Table 4.

epoxide(s) (or unknown metabolites derived from them) are the only microsomal metabolite(s) of benzo[*a*]pyrene leading to observable mutations of the *Salmonella* strains used and that epoxide hydratase is involved in their control.

POSSIBLE CONSEQUENCES WITH RESPECT TO CLINICALLY USED DRUGS

Concern has been expressed that clinically useful drugs which contain moieties that are or might be metabolically epoxidized (5, 36–47) may alter critical biomacromolecules by covalent binding to them, thereby possibly leading to unwanted effects such as mutagenicity, carcinogenicity, cell necrosis, or allergic effects (for a more complete discussion, see refs. 1–5). In this light, structure activity relationships for epoxides as substrates or inhibitors of human epoxide hydratase, an enzyme which is localized in the subcellular fraction where epoxides are formed (i.e., the microsomes) and which transforms such epoxides to much less reactive dihydrodiols, were studied and use of that information for a more rational drug design was proposed (30).

On the other hand, epoxides are a very heterogeneous class of compounds which vary greatly with respect to stability and electrophilic reactivity (for a review, see ref. 48). Some epoxides are quite stable, both chemically and metabolically: 10,11-epoxides are found as stable terminal metabolites in the urine of rat and human after administration of the clinically used drugs carbamazepine (39) or cyproheptadine (43, 45) or the experimental psychotropic agent cyclobenzaprine (Belvedere, Rovei, Pantarotto, and Frigerio, personal communication) (for structures, see Fig. 12). Morselli *et al.* (49) found 1% of the administered dose of carbamazepine to be present as carbamazepine-10,11-oxide in human urines, and Eichelbaum and Bertilsson (50) found this metabolite in the plasma of chronically treated patients in such high concentrations as 0.6–2.3 μg/ml. The present state of knowledge does not permit a prediction of the relationship between the stability of epoxides and their biological effects. The relatively high stability of the above epoxides may be associated with a relatively low electrophilic reactivity. On the other hand, the relatively high chemical and metabolic stability of these epoxides may lead to a high concentration of the epoxide reaching any critical target site.

Figure 13 shows that the 10,11-epoxide which is a known metabolite of the clinically used drug carbamazepine (39) is nonmutagenic to the frameshift-sensitive *Salmonella typhimurium* strain TA 1537. Indeed, it is still ineffective at thousandfold higher concentrations than those of benzo[*a*]pyrene-4,5-oxide leading to significant mutations (Fig. 13). Similar results were obtained when using another frameshift-sensitive strain, TA 1538, and when using TA 1535 designed to detect substitution mutations. The 10,11-epoxide metabolites of the medicinal drugs cyproheptadine and cyclobenzaprine also proved mutagenically inactive with the three tester strains used (data not shown). None of these epoxides was cytotoxic to any of the bacterial strains used, precluding that mutagenicity might have been overshadowed by cytotoxicity. Since the set of tester strains used does not contain all of the theoretically possible permutations of nucleobases in repetitive sequences ("hot spots"), it is conceivable that some mutagenically active species may be negative for all three strains.

The 10,11-epoxides of the three medicinal drugs tested have been observed as metabolites of the corresponding parent compounds. However, all the compounds possess other moieties where epoxidation could possibly also take place (see Fig. 12). These other epoxides, being non-K-region arene oxides, would be expected to be considerably less stable and, consequently, detectable with greater difficulties than the alkene oxides reported so far. In order to detect whether any metabolites other than the known epoxide metabolites might be mutagenic, the histidine-dependent bacteria were incubated in the presence of the parent compounds and an active microsomal monooxygenase system with or without the epoxide hydratase inhibitor 1,1,1-trichloropropene-2,3-oxide. No mutagenic response was elicited by carbamazepine, cyproheptadine, or cyclobenzaprine whether the epoxide hydratase inhibitor was present or not. This was true for a wide range of concentrations of the compounds, up to 3 orders of

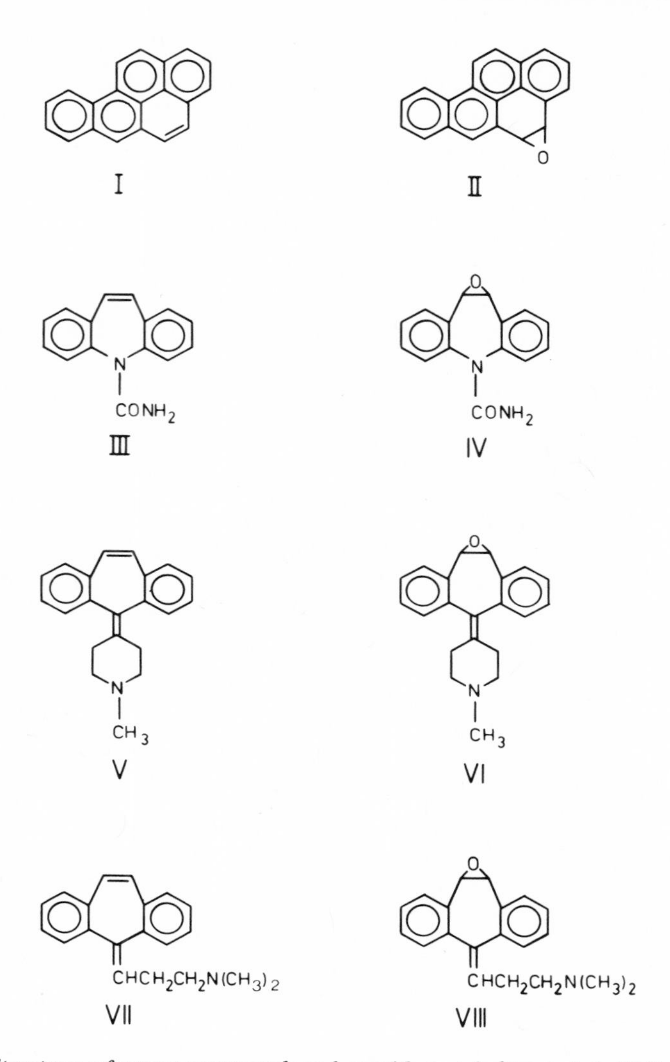

Fig. 12. Structures of parent compounds and epoxide metabolites tested: (I) Benzo[*a*]pyrene, (II) benzo[*a*]pyrene-4,5-oxide, (III) carbamazepine, (IV) carbamazepine-10,11-oxide, (V) cyproheptadine, (VI) cyproheptadine-10,11-oxide, (VII) cyclobenzaprine, (VIII) cyclobenzaprine-10,11-oxide. The 4,5 bond of benzo[*a*]pyrene is depicted as a double bond to indicate its similarity with the 10,11 bonds of compounds III, V, and VII with respect to its simultaneous conjugation with two benzene rings.

magnitude higher than the ones leading to significant mutations with the premutagenic hydrocarbon benzo[*a*]pyrene in the presence of a metabolically active system (data not shown).

Thus it is important to keep in mind that alarming results in terms of biological effects of some epoxide metabolites do not automatically imply

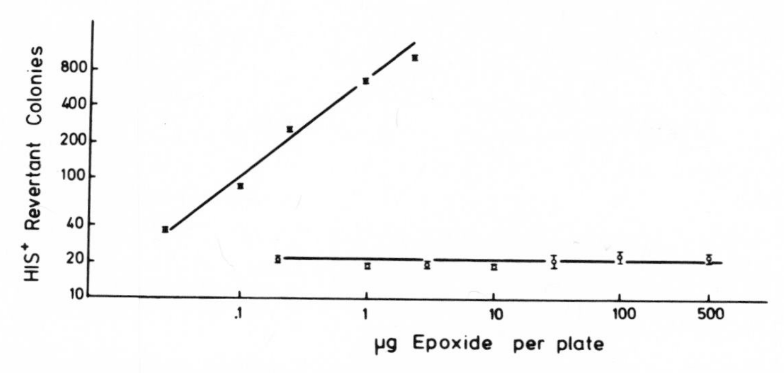

Fig. 13. Mutagenic effect of benzo[a]pyrene-4,5-oxide on the frameshift-sensitive strain TA 1537, and lack of this effect with 10,11-epoxide metabolite of the clinically used drug carbamazepine. 1–2 × 10^8 bacteria were incubated at 37°C for 2 days in histidine-poor medium in the presence of the indicated amounts of the synthetic epoxide, benzo[*a*]pyrene-4,5-oxide (•) or carbamazepine-10,11-oxide (○). Values represent the mean ± SEM of the number of His^+ revertant colonies per plate. Spontaneous mutations led to 21 ± 4 His^+ revertant colonies in control plates with no test compounds added.

similar effects for other epoxides, since they vary greatly among themselves in stability, electrophilic reactivity, and relative activity as substrates for enzymes that transform epoxides. Moreover, they can vary greatly in other parameters not related to the epoxide moiety but potentially important for biological effects in conjunction with the epoxide moiety, such as geometry of the whole molecule.

Yet, even if a battery of biological tests suggests a particular epoxide metabolite to be harmless, it may potentiate effects of other (carcinogenic, mutagenic, and/or cytotoxic) epoxide metabolites derived from ubiquitously present pollutants such as benzo[*a*]pyrene by interfering with enzymes involved in their inactivation. Table 5 shows that the 10,11-epoxides of the three medicinal drugs inhibit epoxide hydratase to very different degrees. Their potentiating effect may therefore also be very different from one to the other. An assessment of this hypothesis is currently in progress.

Table 5. Metabolically Produced Epoxides Derived from Some Medicinal Drugs: Inhibition of Epoxide Hydratase[a]

Inhibitor	Inhibition (%)
Carbamazepine-10,11-oxide	11
Cyclobenzaprine-10,11-oxide	25
Cyproheptadine-10,11-oxide	67

[a]The concentration of both the substrate, styrene oxide, and the inhibitors was 0.5 mM. Male rat liver microsomes were used.

CONCLUSIONS

Since our previous results had allowed us to conclude that metabolically produced epoxides from benzo[*a*]pyrene were mutagenically reactive for *Salmonella typhimurium* TA 1537 and TA 1538 (28) but did not allow us to make conclusions about their relative importance among possibly many other possible reactive metabolites, we purified epoxide hydratase to apparent homogeneity as an epoxide-specific probe. An enzyme preparation was obtained which was homogeneous by gel electrophoretic, analytical ultracentrifugal, and immunological criteria.

The pure enzyme has a minimum molecular weight of 49,000 and, in the absence of detergent, a high tendency to aggregate to very high molecular weight oligomers. This tendency is in agreement with the hydrophobic character of the enzyme. In fact, amino acid analysis showed that 56% of the amino acid residues were hydrophobic. The absorption spectrum of the enzyme has between 235 and 500 nm one single peak at 280 nm and a shoulder at 290 nm. The content in tryptophan and tyrosine residues as determined spectrophotometrically is even slightly lower than if determined chemically, confirming that the preparation is essentially free of the detergent used for solubilizing the enzyme from the microsomal membranes, Cutscum, which strongly absorbs at the same wavelengths. The pH optimum of the pure enzyme is between 8.9 and 9.4 (depending on the buffer used). The ionic strength of the medium appears to have negligible influence on the catalytic activity. The apparent K_m with respect to styrene oxide as substrate is 0.67 mM, the V_{max} is 800 nmol styrene glycol/min/mg protein. Thus the maximal turnover rate is 40 molecules styrene oxide per molecule epoxide hydratase per minute based on the minimum molecular weight of 49,000.

The apparently homogeneous enzyme has a very broad specificity for substrate epoxides, including K-region epoxides derived from carcinogenic polycyclic hydrocarbons. Thus benzo[*a*]pyrene-4,5-(K-region)-oxide is efficiently converted to the corresponding dihydrodiol by the pure enzyme. No indication for the presence of a different enzyme capable of hydrating styrene oxide or benzo[*a*]pyrene-4,5-oxide was obtained in any of the branching-off fractions during purification. Moreover, the relative potencies of inhibitors remained similar for both substrates throughout the purification. Finally, when antibodies raised against the apparently homogeneous enzyme were added to the crudest soluble preparation, the solubilized microsomes, they precipitated the entire activity capable of hydrating styrene oxide or benzo[*a*]pyrene-4,5-oxide. These observations taken together strongly suggest that one single enzyme in microsomal membranes is involved in hydrating styrene oxide and benzo[*a*]pyrene-4,5-oxide and that this is the enzyme purified to homogeneity in this study.

When the premutagen benzo[*a*]pyrene was incubated with rat liver microsomes and NADPH in presence of the tester strain TA 1537, metabolite(s) highly mutagenic for this strain were formed. Addition of increasing amounts of the pure epoxide hydratase led to a dose-dependent decrease of the mutagenic effect. Eight units of pure enzyme prevented the mutagenic effect. Thus, among the great variety of microsomal metabolites of benzo[*a*] pyrene, an epoxide(s) (or unknown metabolites derived therefrom) is the sole species mutagenic in the system used.

However, alarming biological effects of some epoxides do not automatically imply that all epoxides have similar effects. Epoxides vary greatly among themselves in stability, electrophilic reactivity, and relative activity as substrates of epoxide-transforming enzymes. Moreover, they can be very different in parameters of the molecule unrelated to the epoxide moiety, yet possibly of biological importance in conjunction with the epoxide moiety, such as geometry. Indeed, known epoxide metabolites of three medicinal drugs proved mutagenically inactive for *Salmonella typhimurium* TA 1537 and TA 1538 up to concentrations of more than a thousandfold higher than those of benzo[*a*] pyrene-4,5-oxide leading to significant mutations. Moreover, over the whole range of concentrations used, no cytotoxic effects were observed with these strains.

On the other hand, epoxide metabolites of clinically used drugs may lead to potentiation of biological effects of epoxide metabolites derived from ubiquitously present pollutants, such as benzo[*a*] pyrene, by interaction with enzymes responsible for their further biotransformation. Since drugs will normally have to be present at much higher concentrations than are pollutants, they may lead to very effective competitive inhibition of processes which normally inactivate potentially harmful epoxide metabolites of polycyclic hydrocarbons.

ACKNOWLEDGMENTS

The authors thank Mrs. P. Dent for her excellent technical assistance. This work was supported by the Deutsche Forschungsgemeinschaft.

REFERENCES

1. J. W. Daly, D. M. Jerina, and B. Witkop, Arene oxides and the NIH shift: The metabolism, toxicity and carcinogenicity of aromatic compounds, *Experientia* **28**, 1129–1264 (1972).
2. F. Oesch, Mammalian epoxide hydrases: Inducible enzymes catalyzing the inactivation of carcinogenic and cytotoxic metabolites derived from aromatic and olefinic compounds, *Xenobiotica* **3**, 305–340 (1973).
3. D. M. Jerina and J. W. Daly, Arene oxides: A new aspect of drug metabolism, *Science* **185**, 573–582 (1974).

4. P. Sims and P. L. Grover, Epoxides in polycyclic aromatic hydrocarbon metabolism and carcinogenesis, *Adv. Cancer Res.* **20,** 165–274 (1974).
5. M. G. Horning, C. M. Butler, J. Nowlin, and R. M. Hill, Drug metabolism in the human neonate, *Life Sci.* **16,** 651–672 (1975).
6. P. L. Grover and P. Sims, Interactions of the K-region epoxides of phenanthrene and dibenz[*a,h*]anthracene with nucleic acids and histone, *Biochem. Pharmacol.* **19,** 2251–2259 (1970).
7. T. Kuroki, E. Huberman, H. Marquardt, J. K. Selkirk, C. Heidelberger, P. L. Grover, and P. Sims, Binding of K-region epoxides and other derivatives of benz[*a*]anthracene and dibenz[*a,h*]anthracene to DNA, RNA and proteins of transformable cells, *Chem.-Biol. Interactions* **4,** 389–397 (1971/72).
8. P. D. Lawley and N. Jarman, Alkylation by propylene oxide of deoxyribonucleic acid, adenine, guanosine and deoxyguanylic acid, *Biochem. J.* **126,** 893–900 (1972).
9. I. Y. Wang, R. E. Rasmussen, and T. T. Crooker, Isolation and characterization of an active DNA-binding metabolite of benzo[*a*]pyrene from hamster liver incubation systems, *Biochem. Biophys. Res. Commun.* **49,** 1142–1149 (1972).
10. P. L. Grover and P. Sims, K-region epoxides of polycyclic hydrocarbons: Reactions with nucleic acids and polyribonucleotides, *Biochem. Pharmacol.* **22,** 661–666 (1973).
11. M. J. Cookson, P. Sims, and P. L. Grover, Mutagenicity of epoxides of polycyclic hydrocarbons correlates with carcinogenicity of parent hydrocarbons, *Nature (London) New Biol.* **234,** 186–187 (1971).
12. E. L. Huberman, L. Aspiras, C. Heidelberger, P. L. Grover, and P. Sims, Mutagenicity to mammalian cells of epoxides and other derivatives of polycyclic hydrocarbons, *Proc. Natl. Acad. Sci. USA* **68,** 3195–3199 (1971).
13. B. N. Ames, P. Sims. and P. L. Grover, Epoxides of carcinogenic polycyclic hydrocarbons are frameshift mutagens, *Science* **176,** 47–49 (1972).
14. O. G. Fahmy and M. J. Fahmy, Oxidative activation of benz[*a*]anthracene and methylated derivatives in mutagenesis and carcinogenesis, *Cancer Res.* **33,** 2354–2361 (1973).
15. B. L. Van Duuren, L. Langseth, B. M. Goldschmidt, and L. Orris, Carcinogenicity of epoxides, lactones and peroxy compounds. VI. Structure and carcinogenic activity, *J. Natl. Cancer Inst.* **39,** 1217–1228 (1967).
16. P. L. Grover, P. Sims, E. Huberman, H. Marquardt, T. Kuroki, and C. Heidelberger, *In vitro* transformation of rodent cells by K-region derivatives of polycyclic hydrocarbons, *Proc. Natl. Acad. Sci. USA* **68,** 1098–1101 (1971).
17. H. Marquardt, T. Kuroki, E. Huberman, J. K. Selkirk, C. Heidelberger, P. L. Grover, and P. Sims, Malignant transformation of cells derived from mouse prostate by epoxides and other derivatives of polycyclic hydrocarbons, *Cancer Res.* **32,** 716–720 (1972).
18. E. Huberman, T. Kuroki, H. Marquardt, J. K. Selkirk, C. Heidelberger, P. L. Grover, and P. Sims, Transformation of hamster embryo cells by epoxides and other derivatives of polycyclic hydrocarbons, *Cancer Res.* **32,** 1391–1396 (1972).
19. K. Bürki, T. A. Stoming, and E. Bresnick, Effects of an epoxide hydrase inhibitor on *in vitro* binding of polycyclic hydrocarbons to DNA and on skin carcinogenesis, *J. Natl. Cancer Inst.* **52,** 785–788 (1974).
20. B. B. Brodie, W. D. Reid, A. K. Cho, G. Sipes, G. Krishna, and J. R. Gillette, Possible mechanism of liver necrosis caused by aromatic organic compounds, *Proc. Natl. Acad. Sci. USA* **68,** 160–164 (1971).
21. H. L. Falk, P. Kotin, S. S. Lee, and A. Nathan, Intermediary metabolism of benzo[*a*]-pyrene in the rat, *J. Natl. Cancer Inst.* **28,** 699–724 (1962).
22. Ch. Nagata, M. Kodama, and Y. Tagashira, Electron spin resonance study on the interaction between chemical carcinogens and tissue components. II. Free radical

produced by stirring aromatic hydrocarbons with tissue components such as skin homogenates or proteins, *GANN* **58**, 493–504 (1967).
23. A. Dipple, P. D. Lawley, and P. Brookes, Theory of tumour initiation by chemical carcinogens: Dependence of activity on structure of ultimate carcinogen, *J. Cancer* **4**, 493–505 (1968).
24. P. L. Grover, A. Hewer, and P. Sims, Metabolism of polycyclic hydrocarbons by rat-lung preparations, *Biochem. Pharmacol.* **23**, 323–332 (1974).
25. G. Holder, H. Yagi, P. Dansette, D. M. Jerina, W. Levin, A. Y. H. Lu, and A. H. Conney, Effects of inducers and epoxide hydrase on the metabolism of benzo[*a*]pyrene by liver microsomes and a reconstituted system: Analysis by high pressure liquid chromatography, *Proc. Natl. Acad. Sci. USA* **71**, 4356–4360 (1974).
26. R. E. Rasmussen and I. Y. Wang, Dependence of specific metabolism of benzo[*a*]pyrene on the inducer of hydroxylase activity, *Cancer Res.* **34**, 2290–2295 (1974).
27. J. K. Selkirk, R. G. Croy, P. P. Roller, and H. V. Gelboin, High-pressure liquid chromatographic analysis of benzo[*a*]pyrene metabolism and covalent binding and the mechanism of action of 7,8-benzoflavone and 1,2-epoxy-3,3,3-trichloropropane, *Cancer Res.* **34**, 3474–3480 (1974).
28. F. Oesch and H. R. Glatt, in: *Screening Tests in Chemical Carcinogenesis* (R. Montesano, H. Bartsch, and L. Tomatis, eds.), pp. 255–295, International Agency for Research on Cancer, Lyon (1976).
29. F. Oesch and J. Daly, Solubilization, purification and properties of a hepatic epoxide hydrase, *Biochim. Biophys. Acta* **227**, 692–697 (1971).
30. F. Oesch, Purification and specificity of a human microsomal epoxide hydratase, *Biochem. J.* **139**, 77–88 (1974).
31. A. Er-el, Y. Zaidenzaig, and S. Shaltiel, Hydrocarbon coated Sepharoses: Use in the purification of glycogen phosphorylase, *Biochem. Biophys. Res. Commun.* **49**, 383–390 (1972).
32. K. Weber and M. Osborn, The reliability of molecular weight determinations by dodecyl sulfate-polyacrylamide gel electrophoresis, *J. Biol. Chem.* **244**, 4406–4412 (1969).
33. R. A. Capaldi and G. Vanderkooi, The low polarity of many membrane proteins, *Proc. Natl. Acad. Sci. USA* **69**, 930–932 (1972).
34. P. Bentley, F. Oesch, and A. Tsugita, Properties and amino acid composition of pure epoxide hydratase, *FEBS Lett.* **59**, 296–299 (1975).
35. B. N. Ames, F. D. Lee, and W. E. Durston, An improved bacterial test system for the detection and classification of mutagens and carcinogens, *Proc. Natl. Acad. Sci. USA* **70**, 782–786 (1973).
36. T. Chang, A. Savory, and A. J. Glazko, A new metabolite of 5,5-diphenylhydantoin (Dilantin), *Biochem. Biophys. Res. Commun.* **33**, 444–449 (1970).
37. J. T. Matschiner, R. G. Bell, J. M. Amelotti, and T. E. Knauer, Isolation and characterization of a new metabolite of phylloquinone in the rat, *Biochim. Biophys. Acta* **201**, 309–315 (1970).
38. S. F. Sisenwine, C. O. Tio, S. R. Shrader, and H. Ruelius, The biotransformation of protriptyline in man, pig and dog, *J. Pharmacol. Exp. Ther.* **175**, 51–59 (1970).
39. A. Frigerio, R. Fanelli, P. Biandrate, G. Passerini, P. L. Morselli, and S. Garattini, Mass spectrometric characterization of carbamazepine-10,11-epoxide, a carbamazepine metabolite isolated from human urine, *J. Pharm. Sci.* **61**, 1144–1147 (1972).
40. D. J. Harvey, L. Glazener, C. Stratton, D. B. Johnson, R. M. Hill, E. C. Horning, and M. G. Horning, Detection of epoxides of allylsubstituted barbiturates in rat urine, *Res. Commun. Chem. Pathol. Pharmacol.* **4**, 247–260 (1972).
41. D. J. Harvey, L. Glazener, C. Stratton, J. Nowlin, R. M. Hill, and M. G. Horning, Detection of a 5-(2,4-dihydroxy-1,5-cyclohexadien-1-yl) metabolite of phenobarbital

and mephobarbital in rat, guinea pig and human, *Res. Commun. Chem. Pathol. Pharmacol.* **3**, 557–565 (1972).
42. M. G. Horning, D. J. Harvey, J. Nowlin, W. G. Stillwell, and R. M. Hill, The use of gas chromatography mass spectrometry methods in perinatal pharmacology, *Adv. Biochem. Psychopharmacol.* **7**, 113–124 (1973).
43. A. Frigerio, N. Sossi, G. Belvedere, C. Pantarotto, and S. Garattini, Identification of desmethylcyproheptadine-10-11-epoxide and other cyproheptadine metabolites isolated from rat uterine, *J. Pharm. Sci.* **63**, 1536–1539 (1974).
44. P. H. Grantham, L. C. Mohan, E. K. Weisburger, H. M. Fales, E. A. Sokoloski, and J. H. Weisburger, Identification of new water-soluble metabolites of acetanilide, *Xenobiotica* **4**, 69–76 (1974).
45. H. B. Hucker, A. J. Balletto, S. C. Stauffer, A. G. Zacchei, and B. H. Arison, Physiological disposition and urinary metabolites of cyproheptadine in the dog, rat, and cat, *Drug Metab. Dispos.* **2**, 406–415 (1974).
46. H. B. Hucker, A. J. Balletto, J. Demetriades, B. H. Arison, and A. G. Zacchei, Epoxide metabolites of protriptyline in rat urine, *Drug. Metab. Dispos.* **3**, 80–84 (1975).
47. H. Kappus and H. Remmer, Irreversible protein binding of ^{14}C-imipramine with rat and human liver microsomes, *Biochem. Pharmacol.* **24**, 1079–1084 (1975).
48. D. M. Jerina, H. Yagi, and J. W. Daly, Arene oxides-oxepins, *Heterocycles* **1**, 267–299 (1973).
49. P. L. Morselli, P. Biandrate, A. Frigerio, M. Gerna, and G. Tognoni, in: *Gas Chromatographic Determination of Carbamazepine and Carbamazepine-10,11-Epoxide in Human Body Fluids* (J. W. A. Meijer, H. Mienardi, C. Gardner-Thorpe, and E. van der Kleijn, eds.), pp. 169–175, Excerpta Medica, Amsterdam (1973).
50. M. Eichelbaum and L. Bertilsson, Determination of carbamazepine and its epoxide metabolite in plasma by high-speed liquid chromatography, *J. Chromatogr.* **103**, 135–140 (1975).
51. J. McCann, N. E. Spingarn, J. Kobori, and B. N. Ames, Detection of carcinogens as mutagens: Bacterial tester strains with R factor plasmids, *Proc. Natl. Acad. Sci. USA* **72**, 979–983 (1975).

20

Glutathione *S*-Transferases

Donald M. Jerina
Laboratory of Chemistry
National Institute of Arthritis, Metabolism, and Digestive Diseases
National Institutes of Health
Bethesda, Maryland 20014, U.S.A.

and

John R. Bend
Pharmacology Branch
National Institute of Environmental Health Sciences
National Institutes of Health
Research Triangle Park, North Carolina 27709, U.S.A.

Any symposium which deals with the metabolic formation and inactivation of reactive metabolites must necessarily consider the prominent role played by glutathione (GSH) and the glutathione transferases since this simple tripeptide (Fig. 1) and the enzymes which employ GSH as a cosubstrate represent a very important factor in the protection of cellular constituents from the adverse effects of reactive chemicals. Thiols such as GSH are highly reactive as nucleophiles. In addition, thiols readily give up hydrogen atoms on reaction with radicals. Thus the importance of GSH in its spontaneous reactions with electrophiles and free radicals within the cell cannot be overemphasized. The remarkably high concentration of 10^{-4} to 10^{-2} M GSH (1) in aerobic cells cannot be serendipitous.

Given that GSH is critically important in the maintainance of the redox potential of the cell from the standpoint of free thiol groups in proteins as well as the protection of other critical nucleophiles in biopolymers, the next question which arises regards the role of the glutathione transferases. Although the main thrust of this chapter will be a discussion of the enzyme-catalyzed reaction of GSH with arene oxides and simple epoxides, a general discussion of the gluta-

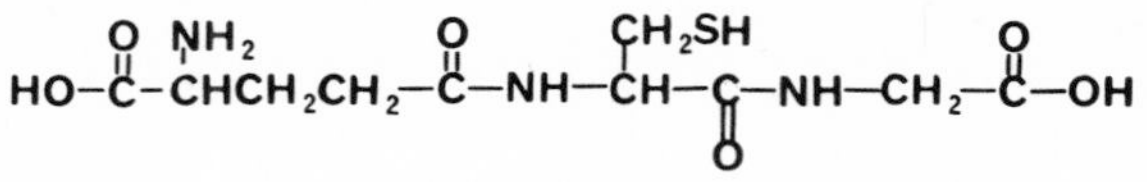

$$HO-\overset{O}{\overset{\|}{C}}-\overset{NH_2}{\overset{|}{C}}HCH_2CH_2-\overset{O}{\overset{\|}{C}}-NH-\overset{CH_2SH}{\overset{|}{C}}H-\underset{O}{\underset{\|}{C}}-NH-CH_2-\overset{O}{\overset{\|}{C}}-OH$$

γ-GLU-CYS-GLY (GSH)

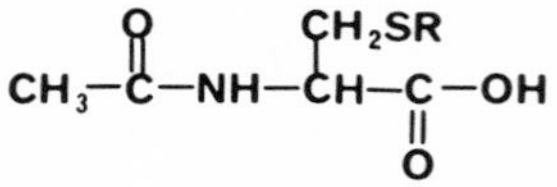

$$CH_3-\overset{O}{\overset{\|}{C}}-NH-\overset{CH_2SR}{\overset{|}{C}}H-\underset{O}{\underset{\|}{C}}-OH$$

S-SUBSTITUTED-N-ACETYLCYSTEINE
(A MERCAPTURIC ACID)

Fig. 1. Structures of the tripeptide glutathione (GSH) and a mercapturic acid metabolite.

thione transferases is essential in order to gain a perspective of the multiple activities which these enzymes possess.

A rich literature exists on the metabolism and excretion of drugs, steroids, pesticides, herbicides, and other environmental chemicals, as *N*-acetylcysteine conjugates by animals. The generally held view on the formation of these metabolites consists of initial conjugation with GSH, loss of the glutamic acid residue via a γ-glutamyltransferase, removal of the glycine residue, and *N*-acetylation of the remaining *S*-substituted cysteine. The resulting conjugates (Fig. 1), commonly known as mercapturic acids, are then excreted via the bile or as urinary metabolites. The remarkably broad spectrum of structural types which undergo conjugation with GSH need not be detailed here as several excellent reviews of this subject are available (2–5). The reactions fall into two basic categories, displacements of good leaving groups from carbon or heteroatoms and addition to activated double bonds. In all of these reactions, the thiol group of GSH functions as the attacking nucleophile. Some typical examples of these reactions are shown in Fig. 2. Displacement of halides, sulfates, sulfonates, phosphates, and nitro groups at saturated carbon atoms are common. The reaction appears to be facilitated if the carbon atom is allylic or benzylic. Displacements of halide and nitro groups occur on aromatic rings provided that the ring system is disposed toward nucleophilic substitution, i.e., has an appropriate number of electron-withdrawing substituents. Thus chloride is displaced from 3,4-dichloronitrobenzene, while chlorobenzene is not a substrate. Strained rings such as epoxides and four-membered lactones are readily opened. Numerous examples of classic Michael additions (6), where the thiol adds to a double bond which bears a strong electron-withdrawing substituent, are known. Conjugation of catechols with GSH via orthoquinone intermediates (2) may also be an example of a Michael addition. Examples of displacement at heteroatoms consist of the release of nitrite from nitrate esters with the production of oxidized glutathione (GSSG) and the similar release of cyanide from organic thiocyanates (3,7). Postulated mechanisms are shown in Fig. 2.

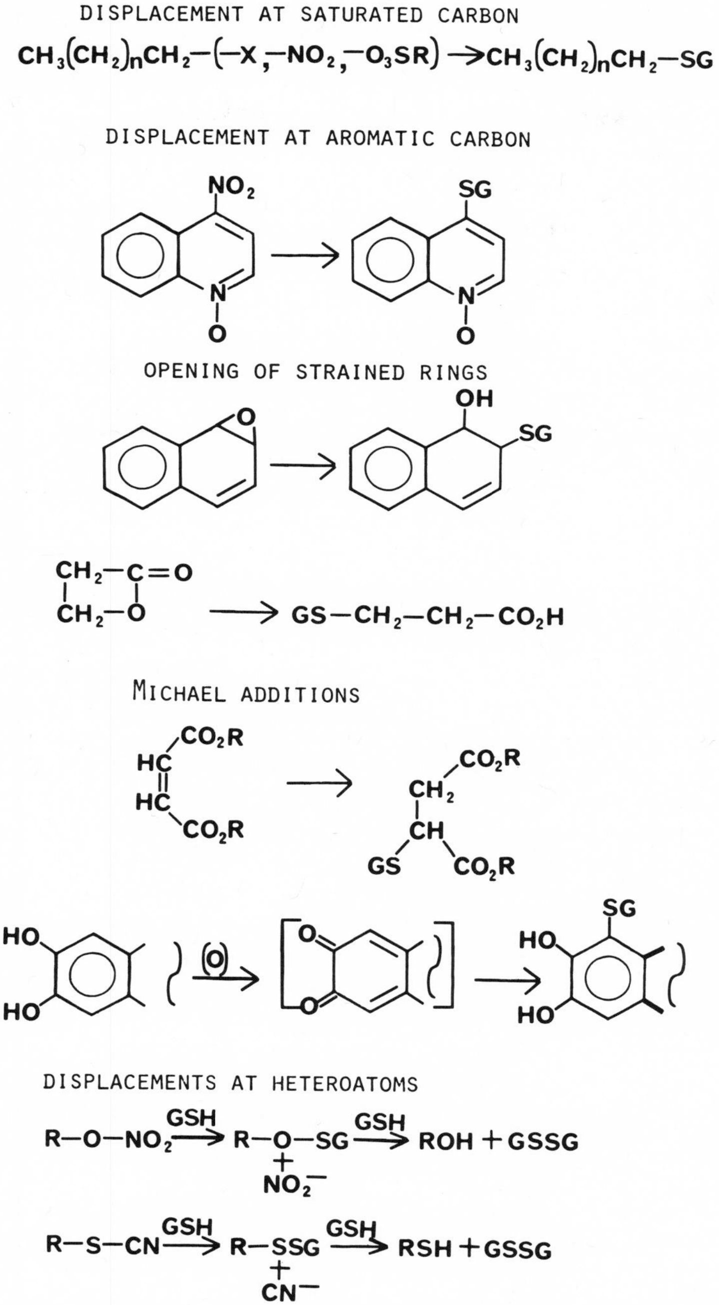

DISPLACEMENT AT SATURATED CARBON

$$CH_3(CH_2)_nCH_2-(-X, -NO_2, -O_3SR) \rightarrow CH_3(CH_2)_nCH_2-SG$$

DISPLACEMENT AT AROMATIC CARBON

NO_2 → SG

N O → N O

OPENING OF STRAINED RINGS

O → OH SG

$$\begin{matrix} CH_2-C=O \\ | \quad\ \ | \\ CH_2-O \end{matrix} \longrightarrow GS-CH_2-CH_2-CO_2H$$

MICHAEL ADDITIONS

$HC(CO_2R)=HC(CO_2R) \longrightarrow GS-CH(CO_2R)-CH_2-CO_2R$

HO, HO $\xrightarrow{(O)}$ [O, O] → HO, HO, SG

DISPLACEMENTS AT HETEROATOMS

$$R-O-NO_2 \xrightarrow{GSH} R-O-SG + NO_2^- \xrightarrow{GSH} ROH + GSSG$$

$$R-S-CN \xrightarrow{GSH} R-SSG + CN^- \xrightarrow{GSH} RSH + GSSG$$

Fig. 2. Examples of reactions which are catalyzed by glutathione S-transferases (2–5).

This remarkable spectrum of substrates which are conjugated with GSH both *in vivo* and *in vitro* with tissue prepreparations has resulted in a series of operational names for these various, or seemingly various, enzyme activities. Attempts at nomenclature and classification have resulted in terms such as "glutathione *S*-aryltransferase," "*S*-alkyltransferase," "*S*-aralkyltransferase," "*S*-epoxidetransferase," and "*S*-alkenetransferase." Although such classifications followed logically from the substrates which had been studied, compelling evidence had not been obtained to indicate that these designations actually represented separate enzymes. Clarification of this problem was forthcoming through the elegant purification studies of Jakoby and co-workers (8,9). The supernatant fraction from rat liver was passed through a diethylaminoethylcellulose column with 90% recovery of transferase activity. The resulting protein fraction was then chromatographed on a carboxymethylcellulose column with an increasing salt gradient. This sequence of purification steps resulted in the identification of seven distinct transferase proteins, which have been designated as AA, A, B, C, D, E, and M (10).

Comparison of the substrate specificities for the several purified glutathione *S*-transferases (Table 1) quickly established that the previous classification of activities was inadequate. Most of the substrates showed detectable activity with most of the enzyme preparations. The substrates in Table 1 are presented in four groups which, in the previous notation, would correspond to substrates for the *S*-epoxidetransferases, *S*-aryltransferases, *S*-alkyl- and aralkyltransferases, and *S*-alkenetransferases. Enzyme E has the highest activity toward 1,2-epoxy-3-(*p*-nitrophenoxy)propane among the six enzymes and three epoxides tested. However, E also has high activity toward several of the substrates in group 3, which react by displacement of leaving group at saturated carbon. In this same group of substrates, enzymes A and C show the highest activity with *p*-nitrobenzyl chloride as the substrate. For displacement of an aromatic substituent, enzyme A was most active toward 1-chloro-2,4-dinitrobenzene in the second group of substrates, and enzyme B was most active toward ethacrynic acid in the last group of substrates. The lack of a clear specificity by any of the enzymes for any of the substrate groups makes a simple classification and nomenclature system difficult. A similar situation has been found for human liver, where several distinct glutathione *S*-transferases have been identified and shown to have overlapping substrate specificities (11).

The problem of classification and nomenclature for the purified glutathione *S*-transferase proteins is even further complicated by the fact that recent studies have established two activities in addition to the catalytic transfer of GSH. One of these activities is the noncovalent binding and transport of endogenous molecules such as bilirubin and metabolites of cortisol. Previously, several laboratories had isolated proteins from rat liver based on their ability to bind a carcinogenic azo dye (12), hemin, bromosulfophthalein (13,14), and the above molecules (15,16). The proteins were established to be identical through the use of an antibody toward one of them, and the term ligandin was coined to

Table 1. **Substrate Specificity of Purified Glutathione S-Transferases from Rat Liver**[a]

Substrate	Glutathione S-transferase[b]					
	A	B	C	E	AA	M
Benzo[a]pyrene 4,5-oxide	0.087	0.011	0.098	0.069	0.004	
Naphthalene 1,2-oxide	0.04	0		0.16		
1,2-Epoxy-3-(p-nitrophenoxy)propane	0.1	0	0	6.7		0
Bromosulfophthalein	0.53	0.006	0.18	0[c]	0.004	
1-Chloro-2,4-dinitrobenzene	62	11	10	0.01	14	
1,2-Dichloro-4-nitrobenzene	4.3	0.003	2.0	0	0.008	0.004
4-Nitropyridine N-oxide	1.7	0		0		0
2,3,5,6-Tetrachloronitrobenzene	3.9	0		0.001		0
Iodomethane	0	0.59	0	8.9	1.4	0
1-Iodopropane	0.39	0.32		1.0		
2-Nitropropane	0.012	0.008	0.014	0	0.01	
Menaphthyl sulfate	0	0.004		0		0.1
p-Nitrobenzyl chloride	11.4	0.1	10.2	4.1	0.09	0.5
p-Nitrophenethyl bromide	0.1			6.1		
trans-4-Phenyl-3-buten-2-one	0.02	0.001	0.40	0		
Ethacrynic acid	0	0.26	0.11	0	0.3	

[a]Data from Jakoby *et al.* (10).
[b]Activity is expressed as μmol conjugate/min/mg protein.
[c]Values of zero indicate that product formation with and without enzyme were not significantly different under the assay conditions.

describe this activity (13). Remarkably, ligandin and glutathione S-transferase B have been shown to be the same protein (17,18). Further studies on the rat transferases AA, A, B, and C have established that each of these proteins binds bilirubin, indocyanine green, and hematin (11). The second of these activities is the covalent binding of small molecules to the protein. Examples are the covalent binding of 4-dimethylaminoazobenzene to ligandin (transferase B) (12) and of 1-chloro-2,4-dinitrobenzene to transferase A (19). At present, the issue is somewhat unclear as to whether or not the soluble "h-protein" in mouse skin, which covalently binds to carcinogenic aromatic hydrocarbons (20), is also a glutathione transferase (21).

The advances in understanding of the glutathione S-transferases, ligandins, and other binding proteins which have occurred in the past few years are truly remarkable. Clearly, terms such as "glutathione S-epoxidetransferase" should no longer be utilized to describe an enzyme activity. However, since the main emphasis of this chapter is on the enzyme-catalyzed reaction between GSH and simple epoxides or arene oxides, we will occasionally use this term with the stipulation that it designates a class of substrates toward which activity is being measured rather than a specific enzyme activity.

FORMATION AND FURTHER METABOLISM OF EPOXIDES AND ARENE OXIDES *IN VIVO*

Metabolism of aromatic hydrocarbons and olefins *in vivo* is now generally accepted to proceed via arene oxides and epoxides. For any given aromatic hydrocarbon, formation of several different arene oxides is possible. In general, monooxygenases in mammals tend to avoid formation of arene oxides at substituted ring positions (22) while such may not be the case in fungi (23). Principal pathways for further reaction of arene oxides include spontaneous isomerization to phenols and solvolysis to dihydrodiols, enzymatic formation of dihydrodiols by epoxide hydrase, and conjugation with GSH to ultimately form mercapturic acids (Fig. 3). As will be discussed later, the structures of many

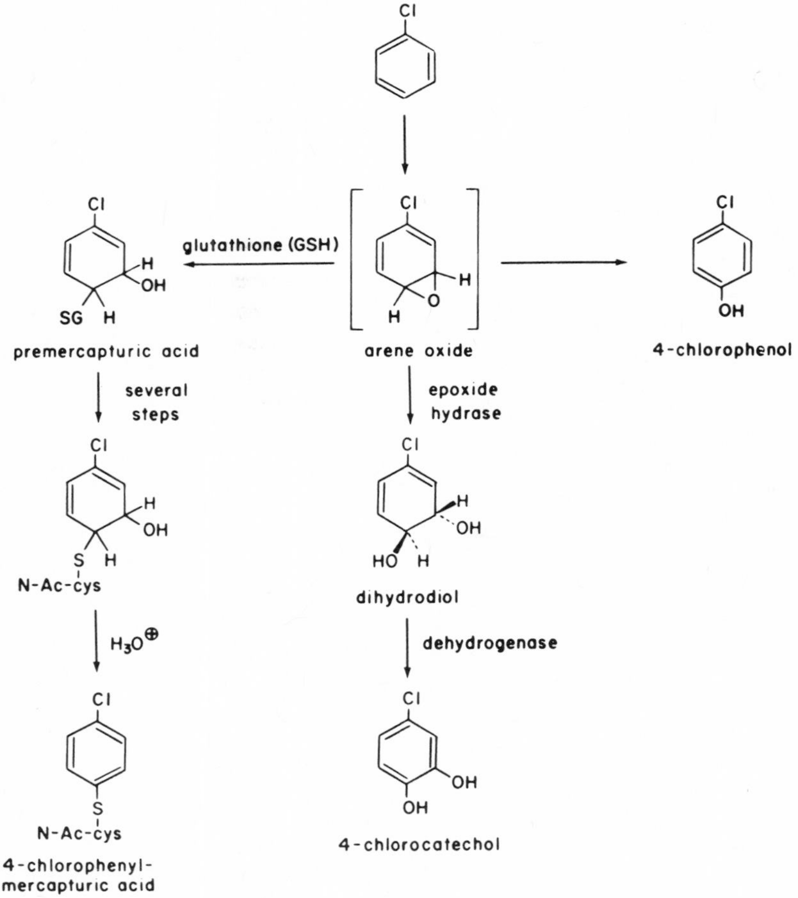

Fig. 3. Typical pathways for metabolism of an aromatic hydrocarbon via an arene oxide.

mercapturic acid precursors from aromatic hydrocarbons may be in error. Catechols result either by dehydrogenation of dihydrodiols (24–26) or by consecutive hydroxylations (27). Phenols, catechols, and dihydrodiols can be detected in urine in the free state or as sulfate and/or glucuronic acid conjugates. For conjugates related to GSH, generally only *N*-acetylcysteine derivatives are found in urine. In contrast, bile and feces often contain GSH, cysteinylglycine, and cysteine conjugates which are precursors to the mercapturic acids. Evidence has been presented which indicates that arene oxides can be metabolically transformed back to the parent hydrocarbon (28) and that direct insertion pathways which do not involve arene oxides are possible routes to phenols (29,30). Extensive reviews on this general area of metabolism are available (31–33).

Although an appreciation of the relative roles played by the glutathione *S*-transferases and epoxide hydrases in the intact animal is of considerable interest in studying the metabolic fate of aromatic hydrocarbons and olefins in the body, relatively few comprehensive metabolism studies have been conducted. Most studies have been qualitative in nature and rarely examine biliary and fecal metabolites which are particularly important for compounds with molecular weights in excess of 250. A comprehensive review of the formation of phenols, catechols, diols, dihydrodiols, mercapturic acids, arene oxides, epoxides, etc., is beyond the scope of this chapter. The *in vivo* studies cited in Table 2 are meant to be representative rather than comprehensive.

METABOLISM OF EPOXIDES AND ARENE OXIDES *IN VITRO*

Toxicity of active metabolites formed from various chemicals *in vivo* is often localized in a single target organ, as in the case of liver necrosis associated with administration of halobenzenes, or pulmonary and mammary carcinomas induced by certain polycyclic aromatic hydrocarbons. For active metabolites which covalently bind to protein, DNA, and RNA, the relationship between the site and method of administration of the chemical, formation of the chemically reactive metabolite in the target organ, and transport of the active metabolite to the target organ from the tissue where it is formed are important factors for consideration. Since, in general, the liver is quantitatively the most important organ for metabolism of nonpolar xenobiotic substances, profiles of urinary metabolites are likely to reflect hepatic metabolism (*cf.* Table 2). Although transport of active metabolites and their metabolic precursors from the liver to remote target tissues where selective uptake might occur is possible, metabolism in extrahepatic tissues must also be considered. Even though drug-metabolizing activity is generally quite low in extrahepatic tissues, metabolic activity which protects against the adverse effects of active metabolites may be even lower. For epoxides and arene oxides as active metabolites, epoxide hydrase and glutathione *S*-transferases are two obvious enzymes capable of catalyzing detoxica-

Table 2. Representative Examples of *in Vivo* Metabolism via Epoxides and Arene Oxides

Arene oxides		Epoxides	
Compound	Reference	Compound	Reference
Benzene and halobenzenes	(34–44)	Styrene	(62–64)
Biphenyl	(45)	Allyl barbiturates	(65)
5-Ethyl-5-phenylbarbituric acid	(46)	2,3-Dimethyl-1-phenyl-4-pyrazolin-5-one (antipyrine)	(66)
5,5-Diphenylhydantoin (Dilantin)	(47)	Carbamazepine, protriptyline, cyproheptadine	(67–69)
[Ethyl 1-(3-cyano-3,3-diphenylpropyl)]-4-phenylisonipecotate (Diphenoxylate)	(48)	Safrole	(70)
2-Ethyl-2-phenylglutanimide (Doriden)	(49)		
N,2-Dimethyl-3-phenylsuccinimide	(50)	Aflatoxin B_1	(71)
Polycyclic hydrocarbons	(51–61)	Phylloquinone	(72)

tion. Moreover, the possibility of inactivation by cytochrome P450 through secondary oxidative metabolism should not be excluded. The emphasis of the present section is on the measurement of epoxide transferase activity in homogenates from liver and extrahepatic tissues. Although purified enzymes clearly provide the best kinetic data, studies of homogenates and isolated, perfused organs provide information on total tissue activity.

Since the overall reaction for glutathione *S*-transferase-catalyzed conjugation of GSH with epoxides and arene oxides involves utilization of both substrates as well as production of a conjugate, assays for enzymatic activity can be based on the disappearance of either substrate or on the formation of product. Assays based on utilization of GSH and formation of conjugate have been described. As with any assay based on disappearance of substrate, maximal sensitivity is seldom possible, because of the difficulty in measuring a small change in concentration of a major reaction component. Nonetheless, disappearance of GSH has been successfully monitored by reaction of residual GSH with iodate (73) and with 5,5′-dithio-bis-(2-nitrobenzoic acid), Ellman's reagent (8). These assays have the advantage that any cosubstrate can be employed, but are best run under anaerobic incubation conditions to reduce destruction of GSH by autooxidation. Several techniques have been employed to monitor product formation from epoxide substrates. Conjugation of 1,2-epoxy-3-(*p*-nitrophenoxy)propane with GSH has been followed spectrophotometrically at 360 nm (8). Unfortunately, few such substrates are available. Radiometric procedures in which either GSH or the epoxide substrate bears the label have proved sensitive and reliable. Use of [^{35}S]GSH allows ready quantitation of product for any cosubstrate, but requires tedious chromatographic separation of the conjugate from unreacted GSH. This method has been employed to assay for enzyme activity (74) as well as to determine substrate specificity (75). Far more rapid and convenient assays based on radioactive epoxides and arene oxides have been developed in which unreacted substrate is simply extracted from the incubation medium into organic solvent, and an aliquot of the aqueous phase containing the water-soluble conjugate is counted. Covalent binding of substrate to water-soluble components of the incubation medium other than GSH has not been a limiting factor. Although this approach is severely limited by availability of labeled substrates, assays based on [^{14}C]styrene oxide (76), [^{3}H]styrene oxide, (77), and several [^{3}H]arene oxides (78,79) have been described. Assay of glutathione transferase activity toward simple epoxides and arene oxides is complicated by the relatively high second-order rate constants for the spontaneous reaction between GSH and these substrates. Manipulation of the GSH concentration and the pH of the incubation medium does, however, permit measurement of the enzymatic reaction. Careful study of the second-order spontaneous reaction between thiols and benzene oxide has established that the rate constants for these reactions are dependent on the pK_a of the thiols and that GSH has a second-order rate constant expected from its pK_a (80).

Glutathione transferase activity has been purified fortyfold from the cytosol of sheep liver. Electrofocusing established that two major and several minor components, with isoelectric points ranging from 6.5 to 7.5, were present in the preparation (74). Apparent K_m values for naphthalene-1,2-oxide (0.11 mM) and styrene oxide (0.13 mM) with this preparation were constant in the range of pH 6.5–8.0, while the apparent K_m for GSH varied from 10 mM to 1.6 mM, respectively. The transferase assay which utilizes [^{35}S]GSH has allowed comparison of the substrate specificity of this preparation toward a wide variety of epoxides and arene oxides (75). With relatively few exceptions, substitution at any position on styrene oxide resulted in decreased substrate activity. Monosubstitution of the phenyl ring by a nitro group or a halide resulted in a 40–70% loss in activity. 4-Phenylstyrene oxide had only 4% of the activity for styrene oxide. Substitution of a methyl or phenyl group at C-7 of the epoxide ring in styrene oxide caused almost complete loss in activity, whereas such 1,1-disubstituted epoxides are relatively good substrates for epoxide hydrase (81). Specificity of the transferase preparation toward *cis* and *trans* 1,2-disubstituted epoxides is the reverse of that known for epoxide hydrase. *trans*-8-Methylstyrene oxide is about 15% more active than styrene oxide with the transferase, while the *cis* isomer has only 20% of the activity found for styrene oxide. Comparison of a series of substituted benzene oxides indicated that electron-withdrawing substituents enhanced the rate of reaction. Thus activity of 3- and 4-carbo-*t*-butoxybenzene oxides was ⩾20-fold better than that of benzene oxide, which had <5% of the activity found for styrene oxide. 4-Carbo-*t*-butoxybenzene oxide was the best substrate tested. In a series of polycyclic arene oxides ranging in size from naphthalene to benzo[*a*]pyrene, there was a trend toward decreasing activity with increasing size. Cross-inhibition studies indicated that naphthalene 1,2-oxide caused a 30% inhibition of the conjugation of styrene oxide at 5% of the styrene oxide concentration. A preliminary study of the homogeneous glutathione *S*-transferases A, B, C, E, and AA from the rat with benzo[*a*]pyrene 4,5-oxide as substrate indicated transferases A and C were most active (78).

Although the cytosol from rat liver (176,000*g* supernatant) clearly contains several glutathione *S*-transferases, study of this subcellular fraction has significance in that estimates of total issue activity are possible. With radioactive styrene oxide, octene 1,2-oxide, and benzo[*a*]pyrene 4,5-oxide as substrates, 10 mM GSH was completely saturating at protein concentrations of 0.10–0.15 mg/ml when incubated at 37°C for 5 min, 10 min, and 10 min, respectively. For all three substrates, bimodal pH profiles were observed with maxima at pH 7.4 and 7.85, the latter being optimal. Apparent kinetic constants for cytosol protein are given in Table 3. The very low K_m of 0.005 mM for benzo[*a*]pyrene 4,5-oxide may reflect the poor solubility of this substrate in water. Specific activity of rat liver cytosol toward styrene oxide, octene 1,2-oxide, and several arene oxides is given in Table 4 along with comparative data for microsomal epoxide hydrase. Structures of the arene oxides are shown in Figure 4. The most

Table 3. Apparent Kinetic Constants for Glutathione S-Transferase Activity in Rat Liver Cytosol[a]

Apparent kinetic constants	Styrene oxide	Octene 1,2-oxide	Benzo[a]pyrene 4,5-oxide
Glutathione K_m (mM)[b]	0.5	0.7	1.4
V_{max}(nmol/min/mg protein)	176.6	42.9	17.9
Epoxide K_m (mM)[c,d]	1.0	2.0	0.005

[a]Data from Bend *et al.* (79).
[b]The concentration of GSH was varied from 0.02 to 20 mM while the cosubstrates styrene oxide (1.0 mM), octene 1,2-oxide (1.0 mM), and benzo[a]pyrene 4,5-oxide (0.10 mM) were used at the indicated concentrations.
[c]The concentration of GSH was 10 mM.
[d]Lineweaver–Burk plots for styrene oxide and octene 1,2-oxide were curvilinear.

striking feature of this comparison of enzyme activities in these crude preparations is that glutathione *S*-transferase activity for each substrate is generally much higher than epoxide hydrase activity. As had been noted previously with the transferase preparation from sheep liver (75), the different enzyme activities complement each other. Epoxide hydrase activity toward benzo[*a*]pyrene 11,12-oxide, 3-methylcholanthrene 11,12-oxide, and dibenzo[*a,h*]anthracene 5,6-oxide is quite low, while glutathione transferase activity toward these substrates is high. For the two non-K-region arene oxides examined, benzo[*a*]pyrene 7,8- and 9,10-oxides, glutathione *S*-transferase activity is low and epoxide hydrase activity is moderate. Interestingly, the two K-region arene oxides of benzo-

Table 4. Specific Glutathione S-Transferase Activity in Rat Liver Cytosol Compared to Specific Epoxide Hydrase Activity in Rat Liver Microsomes with Several Epoxides and Arene Oxides as Substrates[a]

Substrate	Glutathione transferase activity[b]	Epoxide hydrase activity[b]
Styrene oxide	175	6.3
Octene 1,2-oxide	44	13.1
Phenanthrene 9,10-oxide	212	39.3
Benzo[a]anthracene 5,6-oxide	90	12.4
Benzo[a]pyrene 4,5-oxide	10	7.2
Benzo[a]pyrene 7,8-oxide	2.3	8.8
Benzo[a]pyrene 9,10-oxide	0.32	7.4
Benzo[a]pyrene 11,12-oxide	18	0.8
3-Methylcholanthrene 11,12-oxide	101	1.2
Dibenzo[a,h]anthracene 5,6-oxide	35	0.4

[a]Data from Bend *et al.* (79) and Jerina *et al.* (82).
[b]Expressed as nmol product/mg protein/min.

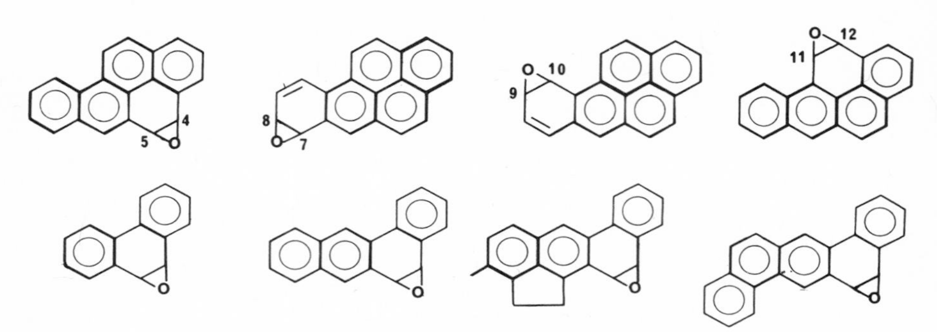

Fig. 4. Structure of the arene oxides utilized in the assay of epoxide hydrase and glutathione S-transferase activities: benzo[a]pyrene oxides (top row); phenanthrene-9,10-oxide, benzo-[a]anthracene-5,6-oxide, 3-methylcholanthrene-11,12-oxide, and dibenzo[a,h]anthracene-5,6-oxide (bottom row).

[*a*] pyrene (4,5 and 11,12 positions) were the best substrates for the glutathione *S*-transferase while the 11,12-K-region arene oxide of benzo[*a*] pyrene was the poorest epoxide hydrase substrate in this set of four arene oxides from benzo-[*a*] pyrene. Either or both of these enzyme activities can play dominant roles in protecting against the biological activity (83) of arene oxides from benzo[*a*] py-rene depending on which arene oxide is considered. Finally, cross-inhibition experiments with rat liver cytosol indicated that neither styrene oxide nor octene 1,2-oxide at 1.0 mM was effective in inhibiting conjugation of benzo[*a*]-pyrene 4,5-oxide at 0.1 mM while conjugation of styrene oxide was inhibited ~30% by an equal concentration of octene 1,2-oxide in the concentration range of 0.2–1.0 mM.

If the total organ epoxide-metabolizing activities are calculated by extrapola-tion from the *in vitro* epoxide hydrase and glutathione *S*-epoxide transferase specific activities, the liver has much more activity than all other tissues com-bined. In the rabbit, for example, the total styrene oxide metabolizing activity (μmol product formed/min/organ) for the two activities was as follows: liver, 251.6; kidney, 3.9; lung, 3.1; small intestinal mucosa, 2.9. The relative contribu-tion of the glutathione *S*-transferase activities was >85% in all four organs. This dominance of the transferase enzymes is only in part due to the much higher protein yield for the soluble fraction compared to the microsomal fraction. These calculations of total *in vitro* activity will not necessarily reflect the *in vivo* situation when styrene is administered to an animal, but do suggest a prominent role for the glutathione *S*-transferases in the *in vivo* metabolism of this substrate. Specific activities of microsomal epoxide hydrase and glutathione *S*-transferase in the cytosol toward styrene oxide are compared for several tissues from rat, rabbit, and guinea pig in Table 5. The highest specific activity of the glutathione *S*-transferase was found in liver, followed by kidney in all cases. For epoxide hydrase, liver had the highest specific activity, followed by either the small

Table 5. Comparison of Epoxide Hydrase and Glutathione S-Transferase Activity Toward Styrene Oxide in Several Tissues from Rat, Rabbit, and Guinea Pig[a]

	Rat		Rabbit		Guinea pig	
Tissue	Epoxide hydrase[b]	Glutathione transferase[c]	Epoxide hydrase	Glutathione transferase	Epoxide hydrase	Glutathione transferase
Liver	4.9	142	5.6	31	14	237
Lung	0.21	12	0.2	6.5	0.50	19
Intestine	0.23	13	2.8	4.4	5.8	25
Kidney	0.80	82	1.4	7.9	1.1	74

[a]Assays conducted with [^{14}C]styrene oxide at 1.0 mM. Data from James *et al.* (76).
[b]Specific activity expressed as nmol styrene glycol/mg protein/min.
[c]Specific activity expressed as nmol conjugate/mg protein/min.

intestinal mucosa or the kidney (rat). Comparison of the substrate specificity of the cytosol fraction from several tissues of the male rat toward epoxides and K-region arene oxides is given in Table 6. Relative to liver, kidney and testis had fairly high glutathione *S*-transferase activities (19–100%) for all substrates tested, whereas activities of the lung and small intestine (5–25%) were appreciably lower. Substrate-specific differences in relative glutathione *S*-transferase rates between liver and other tissues is not surprising since several glutathione *S*-transferases are known. The relatively high activities in testis may be of some physiological importance since arene oxides are known to be mutagenic toward both bacterial and mammalian cells (83).

Substantial concentrations of GSH occur in virtually all mammalian cells (84). The concentration of this cosubstrate for the glutathione *S*-transferases could, however, be rate limiting in selected tissues or become rate limiting under conditions where the rate of formation of glutathione conjugates exceeds the

Table 6. Glutathione S-Transferase Activity toward Several Substrates in Extrahepatic Tissues of Male Rats[a,b]

Substrate	Lung	Kidney	Small intestine	Testis
Styrene oxide	12(7%)	97(55%)	16(9%)	65(37%)
Octene 1,2-oxide	5.7(13%)	36(81%)	6.9(16%)	24(55%)
Benzo[*a*]anthracene 5,6-oxide	4.5(5%)	18(20%)	5.1(5.7%)	17(19%)
Benzo[*a*]pyrene 4,5-oxide	2.3(23%)	10(101%)	2.5(25%)	4.9(50%)
Benzo[*a*]pyrene 11,12-oxide	4.0(22%)	13(69%)	4.0(22%)	16(86%)

[a]Specific activity expressed as nmol conjugate/mg protein/min for the several radioactive substrates. Data from Bend *et al.* (79).
[b]Values in parentheses represent percentage of the specific activity found for the cytosol from liver.

Table 7. Concentration of GSH in Hepatic and Selected Extrahepatic Tissues of Male CD Rats and Male New Zealand Rabbits[a]

Tissue	μg GSH/g wet tissue[b]	
	Rat	Rabbit
Liver	2884 ± 355 (3)	1879 ± 822 (4)
Lung	480 ± 29 (3)	843 ± 107 (4)
Small intestine	620 ± 72 (3)	878 ± 207 (4)
Kidney	216 ± 20 (3)	154 ± 108 (4)

[a]The concentration of GSH was measured by a fluorometric procedure (85). Data from Ben-Zvi *et al.* (86). Fresh homogenates were stored in ice for only a few minutes prior to TCA precipitation.
[b]Results are expressed as mean ± SD (*n*).

rate of resynthesis of GSH. Such is clearly the case in the liver after administration of massive doses of bromobenzene (38). The data on the concentration of GSH in hepatic and extrahepatic tissues of the rat and rabbit in Table 7 are generally consistent with previous reports (84), with the exception of the kidney, which was somewhat lower. In both species, liver had the highest level and was followed by the small intestinal mucosa. Apparent K_m values for GSH as cosubstrate, with styrene oxide at 1.0 mM, were 1.5, 0.21, 0.14, and 0.51 mM for the 176,000*g* supernatant from the liver, lung, small intestine, and kidney of the rabbit, respectively. Thus the endogenous concentrations of GSH are greater than the K_m for GSH in each of these tissues.

The effect of pretreating animals with common inducing agents for drug metabolism on glutathione *S*-transferase activity, with few exceptions (87,88), is generally agreed to result in induction (13,89–91). The effects of pretreatment by phenobarbital, dibenzanthracene, pregnenolone-16α-carbonitrile, and TCDD (2,3,7,8-tetrachlorodibenzo-*p*-dioxin) on hepatic microsomal epoxide hydrase and glutathione *S*-transferase activities toward [^{14}C]styrene oxide in male and female rats are shown in Table 8. Phenobarbital induced both activities in both sexes. In separate experiments, induction of transferase activity by phenobarbital ranged from 25 to 60%. Weak but significant induction of transferase activity by TCDD was also observed. As previously noted (91), pregnenolone-16α-carbonitrile causes a weak induction of transferase activity in the female rat. Extrahepatic epoxide hydrase and glutathione *S*-transferase activities were refractory to pretreatment with all of these compounds, with the following exceptions: epoxide hydrase activity in the kidney of female rats doubled in response to TCDD (0.84 ± 0.07 nmol/min/mg protein vs. 1.65 ± 0.41 nmol/min/mg, mean ± SD, n = 3), whereas no change was observed in the males, and the epoxide hydrase activity in the small intestine increased in both male and female rats as a result of phenobarbital pretreatment (males 0.27 ± 0.01 vs.

Table 8. Effects of Pretreatment of Male and Female Rats with Phenobarbital, Dibenzo[*a,c*]anthracene, Pregnenolone-16α-Carbonitrile, or 2,3,7,8-Tetrachlorodibenzo-*p*-dioxin on Glutathione *S*-Transferase and Epoxide Hydrase Activity

	Specific activity (nmol/min/mg protein)[a]			
	Glutathione *S*-transferase		Epoxide hydrase	
Inducing agent	Males	Females	Males	Females
Control	187	110	5.5	5.4
Phenobarbital[b]	367	157	13.9	14.9
Control	163	122	5.7	6.0
Dibenzo[*a,c*]anthracene[c]	158	126	5.0	7.3
Control	183	126	4.8	6.0
Pregnenolone-16α-carbonitrile[d]	194	179	5.3	9.0
Control	147	106	6.1	4.9
2,3,7,8-Tetrachlorodibenzo-*p*-dioxin[e]	210	151	6.9	7.6

[a]Assayed as described with [^{14}C]styrene oxide (76) except that the concentration of GSH was 10 mM for the measurement of transferase activity. Results are the means of three determinations.
[b]Sodium phenobarbital (80 mg/kg) dissolved in saline and injected i.p. for 3 consecutive days. Sacrificed 24 h after last injection. Controls treated with saline.
[c]Dibenzo[*a,c*]anthracene (80 mg/kg) suspended in 0.5% CMC solution and injected i.p. for 3 consecutive days. Sacrificed 24 h after last injection. Controls treated with CMC.
[d]Pregnenolone-16α-carbonitrile (20 mg/kg) suspended in 0.5% CMC solution and injected for 2 consecutive days. Sacrificed 24 h after last injection. Controls treated with CMC.
[e]2,3,7,8-Tetrachlorodibenzo-*p*-dioxin (10 μg/kg) dissolved in corn oil–acetone (9:1) and injected i.p. as a single dose. Males sacrificed 6 days after the last dose, females sacrificed 10 days after the last dose. Controls treated with corn oil–acetone (9:1).

0.38 ± 0.08 nmol/min/mg protein; females, 0.23 ± 0.04 vs. 0.50 ± 0.17 nmol/min/mg protein).

During development the fetus is exposed to xenobiotic substances and their metabolites which cross the placental barrier. The presence or absence of epoxide-metabolizing enzyme systems in fetal and neonatal animals, and the relative amounts of enzymes in various tissues at various stages of development, may be very important in protecting against epoxide-mediated toxicity in the immature. Consequently, the developmental patterns of glutathione *S*-transferase and epoxide hydrase activities were studied with styrene oxide as substrate in the liver and several extrahepatic tissues of the guinea pig and the rabbit (92,93). Data for the rabbit are given in Tables 9 and 10. Surprisingly, the specific activity of the glutathione *S*-transferases in fetal tissue during the prenatal period is lower in liver compared to other tissues. After birth, the levels of glutathione *S*-transferase in liver increase and finally greatly exceed extrahepatic levels. In contrast, epoxide hydrase is much higher in liver both pre- and

Table 9. Perinatal Development of Glutathione *S*-Epoxidetransferase Activity in 176,000*g* Supernatant Fractions from Liver, Lung, Small Intestine, Kidney, and Placenta of the New Zealand Rabbit[a]

	Glutathione *S*-epoxide transferase activity (nmol/min/mg protein)				
Age (days)	Liver	Lung	Small intestine	Kidney	Placenta
17 (gestation)	1.1	4.8	3.5	4.5	4.3
20 (gestation)	0.87	4.8	3.5	4.5	3.8
25 (gestation)	1.1	5.7	3.4	5.0	4.3
28 (gestation)	0.89	3.8	1.9	3.6	2.5
31 (gestation)	2.1	5.0	2.4	4.3	2.6
1	1.7	2.6	1.3	4.3	
3	5.1	3.3	1.7	5.9	
6	10	3.9	1.6	8.0	
13	10	4.7	1.7	9.5	
20	10	5.5	2.1	9.4	
30	13	4.9	4.0	10	
70	30	7.1	6.1	11	
84	30	6.9	4.6		
98	28	6.4	3.8		
112	27	6.6	3.9	15	
Adults (males)	21	8.0	4.2	9.0	

[a]Assayed with [^{14}C]styrene oxide as previously described (76). Results are expressed as means of two or three litters. Data from James *et al.* (93).

postnatally. Epoxide hydrase levels appear to increase in fetal liver during prenatal development, decrease at birth, and then increase again to adult levels. A similar trend was observed for extrahepatic tissues. Data such as these can be most profitably used to identify target organs which might be especially sensitive or resistant to epoxide-induced toxicity during the developmental period. It appears that the glutathione transferases are quantitatively much more important for epoxide detoxication during the fetal period than is microsomal epoxide hydrase (based on comparisons of specific and total activities toward styrene oxide in liver and extrahepatic tissues).

METABOLISM OF STYRENE OXIDE BY THE ISOLATED, PERFUSED LUNG

Study of metabolism in isolated, perfused organs has advantages in that cellular integrity for the organ is maintained and that a metabolism profile specific for the organ can be obtained. In an attempt to obtain a more physiological measure of the relative roles of the glutathione *S*-transferases and epoxide hydrase, metabolism of [^{14}C]styrene oxide was examined by recirculat-

Table 10. Perinatal Development of Microsomal Epoxide Hydrase Activity in Liver, Lung, Small Intestine, Kidney, and Placenta of the New Zealand Rabbit[a]

	Epoxide hydrase activity (nmol/min/mg protein)				
Age (days)	Liver	Lung	Small intestine	Kidney	Placenta
20 (gestation)	0.29				0.02
25 (gestation)	0.76		0.13		0.22
28 (gestation)	1.4	0.08	0.10	0.14	0.05
31 (gestation)	1.7	0.13	0.11	0.23	0.08
1	0.88	0.11	0.11	0.12	
3	0.91	0.08	0.15	0.12	
6	0.77	0.08	0.13	0.09	
13	0.99	0.24	0.23	0.24	
20	1.8	0.25	0.51	0.37	
30	5.5	0.26	0.83	0.82	
70	6.4	0.18	5.0	1.2	
84	6.4		3.7		
98	6.0		2.5	1.4	
112	8.2	0.15	3.7	0.94	
Adults (males)	6.4	0.37	3.1	0.94	

[a] Assayed with [^{14}C]styrene oxide as previously described (76). Results are expressed as means of two or three litters. Data from James *et al.* (93).

ing perfusion of the isolated lung of the New Zealand rabbit (94), a technique previously used to study mixed-function oxygenase metabolism (95). Production of ethyl acetate nonextractable water-soluble (pH 7.4) metabolites as a function of styrene oxide concentration in the perfusate after 30 min of perfusion is shown in Fig. 5. Perfusion with 66 μmol of styrene oxide for 60 min results in ~50% depletion of the endogenous GSH in the tissue. Very little covalent binding of the styrene oxide to cellular macromolecules occurs under these

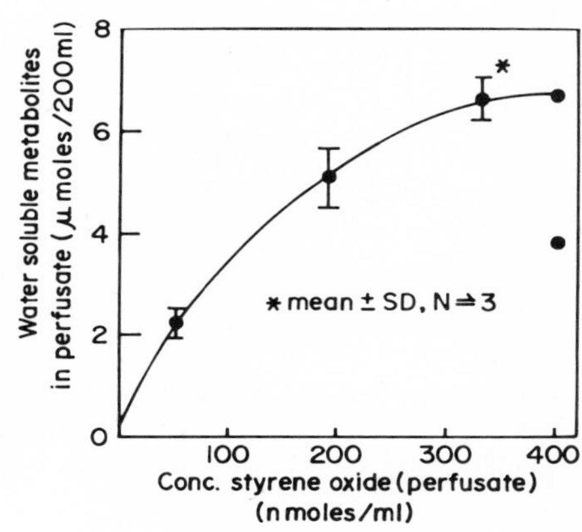

Fig. 5. Formation of water-soluble conjugates (nonextractable into ethyl acetate from pH 7.4 perfusate) on perfusion of the isolated rabbit lung for 30 min (94) as a function of the [^{14}C]styrene oxide concentration in the perfusate (2.7% dextran–Krebs–bicarbonate). At 400 nmol of styrene oxide/ml of perfusate (80 μmol/lung), edema often occurs (lower point at 400 nmol/ml). Approximately 30% of the total metabolism of styrene oxide, at a concentration of 40 μmol/lung, results from initial attack by GSH.

conditions. Of the total amount of styrene oxide (40 mol) perfused for 30 min, about 70% of the products can be accounted for as styrene glycol and mandelic acid. The balance of the metabolism occurs to produce water-soluble conjugates resulting from reaction of styrene oxide with GSH. After the first few minutes of perfusion, the GSH conjugate of styrene oxide (metabolite A) is the only thio ether metabolite found in the lung. After 60 min of perfusion, nonextractable metabolites in the perfusate consisted of: A (21%), B (21%), and C (42%), where metabolite B is a cysteineglycine conjugate and metabolite C is mandelic acid. Time-course experiments suggested that metabolite A serves as a precursor for metabolite B. The dominance of the transferase enzymes over epoxide hydrase, toward styrene oxide in the *in vitro* assays of enzyme activity, is not seen in the perfused lung. Studies are in progress to establish whether this pattern of metabolism is obtained when styrene rather than styrene oxide is employed in the perfusion experiments.

COVALENT BINDING TO CELL COMPONENTS

Metabolism-induced binding of exogeneous substances to various components of the cell has been the underlying theme of this symposium. As applied to arene oxides and epoxides as active metabolites, extensive discussions of the role played by these metabolites in binding to nucleic acid (96–99) and in causing cell death (100–102) have been presented elsewhere. While a clear protective role for GSH and glutathione *S*-transferases has been established for hepatic necrosis, experiments are unavailable to indicate that a similar protective role also exists for chemical carcinogenesis. The present section considers some aspects of binding to protein which have not been treated elsewhere.

Although the thiol residue of cysteine is widely considered as the target site when arene oxides and epoxides bind to protein, several other potential routes for incorporation of hydrocarbon deserve mention. For example, 1,2-epoxy-3-(*p*-nitrophenoxy)propane has been demonstrated to esterify a β-carboxyl group of an aspartyl residue and to alkylate a methionyl residue in pepsin (103). Other enzymes for which epoxides have been utilized as active-site-directed inhibitors include lysozyme, phosphoglucose isomerase, and α- and β-glucosidases (104).

Direct interaction of a cysteine thiol group in a protein with an arene oxide or an epoxide is not the only mechanism by which *S*-substituted proteins can be formed. Conjugates formed from phenathrene 9,10-oxide, benzo[*a*]anthracene 5,6-oxide, and dibenzo[*a,h*]anthracene 5,6-oxide and cysteine are all activated and transferred to tRNA. Subsequent incorporation of these modified cysteines into protein occurs, but not always at a site which normally contains cysteine (105). Thus modified cysteines have been incorporated into β-galactosidase from *Escherichia coli* (106) and into rabbit hemoglobin (107).

STRUCTURE OF GSH–ARENE OXIDE ADDUCTS

Definitive structural studies on mercapturic acids formed from intermediate arene oxides and epoxides *in vivo* and on related GSH conjugates formed *in vitro* are practically unavailable. The occasional high instability of the conjugates as well as the peptide nature of these molecules makes rigorous structural analysis difficult. For an unsymmetrical arene oxide, several modes of attack by GSH are possible (Fig. 6). Addition of the thiol can occur at either carbon of the epoxide group, and the relative stereochemistry of the resulting hydroxyl and thioether groups can be *cis* or *trans*. Homoallylic addition of the thiol is also possible.

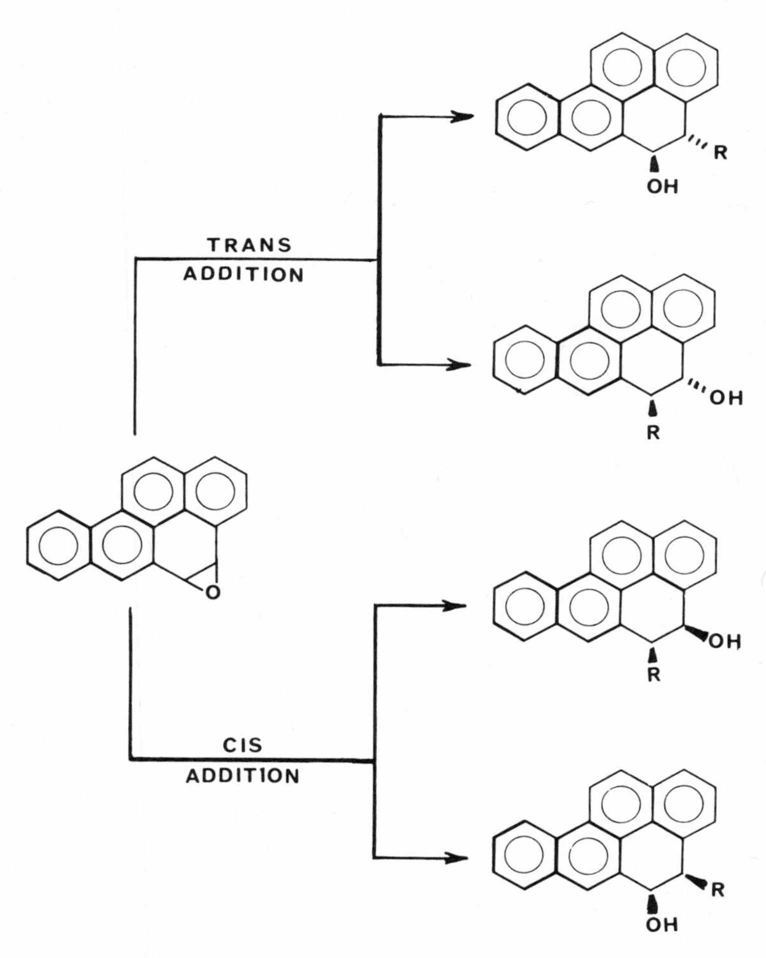

Fig. 6. Four possible adducts resulting from attack of a nucleophile (R^-) on the unsymmetrical benzo[a]pyrene 4,5-oxide.

Although most studies have assumed that the structures of chemically formed adducts between arene oxides and GSH are identical to those formed enzymatically, this need not be the case. Since the spontaneous rate of reaction between GSH and arene oxides can be quite high, samples of enzymatically formed adduct of sufficient size for chemical analysis are difficult to obtain. Although numerous reports have appeared on the structures of GSH–arene oxide adducts, particularly in the polycyclic aromatic hydrocarbon series (see ref. 33 for a review), these assignments must be considered as tentative.

Studies of the reactions between simple nucleophiles and typical arene oxides such as benzene oxide, naphthalene 1,2-oxide, and phenanthrene 9,10-oxide have proved quite valuable in establishing fundamental chemistry involved in these systems (108–111). Reactions between benzene oxide and simple nucleophiles (Fig. 7) occur by direct 1,2 addition as well as by homoallylic 1,4 and 1,6 addition (100) to form both *cis* and *trans* adducts. The preferred point of attack (>90%) on naphthalene 1,2-oxide is at carbon atom 2. For phenanthrene 9,10-oxide, only direct *trans* addition was detected under basic conditions (111). When arene oxides are mixed with aqueous solutions of nucleophiles, at least three competitive reactions occur: spontaneous isomerization to phenols, solvolysis to diols, and attack by the added nucleophile. Thus a comparison of relative reaction rates is of interest (Table 11). A thiol was chosen because soft and polarizable nucleophiles are particularly reactive toward arene oxides in general. Of the five arene oxides examined, phenanthrene 9,10-oxide was the

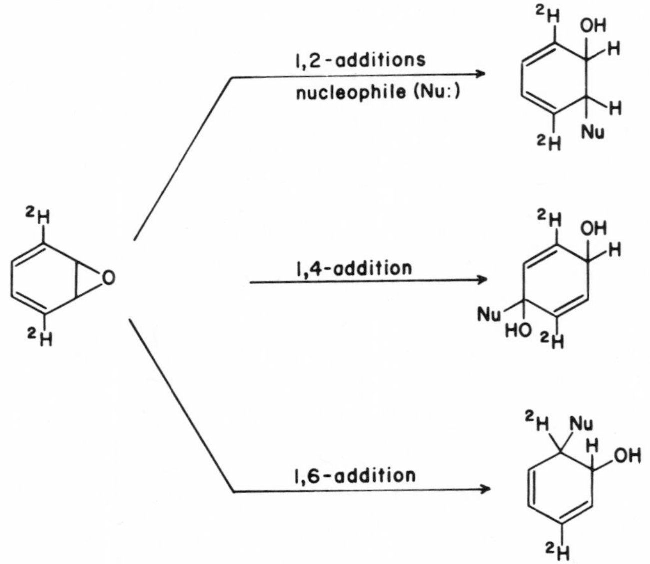

Fig. 7. The use of specifically deuterated benzene oxide to demonstrate 1,6 addition as well as 1,2 and 1,4 addition of nucleophiles (109).

Table 11. Relative Rates of Spontaneous Ring Opening (*A*) and Nucleophilic Attack (*B*) by 2-Mercaptoethanol Anion[a]

Compound	A^b	B^c	B/A = nucleophilic susceptibility
Ethylene oxide	1	1	1
Benzene oxide	2,000	4	0.002
Naphthalene 1,2-oxide	5,000	41	0.008
Phenanthrene 1,2-oxide	50,000	39	0.0008
Phenanthrene 3,4-oxide	88,000	50	0.0006
Phenanthrene 9,10-oxide	300	82	0.3

[a]Data from Bruice *et al.* (111).
[b]Ethylene oxide measured at 25°C, all others at 30°C.
[c]Attack on ethylene oxide measured at 20°C, others at 30°C.

most reactive with the nucleophile, had the lowest rate of spontaneous ring opening, and had the highest susceptibility to attack by the nucleophile. The similarity of the "nucleophile susceptibility index" for phenathrene 9,10-oxide and ethylene oxide, the solvolysis of K-region arene oxides to diols, and the generally high stability of K-region vs. non-K-region arene oxides (111) indicate that K-region arene oxides can be expected to display chemistry typified by simple epoxides.

Extrapolation of the above chemical studies to the study of mercapturic acid formation provides several valuable insights. For any given polycyclic hydrocarbon, the ratio of GSH and related conjugates to phenols should be much higher for a K-region arene oxide since much of the *in vivo* conjugates may be formed by spontaneous reaction (44) because of the high levels of GSH, and since the cytosol transferases have a higher specificity for K-region arene oxides (Table 4). Similarly, for a given polycyclic hydrocarbon, K-region arene oxides can be expected to alkylate cellular components to a greater extent than non-K-region arene oxides. In terms of total arene oxide bound, the fraction due to K-region arene oxides will also be determined by relative rates of enzymatic formation and detoxication as well as by any biological specificity for binding. Chemical examples of homoallylic addition or *cis* addition of GSH to polycyclic arene oxides under basic conditions are presently unknown. The position of attachment of the thioether residue in GSH–arene oxide conjugates, however, is open to serious question.

The structure of the chemically formed adduct between naphthalene 1,2-oxide and GSH has been incorrectly assigned due to an unexpected rearrangement on dehydration (112). Previous assignments were based primarily on the fact that *S*-(1-naphthyl)glutathione forms when the mercapturic acid precursor is dehydrated (Fig. 8). Normally, this would imply the initial attack by the thiol had occurred at C-1 of naphthalene oxide. Since this was not the case for several simple nucleophiles (azide, methide, ethanethiolate) which had been studied (109), the adduct with GSH was reexamined (112). Oxidation of the adduct to

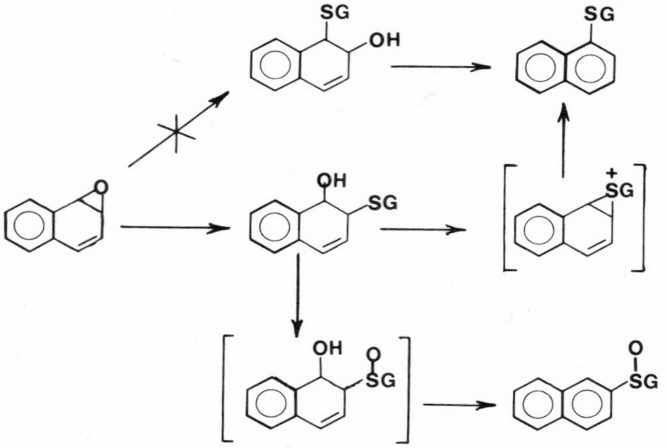

Fig. 8. The predominant attack of GSH on naphthalene 1,2-oxide occurs at C-2. Direct dehydration of the adduct results in migration of the sulfur residue while prior oxidation to a sulfoxide results in dehydration without migration (112).

a sulfoxide prior to dehydration resulted in a 2-substituted naphthalene (Fig. 8). Further examples of these migrations and a discussion of the factors which determine the site at which nucleophiles attack arene oxides are available elsewhere (113). Clearly, the structures of the numerous GSH conjugates derived from simple epoxides and arene oxides will have to be reinvestigated with the aim of establishing the ratio of positional isomers, their relative stereochemistry, and the nature of the products obtained on dehydration.

MECHANISM OF GLUTATHIONE *S*-TRANSFERASES

Relatively little is known about the mechanism by which the glutathione *S*-transferases catalyze the displacement of leaving groups by GSH and the addition of GSH to double bonds. Availability of the several purified transferases has allowed the demonstration of remarkable cross-specificities toward substrates from both classes. Put in other terms, the nature of the leaving group and the site of the displacement seem relatively unimportant, given that the substrate is chemically reactive toward a thiolate. In contrast, the enzymes have a strict requirement for the cosubstrate GSH. Other thiols such as cysteine, dithiothreitol, 2-mercaptoethanol, and thioglycerol are ineffective (73,75,114), possibly because they are not bound by the enzymes. Presumably, the transferases function by common or related mechanisms toward the several types of substrates.

At present, the only transferase for which detailed kinetic studies have been conducted is transferase A (19). With 1,2-dichloro-4-nitrobenzene as substrate, a double-reciprocal plot for saturation with GSH was biphasic. At concentrations of GSH higher than 0.15 mM, an apparent K_m of 0.2 mM was calculated. Below 0.10 mM in GSH, an apparent K_m of 0.01 mM for GSH was found. Studies of product inhibition, competitive substrate inhibition, and covalent binding of radioactive benzyl chloride and 1-chloro-2,4-dinitrobenzene to the enzyme indicated a duality of mechanism in which the major pathway depends on substrate (GSH) concentration. The combined reaction scheme is shown in Fig. 9. Under conditions of high GSH concentration, an ordered sequential pathway predominates in which the enzyme binds GSH, the enzyme–GSH complex (E-GSH) binds the aryl halide to form a second complex (E-GSH-RCl), chloride is released, and the glutathione conjugate (GSR) is released to regenerate resting enzyme. This single displacement pathway was to be anticipated. A more interesting double-displacement pathway predominates under conditions of low GSH concentration. In the ping-pong mechanism, the enzyme binds the organic halide first (E-RCl), chloride is released to form arylated enzyme (ER), GSH is bound (ER-GSH), and the conjugate is released to regenerate resting enzyme.

The same conjugate will be produced by either mechanism when an aryl halide is the substrate. This is not expected to be the case, however, with a cyclic epoxide as the substrate. The single-displacement pathway should produce a conjugate in which the relative stereochemistry of the hydroxyl group and the GSH residue are *trans* while the relative stereochemistry of these groups should be *cis* via the double-displacement pathway (see Figs. 6 and 9). Experiments should be directed at resolution of this point because of its mechanistic implications.

In general, the purified glutathione *S*-transferases are not particularly spectacular in catalytic activity. Where rate data have been obtained as a function of pH, the catalyzed rates are seldom much more than an order of magnitude higher than nonenzymatic rates under the same conditions (8,9,19). Although much of the rate enhancement may be due to the entropy effect of binding both substrates at the "active site," general acid–base catalysis by the protein prob-

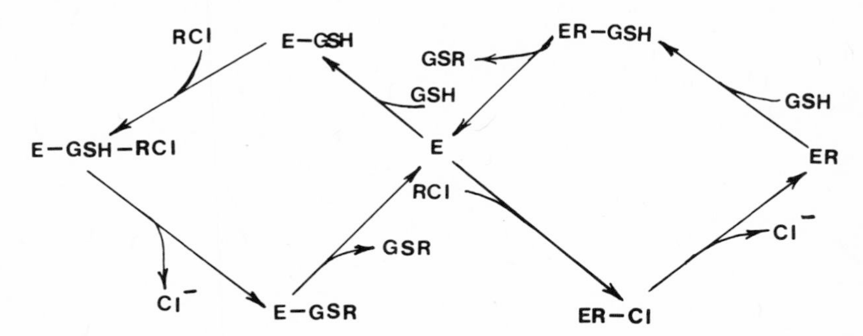

Fig. 9. Expected single-displacement mechanism (left) and unexpected double-displacement mechanism (right) for glutathione transferase A.

ably plays a significant role. Substantial variation in pH optima for different substrates complicates identification of functional groups which may be acting in a catalytic fashion. In the case of the covalent binding of 4-dimethylaminoazobenzene to transferase B, a cysteine residue seems to be involved (115). The nature of the amino acid involved when substrates covalently bind to transferase A in the presence of low concentrations of GSH was not investigated (19). Although attachment via a thioether would seem attractive, amino acid residues other than cysteine could be involved as indicated in the section on covalent binding.

CONCLUDING REMARKS

Glutathione *S*-transferases are a family of related proteins with a broad spectrum of activities ranging from detoxication of xenobiotic substances and their metabolites, to binding and transport of physiologically important molecules such as bilirubin, to covalent binding with carcinogenic azo dyes. Comparison of substrate specificities for homogeneous glutathione *S*-transferases has established that the several proteins have broad specificities for chemically different classes of substrates. In general, the transferases catalyze the conjugation of GSH only with compounds for which there is a significant spontaneous chemical reaction rate with GSH. Known reactions consist of displacement of good leaving groups from saturated carbon atoms, aromatic carbon atoms, and heteroatoms, opening of strained rings such as small-ring lactones and epoxides, and addition to activated double bonds. With regard to the metabolism of simple epoxides and of arene oxides, it is clear that the transferases are at least as important as, if not more important than, epoxide hydrase. From the standpoint of the metabolism of xenobiotic substances to epoxides and arene oxides, it is of interest to note that hydrase and transferase activities complement each other. Although the mechanisms by which the transferases act have yet to be elucidated, both a direct displacement pathway and a mechanism in which the substrate becomes covalently bound to the enzyme have been demonstrated.

REFERENCES

1. E. M. Kosower, Chemical properties of glutathione, in: *Glutathione, Metabolism and Function* (I. M. Arias and W. B. Jakoby, eds.), pp. 1–15, Raven Press, New York (1976).
2. L. F. Chasseaud, Conjugation with glutathione and mercapturic acid excretion, in: *Glutathione, Metabolism and Function* (I. M. Arias and W. B. Jakoby, eds.), pp. 77–114, Raven Press, New York (1976).
3. L. F. Chasseaud, The nature and distribution of enzymes catalyzing the conjugation of glutathione with foreign compounds, *Drug Metab. Rev.* **2,** 185–220 (1973).
4. J. L. Wood, Biochemistry of mercapturic acid formation, in: *Metabolic Conjugation*

and Metabolic Hydrolysis, Vol. II (W. H. Fishman, ed.), pp. 261–299, Academic Press, New York (1970).
5. E. Boyland and L. F. Chasseaud, The role of glutathione and glutathione *S*-transferases in mercapturic acid biosynthesis, *Adv. Enzymol.* **32,** 173–219 (1969).
6. J. B. Hendrickson, D. J. Cram, and G. S. Hammond, *Organic Chemistry,* McGraw-Hill, New York (1970).
7. W. H. Habig, J. H. Keen, and W. Jakoby, Glutathione *S*-transferase in the formation of cyanide from organic thiocyanates and as an organic nitrate reductase, *Biochem. Biophys. Res. Commun.* **64,** 501–506 (1975).
8. T. A. Fjellstedt, R. H. Allen, B. K. Duncan, and W. Jakoby, Enzymatic conjugation of epoxides with glutathione, *J. Biol. Chem.* **248,** 3702–3707 (1973).
9. W. H. Habig, M. J. Pabst, and W. B. Jakoby, Glutathione *S*-transferases: The first enzymatic step in mercapturic acid formation, *J. Biol. Chem.* **249,** 7130–7139 (1974).
10. W. B. Jakoby, W. H. Habig, J. H. Keen, J. N. Ketley, and M. J. Pabst. Glutathione *S*-transferases: Catalytic aspects, in: *Glutathione, Metabolism and Function* (I. M. Arias and W. B. Jakoby, eds.), pp. 189–211, Raven Press, New York (1976).
11. K. Kamisaka, W. H. Habig, J. N. Ketley, L. M. Arias, and W. B. Jakoby, Multiple forms of human glutathione *S*-transferase and their affinity for bilirubin, *Eur. J. Biochem.* **60,** 153–161 (1975).
12. B. Ketterer, P. Ross-Mansell, and J. K. Whitehead, The isolation of a carcinogen-binding protein from livers of rats given 4-dimethylaminoazobenzene, *Biochem. J.* **103,** 316–324 (1967).
13. G. Litwak, B. Ketterer, and I. M. Arias, Ligandin: A hepatic protein which binds steroids, bilirubin, carcinogens, and a number of exogeneous organic anions, *Nature (London)* **234,** 466–467 (1971).
14. J. A. Meuwissen, B. Ketterer, and B. B. Mertens, Binding constants for haem and bilirubin to purified ligandins, *Digestion* **6,** 293 (1972).
15. A. J. Levi, A. Gatamaitan, and I. M. Arias, The role of two hepatic cytoplasmic proteins (Y and Z) in the transfer of sulfobromophthalein (BSP) and bilirubin from plasma into liver, *J. Clin. Invest.* **48,** 2156–2167 (1969).
16. K. S. Morey and G. Litwak, Purification and properties of cortisol metabolite binding proteins of rat liver cytosol, *Biochemistry* **8,** 4813–4821 (1969).
17. W. H. Habig, M. J. Pabst, G. Fleischner, Z. Gatmaitan, I. M. Arias, and W. B. Jakoby, The identity of glutathione *S*-transferase B with ligandin, a major binding protein of liver, *Proc. Natl. Acad. Sci. USA* **71,** 3879–3882 (1974).
18. I. M. Arias, G. Fleischner, R. Kirsch, S. Mishkin, and Z. Gatmaitan, On the structure, regulation, and function of ligandin, in: *Glutathione, Metabolism and Function* (I. M. Arias and W. B. Jakoby, eds.), pp. 175–188, Raven Press, New York (1976).
19. M. J. Pabst, W. H. Habig, and W. B. Jakoby, Glutathione *S*-transferase A. A novel kinetic mechanism in which the major reaction pathway depends on substrate concentration, *J. Biol. Chem.* **249,** 7140–7148 (1974).
20. C. W. Abel and C. Heidelberger, The interaction of carcinogenic hydrocarbons with tissue constituents. VIII. Binding of tritium-labeled hydrocarbons to the soluble proteins of mouse skin, *Cancer Res.* **22,** 931–946 (1962).
21. A. M. Sarrif and C. Heidelberger, On the interaction of chemical carcinogens with soluble proteins of target tissues and in cell culture in: *Glutathione, Metabolism and Function* (I. M. Arias and W. B. Jakoby, eds.), pp. 317–338, Raven Press, New York (1976).
22. N. Kaubisch, D. M. Jerina, and J. W. Daly, Arene oxides as intermediates in the oxidative metabolism of aromatic compounds: Isomerization of methyl substituted arene oxides, *Biochemistry* **11,** 3080–3088 (1972).
23. B. J. Auret, D. R. Boyd, P. M. Robinson, C. Watson, J. W. Daly, and D. M. Jerina, The

NIH shift during hydroxylation of aromatic substrates by fungi, *J. Chem. Soc. (D) Chem. Commun.* 1585–1587 (1971).

24. P. K. Ayengan, O. Hayaisha, M. Nakajima, and J. Tomida, Enzymatic aromatization of 3,5-cyclohexadiene-1,2-diol, *Biochim. Biophys. Acta* **33,** 111–119 (1959).
25. D. M. Jerina, H. Ziffer, and J. W. Daly, The role of the arene oxide–oxepin system in the metabolism of aromatic substrates. IV. Stereochemical considerations of dihydrodiol formation and dehydrogenation, *J. Am. Chem. Soc.* **92,** 1056–1061 (1970).
26. A. M. Jeffrey, H. J. C. Yeh, D. M. Jerina, T. R. Patel, J. F. Davey, and D. T. Gibson, Initial reactions in the oxidation of naphthalene by *Pseudomonas putida, Biochemistry* **14,** 575–584 (1975).
27. P. J. Murphy, J. R. Bernstein, and R. E. McMahon, The formation of catechols by consecutive hydroxylations: A study of the microsomal hydroxylation of butarnoxane, *Mol. Pharmacol.* **10,** 634–639 (1974).
28. J. Booth, A. Hewer, G. R. Keysell, and P. Sims, Enzymic reduction of aromatic hydrocarbon epoxides by the microsomal fraction of rat liver, *Xenobiotica* **5,** 197–203 (1975).
29. J. E. Tomaszewski, D. M. Jerina, and J. W. Daly, Metabolism of aromatic substrates to phenols by animal mono-oxygenases: Evidence for a direct oxidative pathway not involving arene oxide intermediates, *Biochemistry* **14,** 2024–2031 (1975).
30. H. G. Selander, D. M. Jerina, and J. W. Daly, Metabolism of chlorobenzene with hepatic microsomes and solubilized cytochrome P-450 systems, *Arch. Biochem. Biophys.* **168,** 309–321 (1975).
31. J. W. Daly, D. M. Jerina, and B. Witkop, Arene oxides and the NIH shift: The metabolism, toxicity and carcinogenicity of aromatic compounds, *Experientia* **28,** 1129–1149 (1972).
32. D. M. Jerina and J. W. Daly, Arene oxides: A new aspect of drug metabolism, *Science* **185,** 573–582 (1974).
33. P. Sims and P. L. Grover, Epoxides in polycyclic aromatic hydrocarbon metabolism and carcinogenesis, *Adv. Cancer Res.* **20,** 165–274 (1974).
34. T. Sato, T. Fukuyama, T. Suzuki, and Y. Uoshikawa, 1,2,-Dihydro-1,2-dihydroxybenzene and several other substances in the metabolism of benzene, *J. Biochem. (Tokyo)* **53,** 23–27 (1963).
35. B. Spencer and R. T. Williams, Studies in detoxication. 33. The metabolism of halobenzenes. A comparison of the glucuronic acid, ethereal sulphate and mercapturic acid conjugates of chloro-, bromo- and iodobenzenes of the *o*-, *m*-, and *p*-chlorophenols: Biosynthesis of *o*-, *m*- and *p*-chlorophenyglucuronides, *Biochem. J.* **47,** 279–284 (1950).
36. J. R. Lindsay-Smith and B. A. J. Shaw, Mechanisms of mammalian hydroxylation: Some novel metabolites of chlorobenzene, *Xenobiotica* **2,** 215–226 (1972).
37. W. M. Azouz, D. V. Parke, and R. T. Williams, Studies in detoxication. 51. The determination of catechols in urine, and the formation of catechols in rabbits receiving halogenobenzenes and other compounds. Dihydroxylation *in vivo, Biochem. J.* **55,** 146–151 (1953).
38. N. Zampaglione, D. J. Jollow, J. R. Mitchell, B. Stripp, M. Hamrick, and J. R. Gillette, Role of detoxifying enzymes in bromobenzene-induced liver necrosis, *J. Pharmacol. Exp. Ther.* **187,** 218–227 (1973).
39. W. M. Azouz, D. V. Parke, and R. T. Williams, Studies in detoxication. 62. The metabolism of halogenobenzenes, *ortho*- and *para*-dichlorobenzene, *Biochem. J.* **59,** 410–415 (1955).
40. D. V. Parke and R. T. Williams, Studies in detoxication. 63. The metabolism of halgenobenzenes (a). Meta-dichlorobenzene (b). Further observations on the metabolism of chlorobenzene, *Biochem. J.* **59,** 415–422 (1955).

41. W. R. Jondorf, D. V. Parke, and R. T. Williams, Studies in detoxication. 66. The metabolism of halogenobenzenes, 1:2:3-, 1:2:4-, and 1:3:5-trichlorobenzenes, *Biochem. J.* **61**, 512–521 (1951).
42. D. M. Jerina, J. W. Daly, and B. Witkop, Deuterium migration during the acid-catalyzed dehydration of 6-deutero-5,6-dihydroxy-3-chloro-1,3-cyclohexadiene, A nonenzymatic model for the NIH shift, *J. Am. Chem Soc.* **89**, 5488–5489 (1967).
43. F. Oesch, D. M. Jerina, J. W. Daly, and J. Rice, An anomalous prevention of chlorobenzene-induced hepatotoxicity by an inhibitor of epoxide hydrase, *Chem.-Biol. Interactions* **6**, 189–202 (1973).
44. H. G. Selander, D. M. Jerina, D. E. Piccolo, and G. A. Berchtold, Synthesis of 3- and 4-chlorobenzene oxides: Unexpected trapping results during metabolism of ^{14}C-chlorobenzene by hepatic microsomes, *J. Am. Chem. Soc.* **97**, 4428–4430 (1975).
45. H. D. West, J. R. Lawson, I. H. Miller, and G. R. Mathura, The fate of diphenyl in the rat, *Arch. Biochem. Biophys.* **60**, 14–20 (1965).
46. D. J. Harvey, L. Glazener, C. Stratton, J. Nowlin, R. M. Hill, and M. G. Horning, Detection of epoxides of allyl-substituted barbiturates in rat urine, *Res. Commun. Chem. Pathol. Pharmacol.* **3**, 557–565 (1972).
47. M. G. Horning, C. Stratton, A. Wilson, E. C. Horning, and R. M. Hill, Detection of 5-(3,4-dihydroxy-1,5-cyclohexadienyl-1-yl)-5-phenylhydantoin as a major metabolite of 5,5-diphenylhydantoin (Dilantin) in the newborn human, *Anal. Lett.* **4**, 537–545 (1971).
48. A. Karim, G. Garden, and W. Trager, Biotransformation of diphenoxylate in rat and dog, *J. Pharm. Exp. Ther.* **177**, 546–555 (1971).
49. W. G. Stillwell, M. Stafford, and M. G. Horning, Metabolism of glutethimide (Doriden) by the epoxide-diol pathway in the rat and guinea pig, *Res. Commun. Chem Pathol. Pharmacol.* **6**, 579–590 (1973).
50. M. G. Horning, C. Butler, D. J. Harvey, R. M. Hill, and T. E. Zion, Metabolism of *N*,2-dimethyl-2-phenylsuccinimide (Methsuximide) by the epoxide-diol pathway in rat, guinea pig, and human, *Res. Commun. Chem. Pathol. Pharmacol.* **6**, 565–578 (1973).
51. J. Booth and E. Boyland, Metabolism of polycyclic compounds. 5. Formation of 1:2-dihydroxy-1:2-dihydronaphthalenes, *Biochem. J.* **44**, 361–365 (1949).
52. E. Boyland and P. Sims, Metabolism of polycyclic compounds. 12. An acid-labile precursor of 1-naphthylmercapturic acid and naphthol: an *N*-acetyl-*S*-(1,2-dihydrohydroxynaphthyl)-L-cysteine, *Biochem. J.* **68**, 440–447 (1958).
53. H. H. Cornish and W. D. Block, Metabolism of chlorinated naphthalenes, *J. Biol. Chem.* **231**, 583–588 (1958).
54. J. B. Knaak, Biological and nonbiological modifications of carbamates, *Bull. WHO* **44**, 121–131 (1971).
55. J. R. Bend, G. M. Holder, E. Protos, and A. J. Ryan, Water soluble metabolites of carbaryl (1-naphthyl *N*-methylcarbamate) in mouse liver preparations and in the rat, *Aust. J. Biol. Sci.* **24**, 535–546 (1971).
56. P. Sims, Metabolism of polycyclic compounds. 25. The metabolism of anthracene and some related compounds in rats, *Biochem. J.* **92**, 621–631 (1964).
57. E. Boyland and P. Sims, Metabolism of polycyclic compounds. 20. The metabolism of phenanthrene in rabbits and rats: mercapturic acids and related compounds, *Biochem. J.* **84**, 564–570 (1962).
58. E. Boyland and P. Sims, Metabolism of polycyclic compounds. 21. The metabolism of phenanthrene in rabbits and rats: dihydrodihydroxy compounds and related glucosiduronic acids, *Biochem. J.* **84**, 571–582 (1962).
59. E. Boyland and P. Sims, The metabolism of 9,10-epoxy-9,10-dihydrophenanthrene in rats, *Biochem. J.* **95**, 778–792 (1965).

60. E. Boyland and P. Sims, Metabolism of polycyclic compounds. 23. The metabolism of pyrene in rats and rabbits, *Biochem. J.* **90,** 391–398 (1964).
61. E. Boyland and P. Sims, Metabolism of polycyclic compounds. 24. The metabolism of benz[*a*]anthracene, *Biochem. J.* **91,** 493–506 (1964).
62. A. M. El Masri, J. N. Smith, and R. T. Williams, Studies in detoxication. 73. The metabolism of alkylbenzenes: phenylacetylene and phenylethylene (styrene), *Biochem. J.* **68,** 199–204 (1958).
63. N. Ohtsuji and M. Ikeda, The metabolism of styrene in the rat and the stimuıatory effect of phenobarbital, *Toxicol. Appl. Pharmacol.* **18,** 321–328 (1971).
64. S. P. James and D. A. White, The metabolism of phenethyl bromide, styrene and styrene oxide in the rabbit and rat, *Biochem. J.* **104,** 914–921 (1967).
65. D. J. Harvey, L. Glazener, C. Stratton, D. B. Johnson, R. M. Hill, E. C. Horning, and M. G. Horning, Detection of epoxides of allyl-substituted barbiturates in rat urine, *Res. Commun. Chem. Pathol. Pharmacol.* **4,** 247–260 (1972).
66. M. Stafford, G. Kellerman, R. N. Stillwell, and M. G. Horning, Metabolism of antipyrine by the epoxide-diol pathway in the rat, guinea pig and human, *Res. Commun. Chem. Pathol. Pharmacol.* **8,** 593–606 (1974).
67. K. M. Baker, J. Csetenyi, A. Frigerio, P. L. Morselli, F. Parravicini, and G. Pifferi, 10,11-Dihydro-10,11-dihydroxy-5*H*-dibenz[*b,f*]acepine-5-carboxamide, a metabolite of carbamazepine isolated from human and rat urine, *J. Med. Chem.* **16,** 703–705 (1973).
68. H. B. Hucker, A. J. Balletto, J. Dernetriades, B. H. Arison, and A. G. Zacchei, Epoxide metabolites of protriptyline in rat urine, *Drug Metab. Dispos.* **3,** 80–84 (1975).
69. K. L. Hintze, J. S. Wold, and L. J. Fischer, Disposition of cyproheptadine in rats, mice and humans and identification of a stable epoxide metabolite, *Drug Metab. Dispos.* **3,** 1–9 (1975).
70. W. G. Stillwell, M. J. Carman, and M. G. Horning, The metabolism of safrole and 2′,3′-epoxysafrole in the rat and guinea pig, *Drug Metab. Dispos.* **2,** 489–498 (1974).
71. D. H. Swenson, E. C. Miller, and J. A. Miller, Aflatoxin B_1-2,3-oxide: Evidence for its formation in rat liver *in vivo* and by human liver microsomes *in vitro, Biochem. Biophys. Res. Commun.* **60,** 1036–1043 (1974).
72. J. T. Matschiner, R. G. Bell, J. M. Amelotti, and T. F. Knauer, Isolation and characterization of a new metabolite of phylloquinone in the rat, *Biochim. Biophys. Acta* **201,** 299–315 (1970).
73. E. Boyland and D. Williams, An enzyme catalyzing the conjugation of epoxides with glutathione, *Biochem. J.* **94,** 190–197 (1965).
74. T. Hayakawa, R. A. Lemahieu, and S. Udenfriend, Studies on glutathione-*S*-arene oxide transferase: A sensitive assay and partial purification of the enzyme from sheep liver, *Arch. Biochem. Biophys.* **162,** 223–230 (1974).
75. T. Hayakawa, S. Udenfriend, H. Yagi, and D. M. Jerina, Substrates and inhibitors of hepatic glutathione-*S*-epoxide transferase, *Arch. Biochem. Biophys.* **170,** 438–451 (1975).
76. M. O. James, J. R. Fouts, and J. R. Bend, Hepatic and extrahepatic metabolism *in vitro,* of an epoxide (8-^{14}C-styrene oxide) in the rabbit, *Biochem. Pharmacol.* **25,** 187–193 (1976).
77. J. Marniemı and M. G. Parkki, Radiochemical assay of glutathione *S*-epoxide transferase and its enhancement by phenobarbital in rat liver *in vitro, Biochem. Pharmacol.* **74,** 1569–1572 (1975).
78. N. Nemoto, H. V. Gelboin, W. H. Habig, J. N. Kettley, and W. B. Jakoby, K-region benzo[*a*]pyrene 4,5-oxide is conjugated by homogeneous glutathione *S*-transferase, *Nature (London)* **255,** 512 (1975).
79. J. R. Bend, Z. Ben-Zvi, J. Van Anda, P. Dansette, and D. M. Jerina, Hepatic and

extrahepatic glutathione S-transferase activity toward several arene oxides and epoxides in the rat, in: *Polynuclear Aromatic Hydrocarbons* (R. Fruedenthal and P. W. Jones, eds.), pp. 63–79, Raven Press, New York (1976).
80. D. M. Reuben and T. C. Bruice, Relative nucleophilicity of thiols and glutathione towards benzene oxide, *J. Chem. Soc. (D) Chem. Commun.* 113–114 (1974).
81. F. Oesch, N. Kaubisch, D. M. Jerina, and J. W. Daly, Hepatic epoxide hydrase: Structure–activity relationships for substrates and inhibitors, *Biochemistry* **10,** 4858–4866 (1971).
82. D. M. Jerina, P. M. Dansette, A. Y. H. Lu, and W. Levin, Hepatic microsomal epoxide hydrase: A sensitive radiometric assay for hydration of arene oxides of carcinogenic aromatic hydrocarbons, *Mol. Pharmacol.,* in press.
83. A. H. Conney, A. W. Wood, W. Levin, A. Y. H. Lu, R. L. Chang, P. G. Wislocki, R. L. Goode, G. M. Holder, P. M. Dansette, H. Yagi, and D. M. Jerina, Metabolism and biological activity of benzo[*a*]pyrene and its metabolic products, Chap. 37, this volume.
84. A. Meister, Glutathione: Metabolism and function via the γ-glutamyl cycle, *Life Sci.* **15,** 177–190 (1974).
85. V. H. Cohn and J. Lyle, A fluorimetric assay for glutathione, *Anal. Biochem.* **14,** 434–440 (1966).
86. Z. Ben-Zvi, M. O. James, and J. R. Bend, unpublished results.
87. P. L. Grover and P. Sims, Conjugations with glutathione: Distribution of glutathione S-aryltransferase in vertebrate species, *Biochem. J.* **90,** 603–606 (1964).
88. G. Clifton, N. Kaplowitz, J. D. Wallin, and J. Kuhlenkamp, Drug induction and sex differences of renal glutathione S-transferases in the rat. *Biochem. J.* **150,** 259–262 (1975).
89. C. D. Klassen and G. L. Plaa, Studies on the mechanism of phenobarbital-enhanced sulfobromophalein disappearance, *J. Pharmacol. Exp. Ther.* **161,** 361–366 (1968).
90. F. J. Darby and R. K. Grundy, Glutathione S-aryltransferase: The effect of treating male and female rats with phenobarbitone on the apparent kinetic parameters for the conjugation of 1,2-dichloro-4-nitrobenzene and 1-chloro-2,4-dinitrobenzene with glutathione, *Biochem. J.* **128,** 175–177 (1975).
91. N. Kaplowitz, J. Kuhlenkamp, and G. Clifton, Drug induction of hepatic glutathione S-transferases in male and female rats, *Biochem. J.* **146,** 351–356 (1975).
92. J. R. Bend, M. O. James, T. R. Devereux, and J. R. Fouts, Toxication–detoxication systems in hepatic and extrahepatic tissues in the perinatal period, in: *Basic and Therapeutic Aspects of Perinatal Pharmacology* (P. L. Morselli, S. Garattini, and F. Sereni, eds.), pp. 229–243, Raven Press, New York (1975).
93. M. O. James, G. L. Foureman, F. C. P. Law, and J. R. Bend, Perinatal development of epoxide hydrase and glutathione S-transferase in hepatic and extrahepatic tissues of the rabbit and guinea pig, *Drug. Metab. Disp.,* in press.
94. B. R. Smith, Z. Ben-Zvi, F. C. P. Law, and J. R. Bend, unpublished results.
95. F. C. P. Law, T. E. Eling, J. R. Bend, and J. R. Fouts, Metabolism of xenobiotics by the isolated perfused lung: Comparison with *in vitro* incubations, *Drug Metab. Dispos.* **2,** 433–442 (1974).
96. P. Sims, Polycyclic hydrocarbon epoxides as active metabolic intermediates, Chap. 39, this volume
97. P. Brookes, Role of covalent binding in carcinogenicity, Chap. 54, this volume.
98. D. W. Nerbert, A. R. Boobis, H. Yagi, D. M. Jerina, and R. E. Kouri, Genetic differences in benzo[*a*]pyrene carcinogenic index *in vivo* and in mouse cytochrome $P_1$450-mediated benzo[*a*]pyrene metabolite binding to DNA *in vitro,* Chap. 12, this volume.
99. J. A. Miller and E. C. Miller, The concept of reactive electrophilic metabolites in

chemical carcinogenesis: Recent results with aromatic amines, safrole, and aflatonin B_1, Chap. 2, this volume.
100. J. R. Gillette, Kinetics of reactive metabolites and covalent binding *in vivo* and *in vitro,* Chap. 3, this volume.
101. J. R. Mitchell, S. D. Nelson, W. R. Snodgrass, and J. A. Timbrell, Metabolic activation of hydrazines to highly reactive hepatotoxic intermediates, Chap. 27, this volume.
102. D. J. Jollow and C. Smith, Biochemical aspects of toxic metabolites: Formation, detoxication, and covalent binding, Chap. 4, this volume.
103. K. C. Chen and J. Tang, Amino acid sequence around the epoxide-reactive residues in pepsin, *J. Biol. Chem.* **247,** 2566–2574 (1972).
104. A. Quaroni, E. Gershon, and G. Semanza, Affinity labeling of the active sites in the sucrase–isomaltase complex from small intestine, *J. Biol. Chem.* **249,** 6424–6433 (1974), and references therein.
105. E. T. Bucovax, J. C. Morrison, H. L. James, C. F. Dais, and J. L. Wood, Reaction of polycyclic hydrocarbon–cysteine conjugates with the aminoacyl-RNA synthetase system, *Cancer Res.* **30,** 155–161 (1970).
106. S. V. Molinary and J. L. Wood, Phenanthrene bound to a protein by biosynthesis, *Biochem. Biophys. Res. Commun.* **43,** 899–904 (1971).
107. J. Frendo and J. L. Wood, Incorporation of *S*-(9-hydroxy-9,10-dihydro-10-phenanthryl)-L-cysteine into rabbit hemoglobin, *Proc. Soc. Exp. Biol. Med.* **139,** 173–175 (1972).
108. D. M. Jerina, H. Yagi, and J. W. Daly, Arene oxides-oxepins, *Heterocycles* **1,** 267–326 (1973).
109. A. M. Jeffrey, H. J. C. Yeh, D. M. Jerina, R. M. DeMarinis, D. H. Foster, D. E. Piccolo, and G. A. Berchtold, Stereochemical course in reactions between nucleophiles and arene oxides, *J. Am. Chem. Soc.* **96,** 6929–6937 (1974).
110. P. Y. Bruice, T. C. Bruice, P. M. Dansette, H. G. Selander, H. Yagi, and D. M. Jerina, A comparison of the mechanism of solvolysis and rearrangement of K-region vs. non-K-region arene oxides of phenanthrene. Comparative solvolytic rate constants of K-region and non-K-region arene oxides, *J. Am. Chem. Soc.* **98,** 2965–2973(1976).
111. P. Y. Bruice, T. C. Bruice, H. Yagi, and D. M. Jerina, Nucleophilic displacements on the arene oxides of phenanthrene, *J. Am. Chem. Soc.* **98,** 2973–2981 (1976).
112. A. M. Jeffrey and D. M. Jerina, Novel rearrangements during dehydration of arene oxide adducts, *J. Am. Chem. Soc.* **97,** 4427–4428 (1975).
113. D. M. Jerina, Products, specificity, and assay of glutathione *S*-epoxide transferase, in: *Glutathione, Metabolism and Function* (I. M. Arias and W. B. Jakoby, eds.), pp. 267–279, Raven Press, New York (1976).
114. W. B. Jakoby and T. A. Fjellstedt, Epoxidases, in: *The Enzymes,* 3rd ed., Vol. 7 (P. D. Boyer, ed.), pp. 199–212, Academic Press, New York (1972).
115. B. Ketterer and L. Christodoulides, Two specific azodye-carcinogen-binding proteins of the rat liver: The identity of the amino acid residues which bind to the azodye, *Chem.-Biol. Interactions* **1,** 173–183 (1969/1970).

21

Discussion

In response to questions by Snyder on the availability of glutathione in the liver for the conjugation reaction, and whether the availability of glutathione could be rate limiting for the conjugation reaction, Jerina replied that the concentration of glutathione in the liver was 5–10 mM, but that it was not clear whether all this glutathione is available intracellularly for reaction, in that some portion may be tied up in precursor molecules. In regard to the rate-limiting step, Jerina felt that the rate of formation of epoxides in the liver would be the slow step since cytochrome P450 has a low turnover number, whereas both the enzyme-catalyzed and the spontaneous second-order conjugation of glutathione with epoxides are very rapid reactions. However, since the true intracellular concentration of glutathione is not known, the rate-limiting step in the reaction *in vivo* cannot be decided with certainty.

Jacobsson questioned the use of the high dose of 3-methylcholanthrene (40 mg/kg) in rats for induction of epoxide hydratase activity. Studies in Jacobsson's laboratory have indicated that at this high dose some strains of rats become sick and that the induction of epoxide hydratase in the kidney is poor compared with that obtained after low doses. Jerina replied that the rats he had used did not appear to become sick. He reported that these induced animals showed high levels of epoxide hydratase activity in lung, kidney, and liver. The activity in testes was also high and was intermediate between kidney and liver.

In reply to a question as to how one can formally show that the carbonium ion stabilization is higher in the β-position than in the α-position, Jerina replied that there are several lines of reasoning. First, naphthalene oxide opens almost exclusively to α-naphthol. Since this is a carbonium ion reaction, it indicates that the β-carbonium ion is the stable one. Second, electrophilic substitution of naphthalene oxide occurs at C-1, indicating that the carbonium ion is at C-2 and hence must be the more stable. For chlorobenzene oxide, resonance theory can be used to predict the more stable carbonium ion.

Remmer directed attention to the role of Y-protein (ligandin) in glutathione conjugation reactions, recalling that several years ago he had observed what seemed to be an induction of glutathione transferase by phenobarbital. Subsequently, Arias has shown that phenobarbital pretreatment of rats leads to enhanced levels of Y-protein in the liver. Thus the enhancement of glutathione conjugation is associated with an enhancement of binding of cellular glutathione to this protein.

Gelboin raised the question of specificity of glutathione transferase enzymes, referring to the work of Jacoby at NIH. Although the five hepatic glutathione transferase fractions purified by Jacoby are known to show considerable cross-specificity, some of the fractions appear to lack activity toward alkyl epoxides. Gelboin pointed out that this might be important in the toxicity of diol-epoxides. Jerina agreed that the specificity of the glutathione transferases was an important question, but suggested that since the second-order rate constant for the spontaneous conjugation of glutathione with diol-epoxides is very high, it seems unlikely that the absence of the appropriate glutathione transferase would limit this reaction, provided that the concentration of glutathione in the cell was high. In response to a question on the normal concentration of glutathione in the liver, Jollow reported that unpublished experiments in his laboratory indicated that hepatic glutathione undergoes a diurnal variation, ranging from 4 to 10 mM, depending on the time of day and the species under study.

Greim questioned Oesch about testing nonplanar molecules as frameshift mutagens, as planarity of the mutagen is a prerequisite for this type of activity. Oesch responded that the epoxides metabolically produced from the drugs carbamazepine, cyproheptadine, and cyclobenzaprine were tested not only with bacterial strains which undergo frameshift mutations but also with those susceptible to substitution mutations. He pointed out that planarity is not a rigid requirement for compounds causing frameshift mutations. For example, the oxirane ring of epoxides derived from polycyclic hydrocarbons always sticks out of the plane, yet many of these compounds are potent frameshift mutagens—with the epoxide moiety being essential for this activity (see text refs. 13 and 28). Additionally, in the 7,12-dimethylbenz[*a*]anthracene-5,6-oxide, the entire molecule is severely distorted with the angle between the outer rings close to 35° (Glusker *et al., Cancer Biochem. Biophys.* **1**, 43, 1974), yet this compound is a mutagen for the *Salmonella typhimurium* strain TA 1537, which tests for frameshift mutations (Glatt *et al., Int. J. Cancer* **16**, 787, 1975). Oesch stressed that the third important point he wished to make was that alarming biological effects of some epoxides do not justify an epoxide hysteria since they do not imply that all epoxides will have similar effects, for epoxides differ widely from each other in many respects; one such difference is molecular geometry.

22

UDPGlucuronosyltransferase: Substrate Specificity and Reactivation after Partial Separation from Other Membrane Components

Osmo Hänninen, Matti Lang, Ulla Koivusaari, and Tuula Ollikainen
Department of Physiology
University of Kuopio
SF-70101 Kuopio 10, Finland

The oxidation of a foreign compound may lead to the formation of metabolites which are more active both chemically and biologically than the original compound. These metabolites are usually conjugated as carbohydrate, amino acid, or peptide derivatives that usually display low biological activity. Glucuronide formation is a common final step in the metabolism and inactivation of many foreign compounds in mammalian tissues since D-glucuronic acid residues can be attached to hydroxyl, carboxylic, amino, or thiol groups. The enzyme(s) catalyzing the synthesis of glucuronides (UDPglucuronosyltransferase, E.C. 2.4.1.17) are located in the same intracellular structure, i.e., the microsomes, as the enzymes catalyzing the first steps in the metabolism of foreign compounds (1). The glucuronide-forming capacity of a tissue sample may vary considerably depending on the structure and possibly also on the functional state of the membrane. Removing the outer layers of membrane preparations isolated from rat liver, for example, with trypsin (2) and chaotropic agents (3), or the perturbation of membrane structure with surfactants (2,4,5) or phospholipases (6,7) can each increase UDPglucuronosyltransferase activity with a simultaneous decrease in drug-hydroxylating activity.

These findings indicate that some parts of the UDPglucuronosyltransferase protein are probably buried in the membrane, at least in rat liver (2,8). The enzyme obviously has a hydrophobic character. The fact that phospholipases affect the catalytic properties of the enzyme suggests that the enzyme is in intimate contact with the phospholipids of the membrane structure. Because of tight binding of the enzyme to the membrane and a loss of the catalytic properties during processing of the membranes, purification of UDPglucuronosyltransferase has proceeded rather slowly (9–15), although preparations with 100 times the specific activity of the original microsomal fractions have been obtained (13,14). Previously, rather mild treatments have been carried out to remove the enzyme from the membrane structure. These preparations probably still contain a number of other membrane components; membranous structures have been seen with the aid of electron microscopy (9). It is obvious that only after further purification can the total number of different enzymes catalyzing glucuronide biosynthesis be determined and the kinetics of the enzyme(s) understood.

In the present study, we first increased the starting level of hepatic UDPglucuronosyltransferase activity by pretreating rats with 3-methylcholanthrene. The isolated microsomal fraction of the liver was treated with excess Triton X-100 to convert the natural protein–membrane lipid complexes to protein– and membrane lipid–Triton X-100 complexes (16) which were then subjected to gel filtration. The proper lipid environment for the catalytic activity of UDPglucuronosyltransferase and the glucuronidation of a number of aglycones was also studied. Since a method based on the oxidation of NADH was used in UDPglucuronosyltransferase assay with some aglycones, a study of the consumption of reducing equivalents by the membrane components was initiated.

MATERIALS AND METHODS

Male rats (*Rattus norvegicus*) of the Wistar/Af/Han/Mol/(Han 67) strain, about 3 months old, were used. The strain originates from Møllegaard Avlslaboratorier A/S (Ejby, Denmark) and presents the fifth generation outbred by a rotational mating system in the Laboratory Animal Center of the University of Kuopio. 3-Methylcholanthrene (Sigma Chemical Co., St. Louis, Missouri) was given intraperitoneally 20 mg/kg in olive oil for 4 days. The animals were sacrificed 24 h after the last injection and bled; the liver was removed and placed in ice-cold 0.15 M KCl. The subsequent steps were carried out at 0–4°C. The tissue was homogenized with a Potter-Elvehjem glass–teflon homogenizer in 0.15 M KCl to give a 20% (w/v) homogenate. The suspension was first centrifuged at 10,000*g* for 10 min, and then the microsomal fraction was harvested from the supernatant by centrifugation at 105,000*g* for 60 min. The isolated microsomal fraction was resuspended in 0.15 M KCl to give about 30 mg protein/ml.

The isolated microsomal fraction was treated with 25 times as much Triton X-100 as the phospholipid in the sample in order to replace the phospholipid in the lipid–protein complexes of the native membrane. The membrane samples applied to a Sepharose 4B column (diameter 1 cm, height 15.3 cm). The elution were kept at 0°C for 30 min. Aliquots of 5 mg of microsomal protein were fluid was 0.15 M KCl containing 1.5 g Triton X-100/liter. Half-milliliter fractions were collected. In some experiments, lipid mixtures were added back to the fractions immediately after collection. Lysolecithin from egg yolk (type I) and lecithin from soy beans (type II-s) were obtained from Sigma Chemical Co. The membrane lipid from rat liver microsomes was isolated by chloroform–methanol (2:1) extraction (17).

Protein was determined by the biuret (18) and sulfosalicylic (19) methods. The phospholipid concentration of the collected fractions was estimated by a modified Bartlett method (20). The samples were extracted by chloroform–ethanol (1:1). After the solvent was evaporated in a nitrogen stream, samples were hydrolyzed and the released inorganic phosphate was determined.

The activity of UDPglucuronosyltransferase was measured with 4-nitrophenol (E. Merck AG, Darmstadt, Federal Republic of Germany) as the aglycone as described earlier (4) and with 4-methylumbelliferone (British Drug Houses, Poole, England) as described by Aitio (21). With bilirubin (E. Merck AG, Darmstadt, Germany), 3-methyl-2-nitrophenol (Aldrich-Europe, Beerse, Belgium), and ethylmorphine (University Pharmacy, Helsinki, Finland), the NAD-linked assay method developed by Mulder and van Doorn (22) was used. NADH oxidation was monitored in Cary 118 double-beam spectrophotometer at 38°C. The final concentrations of the aglycones were as follows: bilirubin, 0.04 mM; 3-methyl-2-nitrophenol, 0.3 mM; ethylmorphine, 0.15 mM. Phosphoenolpyruvate and pyruvate kinase were obtained from Sigma and the lactate and pyruvate from Boehringer, Mannheim. The endogenous consumption of NADH by the microsomal and other preparations was studied in reaction mixtures lacking phosphoenolpyruvate, pyruvate kinase, and lactate dehydrogenase. Various amounts of Triton X-100, microsomal membrane lipid, lecithin, and lysolecithin were added. In order to inhibit the NADH consumption, potassium cyanide (Merck), rotenone (grade II, Sigma), and DL-α-tocopherol (Merck) were used in various concentrations. The role of oxygen was studied in reaction mixtures which were flushed with a nitrogen or CO stream.

RESULTS

When the microsomal membrane proteins were converted to Triton X-100 complexes, no glucuronidation of 4-nitrophenol could be observed either before or after gel filtration in any of the separated protein fractions. 4-Nitrophenol was, however, conjugated by the gel-filtered enzyme preparations if microsomal membrane lipid preparation, lecithin, lysolecithin, lysolecithin–lecithin, or lyso-

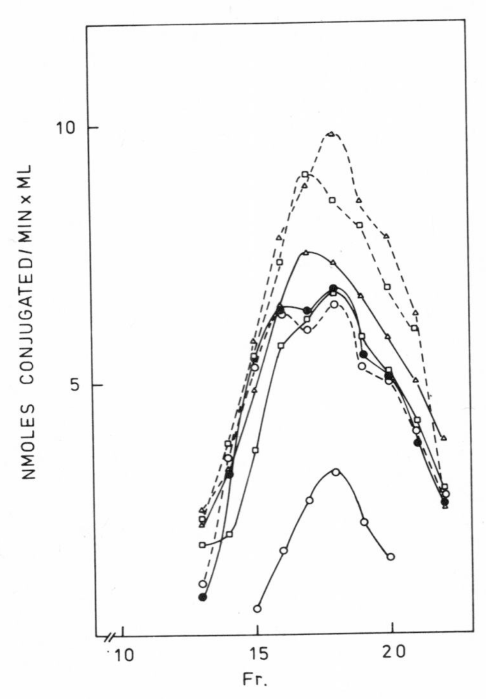

Fig. 1. Reactivation of 4-nitrophenyl UDPglucuronosyltransferase in eluates from Sepharose 4B gel chromatography. Solid lines with circles and black spots indicate the activity after addition (3 mg/ml) of lysolecithin and lecithin, respectively, in the reaction mixture. Solid lines with squares and triangles indicate activity after similar addition of lysolecithin–lecithin mixtures in ratios 1:1 and 3:1, respectively. Broken lines with circles indicate the activity after addition of a microsomal membrane lipid preparation (3 mg/ml) to the reaction mixture, whereas broken lines with squares and triangles represent results of similar addition of lysolecithin–microsomal membrane lipid in ratios 1:1 and 3:1, respectively.

lecithin–microsomal membrane lipid mixtures were added to the reaction mixture or to the collected fractions (Fig. 1). The best reactivation was observed when lysolecithin and microsomal membrane lipid or lecithin mixtures were added.

A slow conjugation could be observed in the presence of gel-filtered UDPglucuronosyltransferase preparations with 3-methyl-2-nitrophenol without any addition of lysolecithin or other lipid mixtures. However, if lysolecithin–membrane lipid mixture was added to the reaction mixture or to the fractions when standing, the UDPglucuronosyltransferase activity was higher. On further standing, the UDPglucuronosyltransferase activity diminished. Also, after an addition of the various lipid mixtures there was a decline of the enzyme activity. The loss of UDPglucuronosyltransferase activity with time varied with different aglycones.

When the solubilized membrane components were subjected to gel filtration on Sepharose 4B columns, some protein came off the column in the void volume, but the main bulk of protein appeared later in one rather broad peak (Fig. 2).

The elution patterns of UDPglucuronosyltransferase activity toward different aglycones were different after gel filtration (Fig. 2). With bilirubin, ethylmorphine, and 3-methyl-2-nitrophenol as aglycones, some UDPglucuronsyltransferase activity could be found in the first protein peak (the void volume), and at least some activity in the later fractions. The void-volume enzyme activity was highest with bilirubin and ethylmorphine as substrates (Figs. 2A, B), whereas with 4-nitrophenol (Fig. 2D) and 4-methylumbelliferone as aglycones no activity was found in the void volume.

With 3-methyl-2-nitrophenol (Fig. 2C), 4-nitrophenol (Fig. 2D), and 4-methylumbelliferone (not shown), the major part of the activity appeared later in the eluate. With 3-methyl-2-nitrophenol, the highest UDPglucuronosyltransferase activity could be observed *after* the main protein peak, and with 4-nitrophenol the main activity *preceded* the main protein peak. With 3-methyl-2-nitrophenol, a third minor activity peak could be observed between the two other peaks in the region where 4-nitrophenol was best conjugated. 4-Methylumbelliferone UDPglucuronosyltransferase activity appeared to be eluted together with 4-nitrophenol transferase activity.

When microsomal UDPglucuronosyltransferase activity was determined by coupling the production of UDP in the glucuronide synthesis with NADH consumption, some NADH was oxidized, even in the absence of an aglycone and UDPglucuronic acid. Similar NADH oxidation was seen even in the absence of phosphoenolpyruvate, pyruvate kinase, and lactate dehydrogenase, and only microsomal preparation, buffer, and NADH were needed.

When Triton X-100 was added to a reaction mixture containing buffer, NADH, and microsomal fraction, the endogenous microsomal consumption of hydrogen was diminished, but adding back either the lipid preparation isolated from the microsomal fraction or lecithin increased the hydrogen consumption. With lysolecithin, no increase was observed.

When gel-filtered UDPglucuronosyltransferase preparation was used, the endogenous hydrogen consumption from NADH was low, but after addition of lecithin or microsomal membrane lipid preparation the NADH oxidation was enhanced. Adding lysolecithin had only a slight stimulatory effect (Fig. 3).

The oxidation of NADH catalyzed by rat liver microsomes was not inhibited by carbon monoxide, DL-α-tocopherol, or rotenone. The rate of NADH oxidation was, however, slightly decreased by 10 mM potassium cyanide. While nitrogen flushing only slightly diminished the initial reaction velocity, NADH oxidation ceased earlier than in air-containing reaction mixtures (Fig. 3).

DISCUSSION

The present results indicate that the effect of phospholipids on glucuronidation of different aglycones varies in preparations derived from rat liver microsomes. When the natural phospholipid environment of UDPglucuronosyltransferase(s) was replaced with an artificial "lipid" environment (Triton X-100), the conjugation of water-soluble aglycones like 4-nitrophenol ceased, although the

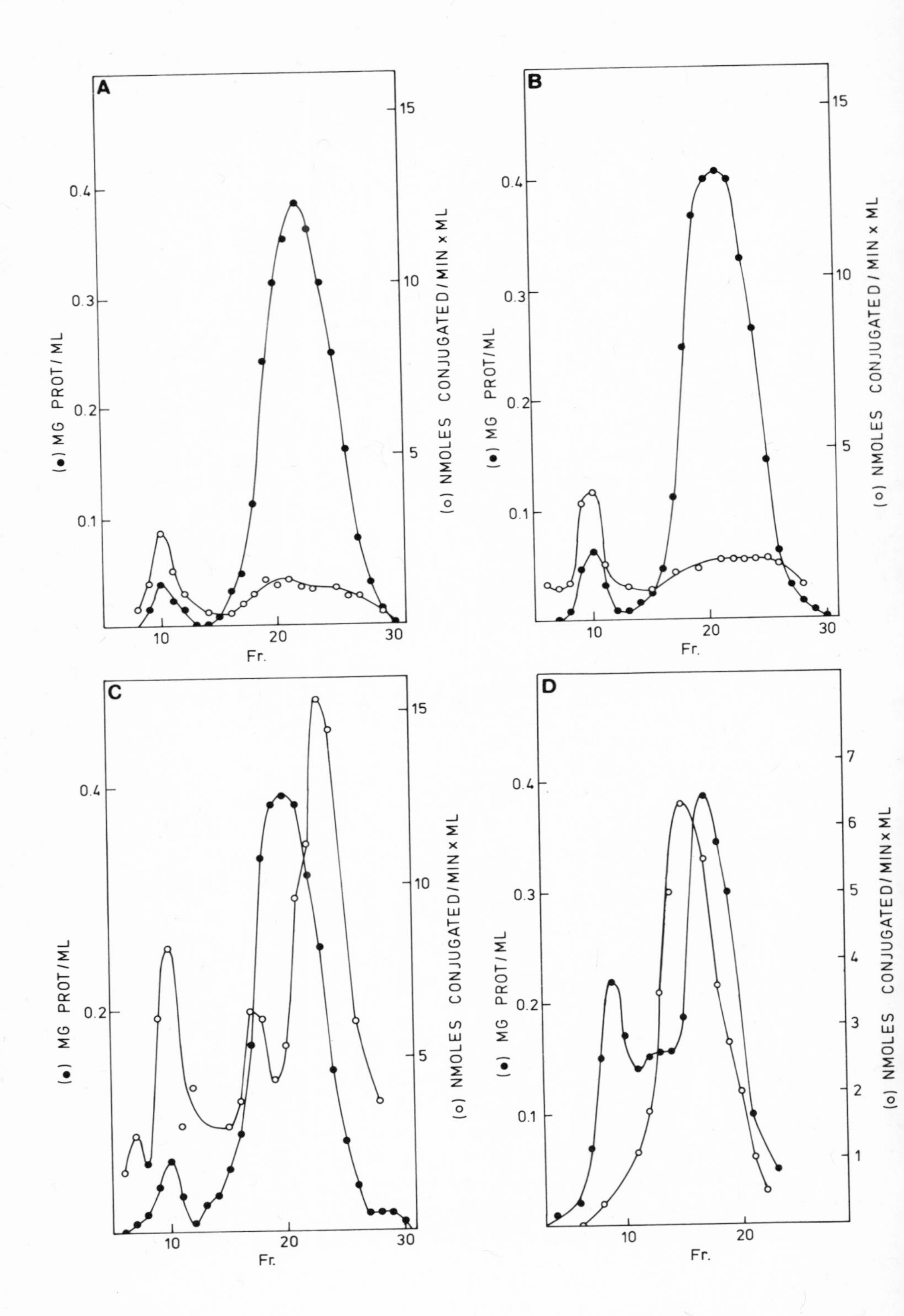

A
B
C
D
(•) MG PROT/ML
(o) NMOLES CONJUGATED/MIN x ML
Fr.

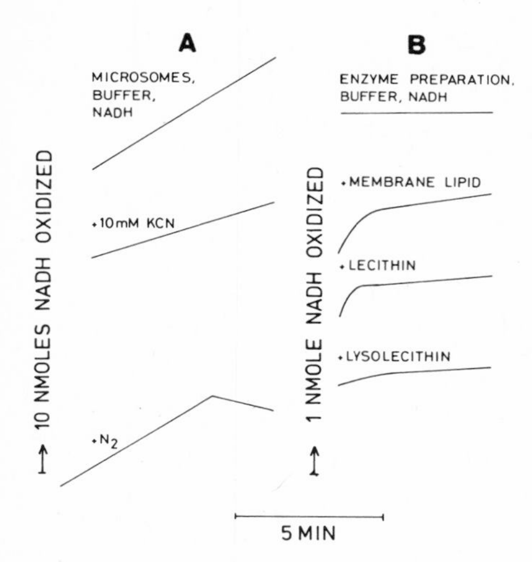

Fig. 3. (A) Microsomal NADH oxidation. One milliliter of a reaction mixture contained rat liver microsomes (0.8 mg protein), 150 mM tris–HCl buffer (pH 7.3), 10 mM $MgCl_2$, and 0.2 mM NADH. The effects on NADH consumption of 10 mM KCN and of flushing the reaction mixture with nitrogen before addition of the microsomes were also determined. *(B) NADH oxidation of a gel-filtered UDPglucuronosyltransferase preparation (fractions 16–20) and the effect of lipids on the reaction.* The reaction mixture contained about 20 μg of protein, 75 μg of Triton X-100, and microsomal membrane lipid, lecithin, or lysolecithin, 0.2 mg/ml. For the other reaction mixture components, see (A).

conjugation of more lipophilic aglycones like 3-methyl-2-nitrophenol continued at a slow rate. Triton X-100 is obviously not an optimal substitute for the membrane lipid. Lecithin, lysolecithin, microsomal membrane lipids, and lysolecithin mixtures provide a better environment for UDPglucuronosyltransferase. Digestion of the phospholipids in rat liver microsomes with phospholipase A increases UDPglucuronosyltransferase activity (6,7) because of a release of lysolecithin and free fatty acids (6). The present results suggest that it is the lysolecithin-containing lipid mixtures that provide the best reactivation of rat liver UDPglucuronosyltransferase activity. Guinea pig liver microsomes differ, since phospholipase A digestion only inactivates UDPglucuronosyltransferase, such inactivation being reversed with lecithin (23).

The separation of UDPglucuronosyltransferase(s) from the proper lipid and protein environment makes the enzyme(s) sensitive to aging, and a rather rapid loss of the catalytic activity is observed. The preparations of transferase made by the method described here are thus different from those obtained using trypsin digestion, digitonin solubilization, and gel filtration (14). These latter preparations are remarkably stable, but contain at least some other membrane components as representatives of the natural environment of the enzyme.

The present results support the idea that different UDPglucuronosyltrans-

←

Fig. 2. Elution of protein and UDPglucuronosyltransferase activity from a Sepahrose 4B gel column using (A) bilirubin, (B) ethylmorphine, (C) 3-methyl-2-nitrophenol, and (D) 4-nitrophenol as substrates. Open circles represent enzymatic activity with the respective substrates, while solid circles represent protein concentration following gel chromatography of Triton X-100-solubilized hepatic microsomal fraction isolated from 3-methylcholanthrene-pretreated rats. Volume of the fractions was 0.5 ml. Determination of UDPglucuronosyltransferase activity was carried out after adding 3 mg of lysolecithin and 3 mg of microsomal membrane lipid per milliliter to the fractions to give a Triton X-100–lipid ratio of 1:4.

ferases may act on different aglycones (1). Bilirubin and ethylmorphine appeared to be conjugated best by the same fractions of the gel-filtered preparation. Different functional groups could accept the D-glucuronic acid residues, however, since ethylmorphine has only one hydroxyl group and bilirubin has both carboxyl and hydroxyl groups. *In vivo,* bilirubin is conjugated via the carboxyl groups. The location of the D-glucuronic acid residue in the product was not determined in the present study. The decline of UDPglucuronosyltransferase activity during storage toward these two aglycones was similar.

3-Methyl-2-nitrophenol seems to be conjugated by a number of protein fractions, since UDPglucuronosyltransferase activity determined with this substrate showed three peaks in the present study. 4-Nitrophenol may be conjugated by only one of these protein fractions. Recently Tephly and coworkers (15) have been able to separate UDPglucuronosyltransferase activity toward morphine and 4-nitrophenol. Their results appear to be in agreement with the present report. Using 4-nitrophenol as substrate, Jansen and Arias (12) could not reactivate UDPglucuronosyltransferase with lecithin in sodium deoxycholate-solubilized rat liver preparations, although activity toward bilirubin was regained, thus providing evidence that different enzymes catalyze the reactions. Since present UDPglucuronosyltransferase preparations are still rather crude and UDPglucuronosyltransferase activity is affected by many factors, only limited conclusions can be drawn.

It has been shown that the affinity of UDPglucuronosyltransferase of rat liver microsomes for substrates increases with the lipophilicity of the compounds. In a series of simple aliphatic alcohols the affinity appears to correlate with the oil–water partition coefficient (8), and lipophilicity also increases glucuronidation in a series of phenolic compounds (22). It is probable that the hydrophobic environment (Triton X-100 or lecithin, lysolecithin, microsomal membrane lipid, and their mixtures) affects the access of the aglycones to the catalytic center of the enzyme(s).

In conjunction with studies on the UDPglucuronosyltransferase assay based on the coupling of UDP production with NADH oxidation, it was noted that liver microsomes as well as the Triton X-100-treated and gel-filtered preparations each consumed reducing equivalents from NADH. In the presence of Triton X-100, NADH oxidation was increased if phospholipids were added to the reaction mixture. Since rotenone was used to inhibit possible mitochondrial oxidation of NADH (24), mitochondrial contamination was probably not the cause of NADH oxidation. The fact that carbon monoxide had no effect indicated that the cytochrome P450 system was also not involved. The lack of effect of α-tocopherol suggests that lipid peroxidation was not using the reducing equivalents (25).

The NADH oxidation by the rat liver microsomes was partially inhibited by cyanide. Nitrogen flushing of the reagents affected the initial velocity of the

NADH oxidation only slightly, but it caused an earlier termination of the reaction, which suggests that oxygen is a factor involved in the NADH consumption by the microsomes. With partially purified membrane preparations, lecithin and microsomal membrane lipid appeared to promote NADH oxidation, which suggests that membrane lipid components are involved in the reaction.

In the determination of UDPglucuronosyltransferase activity using the NADH-coupled enzyme assay (22), substrate-free controls should be employed to measure the endogenous NADH oxidation of the sample even when partially purified enzyme preparations with added lipids are being assayed.

SUMMARY

The microsomal fraction isolated from rat liver after 3-methylcholanthrene pretreatment has been solubilized using an excess of Triton X-100 (25 times the amount of phospholipid in the samples), and the solubilized membrane components have been subjected to gel filtration. Solubilization and gel filtration inactivated UDPglucuronosyltransferase activity toward 4-nitrophenol, although with 3-methyl-2-nitrophenol some activity could be detected even in the absence of added lipids. Addition of lecithin, microsomal membrane lipid, lysolecithin, and mixtures containing lysolecithin reactivated UDPglucuronosyltransferase. On standing, UDPglucuronosyltransferase activity was lost both in the presence of Triton X-100 and with various lipids.

UDPglucuronosyltransferase activity toward the different aglycones was eluted in different fractions from Sepharose 4B. Thus the elution pattern of UDPglucuronosyltransferase which conjugated bilirubin and ethylmorphine was different from that which conjugated 3-methyl-2-nitrophenol and from that which conjugated 4-nitrophenol and 4-methylumbelliferone.

Rat liver microsomes appear to contain an enzyme system oxidizing NADH, and the lipid components of the membrane appear to be involved in the reaction. Microsomal NADH oxidation was not inhibited by carbon monoxide, DL-α-tocopherol, or rotenone. Inhibition of NADH oxidation occurred after addition of potassium cyanide. Nitrogen flushing of the reaction mixture before addition of microsomes had little effect on the initial reaction velocity, but NADH oxidation was terminated earlier than in the presence of air.

ACKNOWLEDGMENT

This study has been supported by a grant from the National Research Council for Natural Sciences, Finland.

REFERENCES

1. G. J. Dutton, in: *Glucuronic Acid, Free and Combined* (G. J. Dutton, ed.), pp. 185–299, Academic Press, New York (1966).
2. O. Hänninen and R. Puukka, Effect of digitonin on UDPglucuronyltransferase in microsomal membranes, *Suom. Kemistil. Finn. Chem. J.* **43**, 451–456 (1970).
3. H. Vainio, Action of chaotropic agents on drug metabolizing enzymes in hepatic microsomes, *Biochim. Biophys. Acta* **307**, 152–161 (1973).
4. O. Hänninen, On the metabolic regulation in the glucuronic acid pathway in the rat tissues, *Ann. Acad. Sci. Fenn. A2,* No. 142, pp. 1–96 (1968).
5. H. Vainio and O. Hänninen, Effect of surfactants on drug metabolism in hepatic microsomes, *Suom. Kemistil.-Finn. Chem. J.* **45**, 56–61 (1972).
6. O. Hänninen and R. Puukka, Activation of microsomal UDPglucuronyltransferase by phospholipases, *Chem.-Biol. Interactions* **3**, 282–284 (1971).
7. D. A. Vessey and D. Zakim, Regulation of microsomal enzymes by phospholipids. II. Activation of hepatic uridine diphosphate–glucuronyltransferase, *J. Biol. Chem.* **246**, 4649–4656 (1971).
8. O. Hänninen and K. Alanen, The competitive inhibition of *p*-nitrophenyl-β-D-glucopyranosiduronic acid synthesis by aliphatic alcohols *in vitro, Biochem. Pharmacol.* **15**, 1465–1467 (1966).
9. A. P. Mowat and I. M. Arias, Partial purification of hepatic UDPglucuronyltransferase: Studies of some its properties, *Biochim. Biophys. Acta* **212**, 65–78 (1970).
10. R. Puukka and M. Laaksonen, Solubilization of UDPglucuronosyltransferase by digitonin from trypsin-digested rat-liver microsomes, *Int. J. Biochem.* **5**, 507–513 (1974).
11. P. L. M. Jansen, Studies on UDPglucuronoyltransferase. The formation of bilirubin mono- and diglucuronide and p-nitrophenyl glucuronide, Ph.D. thesis, University of Nijmegen, Netherlands (1975).
12. P. L. M. Jansen and I. M. Arias, Delipidation and reactivation of UDPglucuronyltransferase from rat liver, Biochim. Biophys. Acta **319**, 23–38 (1975).
13. G. W. Lucier, O. S. McDaniel, and G. E. R. Hook, Nature of the enhancement of hepatic uridine diphosphate glucuronyltransferase activity by 2,3,7,8-tetrachlorodibenzo-*p*-dioxin in rats, *Biochem. Pharmacol.* **24**, 325–335 (1975).
14. R. Puukka, M. Laitinen, H. Vainio, and O. Hänninen, Hepatic UDPglucuronosyltransferase: Partial purification after 3-methylcholanthrene pretreatment of the rats, *Int. J. Biochem.* **6**, 267–270 (1975).
15. E. Del Villar, E. Sanchez, A. P. Autor, and T. R. Tephly, Morphine metabolism. III. Solubilization and separation of morphine and *p*-nitrophenol uridine diphosphoglucuronyltransferase, *Mol. Pharmacol.* **11**, 236–240 (1975).
16. A. Helenius, The effects of detergents on membranes and lipoproteins, Ph.D. thesis, pp. 1–38, University of Helsinki, Finland (1973).
17. J. Folch, M. Lees, and G. H. Sloane Stanley, A simple method for the isolation and purification of total lipids from animal tissues, *J. Biol. Chem.* **226**, 497–509 (1957).
18. A. G. Gornall, C. J. Bardawill, and M. M. David, Determination of serum proteins by means of the biuret reaction, *J. Biol. Chem.* **177**, 751–766 (1949).
19. F. Heepe, H. Karte, and E. Lambrecht, Die Gesamteiweisbestimmung im Liquor mittels lichtelektrischer Messung der Trübund durch Sulfosalicylsäure, Trichloressigsäure and Ferrocyankali Essigsäure, *Z. Kinderheilkd.* **69**, 331–340 (1951).
20. G. R. Bartlett, Phosphorus assay in column chromatography, *J. Biol. Chem.* **234**, 466–468 (1959).
21. A. Aitio, UDPGlucuronosyltransferase of the human placenta, *Biochem. Pharmacol.* **23**, 2203–2205 (1974).

22. G. J. Mulder and A. B. D. van Doorn, A rapid NAD-linked assay for microsomal uridine diphosphate glucuronyltransferase of rat liver and some observations of substrate specificity of the enzyme, *Biochem. J.*, **151**, 131–140 (1975).
23. A. B. Graham and G. C. Wood, The phospholipid-dependence of UDPglucuronyltransferase, *Biochem. Biophys. Res. Commun.* **37**, 567–575 (1969).
24. E. C. Slater, in: *Methods in Enzymology,* Vol. 10 (R. W. Estabrook, ed.), p. 51, Academic Press, New York (1967).
25. B. D. Astill, in: *The Encyclopedia of Biochemistry* (R. J. Williams, ed.), pp. 86–87, Reinhold, New York (1967).
26. D. A. Vessey and D. Zakim, Membrane fluidity and the regulation of membrane bound enzymes, *Horizons Biochem. Biophys.* **1**, 138–174 (1974).

23

Discussion

Zimmerman commented that observations on the Crigler–Najjar syndrome in man support Hänninen's conclusions in that although these affected individuals are deficient in bilirubin glucuronyltransferase activity they are able to conjugate steroids.

In reply to a question as to whether the kinetic constants of the glucuronyl-transferases were altered in the purified preparation, Hänninen replied that partially purified preparations were rather labile and that such kinetic studies had not yet been performed. Schenkman asked if the differences in substrate specificity were maintained in the absence of detergent and lipid or treatment with activators. Hänninen said that this was so, and added that the activating effect of the added lipids was dependent on the aglycone studied, further supporting the idea of the presence of different transferases. It was agreed in further discussion that some of the differences observed experimentally using the different substrates might be due to differences in the physicochemical properties of the substrates rather than to properties of the enzyme(s). Hänninen observed that studies with androsterone tended to support this possibility since it was a good substrate even in the absence of added lipid.

24

Inactivation of Reactive Intermediates in Human Liver

Harri Vainio, Jukka Marniemi, Max Parkki, and Reijo Luoma
Department of Physiology
University of Turku
and
Department of Surgery
City Hospital of Turku
Turku, Finland

The conversion of foreign compounds to oxidized metabolites in mammals has long been considered a means for detoxication and excretion of a variety of harmful toxic substances (1). Recently, however, numerous studies have indicated that the microsomal biotransformation system plays an important role in toxigenesis, e.g., in chemical carcinogenesis (2). It has been estimated that approximately 90% of human cancers are of chemical origin (3). Conjugation with UDPglucuronic acid, a step which follows drug oxidation, is generally regarded as a true detoxication reaction (4).

In man, microsomal monooxygenase and epoxide hydrase activity as well as cytochrome P450 have been found in liver (5–7) and in some extrahepatic tissues such as skin, gastrointestinal tract, and lymphocytes (8–10). The ability of human liver microsomes to alter mutagenicity of drugs has been reported to parallel cytochrome P450 content (11). Microsomes have been isolated from livers as soon as possible after death (12), from needle biopsies (7, 13), or from liver biopsies taken during operations (6, 14–18). The drug-oxidizing enzymes in human liver microsomes have been found to be comparable to those in experimental animals such as the rat (13, 16). Epoxide hydrase activity, measured in samples taken by needle biopsies from human liver, is comparable to that in guinea pig and rat (7). The microsome-bound UDPglucuronosyltransferase has been studied in human liver slices (19). No detailed studies on the properties of UDPglucuronosyltransferase in human hepatic microsomes have been previously performed.

The present study was designed to investigate microsomal monooxygenase, epoxide hydrase, UDPglucuronosyltransferase, and the soluble glutathione *S*-epoxide transferase activities in liver biopsy samples taken during abdominal surgery of patients with cholelithiasis. Correlations between the drug-oxidizing and glucuronidating enzyme activities of individual samples were also studied.

METHODS

Biopsy Samples

Biopsies of adult human liver were taken during abdominal operations under thiopental–nitrous oxide–oxygen anesthesia from patients with cholecystolithiasis. None of the patients was a drug abuser. The samples were forwarded to the pathologist for histological examination and the rest (0.2–1.5 g) was cooled on ice and immediately transferred to the Department of Physiology. The histological pictures were within normal limits on all the samples studied with the exception of slight inflammatory changes. The liver samples were homogenized in a Potter-Elvehjem glass–teflon homogenizer to give a 20% (w/v) homogenate within 1 h of biopsy. The Ca^{2+}-precipitated microsomes were isolated as described by Vainio and Aitio (20). The *in vitro* treatments of microsomes with digitonin, cetylpyridinium chloride, and trypsin were done as described earlier (26). The isolation of microsomes was used for measurements of glutathione *S*-epoxide transferase activity. The biuret method was used for measuring protein content (21). Correlation coefficients were determined by the least squares method with a PDP 8/L computer.

Enzyme Assays

Aryl hydrocarbon hydroxylase activity of human liver microsomes was measured fluorometrically using 3,4-benzpyrene (25 μl of a 0.6 mM solution in acetone) as substrate (22, 23). *p*-Nitroanisole *O*-demethylase activity was determined as described earlier (23). Cytochrome P450 content of liver microsomes was determined by the method of Omura and Sato (24) using a Unicam SP-800 spectrophotometer connected to an external recorder. NADPH cytochrome *c* reductase activity was measured according to Phillips and Langdon (25). Epoxide hydrase was assayed using [7-^{3}H]styrene oxide (specific activity 15.5 mCi/mmol) (NEN Chemicals GmbH, Dreieichenhain, Germany) as substrate (26) as described by Vainio and Parkki (27). UDPGlucuronosyltransferase was assayed using *p*-nitrophenol (0.35 mM) as the aglycone and a concentration of 2.4 mM of UDPglucuronic acid (ammonium salt, 98%) (Sigma), following the method of Isselbacher (28) as modified by Hänninen (29). When 4-methylumbelliferone (0.08 mM) was used as the substrate, the fluorometric method of Aitio (30) was followed. The kinetic studies of UDPglucuronosyltransferase were carried out using ^{14}C-labeled *p*-nitrophenol (16.7 mCi/mmol) (International Chemical and Nuclear Corporation, Irvine, California) as the aglycone substrate (31).

Glutathione *S*-epoxide transferase of human liver supernatant fraction was determined with [7-^{3}H] styrene oxide as the substrate as described by Marniemi and Parkki (32). The water-soluble reaction product was separated from unreacted styrene oxide by extraction with light petroleum and the radioactivity of the aqueous phase was subsequently measured with a liquid scintillation counter (32).

RESULTS

Drug-Oxidizing Monooxygenase System in Individual Liver Biopsies

The data on drug monooxygenase activities in individual liver samples are illustrated in Table 1. Although variations in enzyme activities are rather large, the mean activities are of the same magnitude as those we reported earlier in rats (33). The microsomal Ca^{2+}-precipitated protein content per gram of liver (wet weight) is similar to that obtained in rat liver, i.e., about 20–25 mg/g.

Epoxide Hydrase and Glutathione *S*-Epoxide Transferase in Liver Samples

The epoxide hydrase and glutathione *S*-epoxide transferase activities in individual liver biopsies, both measured with [^{3}H] styrene oxide as substrate, are shown in Table 2. The mean activity of epoxide hydrase is about twice as high as that found in rat livers. Glutathione *S*-epoxide transferase activity was measured only in three individual liver samples. The mean transferase activity was, however, about the same as that obtained from rat liver.

UDPGlucuronosyltransferase Activity in Individual Liver Samples

UDPGlucuronosyltransferase activities, with both *p*-nitrophenol and 4-methylumbelliferone as aglycones, are illustrated in Table 3. The transferase activities as measured in native, nonactivated liver microsomes were about twice as high as those in rat liver microsomes when *p*-nitrophenol was used as the aglycone. With 4-methylumbelliferone, the activities were practically equal in human and rat liver. The measurable UDPglucuronosyltransferase activity in human liver microsomes could be increased by treating the microsomal vesicles *in vitro* with surfactants such as digitonin and cetylpyridinium chloride. The extent of activation was, however, smaller (i.e., three- to fivefold) than that obtained with rat liver microsomes, which increased approximately tenfold. In

Table 1. Components of the Monooxygenase System and Overall Drug Hydroxylation Activities in Samples of Individual Human Liver Microsomes and a Comparison with Values Obtained from Rat Liver Microsomes

Age (yr) and sex of the patient	Protein content (mg/g liver)	NADPH cytochrome *c* reductase (μmol/g wet wt/min)	Cytochrome P450 (nmol/g wet wt/min)	Aryl hydrocarbon hydroxylase (nmol/g wet wt/min)	*p*-Nitroanisole *O*-demethylase (nmol/g wet wt/min)
60 ♀	–	1.20	2.74	0.153	–
61 ♂	18.1	1.56	4.39	0.380	15.0
58 ♀	23.8	1.04	3.84	0.147	12.0
44 ♀	28.4	1.04	6.14	0.203	6.99
22 ♀	25.7	2.08	5.60	0.130	36.0
66 ♀	24.4	1.99	5.05	0.443	30.2
Mean	24.1	1.48	4.63	0.242	20.0
Rat[a]	19.7	2.51	6.2	0.21	8.16

[a]Data from Vainio and Aitio (20).

Table 2. Epoxide Hydrase and Glutathione *S*-Epoxide Transferase Activities in Samples of Individual Human Liver and a Comparison with the Activities Obtained from Rat Liver

Age (yr) and sex of the patient	Epoxide hydrase (nmol/g wet wt/min)	Age (yr) and sex of the patient	Glutathione *S*-epoxide transferase (μmol/g wet wt/min)
60 ♀	286	46 ♀	6.9
61 ♂	455	52 ♀	8.1
58 ♀	276	65 ♀	3.9
44 ♀	645		
22 ♀	383		
66 ♀	322		
Mean	394	Mean	6.3
Rat	170	Rat	6.0

addition, the treatment of liver microsomes with trypsin, which is known to increase the measurable UDPglucuronosyltransferase in rat liver severalfold, did not increase the UDPglucuronosyltransferase activity in human liver biopsy samples.

The kinetic behavior of UDPglucuronosyltransferase in human liver microsomes exhibited atypical Michaelis–Menten kinetics. Lineweaver–Burk plots used to analyze the effect of increasing UDPglucuronic acid concentrations on the reaction rate of UDPglucuronosyltransferase did not yield straight lines with human liver microsomes (Fig. 1). The break in the line was most marked with native microsomes. Kinetic studies with digitonin-treated microsomes also showed break points in the Lineweaver–Burk plots, but the change in slope was less than that seen in nontreated microsomes.

Table 3. Activity of UDPGlucuronosyltransferase of Human Hepatic Microsomes Conjugating *p*-Nitrophenol (PNP) or 4-Methylumbelliferone (MU)[a]

Age (yr) and sex of the patient	Native microsomes		Digitonin treated		CPC treated		Trypsin treated	
	PNP	MU	PNP	MU	PNP	MU	PNP	MU
60 ♀	16.4	28.9	41.2	168	50.0	206	–	–
61 ♂	25.0	28.9	70.4	154	44.9	85.3	–	–
58 ♀	12.6	25.8	46.8	146	–	–	16.5	21.3
44 ♀	28.0	37.3	74.1	142	63.4	146	18.3	37.4
22 ♀	26.1	24.9	70.0	160	52.7	114	25.7	37.3
66 ♀	20.4	25.8	68.2	171	–	–	–	–
Mean	21.4	28.6	61.8	157	52.7	138	20.2	32.1
Rat[b]	12.6	29.9	148	334	166	351	50	64

[a]Activities are expressed as nmol of substrate conjugated/g wet wt/min.
[b]Data of rat liver taken from Vainio and Hänninen (34).

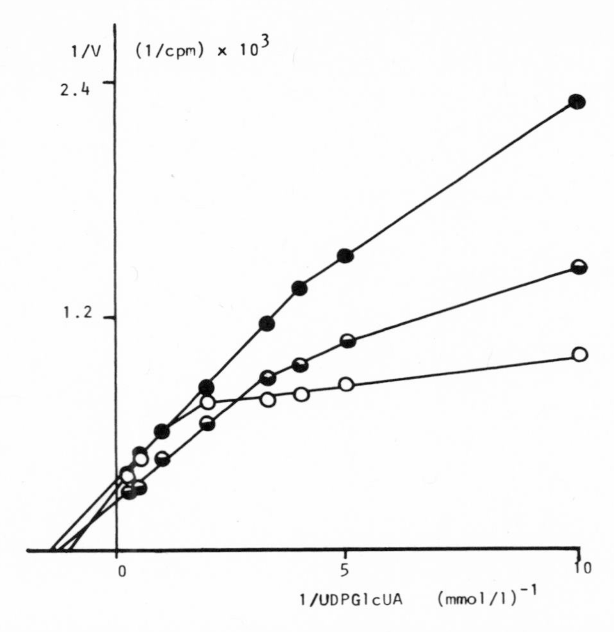

Fig. 1. Double-reciprocal plots of UDPglucuronosyltransferase of human liver microsomes at varying UDPglucuronic and fixed aglycone concentrations. ○, Native microsomes in the presence of 0.3 mM *p*-nitrophenol; ◒, digitonin-treated microsomes in the presence of 0.3 mM *p*-nitrophenol; ●, digitonin-treated microsomes in the presence of 0.6 mM *p*-nitrophenol.

Correlation between Drug-Oxidizing and Glucuronidating Enzymes in Individual Liver Samples

Drug oxidation and glucuronidation are sequential events in drug metabolism. It has been suggested that the enzymes catalyzing these two successive events are also closely associated in the microsomal membrane. A significant correlation was observed between cytochrome P450 levels and UDPglucuronosyltransferase (*p*-nitrophenol) activity in individual liver samples (Fig. 2) with both nontreated and digitonin-treated microsomes. No correlation, however, was seen between drug hydroxylation reactions such as *p*-nitroanisole *O*-demethylation or aryl hydrocarbon hydroxylation and UDPglucuronosyltransferase activity. A positive correlation was also obtained between epoxide hydrase activity and cytochrome P450 levels in individual liver samples (Fig. 3).

DISCUSSION

The enzymes responsible for conversion of lipid-soluble drugs, carcinogens, insecticides, and steroids into their water-soluble metabolites may constitute a

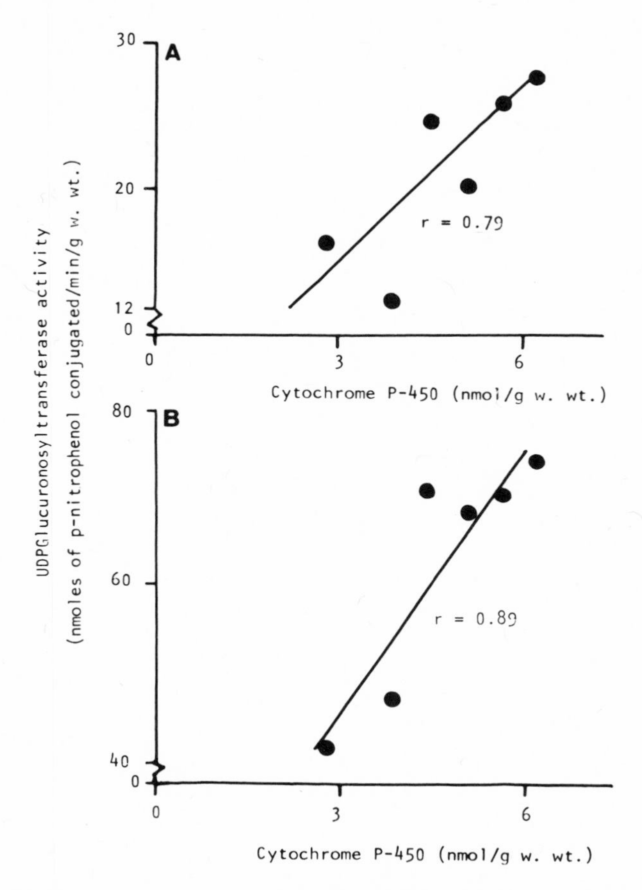

Fig. 2. Correlation between UDPglucuronosyltransferase activity and cytochrome P450 content of liver microsomes of individual patients. UDPGlucuronosyltransferase activity was measured in native (A) or digitonin-treated (B) microsomes.

multienzyme complex in the microsomal membrane (35, 36). Metabolism usually occurs in two steps: a prior oxidation, reduction, or hydrolysis of the parent compound (phase I), followed by conjugation of the phase I product, generally with glucuronic acid in mammals (phase II). In the present study, we found a high correlation in individual liver samples between cytochrome P450, the key component of the monooxygenase complex, and UDPglucuronosyltransferase activity, although interindividual variations were large. This suggests that the enzymes catalyzing the two successive steps might have some regulating factor in common, although the number of individuals was too small to allow conclusive correlations.

The hepatic metabolism of a variety of aromatic compounds by monooxygenases proceeds via intermediate arene oxides which react with tissue macro-

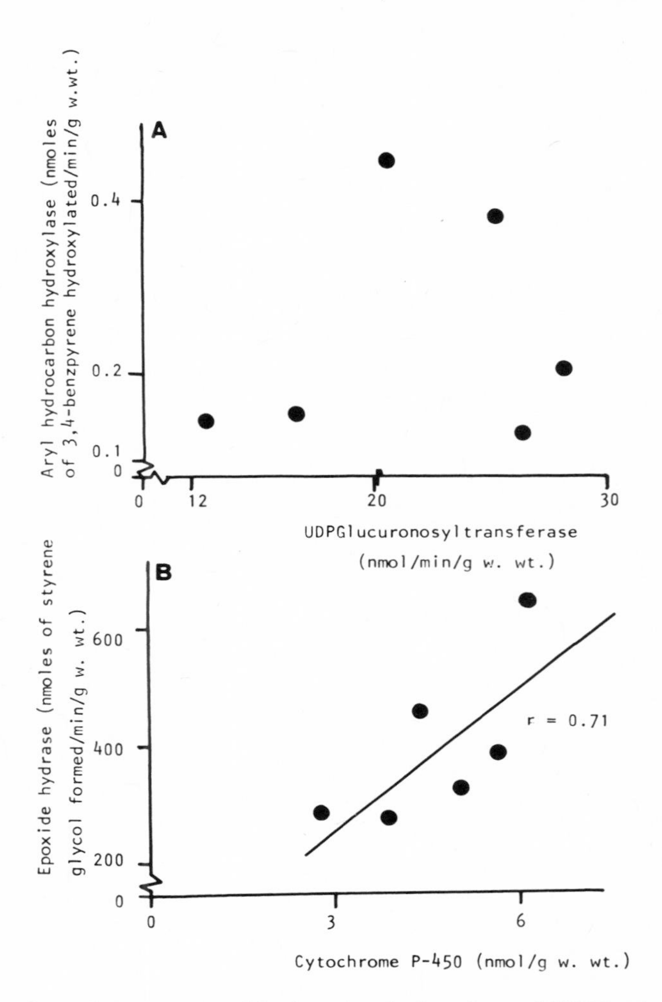

Fig. 3. Correlation between (A) aryl hydrocarbon hydroxylase and UDPglucuronosyltransferase and (B) epoxide hydrase and cytochrome P450 content of untreated human liver microsomes.

molecules, conjugate with glutathione, rearrange nonenzymatically to phenols, or are converted to dihydrodiols by the action of epoxide hydrase (37, 38). A close physical association in the microsomal membrane between cytochrome P450 and epoxide hydrase has been observed (39, 40). However, since their induction with phenobarbital does not occur in parallel (41), it has been suggested that they are not under common genetic control (42). In the present study of human liver microsomes, a very high correlation was also seen between cytochrome P450 and epoxide hydrase activity.

UDPGlucuronosyltransferase in common laboratory animals is a latent enzyme (36), the activity of which is increased three- to tenfold in rodents by agents such as trypsin (43). In these studies, digitonin and cetylpyridinium chloride activated the human liver enzyme but trypsin had no effect. These differences in activation pattern may be the result of structural differences between human and rodent microsomal membranes.

UDPGlucuronosyltransferase of human liver microsomes did not follow typical Michaelis–Menten kinetics; when the data were plotted in the form of a double-reciprocal plot, a decrease in slope was observed at lower UDPglucuronic acid concentrations. This is in agreement with similar findings reported for this enzyme activity in rat liver microsomes (45). The unusual kinetics can also depend on integrity of the microsomal membrane, as reported by Marniemi and Laitinen (46). Thus the interference of other microsomal enzymes such as UDPglucuronic acid pyrophosphatase in the transferase reaction must also be considered. However, Pelkonen (18) has reported that kinetic studies of reactions other than conjugation such as aminopyrine *N*-demethylation, aniline *p*-hydroxylation, and 3,4-benzpyrene hydroxylation by adult human liver samples also resulted in nonlinear double-reciprocal plots of the data. In contrast, straight-line kinetics were observed in studies of fetal liver preparations, which Pelkonen (18) suggested may be explained by the existence of two enzymes in adults responsible for the metabolism of these substrates (aminopyrine, aniline, and 3,4-benzpyrene).

The drug-metabolizing enzymes in human liver were observed in the present study to be similar in many respects to those found in common laboratory animals. Thus studies of these enzymes in laboratory animals appear to have relevance as far as human enzymes are concerned.

ACKNOWLEDGMENTS

This study has been financially supported by a grant from U.S. Public Health Service (AM-06018-13).

REFERENCES

1. R. T. Williams, *Detoxication Mechanisms: The Metabolism and Detoxication of Drugs, Toxic Substances and Other Organic Compounds,* Chapman and Hall, London (1966).
2. C. Heidelberger, Current trends in chemical carcinogenesis, *Fed. Proc.* **32**, 2154–2160 (1973).
3. E. Boyland, The correlation of experimental carcinogenesis and cancer in man, *Prog. Exp. Tumor Res.* **11**, 222–234 (1969).
4. G. J. Dutton, in: *Handbook of Experimental Pharmacology,* Vol. XXVIII (B. B. Brodie and J. R. Gillette, eds.), pp. 378–400, Springer-Verlag, Berlin (1971).
5. R. Kuntzman, L. C. Mark, L. Brand, M. Jacobson, W. Levin, and A. H. Conney,

Metabolism of drugs and carcinogens by human liver enzymes, *J. Pharmacol. Exp. Ther.* **152,** 151–156 (1966).

6. A. P. Alvares, G. Schilling, W. Levin, R. Kuntzman, L. Brand, and L. C. Mark, Cytochromes P-450 and b_5 in human liver microsomes, *Clin. Pharmacol. Ther.* **10,** 655–659 (1969).
7. F. Oesch, H. Thoenen, and H.-J. Fahrlaender, Epoxide hydrase in human liver biopsy specimens: Assay and properties, *Biochem. Pharmacol.* **23,** 1307–1317 (1974).
8. O. Pelkonen, M. Vorne, P. Jouppila, and N. T. Kärki, Metabolism of chlorpromazine and *p*-nitrobenzoic acid in the liver, intestine and kidney of human foetus, *Acta Pharmacol. Toxicol.* **29,** 284–294 (1971).
9. A. P. Alvares, A. Kappas, W. Levin, and A. H. Conney, Inducibility of benzo[*a*]pyrene hydroxylase in human skin by polycyclic hydrocarbons, *Clin. Pharmacol. Ther.* **14,** 30–40 (1973).
10. G. Kellermann, M. Luyten-Kellermann, M. G. Horning, and M. Stafford, Correlation of aryl hydrocarbon hydroxylase activity of human lymphocyte cultures and plasma elimination rates for antipyrine and phenylbutazone, *Drug Metab. Dispos.* **3,** 47–50 (1975).
11. P. Czygan, H. Greim, A. J. Garro, F. Hutterer, J. Rudick, F. Schaffner, and H. Popper, Cytochrome P-450 content and the ability of liver microsomes from patients undergoing abdominal surgery to alter the mutagenicity of a primary and a secondary carcinogen, *J. Natl. Cancer Inst.* **51,** 1761–1764 (1973).
12. F. J. Darby, W. Newnes, and D. A. Price-Evans, Human liver microsomal drug metabolism, *Biochem. Pharmacol.* **19,** 1514–1517 (1970).
13. B. Schoene, R. A. Fleischmann, H. Remmer, and H. F. von Oldershausen, Determination of drug metabolizing enzymes in needle biopsies of human liver, *Eur. J. Clin. Pharmacol.* **4,** 65–73 (1972).
14. E. Ackermann and I. Heinrich, Die Aktivität der *N*- und *O*-demethylase in der Leber des Menschen, *Biochem. Pharmacol.* **19,** 327–342 (1970).
15. A. Auranen, Liver involvement in cholecystitis and in obstructive icterus: A study with special reference to hexobarbital metabolism and glucose-6-phosphatase activity of human liver microsomes, *Acta Chir. Scand. Suppl.* **424,** 1–62 (1972).
16. O. Pelkonen, N. T. Kärki, and T. K. I. Larmi, Liver drug metabolism in tumor-bearing surgical patients, *Chir. Gastroenterol.* **7,** 436–443 (1973).
17. O. Pelkonen, E. H. Kaltiala, T. K. I. Larmi, and N. T. Kärki, Cytochrome P-450-linked monooxygenase system and drug-induced spectral interactions in human liver microsomes, *Chem.-Biol. Interactions* **9,** 205–216 (1974).
18. O. Pelkonen, Developmental change in the apparent kinetic properties of drug-oxidizing enzymes in the human liver, *Res. Commun. Chem. Pathol. Pharmacol.* **10,** 293–302 (1975).
19. E. Hietanen, A. Auranen, O. Hänninen, and B. Savakis, Hepatic drug metabolism and oxygen uptake in biliary diseases, *Scand. J. Gastroentero.* **9,** Suppl. 27, 28 (1974).
20. H. Vainio and A. Aitio, Enhancement of microsomal drug hydroxylation and glucuronidation in rat liver by phenobarbital and 3-methylcholanthrene in combination, *Acta Pharmacol. Toxicol.* **34,** 130–140 (1974).
21. E. Layne, in: *Methods in Enzymology,* Vol. 3 (S. P. Colowick and N. O. Kaplan, eds.), pp. 447–454, Academic Press, New York (1957).
22. D. W. Nebert and H. V. Gelboin, Substrate-inducible microsomal aryl hydroxylase in mammalian cell culture, *J. Biol. Chem.* **243,** 6242–6249 (1968).
23. H. Vainio, Drug hydroxylation and glucuronidation in liver microsomes of phenobarbital treated rats, *Xenobiotica* **3,** 715–725 (1973).
24. T. Omura and R. Sato, The carbon monoxide-binding pigment of liver microsomes. II. Solubilization, purification, and properties, *J. Biol. Chem.* **239,** 2379–2385 (1964).

25. A. H. Phillips and R. G. Langdon, Hepatic triphosphopyridine nucleotide-cytochrome *c*-reductase: Isolation, characterization and kinetic studies. *J. Biol. Chem.* **237**, 2652–2660 (1962).
26. F. Oesch, D. M. Jerina, and J. Daly, A radiometric assay for hepatic epoxide hydrase activity with (7-^{3}H) styrene oxide, *Biochim. Biophys. Acta* **227**, 685–691 (1971).
27. H. Vainio and M. G. Parkki, Protection of microsomal drug biotransformation enzymes against carbon tetrachloride by diethyldithiocarbamate in rat liver, *Res. Commun. Chem. Pathol. Pharmacol.* **9**, 511–522 (1974).
28. K. J. Isselbacher, Enzymatic mechanisms of hormone metabolism. II. Mechanism of hormonal glucuronide formation, *Recent Prog. Horm. Res.* **12**, 134–145 (1956).
29. O. Hänninen, On the metabolic regulation in the glucuronic acid pathway in the rat tissues, *Ann. Acad. Sci. Fenn.*, Ser. AII, No 142, 1–96 (1968).
30. A. Aitio, Glucuronide synthesis in the rat and guinea pig lung, *Xenobiotica* **3**, 13–22 (1973).
31. J. Marniemi and O. Hänninen, Radiochemical assay of UDP glucuronyltransferase (*p*-nitrophenol), *FEBS Lett.* **32**, 273–276 (1973).
32. J. Marniemi and M. G. Parkki, Radiochemical assay of glutathione *S*-epoxide transferase and its enhancement by phenobarbital in rat liver *in vivo, Biochem. Pharmacol.* **24**, 1569–1572 (1975).
33. H. Vainio and A. Aitio, Influence of hematin, carbon tetrachloride and SKF 525-A administration on the enhancement of microsomal monooxygenase and UDPglucuronosyltransferase by 3,4-benzpyrene in rat liver, *Acta Pharmacol. Toxicol.* **37**, 23–32 (1975).
34. H. Vainio and O. Hänninen, A comparative study on drug hydroxylation and glucuronidation in liver microsomes of phenobarbital and 3-methylcholanthrene treated rats, *Acta Pharmacol. Toxicol.* **35**, 65–75 (1974).
35. C. von Bahr, Metabolism of tricyclic antidepressant drugs: Pharmacokinetic and molecular aspects, academic thesis, Department of Pharmacology and Forensic Medicine, Karolinska Institutet, Stockholm, and Department of Clinical Pharmacology, University of Linköping, pp. 1–34 (1972).
36. H. Vainio, On the topology and synthesis of drug-metabolizing enzymes in hepatic endoplasmic reticulum, academic thesis, Department of Physiology, University of Turku, pp. 1–46 (1973).
37. F. Oesch, Mammalian epoxide hydrase: Inducible enzymes catalyzing the inactivation of carcinogenic and cytotoxic metabolites derived from aromatic and olefinic compounds, *Xenobiotica* **3**, 305–340 (1973).
38. D. M. Jerina and J. W. Daly, Arene oxides: A new aspect of drug metabolism, *Science* **185**, 573–582 (1974).
39. F. Oesch, D. M. Jerina, J. W. Daly, A. Y. H. Lu, R. Kuntzman, and A. H. Conney, A reconstituted microsomal enzyme system that converts naphthalene to trans-1,2-dihydroxy-1,2-dihydronaphthalene via naphthalene-1,2-oxide: Presence of epoxide hydrase in cytochrome P-450 and P-448 fractions, *Arch. Biochem. Biophys.* **153**, 62–67 (1972).
40. P. M. Dansette, H. Yagi, D. M. Jerina, J. W. Daly, W. Levin, A. Y. H. Lu, R. Kuntzman, and A. H. Conney, Assay and properties of epoxide hydrase from rat liver microsomes, *Arch. Biochem. Biophys.* **164**, 511–517 (1974).
41. F. Oesch, N. Morris, J. W. Daly, J. E. Gielen, and D. W. Nebert, Genetic expression of the induction of epoxide hydrase and aryl hydrocarbon hydroxylase activities in the mouse by phenobarbital or 3-methylcholanthrene, *Mol. Pharmacol.* **9**, 692–696 (1973).
42. D. W. Nebert, W. F. Benedict, J. E. Gielen, F. Oesch, and J. W. Daly, Aryl hydrocarbon hydroxylase, epoxide hydrase, and 7,12-dimethylbenz[*a*]anthracene-produced skin tumorigenesis in the mouse, *Mol. Pharmacol.* **8**, 374–379 (1972).
43. E. Hietanen and H. Vainio, Effect of administration route of DDT on acute toxicity

and on drug biotransformation in various rodents, *Arch. Environ. Contam. Toxicol.* **4**, 201–216 (1976).

44. A. Winsnes, Kinetic properties of different forms of hepatic UDPglucuronyltransferase, *Biochim. Biophys. Acta* **284**, 394–405 (1972).
45. J. Marniemi and O. Hänninen, Kinetic properties of liver UDPglucuronosyltransferase determined with ^{14}C-labeled *p*-nitrophenol as substrate, *Acta Physiol. Scand. Suppl.* **396**, 136 (1973).
46. J. Marniemi and M. Laitinen, Regulatory properties of partially purified UDPglucuronosyltransferase (*p*-nitrophenol), *Int. J. Biochem.* **6**, 345–353 (1975).

25

Discussion

Conney briefly described similar studies in his laboratory aimed at the question of how many cytochrome P450 enzymes are present in human liver. The metabolism of six different substrates was estimated in each of ten human livers sampled at autopsy, to determine if there was the same variability in the metabolic rate of all substrates. He reported a thirtyfold variation in benzpyrene hydroxylation and a high correlation with zoxazolamine and coumarin hydroxylation. In contrast, antipyrine metabolism varied independently, suggesting that antipyrine is not a good indicator of general drug metabolism in man. In other studies, a variety of epoxides were used as substrates, and suggested that either human liver contains a single epoxide hydratase or that multiple epoxide hydratases are under similar regulatory control in the different individuals. Ullrich commented on experiments in his laboratory also aimed at determination of the multiplicity of P450-dependent drug-metabolizing enzymes in human liver. His studies indicate that man shows high differences in activity of the P450 enzymes and in their response to inhibitors. Ullrich suggested that each individual may have a completely different pattern of drug metabolism, which would make predictions of toxicity even more difficult. Remmer asked what was the time delay between death and obtaining the autopsy liver samples. Conney replied that in his studies the time interval was longer than he would like. He pointed out, however, that if a single enzyme oxidized two substrates, loss of the enzyme during the delay should be reflected in similar loss of both substrate oxidations.

26

Nucleotide Pyrophosphatase as a Competitor of Drug Hydroxylation and Conjugation in Rat Liver Microsomes

Eino Puhakainen
Department of Physiology
University of Kuopio
SF-70101 Kuopio 10, Finland

Many drug metabolism reactions take place in the endoplasmic reticulum of hepatocytes in mammals. The oxidative reactions may lead to either activation or inactivation of the drug, but the conjugation with glucuronic acid inactivates the biological effects of the aglycone (1). In the monooxygenase system, NADPH gives electrons to cytochrome P450, although NADH can also serve as an electron donor (2). The conjugation reaction, glucuronide formation, is catalyzed by UDPglucuronosyltransferase (E.C. 2.4.1.17). In this reaction, aglycones are conjugated with a D-glucuronic acid residue donated by UDPglucuronic acid (3). In mammalian liver there exists also a nucleotide pyrophosphatase (E.C. 3.6.1.9). This enzyme hydrolyzes several nucleotides including UDPglucuronic acid, NADP, NADPH, NAD, and NADH (4). This nucleotide pyrophosphatase competes with drug metabolism by consuming the same substrates and coenzymes. However, little is known about microsomal nucleotide pyrophosphatase compared to drug-metabolizing enzymes. In this work, the rat liver microsomal nucleotide pyrophosphatase and its effects on substrates and coenzymes needed in drug hydroxylation and conjugation have been studied *in vitro*.

MATERIALS AND METHODS

Animals

Male rats (*Rattus norvegicus*) of the Wistar/Af/Han/Mol/(Han 67) strain, about 3 months old, were used in these experiments. The rats were killed by a blow on the head and the cervical vessels were cut. Microsomes were prepared as described earlier (5). Protein was determined by the method described by Lowry *et al.* (6) using bovine serum albumin as the reference protein.

Determination of Enzyme Activities

Nucleotide pyrophosphatase activity was measured by radiochemical, coupled phosphatase, and spectrophotometric methods.

In the radiochemical method, the incubations were carried out at 38°C in 70 μl of reaction mixture containing 0.2 M tris-HCl buffer (pH 7.4), 2 mM UDPglucuronic acid (75 nCi labeled UDPglucuronic acid, ^{14}C in D-glucuronic acid residue), and microsomes corresponding to about 0.1 mg of microsomal protein. When conjugation was also studied, 0.35 mM *p*-nitrophenol and 10 mM K_2EDTA were added to the reaction mixture. The reactions were terminated after a 10-min incubation by adding 100 μl of absolute ethanol and immersing the tubes in an ice bath. The radioactive metabolites were separated and analyzed as described earlier (7).

In the coupled phosphatase method, the nucleotide pyrophosphatase activity was determined by measuring the production of inorganic phosphate by a modification of Fiske and Subbarow's method (8). An excess of purified alkaline phosphatase was added to the reaction mixture to release inorganic phosphate from the hydrolysis products in the reaction catalyzed by nucleotide pyrophosphatase. The incubations were carried out in a final volume of 150 μl in the presence of 67 mM tris-HCl (pH 8.9) containing 2 mM UDPglucuronic acid or other nucleotide substrates, 2000 U/liter alkaline phosphatase, and microsomes corresponding to about 25 μg of microsomal protein.

Aniline hydroxylase activity was measured by a modification of the method of Brodie and Axelrod (9). Hydrolysis of NADP under aniline hydroxylation assay conditions was measured by a spectrophotometric method. The reaction mixtures were incubated at 38°C with and without microsomes. The amount of NADP left in the reaction mixture after varying time periods was determined by adding an excess amount of isocitric acid and isocitric acid dehydrogenase to reduce unhydrolyzed NADP, and the absorbance at 340 nm was measured by a Cary 118C spectrophotometer.

RESULTS AND DISCUSSION

The radiochemical method makes it possible to measure simultaneously the hydrolysis of UDPglucuronic acid and the conjugation of *p*-nitrophenol with D-glucuronic acid. The pH optimum of microsomal nucleotide pyrophosphatase is 8.9, but even at pH 7.4 in 0.2 M tris-HCl buffer containing 0.35 mM *p*-nitrophenol a specific activity of 15 nmol UDPglucuronic acid hydrolyzed/min/mg protein was found. Under the same conditions, only 0.4 nmol *p*-nitro-

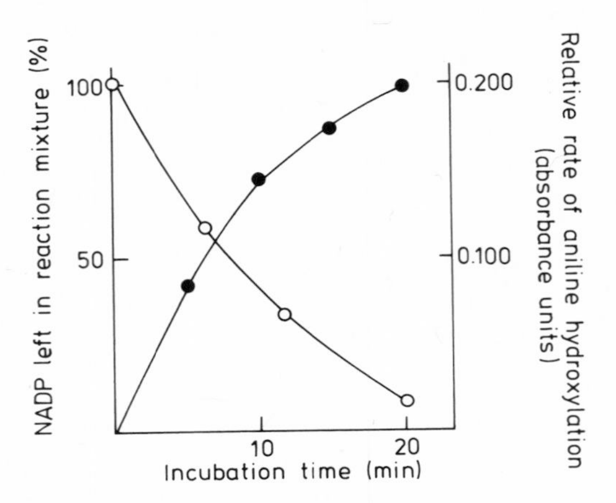

Fig. 1. Decrease of NADP (○) and increase of aniline hydroxylation (•) vs. incubation time in 37.5 mM potassium phosphate buffer (pH 7.5) containing 2.5 mM $MgCl_2$, 2.5 mM $MnCl_2$, 1.25 mM Na_3-isocitrate, 0.49 mM NADP, 4 mM aniline-HCl, and rat liver microsomes corresponding to 2 mg of protein in a final volume of 0.5 ml.

phenol/min/mg protein was conjugated. Even in the presence of 10 mM K_2EDTA, hydrolysis still predominated over conjugation, the corresponding specific activities being 3.5 and 0.38 nmol UDPglucuronic acid hydrolyzed and conjugated, respectively, per min/mg protein.

When the hydrolysis of NADP under aniline hydroxylation assay conditions was measured, more than 90% of the coenzyme was hydrolyzed during the 20-min incubation and the aniline hydroxylation reaction was linear only during the first 5 min (Fig. 1). At least part of the nonlinearity of the reaction may be caused by a rapid hydrolysis of the coenzyme, especially since NADPH, which is produced by the generating system and is finally used as a hydrogen donor, is hydrolyzed twice as fast as NADP (Fig. 2). Under hydroxylation assay conditions where the NADPH-generating system is used, chelators cannot be added to inhibit nucleotide pyrophosphatase, since divalent cations are essential for the generating system.

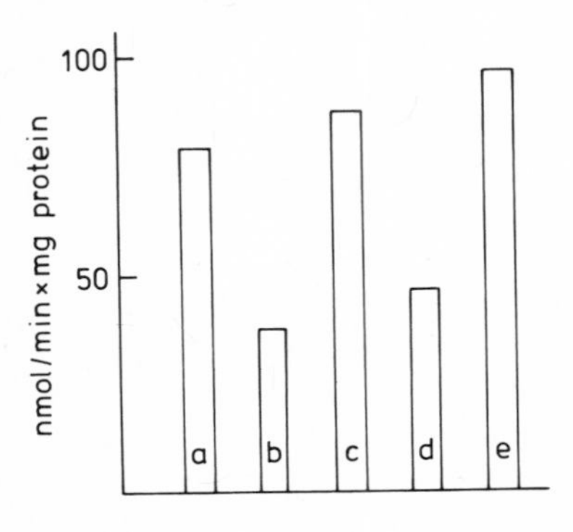

Fig. 2. Hydrolysis of NADPH (a), NADP (b), NADH (c), NAD (d) and UDPglucuronic acid (e) in 67 mM tris-HCl buffer (pH 8.9) containing 25 mM $MgCl_2$ catalyzed by rat liver microsomes.

Nucleotide pyrophosphatase is a powerful competitor of drug conjugation and hydroxylation, hydrolyzing the substrates and coenzymes used in these reactions. Special care must be taken to ensure substrate and coenzyme saturation in drug hydroxylation and conjugation assay conditions *in vitro*, especially in end-point methods, when rat liver microsomes are used as enzyme source.

ACKNOWLEDGMENTS

This work was supported by a grant from Orion Research Foundation, Finland.

REFERENCES

1. H. G. Mandel, in: *Fundamentals of Drug Metabolism and Drug Disposition* (B. N. La Du, H. G. Mandel, and E. L. Way, eds.), pp. 149–186, Williams and Wilkins, Baltimore (1971).
2. J. R. Gillette, D. C. Davis, and H. A. Sasame, Cytochrome P-450 and its role in drug metabolism, *Annu. Rev. Pharmacol.* **12**, 57–84 (1972).
3. G. J. Dutton, in: *Glucuronic Acid Free and Combined*, pp. 186–299, Academic Press, New York (1966).
4. H. Ogawa, M. Sawada, and M. Kawada, Purification and properties of a pyrophosphatase from rat liver microsomes capable of catalyzing the hydrolysis of UDPglucuronic acid, *J. Biochem. (Tokyo)* **59**, 126–134 (1966).
5. M. Marselos, T. Rantanen, and E. Puhakainen, Comparison of the effect of phenobarbital on the D-glucuronic acid pathway in euthyroid and hypothyroid rats, *Acta Pharmacol. Toxicol.* **37**, 415–424 (1975).
6. O. H. Lowry, N. J. Rosebrough, A. L. Farr, and R. J. Randall, Protein measurements with the Folin phenol reagent, *J. Biol. Chem.* **193**, 265–275 (1951).
7. E. Puhakainen and O. Hänninen, Lipid acceptor in UDPglucuronic acid metabolism in rat liver microsomes, *FEBS Lett.* **39**, 144–148 (1974).
8. A. E. Harper, in: *Methods in Enzymatic Analysis* (H. -U. Bergmeyer, ed.), pp. 788–792, Verlag Chemie, Weinheim (1963).
9. B. B. Brodie and J. Axelrod, The estimation of acetanilide and its metabolic products, aniline, *N*-acetyl-*p*-aminophenol and *p*-aminophenol (free and total conjugated) in biological fluids and tissues, *J. Pharmacol. Exp. Ther.* **94**, 22–28 (1948).

IV

Specific Reactive Intermediates

27

Metabolic Activation of Hydrazines to Highly Reactive Hepatotoxic Intermediates

Jerry R. Mitchell, Sidney D. Nelson, Wayne R. Snodgrass, and John A. Timbrell

Laboratory of Chemical Pharmacology
National Heart and Lung Institute
National Institutes of Health
Bethesda, Maryland 20014, U.S.A.

Substituted hydrazine derivatives are receiving considerable attention in chemical toxicology because of their widespread use as therapeutic agents for the treatment of depression, tuberculosis, and cancer, as herbicides, as intermediates in chemical syntheses, and even as rocket fuels. Hydrazines can produce many toxic responses, including methemoglobinemia, cell necrosis, mutagenesis, and carcinogenesis. We would like to report on our work concerning the possible mechanism of hepatotoxicity caused by the tuberculostatic drug isoniazid and related hydrazine compounds.

RESULTS

The conversion of isoniazid to its principal urinary metabolite, acetylisoniazid (1), is known to be under genetic control in humans (2). Three of our studies have suggested a possible correlation between susceptibility to the serious hepatitis that occurs in 1% of isoniazid recipients and rapid acetylation of the drug by patients. A prospective study was first carried out in 1972 (3); SGOT and bilirubin concentrations were examined monthly in 250 patients receiving isoniazid for 1 year. Isoniazid was hepatotoxic in a large proportion of individuals, but most adapted to the insult and recovered rather than progressing to severe hepatitis. Measurement of the 6-h plasma concentrations of isoniazid in these patients failed to show a correlation between plasma levels of isoniazid and liver injury, suggesting that a toxic metabolite rather than isoniazid itself might be responsible for the hepatotoxicity. In this study, no anti-isoniazid antibodies were found and no correlation was seen between hepatic injury and antinuclear antibodies measured at the end of the study.

The second study was a retrospective analysis of 114 patients with isoniazid-related hepatitis (4). Some of the important findings include the following: (1) isoniazid-related liver injury was indistinguishable biochemically and morphologically from iproniazid-induced liver damage or from other causes of acute hepatocellular injury such as viral hepatitis; (2) no clinical evidence for a hypersensitivity mechanism such as rash, fever, arthralgia, or eosinophilia was found; and (3) about 30% of the patients with hepatic reactions were residents of Honolulu and were of Oriental ancestry; on a genetic basis, 90% or more of these patients would be expected to be rapid acetylators of isoniazid, in contrast to black and white populations, in which 45% are rapid acetylators (2).

In the third study (5), 21 non-Oriental patients who had recovered from isoniazid hepatitis were genetically phenotyped as rapid or slow acetylators of isoniazid using the sulfamethazine method. Eighty-six percent of them displayed the rapid acetylator phenotype for isoniazid metabolism.

We recently reexamined the metabolism of isoniazid and acetylisoniazid and identified the structure of the metabolites by cochromatography and reverse isotope dilution with synthesized standards and by mass spectral analysis (5). ^{3}H-Ring-labeled isoniazid and acetylisoniazid were given to human volunteers in single 300 mg doses and urinary metabolites were collected for 24 h. About 55% of a dose of acetylisoniazid was metabolized by hydrolysis to isonicotinic acid and free acetylhydrazine regardless of the genetic phenotype of the patients for acetylating isoniazid. In contrast, the pattern of metabolites after the administration of isoniazid was very dependent upon the rate at which isoniazid was acetylated. On the basis of the relative amounts of acetylisoniazid and isonicotinic acid excreted into the urine, we calculated that almost all of the isonicotinic acid was formed by way of acetylisoniazid. We also calculated that patients who were fast metabolizers of isoniazid converted about 94% of the dose of isoniazid to acetylisoniazid; only 2.8% of the drug was excreted unchanged in the urine and 3.6% was excreted as hydrazone conjugates. Slow acetylators, on the other hand, excreted almost 37% of the drug in the urine either free or as a hydrazone. Thus only 63% was converted to acetylisoniazid and subsequently to isonicotinic acid and acetylhydrazine. We concluded, therefore, that fast acetylators are exposed to much more acetylisoniazid and acetylhydrazine than are slow acetylators (5).

Acetylisoniazid and isoniazid were given to rats, mice, and hamsters to see if they could produce hepatic necrosis (6). These hydrazines were given in a dose–response manner to several hundred animals. Isoniazid did not cause necrosis in any of the animals. In contrast, acetylisoniazid produced occasional single-cell necrosis in rats and mice, and phenobarbital pretreatment, which is known to increase drug-metabolizing enzymes, greatly potentiated the necrosis.

The liver damage was prevented by pretreatment of animals with cobalt chloride, which inhibits the synthesis of cytochrome P450 drug-metabolizing enzymes. Similarly, when the hydrolysis of acetylisoniazid was inhibited by pretreatment of animals with bis-*p*-nitrophenyl phosphate (BNPP), the necrosis was prevented.

The effect of acetylhydrazine on the liver was also examined. This hydrazine is a very potent hepatotoxin, which produces hepatic necrosis in phenobarbital-pretreated rats after single doses of only 10 mg/kg. The necrosis was potentiated by pretreatment with phenobarbital and prevented both by cobalt chloride and by another inhibitor of drug metabolism, aminotriazole. However, BNPP, which inhibited the hydrolysis of acetylisoniazid and prevented the necrosis, had no effect on the necrosis produced by acetylhydrazine. Thus the metabolic activation of the liberated hydrazino moiety of isoniazid to a toxic metabolite satisfactorily accounts for the hepatic necrosis produced by isoniazid (6).

As further proof of the hypothesis that acetylisoniazid is converted in the body to a chemically reactive form via activation of acetylhydrazine, [^{14}C]-acetylisoniazid radiolabeled in the acetyl moiety was given to rats, and evidence for covalent binding to tissue macromolecules was sought (6–8). We have previously used this approach to implicate toxic metabolites as mediators of hepatic necrosis produced by other commonly administered drugs such as acetaminophen, phenacetin, and furosemide (8). A large amount of covalent binding, about 0.5 nmol/mg protein, was found upon digestion of the proteins in the liver, the target organ for toxicity, but none was found in other tissues. This binding was proportional to dose, was increased by pretreatment with phenobarbital, and was markedly decreased by pretreatment with cobalt chloride or aminotriazole. However, no covalent binding occurred after acetylisoniazid when the radiolabel was in the pyridine ring, demonstrating that the reactive metabolite came only from the acetylhydrazine moiety.

Subsequently, isoniazid itself was shown to produce acute hepatic necrosis in pheonobarbital-treated rats. The proportion of the isoniazid that is acetylated in rats decreases markedly at doses above 100 mg/kg. Thus a single large dose does not cause liver necrosis, but by administering isoniazid in six single doses of 100 mg/kg/h it was possible to produce acute hepatic necrosis with isoniazid itself.

Recent double-isotope studies with acetylisoniazid and acetylhydrazine labeled with tritium and ^{14}C in the acetyl moiety demonstrate that each is converted to a reactive metabolite that becomes covalently bound to liver proteins in rats (6–8). The ^{3}H/^{14}C ratio of the covalently bound metabolite from either acetylisoniazid or acetylhydrazine was 0.94, which indicates that all hydrogen atoms of the acetyl group are bound and makes ketene an unlikely intermediate (Fig. 1) (7). The data are consistent with the reactive intermediate being either an acetyl radical or an acetylonium ion that acylates hepatic macromolecules and leads to hepatic necrosis. These acylating intermediates can also react in part with the aqueous medium to yield acetate. This will be incorporated into the acetate pool and eventually be expired as carbon dioxide

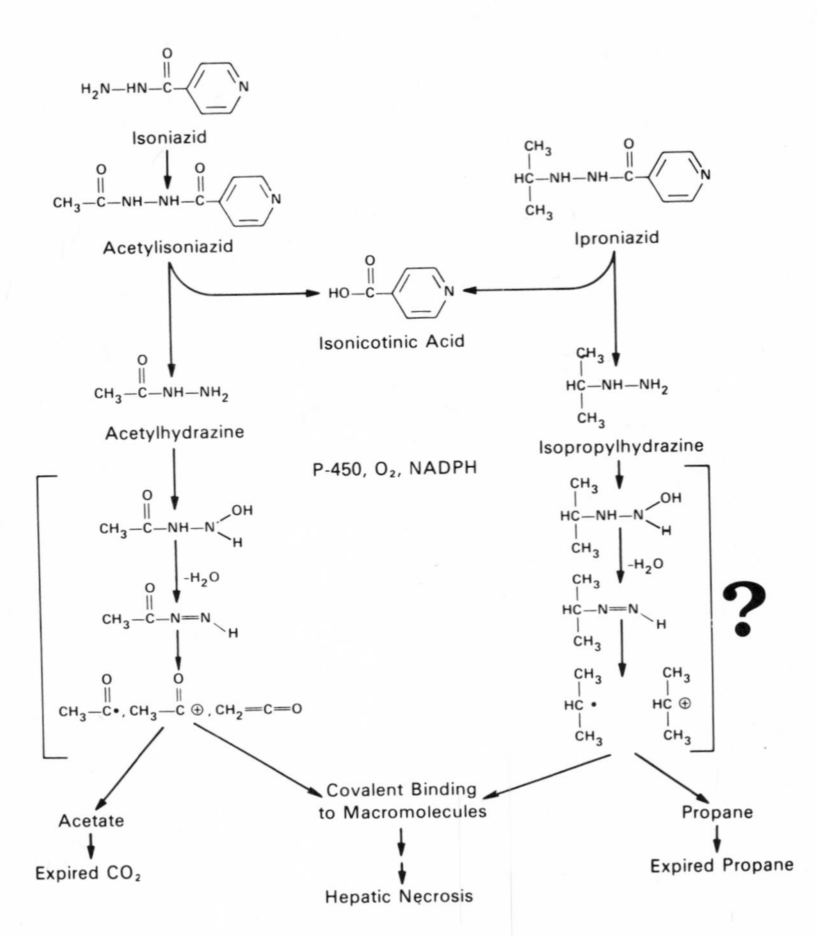

Fig. 1. Proposed metabolic activation pathways for isoniazid, acetylisoniazid, and isopropylisoniazid (iproniazid).

(Fig. 1). Subsequent experiments in rats have shown a direct proportionality between covalent binding, hepatic necrosis, and $^{14}CO_2$ production after administration of [^{14}C]acetylisoniazid. Thus the measurement of expired $^{14}CO_2$ after administration of [^{14}C]acetylisoniazid can be used as an index of the amount of acetylhydrazine (AcHz) that is converted to a chemically reactive acylating species *in vivo.* This was confirmed by analysis of urinary metabolites in these experiments. Cobalt chloride or piperonyl butoxide pretreatment markedly decreased expiration of $^{14}CO_2$ and increased the total urinary excretion of radioactivity as well as the urinary excretion of free AcHz, diacetylhydrazine, and the hydrazones of AcHz, presumably because the treatments inhibited the oxidation of AcHz by cytochrome P450. The excretion of these metabolites is proportional to the total amount of AcHz in the body. As expected, phenobarbital pretreatment had the converse effect; it decreased the urinary excretion of AcHz, diacetylhydrazine, and the hydrazones of AcHz.

It is important to emphasize that the covalent binding measured in the above experiments results primarily from acylation and not from acetate incorporation into proteins. Pretreatment of rats with doses of cycloheximide that decreased by $> 98\%$ the incorporation of sodium acetate[^{14}C] into tissue proteins only slightly reduced the amount of bound metabolite. Similarly, trapping experiments *in vitro* show that trideuteromethyl-labeled acetylhydrazine is activated by microsomal cytochrome P450 to an electrophilic metabolite that reacts with cysteine to yield trideuteromethyl-*N*-acetylcysteine (7).

These data are not unique to acetylisoniazid but can be extrapolated to other hydrazide and hydrazine drugs.

Animal studies with iproniazid (isopropylisoniazid or IpINH), an antidepressant removed from clinical use because of a high incidence of isoniazidlike hepatic injury (9), provided results similar to those obtained with acetylisoniazid. Isopropylhydrazine (IpHz) is liberated *in vivo* by enzymatic hydrolysis of IpINH and is an extremely potent liver toxin (6–8).

In rats, IpHz is converted to propane, which is expired and can be measured by gas chromatography–mass spectrometry (7). The technique of administering mixtures of isotopes of IpINH or IpHz and comparing the $^{3}H/^{14}C$ ratios of covalently bound metabolite and expired propane was again used to demonstrate that the entire isopropyl group was covalently bound and therefore that the same pathway led to both alkylation of protein and formation of propane. Both the covalently bound material and expired propane were shown to have a ratio of $^{3}H/^{14}C$ of 0.92–0.98 after administration of a similar mixture of [2-^{3}H]-IpINH and [2-^{14}C]IpINH or [1,3-^{14}C]IpINH (or [2-^{3}H]IpHz and [2-^{14}C]-IpHz).

The covalent binding of both AcHz and IpHz to rat hepatic microsomal enzymes *in vitro* required oxygen and NADPH and was inhibited by a $CO–O_2$ (9:1) atmosphere, by an antibody against NADPH-cytochrome *c* reductase, by SKF 525A, and by piperonyl butoxide (7). Thus the enzyme system that activated the hydrazines was a P450 oxidase and not a flavoprotein-amine oxidase. Neither superoxide dismutase nor ascorbic acid inhibited binding. Kinetic analyses, using covalent binding as an index of reactive product formation, demonstrated that phenobarbital pretreatment markedly increased and cobalt chloride decreased the V_{max} without altering the apparent K_m. Formation of propane from IpHz by rat microsomes was proportional to the rate of covalent binding at various substrate concentrations. Moreover, both were increased by pretreatment of animals with phenobarbital and both were decreased by pretreatment with cobalt chloride. Experiments with human microsomes showed the presence of a similar enzyme system for activation of monosubstituted hydrazines.

Discussion

From these results, we postulate that monoacyl and monoalkyl hydrazines are oxidized by cytochrome P450 to *N*-hydroxy intermediates followed by dehydration of the respective diazenes (Fig. 1). Monoalkyl diazenes are known to fragment homolytically to yield radicals (10) or heterolytically to produce electrophilic cations. Our studies showing correlations between propane formation and binding to tissue macromolecules suggest to us that radicals are the proximate reactive species. Another mechanism that cannot be ruled out at present is a second oxidation of the postulated diazene intermediate to form a diazohydroxide. This intermediate would be similar to that envisioned by Magee (11) for carcinogenic nitrosamines and by Preussman (12) and Druckrey (13) for activating 1,2-dialkylhydrazines. However, our kinetic data do not support this possibility for AcHz and IpHz. Whatever the intermediate may be, the implications are clear that metabolic activation by microsomal enzymes is required to elicit the toxicity of many hydrazine compounds, including the serious and occasionally lethal hepatitis seen with isoniazid and iproniazid therapy. Since these enzyme systems were shown to occur in human tissues, the possible risk of neoplasia after chronic therapy with such drugs needs to be continually monitored. Fortunately, careful studies to date have found no evidence that isoniazid is carcinogenic in man (14).

REFERENCES

1. J. H. Peters, K. S. Miller, and P. Brown, Studies on the metabolic basis for the genetically determined capacities for isoniazid inactivation in man, *J. Pharmacol. Exp. Ther.* **150,** 298–304 (1965).
2. W. P. Kalow, *Pharmacogenetics, Heredity and the Response to Drugs,* p. 99, Saunders, Philadelphia (1962).
3. J. R. Mitchell, M. W. Long, U. P. Thorgeirsson, and D. J. Jollow, Acetylation rates and monthly liver function tests during one year of isoniazid preventive therapy, *Chest* **68,** 181–190 (1975).
4. M. Black, J. R. Mitchell, H. J. Zimmerman, K. G. Ishak, and G. R. Epler, Isoniazid-associated hepatitis in 114 patients, *Gastroenterology* **69,** 289–302 (1975).
5. J. R. Mitchell, U. P. Thorgeirsson, M. Black, J. A. Timbrell, W. R. Snodgrass, W. Z. Potter, D. J. Jollow, and H. R. Keiser, Increased incidence of isoniazid hepatitis in rapid acetylators: Possible relation to hydrazine metabolites, *Clin. Pharmacol. Ther.* **18,** 70–79 (1975).
6. W. R. Snodgrass, W. Z. Potter, J. A. Timbrell, D. J. Jollow, and J. R. Mitchell, Possible mechanisms of isoniazid-related hepatic injury, *Clin. Res.* **22,** 323A (1975).
7. S. D. Nelson, W. R. Snodgrass, and J. R. Mitchell, Metabolic activation of hydrazines to reactive intermediates: Mechanistic implications for isoniazid and iproniazid hepatitis, *Fed. Proc.* **34,** 784 (1975).
8. J. R. Mitchell and D. J. Jollow, Metabolic activation of drugs to toxic substances, *Gastroenterology* **68,** 392–410 (1975).
9. L. E. Rosenblum, R. J. Korn, and H. J. Zimmerman, Hepatocellular jaundice as a complication of iproniazid therapy, *Arch. Intern. Med.* **105,** 583–593 (1960).

10. R. A. Prough, J. A. Witkop, and D. J. Reed, Further evidence on the nature of microsomal metabolism of procarbazine and related alkylhydrazines, *Arch. Biochem. Biophys.* **140,** 450–458 (1970).
11. P. N. Magee and J. M. Barnes, Carcinogenic nitroso compounds, *Adv. Cancer Res.* **10,** 163–246 (1967).
12. R. Preussman, Chemical structure and carcinogenicity of aliphatic hydrazo- and azoxy-compounds and of triazenes: Potential *in vivo* alkylating agents, *Ann. N.Y. Acad. Sci.* **163,** 697–806 (1969).
13. H. Druckery, Specific carcinogenic and teratogenic effects of indirect alkylating methyl and ethyl compounds and their dependency on stages of ontogenic development, *Xenobiotica* **3,** 271–303 (1973).
14. S. H. Ferebee, Controlled chemoprophylaxis trials in tuberculosis: A general review, *Adv. Tuberc. Res.* **17,** 28–106 (1969).

28

Discussion

Conney observed that the inhibition studies presented on microsome-catalyzed covalent binding of acetylhydrazine to microsomal protein indicated that carbon monoxide and the antibody to NADPH-cytochrome *c* reductase only partially inhibited the covalent binding reaction. Conney speculated that non-cytochrome P450-dependent metabolism might also be involved in this reaction. Mitchell agreed that other pathways might be involved in the metabolism of acetylhydrazine but did not think that the evidence indicated their involvement in the formation of the chemically reactive alkylating species. Mitchell pointed out that a variety of P450-dependent nitrogen oxidations such as that of 2-acetylaminofluorene are similarly not well inhibited by carbon monoxide. In regard to the incomplete inhibition by the cytochrome *c* reductase antibody, Mitchell observed that incomplete inhibition of cytochrome P450 metabolism was observed even with classical P450-catalyzed reactions such as the *N*-demethylation of ethylmorphine and aliphatic carbon oxidation. Mitchell added that the evolved carbon dioxide is due to the alkylating species and not due to incorporation into the acetate pool, as evidenced by isotope ratio experiments, trapping experiments with glutathione, and the fact that microsomes lack the enzymes to metabolize acetate to carbon dioxide.

Netter asked if there were reason to hope that man might rapidly excrete the acetylated compounds in the kidneys and be less at risk than the experimental animals. Mitchell replied that this seemed unlikely since acetylated compounds tend to be more lipophilic than the parent compounds, and added that in human metabolic studies at least 40% of a dose of isoniazid is metabolized to acetylhydrazine. This indicates that acetylisoniazid is not being excreted sufficiently rapidly to spare man from acetylhydrazine's potential hepatotoxicity.

29

Activation of Nitrite

William Lijinsky
Biology Division
Oak Ridge National Laboratory
Oak Ridge, Tennessee 37830, U.S.A.

Nitrites and nitrous acid are well-known mutagens in bacteria and could be considered a hazard to man, even though nitrites are widely distributed in nature because they are formed by bacterial reduction of nitrates, which are almost ubiquitous. Nitrite is present in saliva because of this process. Nitrites have also been reported at unusually high concentrations in plants growing on mineral-deficient soil. This leads to deficiency of enzymes that convert nitrite to hydroxylamine and ammonia, causing nitrite to accumulate (1). The bacteria in the stomachs of infants can reduce nitrate to nitrite when occasionally present at high concentrations in drinking water. All of these reactions, including formation of nitrite from nitrate in vegetables which have been allowed to stand in the presence of bacteria (and some vegetables have high contents of nitrate), have been associated with methemoglobinemia, the best-known toxic effect of nitrite in man, which is occasionally fatal. Methemoglobinemia in animals has been associated with nitrite from plants (2). There is no evidence, however, that nitrites are mutagenic in man or any other animal.

At low doses, nitrite has no obvious deleterious effect, even when ingested for long periods (3), which is why nitrites have been used for a long time as flavoring and color additives in meat and fish; they are also valuable preservatives in food in which there is a danger of botulism.

In recent years, there has been increasing concern that one form of activation of nitrite might be related to cancer. That is the property of interaction with amino compounds to form *N*-nitroso compounds, which comprise the most broadly acting and one of the most potent group of chemical carcinogens. Although no overt connection has been made between exposure to nitrite and any human cancer, this might be difficult to do precisely because ingestion of nitrite and nitroso compounds is so common.

More than 100 *N*-nitroso compounds have been tested for carcinogenic activity since the discovery of this activity of dimethylnitrosamine by Barnes and Magee (4) in 1954. About 80% of those tested have induced tumors and have covered a wide range of activity from very potent to quite weak. They are most effective when administered as small continuous doses over a long period and the oral route of administration evokes the best response. A total dose of 0.2 mmol of dimethylnitrosamine or nitrosomorpholine has induced liver tumors in 30–40% of a group of 30 rats (5), so that these compounds are among the most potent group of carcinogens. Since most nitroso compounds act systematically, administration by any route can induce tumors in those organs and tissues that are susceptible, and almost all organs of the rat are susceptible to one nitroso compound or another. The chemical structure of the nitroso compound has great bearing on its carcinogenic activity. Substituents at various positions can inhibit or enhance the potency, or change the organ in which tumors are induced, depending on the nature of the substituent and the position of substitution. The carbon atoms alpha to the nitroso group appear to be the first points of activation of the molecule, since substitution there markedly reduces carcinogenic activity, but we do not know what form this activation takes or the enzyme or enzymes involved. Answers to these questions might be gained through more studies of the chemistry and metabolism of a variety of *N*-nitroso compounds, too few of which have been examined until now to give a comprehensive picture.

Since treatment of animals with nitroso compounds provides good models of so many human cancers, including lung, esophagus, bladder, stomach, pancreas, kidney, and liver, it is tempting to believe that nitroso compounds are involved in the causation of cancer in man. However, there is no evidence that any nitroso compound occurs in the environment at anything but trace concentrations. Dimethylnitrosamine has been found in a few samples of food treated with nitrite at levels of 10–100 parts per billion; and nitrosopyrrolidine, probably formed by decarboxylation of nitrosoproline (6), has been found in cooked (but not raw) bacon at levels of 100 ppb. Even though these levels are low, they are not without significance. But a more important exposure to nitroso compounds is through their formation from amino compounds and nitrite in the stomach, which provides the mildly acid medium favorable for nitrosation, above pH 3 during digestion of a meal.

The pH of the medium is important because, while secondary amines are nitrosated reasonably easily at low pHs, reaction of tertiary amines with nitrous acid is negligible below pH 3 and proceeds well between pH 3 and 6. Since a large proportion of commonly used drugs and agricultural chemicals are secondary or tertiary amines, the presence of nitrite in the stomach together with any of them provides a source of nitroso compounds. While it has not been possible to estimate the extent of exposure of people to nitroso compounds formed in this way, carcinogenesis by nitroso compounds formed in this way in experimental animals has been amply demonstrated. Extrapolation of these results to man has been difficult because, although the individual dose for a rat has been much higher than that which man could receive in most cases, the duration of exposure of a rat is very much shorter than the 70-year lifespan of a man or woman. Similarly, we have no way of knowing whether man is more or less sensitive than a rat to nitroso compounds (or to any other carcinogen), nor do we know which organ in man is susceptible to any particular nitroso compound; there are many examples of pronounced differences in organ specificity of a nitroso compound between different species.

Many factors apart from the nature of the amine and the concentration of nitrite can affect the extent of formation of nitroso compounds in the stomach. There are inhibitors, catalysts, and accelerators of nitrosation, all of which can affect the reactions in a manner which is at present unpredictable. There are probably other factors that play a part, but are still unknown.

Our animal studies of administration of nitrite and amines, then, can serve as a model of cancer induction in man, and can no more be dismissed than any other carcinogenesis studies in model systems. Several secondary amines have been administered to rats together with nitrite and have given rise to the tumors induced by the nitrosamine product itself; the amines include methylbenzylamine, morpholine (7), heptamethyleneimine, ethylurea, and methylurea. One tertiary amine, aminopyrine, has been tested in the same way and has formed a highly carcinogenic combination with nitrite (8). The ingestion of morpholine by man is probable, and that of aminopyrine certain, since the latter has been used as a drug for many decades. Another common drug that is a tertiary amine, oxytetracycline, has been fed to rats together with nitrite (each at 0.1% in drinking water). Four of 30 rats fed this combination for 60 weeks developed tumors in the liver. While this cannot be considered a definitely significant incidence of tumors, it was noticed that neither control group, oxytetracycline nor sodium nitrite, developed any liver tumors. Since the use of tetracyclines is so widespread and their reaction with nitrous acid can be demonstrated in a flask, it would seem that further investigation of their interaction with nitrite *in vivo* is warranted.

Our tests of several other amino compounds together with nitrite *in vivo* are still in progress, although some are almost certain to be negative. This is not to say that no nitroso compound was formed, but only that insufficient compound was formed under these conditions to give rise to a detectable incidence of tumors during the lifetime of the animals. With other compounds, there have been one or two unusual tumors, but not enough yet to assume that the treatment was carcinogenic. Trimethylamine oxide, chlorpromazine, piperidine, hexamethylenetetramine, arginine, and methylguanidine appear to be negative. Tests of lucathone, tolazamide, chlordiazepoxide, methapyrilene, dimethyl-phenylurea, and cyclizine are still in progress and many animals are alive. Carbaryl, a representative of a large group of insecticides that are *N*-methylcarbamic esters, forms a nitroso derivative by reaction with nitrite (9), and this is analogous to the highly carcinogenic and mutagenic nitrosomethylurethane. Nitrosocarbaryl itself is carcinogenic, although somewhat weaker than nitrosomethylurethane. Mixtures of carbaryl and nitrite have been fed to rats, including pregnant females, and the progeny have been kept. Although the latter test is considered very sensitive for carcinogens, there have been no tumors in the offspring that have so far died that can be attributed to the treatment.

It is obviously advisable to discover whether the nitroso product of an amine is carcinogenic before feeding a combination of it with nitrite to animals to test for carcinogenicity of the combination. In two such cases the drugs phenmetrazine and methylphenidate, derivatives of morpholine and piperidine, respectively, the nitroso products have not induced any tumors in rats. This demonstrates the profound effect of certain substituents on the potency of such strong carcinogens as nitrosomorpholine and nitrosopiperidine. Similarly, nitrosoguvacoline, prepared by reaction of guvacoline (present in betel nut) with nitrite, is noncarcinogenic, although its parent, nitrosotetrahydropyridine, is a potent carcinogen in rats.

Since formation of nitroso compounds in the stomach is so much more common than in other parts of the body, this is the main source of nitroso compounds that has been discussed. However, there have been reports of similar reactions of amino compounds with nitrite at other sites in the body, such as the small intestine and the urinary bladder (especially after infection with bacteria that reduce nitrate to nitrite in urine), and these could undoubtedly provide additional contributions of the systematically acting *N*-nitroso compounds.

ACNOWLEDGMENT

The Oak Ridge National Laboratory is operated for the Energy Research and Development Administration by Union Carbide Corporation.

REFERENCES

1. R. J. W. Burrell, W. A. Roach, and A. Shadwell, Esophageal cancer in the Bantu of the Transkei associated with mineral deficiency in garden plants, *J. Natl. Cancer Inst.* **36**, 201–214 (1966).
2. J. I. Quin and C. Rimington, Tribulosis and methaemoglobinaemia in South African sheep, *Nature (London)* **130**, 926–927 (1932).
3. H. W. Taylor and W. Lijinsky, Tumor induction in rats by feeding heptamethyleneimine and nitrite in water, *Cancer Res.* **35**, 812–815 (1975).

4. J. M. Barnes and P. N. Magee, Some toxic properties of dimethylnitrosamine, *Br. J. Ind. Med.* **11**, 167–174 (1954).
5. L. K. Keefer, W. Lijinsky, and H. Garcia, Deuterium isotope effect on the carcinogenicity of dimethylnitrosamine in rat liver, *J. Natl. Cancer Inst.* **51**, 299–302 (1973).
6. W. Lijinsky and S. S. Epstein, Nitrosamines as environmental carcinogens, *Nature (London)* **225**, 21–23 (1970).
7. J. Sander and G. Bürkle, Induktion maligner Tumoren bei Ratten durch gleichzeitige Verfütterung von Nitrit und secondären Aminen, *Z. Krebsforsch.* **73**, 54–66 (1969).
8. W. Lijinsky, H. W. Taylor, C. Snyder, and P. Nettesheim, Malignant tumors of liver and lung in rats fed aminopyrine or heptamethyleneimine together with nitrite, *Nature (London)* **244**, 176–179 (1973).
9. R. K. Elespuru and W. Lijinsky, The formation of carcinogenic nitroso compounds from nitrite and some types of agricultural chemicals, *Food Cosmet. Toxicol.* **11**, 807–817 (1973).

30

Discussion

Referring to Lijinsky's suggestion that lifetime carcinogenicity testing of drugs and environmental chemicals in rats might not be predictive of their carcinogenicity in longer living species such as man, Miller observed that all our present information on the natural history of cancer, both spontaneous and induced, in the experimental animal matches that of man. Lijinsky agreed that this was so, but pointed out that there are at present no good experimental data to indicate whether the latent period is related to dose of the carcinogen or to life span of the species. He suggested that the following experiment might be informative: administration of the same carcinogen to species with widely differing lifespans and determination of whether the tumors appear after a fixed time interval or at the same stage in the lifespan. In the ensuing discussion, Rentsch observed that rats reared under strictly germ-free conditions live appreciably longer (4–5 years) than rats maintained under the usual laboratory conditions (2–3 years), and suggested that this fact could be used experimentally to explore this problem. In reply, Lijinsky agreed but noted that the maintenance of appreciable numbers of rats under germ-free conditions would be very expensive.

Miller asked what kind of experimental evidence was available to indicate that the cyclic nitrosamines are activated in a manner different from that so far discovered for dimethylnitrosamine. Lijinsky replied that his suggestion was based on quantitative studies comparing the carcinogenity of a variety of nitrosamines and their interaction with macromolecules. These studies indicated that compounds such as nitrosomorpholine and dimethylnitrosamine which show similar carcinogenicity can vary by 3 orders of magnitude in their alkylation of cell macromolecules, suggesting that their activation might be different. Miller replied that it is no longer profitable to compare total binding since recent studies have shown the importance of specific binding sites on DNA in the development of tumors.

In response to a question on whether the animals showed deleterious effects due to methemoglobinemia, Lijinsky replied that the rats receiving 10% sodium nitrite in their diet did not show evidence of methemoglobinemia. He noted that this was in marked contrast to an experiment in which 40 mg of sodium nitrite was dissolved in a small volume of water and injected directly into the stomach. Under these conditions, appreciable methemoglobinemia was induced. Lijinsky observed that although the animals ingest grams of sodium nitrite in the course of the carcinogenicity experiments, it is taken gradually over a long period of time, and no deleterious effects due to methemoglobinemia develop.

Mitchell noted that although the amount of nitrite ingested by the animals (10% of diet) was only slightly higher than that which might be in the human diet, the amount of Librium fed to the animals (2% of diet) was appreciably greater than the usual human dose (15–25 mg). Mitchell suggested that since the formation of a nitrosamine in the stomach would depend on the concentration of both reactants (nitrous acid and Librium) the unusually high dose of Librium should be taken into account in the interpretation of the biological significance of the animal model observations. In reply, Lijinsky pointed out that the animal model studies indicated only that carcinogenic nitrosamines could be formed *in vivo* by reaction of amine drugs with dietary nitrite, and not that this was of necessity a cause of human cancer. He further pointed out that exaggeration of dose was an essential feature of chemical carcinogenicity studies and that if this principle had not been accepted no compound would have been shown to be a carcinogen. In the ensuing discussion, Remmer noted that primary carcinoma of the liver was rare in man and suggested that an epidemiological study on the incidence of this tumor might reveal whether there was an increase in its incidence associated with increase in widespread use of amine drugs. Zimmerman observed that some epidemiological evidence was available to suggest that the incidence of primary carcinoma of the liver in the United States had increased by about twofold and up to about tenfold in one area of the country. However, he felt that this could not be clearly related to chemical carcinogenicity since other oncogenic factors such as viral B hepatitis might be involved. Lijinsky added that it was not known if a nitrosamine carcinogen arising by interaction of a dietary amine and nitrite would cause primary carcinoma of the liver in man. He pointed out that, although rodents develop liver tumors after dimethylnitrosamine, other species may show different target organ specificity. For example, in one study, the feeding of dimethylnitrosamine to pigs resulted in the development of kidney tumors.

Lenk asked if the reaction that converts aminopyrine to its carcinogenic derivative is known. Lijinsky replied that the reaction has been studied in detail by several laboratories and that it is complex. Dimethylnitrosamine and possibly other as yet unrecognized carcinogens are produced in the reaction. He added that the reaction of aminopyrine with nitrous acid is very rapid and appreciably faster than the reaction of dimethylamine and nitrous acid, suggesting that the reaction does not proceed via the secondary imino ion. Conney added that studies by Kamm in his laboratory have shown that dimethylnitrosamine can be measured in the blood of animals fed aminopyrine and sodium nitrite. Coadministration of ascorbic acid with aminopyrine and nitrite blocked the formation of dimethylnitrosamine and prevented tumor formation.

31

Benzene Metabolism and Toxicity

R. Snyder, L. S. Andrews, E. W. Lee, C. M. Witmer, M. Reilly, and J. J. Kocsis
Department of Pharmacology
Thomas Jefferson University
Philadelphia, Pennsylvania 19107, U.S.A.

Chronic exposure to benzene leads to blood dyscrasias characterized by progressive depression in the levels of circulating leukocytes, thrombocytes, and/or erythrocytes, eventually leading to pancytopenia and aplastic anemia (1). The literature contains many descriptions of human benzene toxicity following chronic inhalation (2–4) and the process has been reproduced in several animal species either by exposing animals to atmospheric benzene (5,6) or by parenteral administration of benzene (7,8). The mechanism by which benzene produces bone marrow depression has been explored in a series of studies by Kissling and Speck (9,10) and by Boje *et al.* (11), who demonstrated that nucleic acid synthesis in bone marrow was inhibited in chronic benzene toxicity, but the molecular site of action is as yet unknown.

Benzene metabolism leads to the formation of phenol and in lesser quantity to a series of related phenolic compounds which appear in the urine largely as ethereal sulfate or glucuronide conjugates (12–15). Studies of benzene metabolism *in vitro* suggest that benzene is a substrate for the hepatic mixed-function oxidase (16). Benzene metabolism can be increased by treating animals with benzene, phenobarbital,3-methylcholanthrene, or DMSO (17–20).

The growing awareness in recent years that the formation of toxic intermediates during metabolism of xenobiotic compounds is not a rare or unusual event but a common mechanism of toxicity suggests reevaluation of the few reports in which the relationship between benzene metabolism and benzene toxicity was studied. Dustin (21), who observed that metabolites of benzene interfered with mitosis, suggested several mechanisms by which phenolic or quinone metabolites of benzene might produce toxicity. Nomiyama's observations (22–25) that young rats which were more susceptible to benzene toxicity than old rats also metabolized benzene more rapidly are consistent with the postulate that metabolism is required to produce benzene toxicity. This view was supported by the demonstration that inhibition of benzene metabolism by 3-amino-1,2,4-triazole is accompanied by decreased susceptibility to benzene-induced leukopenia in rats. In apparent contrast to these results, studies by Ikeda (26,27) and Drew *et al.* (19) showed that treating rats with phenobarbital increased the rate of benzene metabolism and protected them against leukopenia. Ikeda *et al.* (28) also suggested that inhibition of benzene metabolism by toluene would be expected to intensify benzene toxicity by preventing its detoxication. Thus the role of benzene metabolism either in detoxication of benzene or in the production of an active intermediate remains to be clarified.

The studies reported here differ from previous investigations in which both benzene metabolism and toxicity were evaluated, since earlier studies were based on production of leukopenia by benzene in rats. Benzene metabolism was determined either by chemically assessing changes in urinary constituents after giving benzene or *in vitro* by using subcellular fractions of liver. We elected to evaluate benzene toxicity by studying the inhibitory effect of benzene on hematopoiesis in the mouse using ^{59}Fe incorporation into hemoglobin as a measure of red cell production. Benzene metabolism in mice was evaluated by studying the appearance of labeled benzene metabolites in the urine after treating mice with [^{3}H]benzene.

RESULTS

One of the difficulties facing investigators who have studied the relationship between benzene toxicity and metabolism has been the inability to detect toxic effects early after giving benzene at the time when benzene is undergoing metabolism. Since benzene retards development of blood cells in bone marrow rather than destroying circulating blood cells, the normal cell complement must undergo turnover before impaired functioning of the marrow can be detected by failure to replace blood components. In the past the development of leukopenia has most frequently been used to measure benzene toxicity because leukocytes turn over in a matter of days and must be rapidly replaced. Since total leukocyte numbers are relatively small (5000–15,000/μl), losses can be rapidly detected. Thrombocytes, which occur in greater numbers (approximately 250,000/μl), have been more difficult to determine. Red blood cells (approximately 5 million/μl) have a much longer half-life, and anemia takes much longer to detect than leukopenia in experimental animals after giving benzene.

In an attempt to measure benzene toxicity soon after giving benzene, we developed a method which does not require cell counting but is based on new cell formation. The procedure involves the administration of ^{59}Fe to mice; this

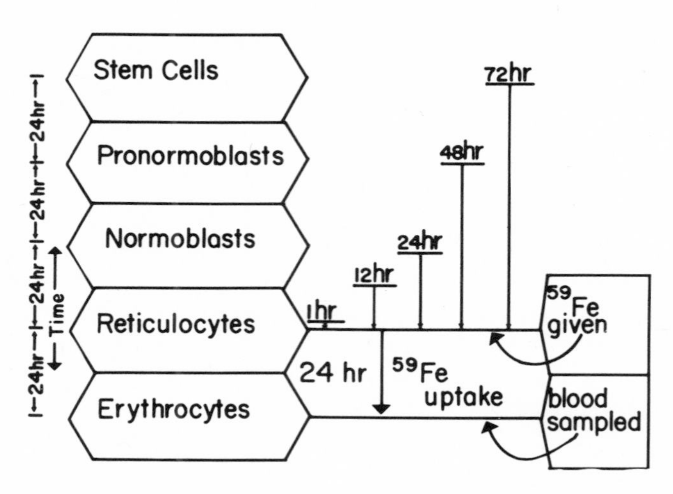

Fig. 1. Experimental design for studying the effect of benzene on the uptake of ^{59}Fe into red cells in mice.

is then taken up into erythrocyte precursor cells in the bone marrow and used to make hemoglobin. Blood levels of radioactive iron return to control levels in a few hours and remain depressed until mature red cells carrying ^{59}Fe-labeled hemoglobin are released into the blood. The maturation sequence in mice involves differentiation of the stem cell to the pronormoblast stage, followed successively by maturation to the normoblast, reticulocyte, and finally mature erythrocyte. Each stage requires 24 h, thus allowing fairly rapid determination of ferrokinetics and early detection of the effects of benzene on red cell formation.

If ^{59}Fe is given to a mouse and the circulating blood is sampled for radioactivity 24 h later, the radioactivity found therein is a measure of the reticulocyte pool at the time that the iron was given, since it requires 24 h for a reticulocyte to mature to a red cell in the mouse. Figure 1 shows that a series of experiments can be designed to measure the effect of benzene on red cell formation by giving benzene at various times prior to giving iron and then performing 24-h ^{59}Fe uptake studies to measure the effect of benzene on the reticulocyte pool. For example, if benzene and ^{59}Fe were given at the same time, a reduction in ^{59}Fe uptake 24 h later would indicate either that benzene had inhibited hemoglobin synthesis or that benzene impaired reticulocyte maturation. In contrast, reduction in 24-h iron uptake if benzene were given 24 h before iron suggests that benzene inhibited the maturation of normoblasts to reticulocytes because that process, like each of the other maturation steps, requires 24 h. Accordingly, benzene was given at each of the time periods shown in Fig. 1 prior to measuring 24-h ^{59}Fe uptake.

The results shown in Table 1 indicate that the greatest reduction in ^{59}Fe uptake into red cells occurred when benzene (administered in oil) was given 48 h

Table 1. Effect of Benzene on 24-h Erythrocyte ^{59}Fe Utilization

Benzene dose (mg/kg)	24-h[a] mean ± SD (% control)	48-h mean ± SD (% control)
Olive oil control	100[b] (19)[c]	100[b] (12)
88	99.5 ± 20.2 (26)	91.9 ± 23.1 (11)
440	77.3 ± 18.2 (23)[d]	65.3 ± 24.1 (15)[d,]
2200	57.0 ± 25.1 (19)[d,e]	37.6 ± 36.6 (15)[d,e]

[a]Time after benzene administration at which 0.5 μCi ^{59}Fe (equivalent to 20–40 ng) was given.
[b]Each control value (100%) represents approximately a 20% uptake of the administered ^{59}Fe into red cells.
[c]Number of animals is given in parentheses.
[d]Significantly different from control ($p < 0.001$).
[e]Significantly different from 440 mg/kg dose ($p < 0.01$).

prior to iron. Since it requires 48 h for pronormoblasts to mature to reticulocytes, it appears that pronormoblasts were most sensitive to benzene. A similar though smaller decrease occurred when benzene was given 24 h prior to iron, suggesting that normoblasts were also affected by benzene. The fact that there was no decrease when benzene was given 72, 12, or 1 h prior to iron shows that stem cells and reticulocytes are less sensitive to benzene and that benzene does not appear to inhibit hemoglobin synthesis.

The reduction in ^{59}Fe uptake is related to the dose of benzene (Table 1). While the lowest dose (88 mg/kg) produced no effect, a significant depression of iron uptake was observed at both 440 mg/kg and 2200 mg/kg doses, with the latter producing the greater effect.

To parallel the studies of benzene toxicity, we also investigated benzene

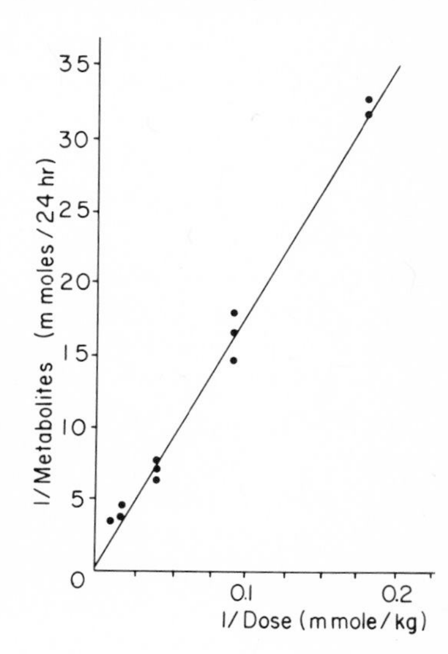

Fig. 2. Urinary excretion of [^{3}H]benzene metabolites as a function of dose. The radioactivity in the 24-h urine sample collected following 440, 880, 2200, 4400, and 8800 mg/kg subcutaneous doses of [^{3}H]benzene was determined. The calculation of mmol equivalency values involved an estimated correction factor of 1/6 for loss of tritium due to hydroxylation. Each value was obtained from the analysis of pooled urine from six or more animals. Least squares analysis of the data yielded a linear regression coefficient of $r = 0.995$.

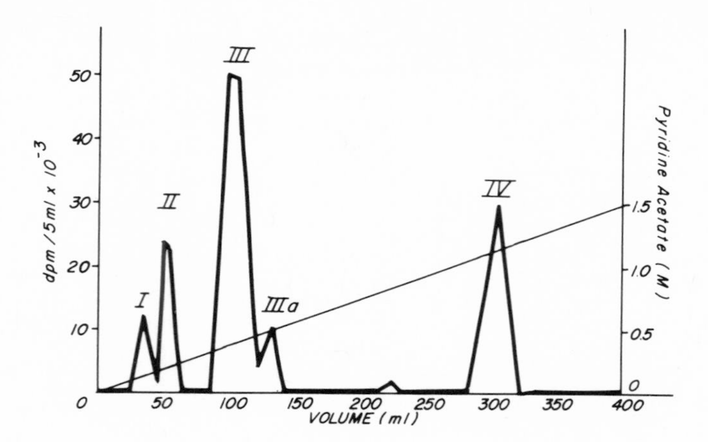

Fig. 3. Anion-exchange chromatography of [^{3}H]benzene metabolites in mouse urine. Untreated urine was applied to DEAE-Sephadex A25 columns and eluted with a linear pyridine acetate gradient (005–1.5 M). The column effluent was collected in fractions and assayed for radioactivity. Peaks I, II, III, IIIa, and IV are identified in the text.

metabolism in mice *in vivo.* Following the subcutaneous administration of [^{3}H]benzene to mice at several dose levels between 440 and 8800 mg/kg, labeled benzene metabolites were collected in the urine over a 24-h period. Figure 2 shows a double-reciprocal plot of the dose of benzene vs. total metabolites of benzene in the urine. The intercept at the ordinate suggests that the mouse is capable of metabolizing approximately 2 mmoles of benzene per day.

When the urine from mice given [^{3}H]benzene was chromatographed on DEAE-Sephadex using a linear gradient of pyridine acetate (0.05–1.5 M) for elution, four major peaks of radioactivity (Fig. 3 and Table 2) were isolated. The glucuronide conjugates (III and IIIa) of phenolic benzene metabolites comprised 50–70% of the total radioactivity and the ethereal sulfates (IV) comprised

Table 2. Excretion of [^{3}H]Benzene Metabolites as a Function of Dose[a]

	Percent sample composition		
Dose (mg/kg)	Free phenol	Glucuronides	Sulfates
440	33.3	26.8	39.8
880	7.0	52.7	40.4
2200	0.75	68.4	30.9
4400	6.1	58.4	35.7
8800	5.1	67.8	27.1

[a]Analyses were performed on two pooled 24-h urine samples. Each sample was collected from three mice.

Table 3. Thin-Layer Chromatography of Urinary $[^3H]$Benzene Metabolites

	R_f	Glucuronides (dpm)	Sulfates (dpm)
Radioactivity applied[a]		3042	695
Metabolites			
Catechol	0.20	135	51
Phenol	0.52	2718	451
Recovery		3202	660
		(105% of applied)	(95% of applied)

[a]Glucuronide and sulfate conjugates isolated from pooled 24-h urine samples collected from mice injected with $[^3H]$benzene (880 mg/kg) were hydrolyzed and the hydrolysates extracted with ether. The concentrated ether extracts were spotted on silica gel TLC plates and run in benzene–methanol–water–ammonium hydroxide (17:2.7:0.1:0.1).

30–40%. Approximately 5–10% was free phenol (II), and tritiated water comprised part of peak I. No *free* phenolic metabolite other than phenol was detected. The glucuronide fractions were pooled and hydrolyzed with β-glucuronidase and the ethereal sulfate fraction was hydrolyzed in dilute acid. Thin-layer chromatography of the released phenolic metabolites on silica gel (Table 3) showed that phenol was the major metabolite and that a small quantity of catechol was also present.

The effect of benzene metabolism on benzene toxicity can be studied by treating animals either with a microsomal stimulant to increase the rate of

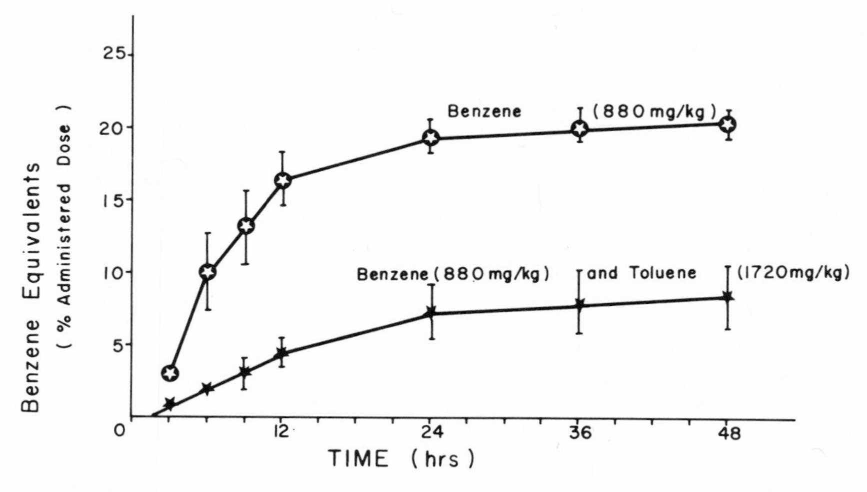

Fig. 4. Effects of toluene on the urinary excretion of $[^3H]$benzene metabolites in the mouse. Each data point represents the mean percentage of administered dose ± SD recovered as urinary $[^3H]$-metabolites from 15 or more mice given $[^3H]$benzene (880 mg/kg) or $[^3H]$benzene (880 mg/kg) plus toluene (1720 mg/kg) subcutaneously.

metabolism or with an inhibitor to decrease metabolism and then studying the effects on benzene toxicity. The most fruitful studies to date have been those in which the latter approach was used. Toluene was chosen as the inhibitor for benzene metabolism because of its structural similarity to benzene and because Ikeda *et al.* (28) had previously suggested that toluene inhibited benzene metabolism in the rat. Figure 4 shows the time course of the urinary excretion of benzene metabolites when benzene was given either alone or with toluene. Analysis of the data showed that all of the metabolites were reduced to an equal extent and Fig. 5 shows a Dixon plot in which it can be seen that toluene is a competitive inhibitor of benzene metabolism in mouse liver microsomes *in vitro*. Thus it appears that, in the mouse, toluene is indeed an inhibitor of benzene metabolism.

The fact that there was reduction in benzene metabolites in the urine raises questions concerning the fate of benzene in the body when the two structurally related hydrocarbons were given together. Figure 6 shows the time course for the pulmonary excretion of benzene when given alone or in combination with toluene. The net effect of toluene was to increase the excretion of benzene through the lungs. Table 4 shows that in either case 92% of the administered dose was recovered from the breath and urine. The reduction in urinary metabolites after giving toluene was compensated for by the increase in pulmonary excretion.

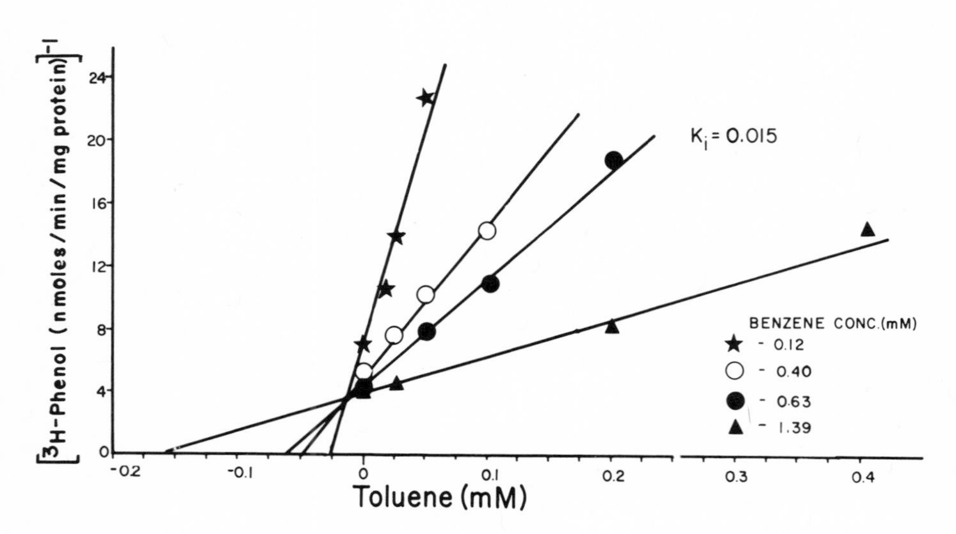

Fig. 5. Effects of toluene on [^{3}H]benzene metabolism in mouse liver microsomes. The indicated concentrations of [^{3}H]benzene were incubated with varying concentrations of toluene and mouse liver microsomes from untreated animals. The data represent the mean of two experiments. Details of microsome preparation, incubation, and assay procedures are described by Gonasun *et al.* (16).

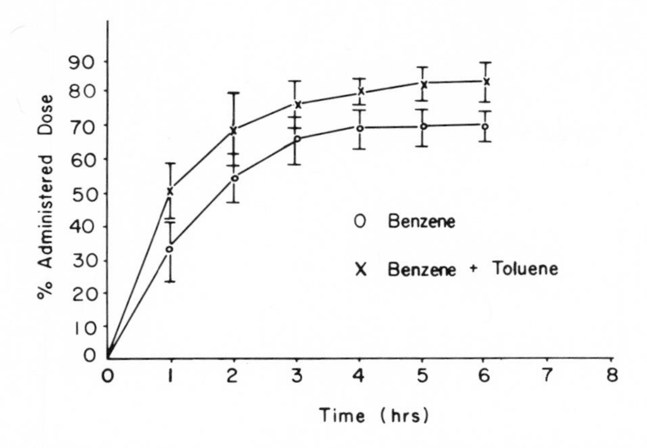

Fig. 6. Effects of toluene on the pulmonary excretion of unchanged [^{3}H] benzene in the mouse. Each point represents the mean percentage of the administered dose ± SD excreted as unchanged [^{3}H]benzene following subcutaneous injection of [^{3}H]benzene (880 mg/kg) or [^{3}H]benzene (880 mg/kg) plus toluene (1720 mg/kg) to six or more mice.

The toxicity and metabolism of benzene were then evaluated in the same animals. [^{3}H] Benzene was given at doses of either 440 or 880 mg/kg and the metabolites in urine were measured over a 24-h period. Forty-eight hours after giving benzene, ^{59}Fe was administered to determine the degree of benzene toxicity. Controls for the iron uptake studies received the oil vehicle but no benzene. The effect of toluene on benzene metabolism and toxicity was measured by giving toluene (1720 mg/kg) with either dose of benzene. Controls received the toluene in oil without benzene. Table 5 shows that toluene reduced the appearance of benzene metabolites in the urine to 45% of controls at the higher dose of benzene and to 30% of controls at the lower dose of benzene.

Table 4. Effects of Toluene on Respiratory and Urinary Excretion of [^{3}H] Benzene and Its Metabolites in the Mouse

	Percent administered dose recovered in 24 h	
	[^{3}H] Benzene (880 mg/kg) (mean ± SD)	[^{3}H] Benzene (880 mg/kg) and toluene (1720 mg/kg) (mean ± SD)
Exhaled [^{3}H]benzene	71.5 ± 6.6 (4.1 h)[a]	85.0 ± 5.5[b] (5.3 h)
Urinary [^{3}H]benzene equivalents	20.1 ± 4.5	7.9 ± 2.7[b]
Total recovery	92.5 ± 5.2	92.8 ± 4.5
$n = 6$		

[a]Time to complete exhalation of [^{3}H]benzene.
[b]Significantly different from group receiving only benzene, $p < 0.01$.

Table 5. Effects of Toluene on Benzene Metabolism and on Red Cell ^{59}Fe Uptake in the Mouse

Treatment	[^{3}H] Benzene metabolism/24 h		Percent [^{59}Fe]utilization/24 h (mean ± SD)
	Percent administered dose (mean ± SD)	Benzene equivalents[a] (μmol)	
Benzene (880 mg/kg)	22.6 ± 5.7 (17)[b]	72.6	4.9 ± 3.4[c] (19)
Benzene + toluene	9.9 ± 3.9 (17)	33.2	9.9 ± 4.5[d] (17)
Toluene (1720 mg/kg)			15.8 ± 5.5 (19)
Control			18.4 ± 6.2 (22)
Benzene (440 mg/kg)	35.8 ± 1.1 (2)[e]	60.1	15.7 ± 4.8[c] (21)
Benzene + toluene	10.9 ± 7.5 (2)[e]	18.3	22.0 ± 5.9[d] (28)
Toluene (1720 mg/kg)			23.3 ± 4.6 (25)
Control			24.2 ± 4.7 (27)

[a]Calculated as μ mol equivalents of [^{3}H]benzene metabolism by a 30-g mouse.
[b]Number of animals or number of groups is given in parentheses.
[c]Significantly different from both control and toluene groups, $p < 0.05$.
[d]Significantly different from group receiving benzene alone, $p < 0.05$.
[e]Two groups of animals, three animals per group.

Toluene did not inhibit ^{59}Fe uptake into red cells but did alleviate the reduction in ^{59}Fe uptake caused by benzene. Taken together, these results could be interpreted to suggest that inhibition of benzene metabolism was accompanied by alleviation or prevention of benzene toxicity and that benzene toxicity is probably caused by a toxic metabolite of benzene.

An alternative point of view suggested by these data is that while benzene metabolism may well be inhibited by toluene *in vitro,* the more significant effect is a change in the distribution of benzene *in vivo* when given with toluene which would prevent sufficient benzene from reaching the site of toxicity for it to act. It might also be argued that even if a metabolite were responsible for toxicity, insufficient quantities of the metabolite might be present at the active site if benzene distribution were altered. Therefore, the following experiments were performed to determine the effect of toluene on the distribution of benzene and its metabolites.

Figure 7 shows the distribution of [^{3}H]benzene (880 mg/kg) in several organs of the mouse following a single injection. Benzene was rapidly absorbed from the injection site, reached blood levels close to 800 nmol/ml, and fell progressively over the subsequent 8-h period. High levels of benzene were

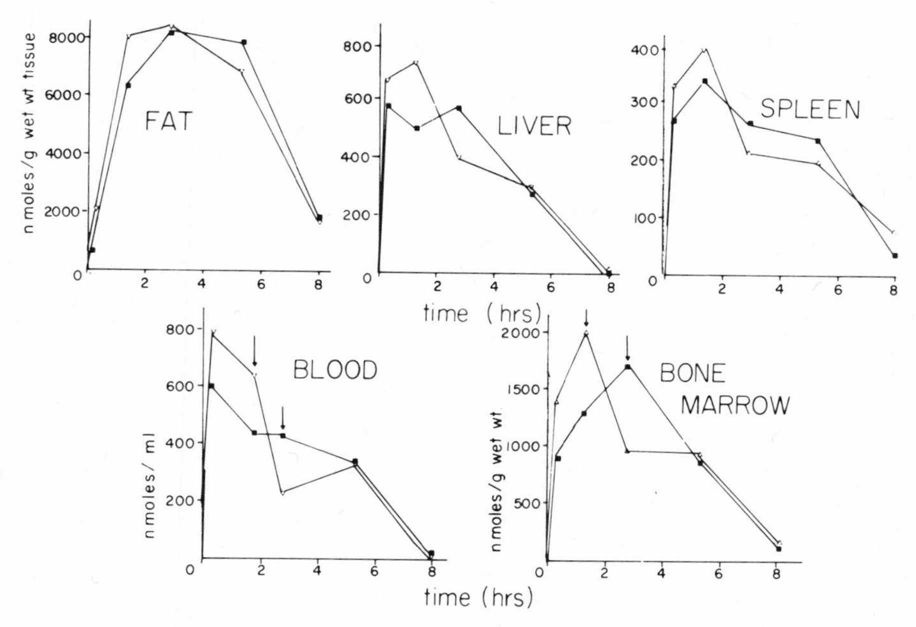

Fig. 7. Concentrations of [^{3}H] benzene in mouse tissues. [^{3}H]Benzene (880 mg/kg) (△) or [^{3}H]benzene (880 mg/kg) plus toluene (1720 mg/kg) (■) was given subcutaneously in olive oil solutions. Tissues were removed and weighed, toluene-extractable radioactivity was determined, and the amount of [^{3}H]benzene present was assayed. Each point represents a mean value obtained from six mice.

observed in the fat and bone marrow with slightly lower levels apparent in liver and spleen. As might be expected, the tissues having the highest fat content contained the most benzene. When toluene was given with benzene in the ratio of two parts toluene to one of benzene there appeared to be a slight delay in absorption from the injection site into the blood since the level at 1 h 15 min in the absence of toluene is significantly greater than that in the presence of toluene. Conversely, at 2 h 45 min the level was lower in the absence of toluene than when toluene was given with benzene. Similar observations were made in bone marrow. Thus attainment of maximum benzene levels in bone marrow was delayed by no more than an hour by toluene but maximum levels reached were not significantly different. Administration of toluene with benzene led to no difference in benzene levels in fat, liver, or spleen.

In addition to [^{3}H]benzene, these organs contained significant amounts of radioactivity associated with polar compounds which appear to be benzene metabolites (Fig. 8). Maximum levels were observed in bone marrow, the site of benzene toxicity. Hepatic levels of radioactivity underwent biphasic changes: At first, high levels of metabolites were observed, which then seemed to disappear.

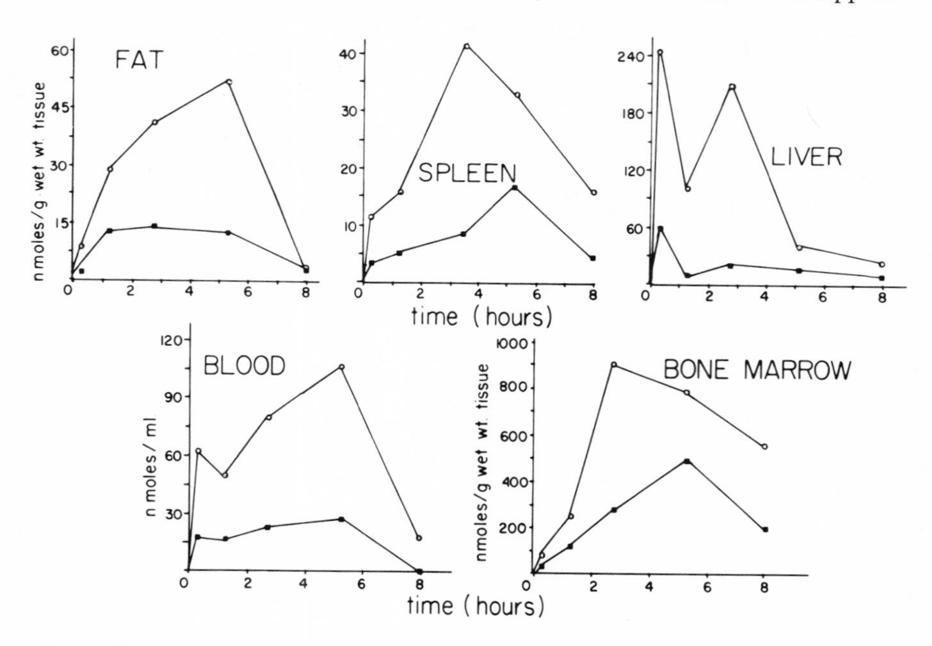

Fig. 8. Concentrations of metabolites of [^{3}H] benzene in mouse tissues. Following subcutaneous administration of [^{3}H]benzene (880 mg/kg) (○) or [^{3}H]benzene (880 mg/kg) plus toluene (1760 mg/kg) (■) to mice, the water-soluble radioactivity present in various tissues was determined and nmol [^{3}H]benzene metabolites present was calculated using the specific activity of the injected [^{3}H]benzene. Each point represents a mean value obtained from six mice.

Thereafter, a second peak level occurred, followed by a final decrease. Blood levels were a reflection of liver levels but at reduced concentrations, while fat and spleen, which were low, appeared to follow the blood in a general way. Toluene reduced metabolite levels without altering the pattern of their appearance.

A comparison of the areas under the curves for benzene and metabolites suggests that, in general, liver and bone marrow either metabolize or contain in the form of metabolites approximately 25–30% of the benzene with which they are presented, whereas for spleen and fat the corresponding figures are about 10% and 0.5%, respectively.

[^{3}H] Benzene metabolites recovered from liver and bone marrow consisted of glucuronide and ethereal sulfate conjugates as shown by chromatography on silica gel TC plates (Table 6). Further chromatography of the products released upon treatment of the metabolites with β-glucuronidase and sulfatase showed that over the 5-h monitoring period the conjugated products of benzene hydroxylation were phenol (96–97%) and catechol (3–4%).

The fact that benzene metabolism is known to occur in liver probably explains why there were higher concentrations of benzene metabolites in liver than in blood. The concentration of metabolites in bone marrow, however, far exceeded the level in blood (Fig. 8). These observations could be explained either by the action of a benzene-metabolizing enzyme system in bone marrow or by the active uptake and/or sequestration of benzene metabolites by the bone marrow. One approach to this problem has been to isolate metabolites from

Table 6. Levels of Benzene Metabolites Accumulating in the Bone Marrow and Liver of Mice Administered [^{3}H] Benzene[a]

Tissue	Time (h)	Free phenols	Conjugated phenol	Conjugated catechol
Femoral bone marrow	0.25	21	140	–
	1.0	34	245	–
	3.0	24	760	34
	5.0	105	1170	38
Liver	0.25	21	120	3.3
	1.0	54	598	3.4
	3.0	17	512	31
	5.0	32	973	59

[a]Mice (n = 12) were given [^{3}H]benzene specific activity (1500 dpm/nmol) (880 mg/kg) s.c. and sacrificed (0.25, 1.0, 3.0, and 5.0 h) following injection. Liver and femoral bone marrow samples were obtained and extracted with toluene in order to determine amounts of unconjugated phenols. Tissue samples were pooled, and water-soluble [^{3}H]benzene metabolites were isolated and treated with an enzyme preparation containing β-glucuronidase and arylsulfatase activity. The phenols liberated were identified by silica gel thin-layer chromatography using benzene–dioxane–acetic acid (19:5:1) as the solvent system. Standard deviation of values obtained for free phenols ranged from 25 to 55% of the mean. The limits of detection of conjugated phenols were approximately 2 nmol/g wet weight.

[b]Values are expressed as nmol/g wet weight liver or femoral bone marrow.

Table 7. Uptake of Injected [^{3}H]Benzene Metabolites from Blood into Mouse Bone Marrow[a]

	Blood concentration[b] (nmol/ml) (mean ± SD)		Marrow concentration (nmol/g) (mean ± SD)	
Metabolite	t = 15 min	t = 30 min	t = 15 min	t = 30 min
[^{3}H]Phenol	66.9 ± 15.9	30.2 ± 11.2	19.4 ± 8.4	12.0 ± 3.6
[^{3}H]Phenylglucuronide[c]	60.6 ± 16.4	57.0 ± 25.2	14.0 ± 4.6	17.8 ± 2.9
[^{3}H]Phenylsulfate[d]	50.1 ± 12.2	84.2 ± 13.8	8.1 ± 2.9	13.9 ± 7.5

[a]Mice (n = 5) were injected i.p. with [^{3}H]phenol (12.5 mg/kg), [^{3}H]phenylglucuronide (15.3 mg/kg), or [^{3}H]phenylsulfate (9.9 mg/kg). Concentrations of benzene metabolites in blood and bone marrow were not detectable 3 h following injection.
[b]Blood concentrations were greater than respective marrow concentrations, $p < 0.05$.
[c]Contained < 5% [^{3}H]catechol glucuronide.
[d]Contained <5% [^{3}H]catechol sulfate.

mouse urine and reinject them into mice to attain blood levels approximately approaching those observed after benzene administration. Table 7 shows blood and marrow concentrations of tritium-labeled phenol, phenyl glucuronide, and phenylsulfate at 15 and 30 min after these metabolites were given to mice. No radioactivity was observed in these organs 3 h after injection. Blood concentrations were in each case far in excess of marrow concentrations and there was no evidence for active concentration of these metabolites by bone marrow.

DISCUSSION

Benzene toxicity, which is principally a function of depressed bone marrow activity, is prevented or alleviated in mice treated with toluene. Since toluene inhibits benzene metabolism but does not reduce benzene levels in bone marrow, it is postulated that a metabolite of benzene mediates benzene toxicity. To further investigate the relationship between the metabolism of benzene and its toxicity to bone marrow, the time course of the distribution of [^{3}H]benzene and its metabolites was studied in blood, spleen, fat, bone marrow, and liver. These organs were chosen because (1) it is known that benzene is metabolized in liver, (2) benzene is nonpolar and would be expected to accumulate in fat, and (3) bone marrow, blood, and spleen are critical organs affected in benzene toxicity.

The most interesting result of these studies was the finding that the level of benzene metabolites in bone marrow equaled or exceeded that in liver and that these were each far in excess of blood levels. The high level of benzene metabolites in liver is to be expected since benzene is metabolized in liver. The concentration of metabolites in bone marrow may have resulted from either the active uptake into marrow or the actual production of the metabolites in marrow. The demonstration of Tschernikow (29) that hepatectomized frogs given benzene excreted phenol and of Harper *et al.* (30) showing benzene metabolism in lung microsomes argues for extrahepatic benzene metabolism. Furthermore, Gadaskina *et al.* (31) reported benzene hydroxylation in bone marrow homogenates and leukocyte suspensions. The demonstration here that benzene metabolites administered to mice *in vivo* did not accumulate in bone marrow even though high levels of metabolites were found in marrow after giving benzene supports the contention that benzene is in part metabolized in bone marrow.

The data collected on bone marrow in these studies were based on radioactivity washed out of two femurs in each mouse. It has been reported, however, that total active marrow in all of the bones of the body is about equal in volume to the size of the liver (32). The finding that metabolite levels in liver and marrow were similar on a weight basis suggests that if bone marrow metabolizes benzene it may be a significant source of benzene metabolites. Furthermore, the metabolites appeared to be ethereal sulfate and glucuronide conjugates, which suggests that the bone marrow may display some enzyme activities usually associated with liver. If bone marrow depression by benzene depends on local formation of an active intermediate, bone marrow drug metabolism leading to the formation of toxic metabolites may help to explain the toxicity of chloramphenicol, phenylbutazone, chlorpromazine, and other drugs known to produce blood dyscrasias.

Jerina and Daly (33) and their associates have suggested that benzene hydroxylation probably occurs via the formation of an epoxide intermediate. Benzene oxide can then rearrange nonenzymatically to form phenol or can serve as a substrate for epoxide hydrase to yield a dihydrodiol which then may be enzymatically converted to catechol. An alternative mechanism is the reaction of glutathione with benzene oxide to form a premercapturic acid. It has been suggested that because of great chemical reactivity benzene oxide may be the active intermediate that reacts with proteins, nucleic acids, etc., to produce benzene toxicity. Irreversible binding of catechol or semiquinone derivatives of estrogens (34,35) and methyldopa (36) suggests another form of active intermediate. Indeed, there are reports of leukopenia and anemia in cats (37) and rats (25) following treatment with catechol, a benzene metabolite. Thus, although its exact structure has yet to be determined, the evidence presented here suggests that benzene toxicity results from the formation of an active toxic metabolite of benzene.

REFERENCES

1. R. Snyder and J. J. Kocsis, Current concepts of chronic benzene toxicity, *Crit. Rev. Toxicol.* 3, 265–288 (1975).

2. E. Browning, *Toxicity and Metabolism of Industrial Solvents,* Chap. 1, Elsevier, Amsterdam (1965).
3. G. Saita, Benzene induced hypoplastic anemias and leukemias, in: *Blood Disorders Due to Drugs and Other Agents* (R. H. Girdwood, ed.), pp. 127–146, Excerpta Medica, Amsterdam (1973).
4. M. Aksoy, K. Dincol, S. Erdem, T. Akgun, and G. Dincol, Details of blood changes in 32 patients with pancytopenia associated with long term exposure to benzene, *Br. J. Ind. Med.* **29**, 56–64 (1972).
5. H. G. Weiskotten, C. B. F. Gibbs, E. O. Boggs, and E. R. Templeton, The action of benzol. VI. Benzol vapor leucopenia (rabbit), *J. Med. Res.* **41**, 425–438 (1920).
6. W. B. Deichmann, W. E. MacDonald, and E. Bernal, Hemopoietic tissue toxicity of benzene vapors, *Toxicol. Appl. Pharmacol.* **5**, 201–224 (1963).
7. L. Selling, Benzol as a leucotoxin: Studies on the degeneration and regeneration of the blood and haematopoietic organs, *Johns Hopkins Hosp. Rep.* **17**, 83–148 (1916).
8. J. S. Latta and L. T. Davies, Effects on the blood and hemopoietic organs of the albino rat of repeated administration of benzene, *Arch. Pathol.* **31**, 55–67 (1941).
9. M. Kissling and B. Speck, Chromosome aberrations in experimental benzene intoxication, *Helv. Med. Acta* **36**, 59–66 (1969).
10. M. Kissling and B. Speck, Further studies on experimental benzene induced aplastic anemia, *Blut* **25**, 97–103 (1972).
11. V. H. Boje, W. Benkel, and H. J. Heiniger, Untersuchungen zur Leukopoese in Knochenmark der Ratte nach chronischer Benzol Inhalation, *Blut* **21**, 250–257 (1970).
12. J. W. Porteous and R. T. Williams, Studies in detoxication. 19. The metabolism of benzene I. (a) The determination of phenol in urine with 2:6-dichloroquinonechloroimide. (b) The excretion of phenol glucuronic acid and ethereal sulfate by rabbits receiving benzene and phenol. (c) Observations on the determination of catechol, quinol and muconic acid in urine, *Biochem. J.* **44**, 46–55 (1949).
13. J. W. Porteous and R. T. Williams, Studies in detoxication. 20. The metabolism of benzene. II. The isolation of phenol, catechol, quinol and hydroxyquinol from the ethereal sulfate fraction of the urine of rabbits receiving benzene orally, *Biochem. J.* **44**, 56–61 (1949).
14. D. V. Parke and R. T. Williams, Studies in detoxication. 49. The metabolism of benzene containing $^{14}C_1$ benzene, *Biochem. J.* **54**, 231–238 (1953).
15. D. V. Parke and R. T. Williams, Studies in detoxication. 54. The metabolism of benzene. (a) The formation of phenyl glucuronide and phenylsulfuric acid from ^{14}C benzene. (b) The metabolism of ^{14}C phenol, *Biochem. J.* **55**, 337–340 (1953).
16. L. M. Gonasun, C. M. Witmer, J. J. Kocsis, and R. Snyder, Benzene metabolism in mouse liver microsomes, *Toxicol. Appl. Pharmacol.* **26**, 398–406 (1973).
17. R. Snyder, F. Uzuki, L. Gonasun, E. Bromfeld, and A. Wells, The metabolism of benzene *in vitro, Toxicol. Appl. Pharmacol.* **11**, 346–360 (1967).
18. J. J. Kocsis, S. Harkaway, M. C. Santoyo, and R. Snyder, Dimethyl sulfoxide: Interactions with aromatic hydrocarbons, *Science* **160**, 427–428 (1968).
19. R. T. Drew, J. R. Fouts, and C. Harper, The influence of certain drugs on the metabolism and toxicity of benzene, in: *Symposium on Toxicology of Benzene and Alkylbenzenes,* (D. Braun, ed.), pp. 17–31, Industrial Health Foundation, Pittsburgh (1974).
20. R. T. Drew and J. R. Fouts, The lack of effects of pretreatment with phenobarbital and chlorpromazine on the acute toxicity of benzene in rats, *Toxicol. Appl. Pharmacol.* **27**, 183–193 (1974).
21. P. Dustin, Jr., The action of mitotic poisons on normal and pathological blood cell formation, *Sang* **21**, 297–330 (1950).

22. T. Hirokawa and K. Nomiyama, Studies on poisoning by benzene and its homologues. (5) Oxidation rate of benzene in rat liver homogenates, *Med. J. Shinshu Univ.* **7**, 29–39 (1962).
23. K. Nomiyama, Studies on poisoning by benzene and its homologues. (6) Oxidation rate of benzene and benzene poisoning, *Med. J. Shinshu Univ.* **7**, 41–48 (1962).
24. K. Nomiyama, Experimental studies on benzene poisoning, *Bull. Tokyo Med. Dent. Univ.* **11**, 297–313 (1964).
25. K. Nomiyama, Studies on poisoning by benzene and its homologues. (7) Toxicity of benzene metabolites to hemopoiesis, *Ind. Health* **3**, 53–57 (1965).
26. M. Ikeda, Enzymatic studies on benzene intoxication, *J. Biochem. (Tokyo)* **55**, 231–243 (1964).
27. M. Ikeda and H. Ohtsuji, Phenobarbital-induced protection against toxicity of toluene and benzene in the rat, *Toxicol. Appl. Pharmacol.* **20**, 30–43 (1971).
28. M. Ikeda, H. Ohtsjui, and T. Imamura, *In vivo* suppression of benzene and styrene oxidation by co-administered toluene in rats and effects of phenobarbital, *Xenobiotica* **2**, 101–106 (1972).
29. A. M. Tschernikow and I. D. Gadaskina, Oxidation of benzene in the liver of warm and cold-blooded animals, *Naunyn-Schmiedebergs Arch. Pharmakol.* **154**, 222–227 (1930).
30. C. Harper, R. T. Drew, and J. R. Fouts, Species differences in benzene hydroxylation to phenol by pulmonary and hepatic microsomes, *Drug Metab. Dispos.* **3**, 381–388 (1975).
31. I. D. Gadaskina, Z. I. Abramova, and Z. I. Bikerskaya, Participation of bone marrow in the oxidation of benzene, *Gig. Tr. Prof. Zabol.* **12**, 2–7 (1963).
32. A. J. Erslev, Pathophysiology of aplastic anemia, in: *Drugs and Hematologic Reactions* (N. V. Dimitrov and J. H. Nodine, eds.), pp. 65–78, Grune and Stratton, New York (1974).
33. D. Jerina and J. W. Day, Arene oxides: A new aspect of drug metabolism, *Science* **185**, 573–582 (1974).
34. F. Marks and E. Hecker, Metabolism and mechanism of action of oestrogens. XII. Structure and mechanism of formation of water soluble and protein-bound metabolites of oestrone in rat liver microsomes *in vitro* and *in vivo, Biochim. Biophys. Acta* **187**, 250–265 (1969).
35. H. Kappus, H. M. Bolt, and H. Remmer, Irreversible protein binding of metabolites of ethynylestradiol *in vivo* and *in vitro, Steroids* **22**, 203–225 (1973).
36. E. Dybing, J. R. Mitchell, S. D. Nelson, and J. R. Gillette, Metabolic activation of methyldopa by cytochrome P450-generated superoxide anion, Chap. 17, this volume.
37. W. B. Dietering, Über Brenzcatechin-vergiftung, *Naunyn-Schmiedebergs Arch. Pharmakol.* **188**, 493–499 (1938).

32

Benzene and *p*-Xylene: A Comparison of Inhalation Toxicities and *in Vitro* Hydroxylations

C. Harper, R. T. Drew, and J. R. Fouts

Pharmacology Branch
National Institute of Environmental Health Sciences
Research Triangle Park, North Carolina 27709, U.S.A.

Aromatic hydrocarbons such as benzene, cumene, mesitylene, toluene, and *p*-xylene are widely used as solvents in the chemical industry. They are also found in common household supplies such as paint removers, degreasing agents, lacquers, insecticides, and pesticides. In addition to the individual solvents, these solvents are often found as contaminants of one another.

Many of the aromatic hydrocarbons produce similar toxic reactions. The most common symptom of acute toxicity is central nervous excitation followed by depression. However, chronic exposure to benzene is known to produce additional toxic reactions which are not encountered with pure samples of toluene or xylenes (1).

Acute exposure to moderate doses of the aromatic hydrocarbons is not known to be fatal, but symptoms which resemble ethyl alcohol inebriation have been observed. In some cases, anesthesia followed the inebriation (2).

The comparative differences in acute toxicities among the aromatic hydrocarbons are not well documented. It is widely accepted, however, that chronic poisoning with benzene is much more severe and represents considerably greater danger to health than chronic poisoning with the benzene alkyl derivatives. The most significant difference between chronic benzene toxicity and toxicity from the other aromatic solvents is the bone marrow damage caused by chronic exposure to benzene (3,4). Bone marrow and blood abnormalities resulting from chronic exposure to benzene have been reported by several investigators (4–8). These bone marrow and blood abnormalities have not been observed after acute exposures to benzene.

The acute toxicity of xylene is believed to be similar to toluene and benzene toxicities (9). Tremors, restlessness, hypertonus, and hyperactive reflexes are the usual symptoms observed after exposure to toxic doses of these aromatic hydrocarbons (10). Benzene is believed to cause greater adverse neural responses than xylene or toluene, but the latter solvents are most irritating to the mucous membranes (11).

The excretion of aromatic solvents takes place primarily through the lungs (9). The lungs are also rich in microsomal mixed-function oxidase enzymes which hydroxylate benzene and *p*-xylene (12–14).

Unlike the halogenated benzenes (15), no definitive relationships between the toxicity and metabolism of benzene and alkyl benzenes have been established. We have studied benzene and *p*-xylene with respect to their toxicities and metabolism *in vitro* by pulmonary and hepatic microsomes.

RESULTS AND DISCUSSION

Benzene Toxicity

Acute inhalation toxicity of benzene was studied with female Sprague–Dawley rats (Charles River CD). Treatment with benzene was carried out as previously described (13). The rats were subjected to a single 4-h exposure to benzene vapors. The benzene toxicities as measured by the LC_{50} are shown in Table 1. Some of the animals were pretreated with phenobarbital (75 mg/kg) or chlorpromazine (15 mg/kg) by intraperitoneal injections for 3 consecutive days prior to the exposures. Animals which did not survive because of the exposure died during the exposure or within the first 24-h postexposure period. The only other index of toxicity observed was lung and liver congestion (increase in the number of red blood cells). All of the animals which died because of the exposure had highly congested livers and most of them had highly congested lungs; conversely, the animals which did not die from the exposure but were subsequently sacrificed had a very low degree of congestion in the liver and/or lung. Pretreatment of the animals with phenobarbital or chlorpromazine had no significant effect on the LC_{50} or the severity of congestion. When the rats were given benzene by intraperitoneal injection, the acute toxicity (LD_{50}) was not

Table 1. Acute Inhalation Toxicity of Benzene

Treatment	Dose (mg/kg)	LC_{50} (ppm)
Saline	–	13,700
Phenobarbital	75	13,770
Chlorpromazine	15	14,370

Table 2. Acute Toxicity of Benzene Administered by Intraperitoneal Injection

Treatment	Dose (mg/kg)	LD_{50} (mg/kg)
Saline (control)	–	2.94
Phenobarbital	75	2.46
Chlorpromazine	15	3.32
Corn oil (control)	–	2.89
3-Methylcholanthrene	20	3.12

significantly changed by pretreatment with phenobarbital, chlorpromazine, or 3-methylcholanthrene (Table 2). Unlike the inhalation toxicity, no histological damage was observed in any of the rats treated with benzene by intraperitoneal injection.

Toxicity caused by chronic inhalation of benzene was studied by exposing adult female rats to 1650 ppm benzene vapors 6 h/day, 5 days/week for 12 weeks (16). A second group of rats was given phenobarbital (1 mg/ml in drinking water) throughout the exposure to benzene. Rats from each group were removed and sacrificed for histological examination and biochemical studies after 2, 4, 8, and 12 weeks exposure to benzene vapors.

The red blood cells of the exposed animals were examined for the erythrocyte count, hematocrit, hemoglobin concentration, reticulocyte count, mean corpuscular hemoglobin concentration, and mean corpuscular volume. No sig-

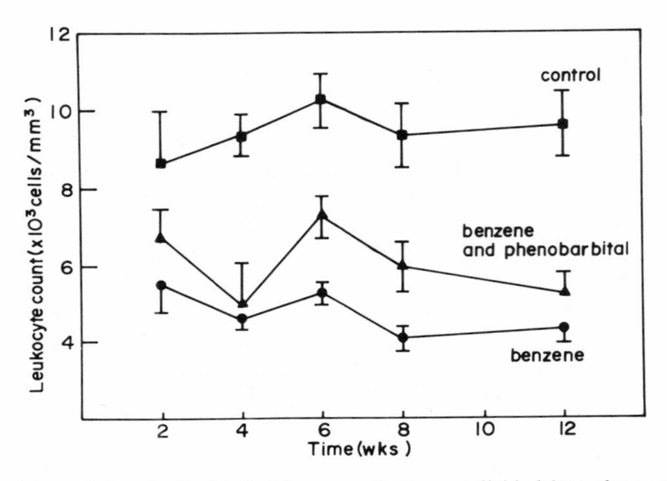

Fig. 1. Toxicity of chronically inhaled benzene. Rats were divided into three groups. The control group was exposed to air only. The first experimental group was exposed to 1650 ppm benzene vapors for 6 h/day, 5 days/week. The second experimental group was exposed to benzene exactly as the first and maintained on drinking water containing phenobarbital (10 g/liter) starting 3 days before the first exposure.

nificant difference between rats exposed to benzene and the control rats was observed in any of the red blood cell examinations.

Examination of the white blood cells revealed significant differences between the control and benzene-exposed rats. The leukocyte counts of the benzene-exposed rats were considerably lower than the leukocyte counts of control rats after 4 weeks of exposure and remained low through 12 weeks. The benzene-induced leukopenia was partially reversed by administration of phenobarbital before and during the exposure (Fig. 1). Maximum protection by phenobarbital administration was observed after 6 weeks of exposure to benzene. The effects of phenobarbital against benzene-induced leukopenia gradually disappeared between 6 and 12 weeks of benzene exposure.

Benzene Metabolism *in Vitro*

Benzene is hydroxylated primarily to phenol when incubated with pulmonary or hepatic microsomes of the rat (12,17). Pulmonary tissue was studied because the lung plays a major role in both benzene inhalation and benzene excretion (9). Metabolism by hepatic microsomes was studied because the liver is the primary site of detoxication in mammalian species.

The rate of hydroxylation of benzene to phenol by hepatic microsomes was elevated when rats were preexposed to high concentrations of benzene vapors 4 h/day for 3 days (Table 3). The rate of hydroxylation by pulmonary microsomes was slightly lower after the exposure to benzene vapors. The effects of acute exposures on rates of microsomal hydroxylation were absent when the concentration of benzene in the exposure chambers was 2000 ppm or below.

Chronic inhalation of benzene (1650 ppm, 6 h/day, 5 days/week) did not significantly modify the rates of benzene hydroxylation by pulmonary or hepatic microsomes (Table 4). The increased rate of hepatic microsomal hydroxylation in phenobarbital-induced animals was maintained throughout the 12 weeks of exposure, although the effect seemed to diminish with time.

Table 3. Effects of Benzene Inhalation on Benzene Hydroxylase in Rat Liver and Lung Microsomes

	nmol metabolized/min/mg protein			
	Liver		Lung	
Average concentration (ppm)	Control	Exposed[a]	Control	Exposed[a]
4050	0.81 ± 0.02	1.58[b] ± 0.03	0.55 ± 0.07	0.32 ± 0.04
2230	0.74 ± 0.03	0.87 ± 0.03	0.43 ± 0.04	0.33 ± 0.06
2030	0.44 ± 0.01	0.49 ± 0.02	0.28 ± 0.02	0.24 ± 0.02

[a]Rats exposed 4 h/day for 3 days prior to sacrifice.
[b]Significantly different from control, $P < 0.01$.

Table 4. Effects of Benzene Inhalation[a] and Phenobarbital Ingestion[b] on Benzene Hydroxylase in Rat Liver Microsomes

	nmol metabolized/min/mg protein[c]		
Time (weeks)	Control	Benzene	Benzene and phenobarbital
2	0.48 ± 0.03	0.60 ± 0.01	8.63[d] ± 0.28
4	0.43 ± 0.01	0.64[d] ± 0.02	6.48[d] ± 0.04
8	0.63 ± 0.03	0.77 ± 0.03	6.08[d] ± 0.22
12	0.62 ± 0.03	0.74 ± 0.04	5.34[d] ± 0.09

[a] 1640 ppm, 6 h/day, 5 days/week.
[b] 1 mg/ml in drinking water.
[c] Standard error of the mean. $N = 4$ in each case.
[d] Significantly different from other two groups, $P<0.01$.

Whether or not there is a relationship between phenobarbital-induced increases in benzene-hydroxylating activity of hepatic microsomes and the phenobarbital protection against benzene-induced leukopenia is not clear. Phenobarbital increases the activity of several enzymes which metabolize xenobiotic substances as well as intermediates and products of xenobiotic substances. It is not clear how phenobarbital affects the balance of these reactions relative to benzene metabolism. Unless it can be shown that phenobarbital induces detoxication pathways in other tissues more directly related to leukocyte production, these results can be interpreted to suggest that leukopenia is caused by benzene itself or a toxic intermediate produced in the liver. A less likely possibility is that the active intermediate is formed in the leucocyte-producing tissues and transported to the liver for detoxication. Indeed, in the unlikely event that the benzene-induced leukopenia and/or benzene detoxication involves the transport of a toxic intermediate, the intermediate would have to be considerably less reactive than the highly unstable benzene oxide. The metabolism of benzene in tissues more directly related to leukocyte production needs to be explored.

Several other factors cause the correlation between hepatic microsomal benzene hydroxylation and benzene toxicity to be problematical. The differences between *in vitro* and *in vivo* metabolism of benzene are numerous.

Phenol accumulates in the *in vitro* reaction mixtures because of the absence of phenol-conjugating systems. In addition, the absence of the conjugating systems allows phenol to be converted to other unidentified nonpolar metabolites unless sufficient benzene is present to inhibit the reaction in hepatic microsomes (12). This alternative pathway for phenol disposition by hepatic microsomes has been observed in the rat, rabbit, and hamster (Fig. 2). Further consideration must be given to the fact that phenol is a competitive inhibitor of benzene hydroxylation *in vitro* (unpublished results).

The biotransformation of phenol to nonconjugated nonpolar products and competitive inhibition of benzene metabolism by phenol are characteristic of

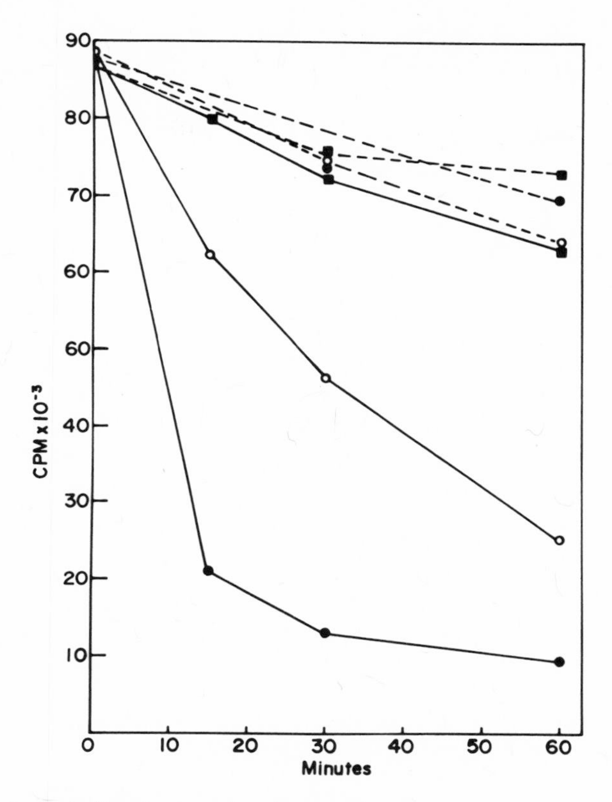

Fig. 2. Phenol metabolism by hepatic microsomal preparations. Each incubation mixture contained 2.5 μmol [^{14}C]phenol, 2.5 mg microsomal protein, 12.5 μmol glucose-6-phosphate, 12.5 μmol magnesium chloride, 2.5 μmol $NADP^+$, 1.0 unit glucose-6-phosphate dehydrogenase, and 250 μmol HEPES buffer (pH 7.6). – – –, 28 μmol benzene; ——, no benzene. ●, Hamster liver; ■, rat liver; ○, rabbit liver. The vertical axis represents the amount of unmetabolized [^{14}C]phenol remaining after incubation.

the *in vitro* assay for benzene hydroxylation by hepatic microsomes. It is not known to what extent these characteristics influence benzene metabolism *in vivo* or if they occur at all. It is possible and likely that some of these characteristics occur only in reactions *in vitro* and are not consequential to benzene metabolism in vivo. These phenomena must be accounted for before any modification of benzene metabolism rates *in vitro* can be extrapolated to explain modifications of benzene toxicity *in vivo.*

p-Xylene Toxicity

Unlike the parent compound benzene, the 1,4-dialkyl derivative of benzene (*p*-xylene) has not been known to cause leukopenia as a consequence of chronic exposure.

Table 5. Effect of Phenobarbital, Chlorpromazine, and 3-Methylcholanthrene on the LC_{50} of *p*-Xylene

Treatment	Dose (mg/kg)	LC_{50} (ppm)	95% CL
Saline	–	4740	4520–4960
Phenobarbital	75	5810	5460–6160
Chlorpromazine	15	4970	4700–5240
Corn oil	–	4550	3850–4750
3-Methylcholanthrene	20	4960	4710–5200

The acute toxicity of *p*-xylene was studied with adult female rats (CD strain). The exposures to *p*-xylene were carried out in chambers exactly as described for benzene exposure (14). As shown in Table 5, the LC_{50} for female rats exposed to *p*-xylene vapors is influenced by the administration of certain drugs. Pretreatment with phenobarbital increased the LC_{50} for *p*-xylene by more than 20%. Pretreatment with 3-methylcholanthrene caused only a 9% increase in the LC_{50}, while chlorpromazine was even less effective, causing only a 5% increase in the LC_{50} for *p*-xylene.

The acute toxicity of *p*-xylene when administered by intraperitoneal injection is shown in Table 6. A 28% increase in the LD_{50} was observed when the *p*-xylene-treated animals were pretreated with 3-methylcholanthrene. Both chlorpromazine and phenobarbital caused insignificant changes in the LD_{50}. A summary of these toxicities is given in Table 7. Histological examination of tissues treated with *p*-xylene (0.1 mg/kg, 3 days, sacrificed 24 h after the final intraperitoneal injection) revealed only moderate fatty infiltration of the liver and dilation of the smaller blood vessels of the lung.

p-Xylene Metabolism

Hydroxylation was found to occur primarily on the methyl group of *p*-xylene when incubated with hepatic or pulmonary microsomes from adult female rats (18). The product, *p*-methylbenzyl alcohol, is converted to *p*-methyl-

Table 6. Effect of Phenobarbital, Chlorpromazine, and 3-Methylcholanthrene on the LD_{50} of *p*-Xylene

Treatment	Dose (mg/kg)	LD_{50} (ml/kg)	95% CL
Saline	–	3.81	3.35–4.27
Phenobarbital	75	3.62	3.39–3.86
Chlorpromazine	15	3.75	3.43–4.07
3-Methylcholanthrene	20	4.81	4.51–5.11

Table 7. Protection from *p*-Xylene Toxicity

Xenobiotic	Inhalation	Injection
Phenobarbital	++++	0
Chlorpromazine	+	+
3-Methylcholanthrene	++	++++

benzoic (toluic) acid if hepatic cytosol fraction or some other source of aldehyde dehydrogenase and liver alcohol dehydrogenase is added.

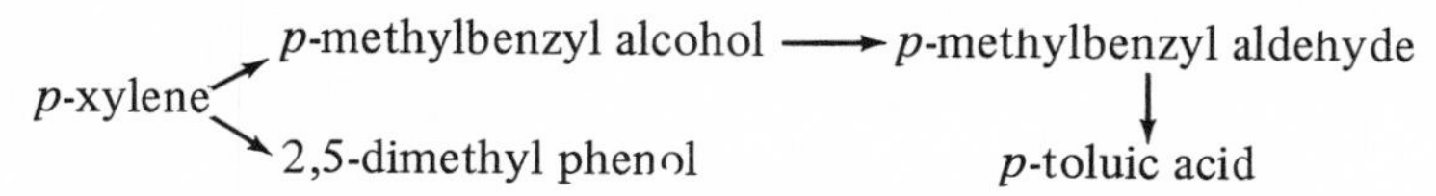

Less than 1% of the product formed with the hepatic or pulmonary microsomes was identified as 2,5-dimethyl phenol.

Neither phenobarbital nor 3-methylcholanthrene was found to modify the rate of *p*-xylene hydroxylation by pulmonary microsomes. Both phenobarbital, which reduced the inhalation toxicity of *p*-xylene, and 3-methylcholanthrane, which reduced the injection toxicity of *p*-xylene, were found to increase the rate of *p*-xylene hydroxylation by hepatic microsomes (Table 8).

It is not clear whether or not there is a relationship between the modest reduction in toxicities of *p*-xylene when rats are pretreated with 3-methylcholanthrene or phenobarbital and the increased rate of hepatic microsomal hydroxylation of *p*-xylene caused by these drugs. If both drugs caused increased rates of hepatic metabolism *in vivo* and neither caused a change in the rate of pulmonary metabolism *in vivo,* as observed *in vitro,* it is difficult to explain their modifications of *p*-xylene toxicities by increased metabolism. If phenobarbital reduces inhalation toxicity and not injection toxicity by increasing hepatic metabolism of *p*-xylene, then other drugs which increase hepatic metabolism should modify inhalation and injection toxicities in the same manner as phenobarbital.

The differences between *in vitro* and *in vivo* metabolism of *p*-xylene need to

Table 8. Effects of Chemical Agents on Hepatic *p*-Xylene Metabolism

Treatment	nmol metabolized/min/mg protein
Saline	0.31 ± 0.01
Phenobarbital	1.03 ± 0.06
Chlorpromazine	0.49 ± 0.01
Cottonseed oil	0.36 ± 0.02
3-Methylcholanthrene	1.53 ± 0.05

be studied. We treated adult female rats with *p*-xylene (0.1 mg/kg) and examined the urinary metabolites. Although no quantitative ratios were established, these preliminary results suggested that the urinary fraction of 2,5-dimethyl phenol produced *in vivo* from *p*-xylene was considerably greater than the fraction formed *in vitro* in pulmonary and hepatic microsomal reactions. The effects of 3-methylcholanthrene and phenobarbital on 2,5-dimethyl phenol production *in vivo* from *p*-xylene need to be studied and compared with their effects on inhalation and injection toxicities. Attention also should be given to those organs which metabolize *p*-xylene to *p*-methylbenzyl alcohol but do not have alcohol and aldehyde dehydrogenase activities.

Although benzene and *p*-xylene have similar physical–chemical properties which allow them to be used interchangeably in certain industrial and laboratory situations, they interact quite differently with biological systems. The reactive alkyl groups of *p*-xylene cause its transformation in biological systems to be quite different from benzene. It cannot be stated whether the differences in toxic symptoms produced by the two substances are related to the differences in biotransformation of *p*-xylene as compared to benzene or differences in physical–chemical interactions of biological structures with benzene as compared to *p*-xylene.

ACKNOWLEDGMENTS

We wish to express our appreciation to the following investigators who participated in the study: Drs. Anna Ottolenghi, Bhola N. Gupta, Michael Hogan, and Joseph Zinkl.

We also wish to express our appreciation for the technical assistance given by Ms. Roberta J. Pohl, Mr. Emil M. Lores, Ms. Joanna L. Schroeder, Mr. Michael T. Riley, and Mr. Frank W. Harrington.

REFERENCES

1. M. A. Wolf, V. K. Rowe, D. D. McCollister, R. L. Hollingworth, and F. Oyen, Toxicological studies of certain alkylated benzenes and benzene, *Arch. Ind. Health* **14**, 387–389 (1956).
2. E. O. Longley, A. T. Jones, R. Welch, and O. Lomaess, Two acute toluene episodes in merchant ships, *Arch. Environ. Health* **14**, 481–487 (1967).
3. J. L. Svirbely, R. C. Dunn, and W. F. von Oettingen, The chronic toxicity of moderate concentrations of benzene and mixtures of benzene and its homologues for rats and dogs, *J. Ind. Hyg. Toxicol.* **26**, 37–46 (1965).
4. E. C. Vigliani and G. Saita, Benzene and leukemia, *N. Engl. J. Med.* **271**, 872–876 (1964).
5. D. H. Anderson, Benzene with hyperplasia of the bone marrow, *Am. J. Pathol.* **10**, 101–111 (1934).

6. R. L. DeGowen, Benzene exposure and aplastic anemia followed by leukemia 15 years later, *J. Am. Med. Assoc.* **185**, 748–751 (1963).
7. E. R. Hayhurst and B. E. Neiswander, A case of chronic benzene poisoning, *J. Am. Med. Assoc.* **96**, 269–270 (1931).
8. F. T. Hunter and S. S. Hanflig, Chronic benzene poisoning. A report of four case histories, *Boston Med. Surg. J.* **197**, 292–299 (1927).
9. H. W. Gerade, *Toxicology and Biochemistry of Aromatic Hydrocarbons,* Elsevier, New York (1960).
10. W. F. von Oettingen, *Toxicity and Potential Dangers of Aliphatic and Aromatic Hydrocarbons: A Critical Review of the Literature,* U.S. Public Health Service, Public Health Bull. No. 255 (1940).
11. American Petroleum Institute, New York, *Cumene,* A.P.I. Toxicological Review (1948).
12. C. Harper, R. T. Drew, and J. R. Fouts, Species differences in benzene hydroxylation to phenol by pulmonary and hepatic microsomes, *Drug Metab. Dispos.* **3**, 381–388 (1975).
13. R. T. Drew and J. R. Fouts, The lack of effects of pretreatment with phenobarbital and chlorpromazine on the acute toxicity of benzene in rats, *Toxicol. Appl. Pharmacol.* **27**, 183–193 (1974).
14. M. F. Carlone and J. R. Fouts, *In vitro* metabolism of *p*-xylene by rabbit lung and liver, *Xenobiotica* **11**, 705–715 (1974).
15. J. N. Zampaglione, D. J. Jollow, J. R. Mitchell, B. Stripp, M. Hamrick, and J. R. Gillette, Role of detoxifying enzymes in bromobenzene-induced liver necrosis, *J. Pharmacol. Exp. Ther.* **187**, 218–227 (1973).
16. R. T. Drew, J. R. Fouts, and C. Harper, The influence of certain drugs on the metabolism and toxicity of benzene, in: *Proceedings of Symposium on Toxicity of Benzene and Alkylbenzenes* (D. Braun, ed.), pp. 17–31, Industrial Hygiene Foundation, Pittsburgh (1974).
17. R. Snyder, R. Uzuki, L. Gonasun, E. Bromfeld, and A. Wells, The metabolism of benzene *in vitro, Toxicol. Appl. Pharmacol.* **11**, 346–360 (1967).
18. C. Harper, *p*-Xylene metabolism by rat pulmonary and hepatic microsomes, *Fed. Proc.* **34**, 785 (1975).

33

Discussion

Nebert stated that his group has shown that benzpyrene is metabolized by bone marrow, thus indicating that bone marrow contains the AHH enzyme system. When both "responsive" and "nonresponsive" mice received a chronic dose of benzpyrene (1 mg/kg) during the duration of their life span, the "nonresponsive" mice died within 1–3 weeks, whereas the "responsive" animals lived as long or longer than control mice. AHH activity as measured by benzpyrene hydroxylation was generally depressed in "nonresponsive" mice, but was higher per milligram of tissue in bone marrow than in lung or kidney. Their conclusions were that the "nonresponsive" allele is probably associated with the death of these mice, that it is found as a single chromosomal locus, and that death does not appear to be directly related to the amount of overall benzpyrene metabolism.

In reply to a question, Snyder reported that his group did not study the uptake of tritiated thymidine into DNA as a measure of damage to stem cells or other erythrocyte precursors. Using the ^{59}Fe uptake technique, it was found that the activities of only pronormoblasts and normoblasts were depressed by the benzene treatment employed here. If stem cells had been damaged by benzene, iron uptake would have been depressed 72 h after administration of benzene. The fact that no decrease in iron uptake was observed at that time plus previous reports by Moeschlein and Speck (*Acta Haematol.* **38**, 104, 1967) and Steinberg (*Blood* **4**, 550, 1949) using radiolabeling and morphological techniques, respectively, supports the concept that early blast cell stages are more susceptible to benzene than stem cells.

Although Snyder and co-workers did not attempt to induce leukemia by benzene treatment in AKR mice which harbor endogenous leukemia viruses or in any other strains of mice susceptible to leukemia, Amiel (*Rev. Fr. Etud. Clin. Biol.* **5**, 198, 1960) reported that leukemia was not observed more frequently in benzene-treated AKR mice than in controls.

There is no evidence to suggest that the increase in radioactivity in the exhaled air after administration of toluene with benzene is the result of an alteration in the pathway of benzene metabolism. The increase in expired radioactivity appears to result from a longer plasma half-life for benzene because less benzene is metabolized, with the result that more benzene remains available for expiration.

Rentsch questioned whether red cell survival time was studied after benzene treatment. Although not studied by Snyder and co-workers, it has been previously reported (Erf and Rhoads, *J. Ind. Hyg.* **21**, 421, 1939; Askoy *et al., Blut* **13**, 85, 1966) that advanced benzene toxicity results in increased red cell hemolysis.

When Jerina was asked to comment on the possible toxicity of benzene epoxide, he replied that not only was the arene oxide a possible toxic intermediate but also the dihydrodiol should not be overlooked as the causative agent in benzene toxicity. Snyder commented that Remmer had suggested that semiquinones might act as reactive intermediates, and since catechol and hydroquinone have been reported to be metabolites of benzene and can exist in the semiquinone form they should be considered along with the arene oxides in future studies of potential active intermediates in benzene toxicity.

34

Hepatotoxicity of Carbon Disulfide and of Other Sulfur-Containing Chemicals: Possible Significance of Their Metabolism by Oxidative Desulfuration

Francesco De Matteis
Biochemical Mechanisms Section
MRC Toxicology Unit
Carshalton, Surrey, England

When carbon disulfide (CS_2) is administered to starved male rats, there is a rapid loss of cytochrome P450 from the liver microsomes and reduction in the activity of the drug-metabolizing enzymes of the liver (1, 2). These changes are observed in the absence of any histological lesion. However, when the rats are pretreated with phenobarbital in order to stimulate the activity of the liver drug-metabolizing enzymes, administration of CS_2 causes a greater loss of cytochrome P450, accumulation of water in the liver (1), and, histologically, marked hydropic degenerative changes in the centrilobular zones of the liver (3, 4). The potentiation of the liver toxicity of CS_2 by pretreatment with phenobarbitone has been interpreted to indicate that a metabolite of CS_2 rather than CS_2 itself may be the real toxic agent responsible for the liver changes considered above (1), a concept in whose favor a number of experimental findings have been obtained in the last 2–3 years.

The purpose of this chapter is to discuss in the light of these findings the possible significance of the metabolism of CS_2 for its toxicity to the liver and also to consider briefly the effects of other sulfur-containing chemicals which share with CS_2 some characteristic features of liver toxicity and may be acting by a similar mechanism.

OXIDATIVE DESULFURATION OF CS_2

Evidence has recently been obtained (5) that CS_2 is partially metabolized in the intact rat to CO_2 and that the extent of conversion to CO_2 is directly related, under a number of experimental conditions, to the amount of cytochrome P450 present in the liver at the time of CS_2 administration. In addition, a correlation has been found between the extent of conversion of CS_2 to CO_2 and the degree of toxic liver changes: phenobarbitone pretreatment, designed to enhance the severity of the liver injury, also significantly increases the oxidation of CS_2 to CO_2; both toxicity and oxidative metabolism are decreased by pretreatment with a small "ineffective" dose of CS_2, and both of them are greater in male than in female rats. CS_2 has also been found to cause loss of cytochrome P450 *in vitro* when microsomes are incubated aerobically in its presence, but can do this markedly only when NADPH (the cofactor essential for metabolism of foreign chemicals by the mixed-function oxidase system) is also present. All of these findings (5) are compatible with the hypothesis of a toxic metabolite of CS_2, even though they do not conclusively prove it. Some more direct evidence for a toxic derivative has been provided by studies carried out in the last 2 years; these studies, which also give some insight into the possible nature of the toxic metabolite, are discussed below.

The conversion of CS_2 to CO_2 can be considered as a two-stage desulfuration reaction (Fig. 1) analogous to other oxidative desulfurations involving conversion of either P=S to P=O or of C=S to C=O (6). A typical example is the conversion of parathion to paraoxon. This takes place mostly in the liver, is also stimulated by previous treatment with phenobarbitone (7), and is catalyzed by the microsomal mixed-function oxidase system (8–10). Ptashne *et al.* (11) have recently suggested a mechanism for the oxidative metabolism of parathion involving the production of free reactive sulfur in the atomic state. By analogy, it has been suggested that CS_2 could give rise to elemental sulfur in two stages through the monooxygenated intermediate COS (Fig. 1) and that reactive sulfur liberated at the two stages of the oxidative desulfuration may become bound to cellular components and initiate toxic changes in the liver (5)

The following recent findings provide support for this hypothesis concerning both the pathway of CS_2 oxidation and also the possible involvement of reactive sulfur in its liver toxicity:

1. When liver microsomes are incubated with either ^{14}C- or ^{35}S-labeled CS_2 in the presence of NADPH (under conditions where loss of cytochrome P450 is observed), labeled sulfur becomes covalently bound to the microsomes (12, 13). Although some increase of ^{14}C binding (over the values obtained on incubation without NADPH) is also observed, the binding of ^{35}S is greatly in excess of that of ^{14}C, indicating that sulfur itself becomes bound. An even greater excess of ^{35}S over ^{14}C radioactivity is found in the trichloroacetic acid

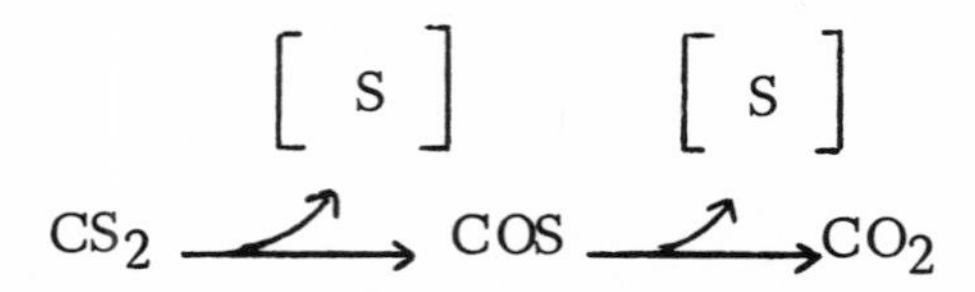

Fig. 1. Proposed pathway of CS_2 oxidative desulfuration. Both sulfur atoms released during the oxidative process are shown in brackets to indicate that they are highly reactive. After De Matteis and Seawright (5).

washings from microsomes incubated with NADPH, and some excess of ^{35}S radioactivity is also seen in the organic solvent washings (Table 1). In contrast, when fraction V from bovine plasma is substituted for the microsomes, no ^{35}S radioactivity in excess of ^{14}C radioactivity can be demonstrated in either the precipitate (12) or the washings thereof (Table 1). These observations clearly indicate that, on incubation with liver microsomes and NADPH, CS_2 is actively desulfurated and that a portion of the sulfur liberated becomes covalently bound to the microsomes. Since ^{35}S radioactivity is found in excess of ^{14}C radioactivity in all fractions, it can also be concluded that most of the corresponding carbon must have been lost from the incubation mixture, probably as the volatile oxidation products COS and CO_2 (see below). Also, since less than 20% of the total recovered "free" sulfur is found bound to the microsomal precipitate, it is clear that, at least under the conditions of our experiments, covalent binding of sulfur to the microsomal precipitate cannot be taken as a quantitative measure of the total desulfuration of CS_2.

The nature of the "free" sulfur in the trichloroacetic acid washings (whether covalently bound to acid-soluble organic molecules or present in an inorganic form) as well as the nature of target sites in the macromolecules of the microsomes remain to be elucidated.

2. Also in favor of the hypothesis outlined in Fig. 1 is the finding that when microsomes are incubated with CS_2 in the presence of NADPH, the production of COS, the postulated monooxygenated intermediate in CS_2 oxidation, can be demonstrated (13).

3. Finally, COS can itself cause loss of cytochrome P450 when incubated with liver microsomes in presence of NADPH. Under these conditions, some of its sulfur becomes covalently bound to the microsomes (14), while presumably the corresponding carbon is fully oxidized to CO_2. These observations are compatible with the two-stage desulfuration of CS_2 outlined in Fig. 1 and demonstrate that, as originally postulated, both atoms of sulfur present in CS_2 can be liberated in a reactive (and potentially toxic) form.

Table 1. Oxidative Desulfuration of CS_2 by Rat Liver Microsomes in the Presence of NADPH: Radioactivity Recovered in Either the Trichloroacetic Acid or the Organic Solvent Washings of the Microsomes after Incubation with Radioactive CS_2[a]

Source of protein	Addition		Fraction	Radioactivity recovered in fraction[b]		
				Total radioactivity recovered		
	CS_2	NADPH		^{35}S	^{14}C	^{35}S in excess of ^{14}C
Rat liver microsomes	+	–	Trichloroacetic acid washings	3.3 ± 1.9	nil	3.3
			Organic solvent washings	25.2 ± 4.6	27.7 ± 5.8	nil
	+	+	Trichloroacetic acid washings	74.8 ± 5.5	trace	74.8
			Organic solvent washings	39.5 ± 5.3	24.3 ± 4.9	15.2
Fraction V from bovine plasma	+	–	Trichloroacetic acid washings	nil	nil	nil
			Organic solvent washings	11.4	9.0	2.4
	+	+	Trichloroacetic acid washings	nil	nil	nil
			Organic solvent washings	12.1	12.8	nil

[a] Rat liver microsomes or fraction V from bovine plasma (both at a concentration of 2 mg protein/ml of incubation mixture) were incubated aerobically in stoppered screw-capped bottles for 15 min in an incubation mixture, the composition of which has been given (12), with 33.3 μmol of either $^{14}CS_2$ or $C^{35}S_2$. At the end of the incubation, trichloroacetic acid was added to a final concentration of 7%, and air was gently blown on the surface of the incubation mixture for 30 min to displace any volatile radioactivity. The precipitates were then extensively washed (until no more radioactivity could be removed from them), first with 7% trichloroacetic acid, then with organic solvents (chloroform–methanol, 2:1 by volume; methanol and ether), and the radioactivity present in the washings was determined by liquid scintillation counting. Details of the methods followed for counting are given in De Matteis (12), where the radioactivity recovered in the precipitates is also given.

[b] nmol CS_2 Eq/bottle.

LOSS OF CYTOCHROME P450 DURING OXIDATIVE DESULFURATION OF PARATHION AND OF OTHER CHEMICALS: LIVER TOXICITY OF PHOSPHOROTHIONATES

In previous work (7, 15) it had been reported that during the oxidative desulfuration of parathion, sulfur becomes covalently bound to the microsomes, but the possible toxicological significance of this finding had not been appreciated. Experiments have therefore been carried out recently to discover whether a loss of microsomal cytochrome P450 would also occur during the desulfuration of parathion and of other chemicals. Three pairs of drugs were studied, each pair consisting of a compound containing sulfur (as either P=S or C=S) and the corresponding oxygen analogue (12). All three sulfur-containing drugs (parathion, phenylthiourea, and 1-naphthylisothiocyanate) caused loss of cytochrome P450 in the presence of NADPH, whereas the oxygen-containing analogues (paraoxon, phenylurea, and 1-naphthylisocyanate) were all inactive. In addition, the loss of cytochrome P450 caused by parathion could be inhibited by piperonyl butoxide and by replacement of air with N_2, and stimulated by either NADH or, to a larger extent, NADPH (12). These same factors have been reported to inhibit and stimulate, respectively, the oxidative desulfuration of parathion to paraoxon *in vitro* and the covalent binding of sulfur to microsomes which accompanies it (15, 16).

The liver effects of an analogue of parathion such as *O,O*-diethyl, *O*-phenyl phosphorothionate (SV_1), which possesses a low general toxicity and can therefore be administered to rats in large doses, have recently been studied *in vivo* (17). When SV_1 is administered to rats that have previously been dosed with phenobarbitone, it, like CS_2, causes loss of cytochrome P450 and centrilobular changes in the liver. Phenobarbitone pretreatment is essential for the histological liver lesion to appear and the lesion is characterized by marked hydropic changes. Necrotic cells are uncommon and fat infiltration is only modest. In all these respects, the SV_1-induced liver lesion is similar to that caused by CS_2 and different from that produced by CCl_4.

It can therefore be concluded that a similar lesion characterized by loss of cytochrome P450 and centrilobular hydropic degeneration is produced in rat liver by either CS_2 (a compound containing C=S) or by SV_1, a phosphorothionate (which contains P=S). Reactive sulfur liberated during the oxidative desulfuration of CS_2 and of a phosphorothionate becomes bound to cellular components and may thereby initiate toxic changes in the liver. Oxidative desulfuration had already been suspected to be implicated in the toxic effects of phenylthiourea in the rabbit (18).

REFERENCES

1. E. J. Bond and F. De Matteis, Biochemical changes in rat liver after administration of carbon disulphide, with particular reference to microsomal changes, *Biochem. Pharmacol.* **18,** 2531–2549 (1969).
2. K. J. Freundt and W. Dreher, Inhibition of drug metabolism by small concentrations of carbon disulphide, *Naunyn Schmiedeberg's Arch. Pharmacol. Exp. Pathol.* **263,** 208–209 (1969).
3. E. J. Bond, W. H. Butler, F. De Matteis, and J. M. Barnes, Effects of carbon disulphide on the liver of rats, *Br. J. Ind. Med.* **26,** 335–337 (1969).
4. L. Magos and W. H. Butler, Effect of phenobarbitone and starvation on hepatotoxicity in rats exposed to carbon disulphide vapours, *Br. J. Ind. Med.* **29,** 95–98 (1972).
5. F. De Matteis and A. A. Seawright, Oxidative metabolism of carbon disulphide by the rat; effect of treatments which modify the liver toxicity of carbon disulphide, *Chem.-Biol. Interactions* **7,** 375–388 (1973).
6. R. D. O'Brien, Desulfuration, in: *Proceedings of the 1st International Pharmacology Meeting,* Vol. 6 (B. B. Brodie and E. G. Erdös, eds.), pp. 111–119, Pergamon Press, New York (1961).
7. R. E. Poore and R. A. Neal, Evidence for extrahepatic metabolism of parathion, *Toxicol. Appl. Pharmacol.* **23,** 759–768 (1972).
8. A. N. Davison, The conversion of Sehradan (OMPA) and parathion into inhibitors of cholinesterase by mammalian liver, *Biochem. J.* **61,** 203–209 (1955).
9. R. D. O'Brien, Activation of thionophosphates by liver microsomes, *Nature (London)* **183,** 121–122 (1959).
10. R. A. Neal, Studies on the metabolism of diethyl-4-nitrophenyl phosphorothionate (parathion) *in vitro, Biochem. J.* **103,** 183–191 (1967).
11. K. A. Ptashne, R. M. Wolcott, and R. A. Neal, Oxygen-18 studies on the chemical mechanisms of the mixed function oxidase catalysed desulfuration and dearylation reactions of parathion, *J. Pharmacol. Exp. Ther.* **179,** 380–385 (1971).
12. F. De Matteis, Covalent binding of sulfur to microsomes and loss of cytochrome P-450 during the oxidative desulfuration of several chemicals, *Mol. Pharmacol.* **10,** 849–854 (1974).
13. R. R. Dalvi, R. E. Poore, and R. A. Neal, Studies of the metabolism of carbon disulphide by rat liver microsomes, *Life Sci.* **14,** 1785–1796 (1974).
14. R. R. Dalvi, A. L. Hunter, and R. A. Neal, Toxicological implications of the mixed-function oxidase catalysed metabolism of carbon disulphide, *Chem.-Biol. Interactions* **10,** 347–361 (1975).
15. T. Nakatsugawa and P. A. Dahm, Microsomal metabolism of parathion, *Biochem. Pharmacol.* **16,** 25–38 (1967).
16. T. Nakatsugawa, N. M. Tolman, and P. A. Dahm, Degradation and activation of parathion analogs by microsomal enzymes, *Biochem. Pharmacol.* **17,** 1517–1528 (1968).
17. A. A. Seawright, J. Hrdlicka, and F. De Matteis, *Brit. J. Exp. Path.* **57,** 16–22 (1976).
18. R. T. Williams, *Detoxication Mechanisms,* 2nd ed., p. 181, Chapman and Hall, London (1959).

35

Studies of the Formation of Reactive Intermediates of Parathion

Robert A. Neal, Tetsuya Kamataki, Marie Lin, Kay A. Ptashne, Ramesh R. Dalvi, and Raymond E. Poore
Center in Toxicology
Department of Biochemistry
Vanderbilt University School of Medicine
Nashville, Tennessee 37232, U.S.A.

Parathion is one of a class of phosphorothionate triesters widely used as insecticides. These compounds exert their toxic effects in insects and mammals by inhibiting the enzyme acetylcholinesterase. The phosphorothionates, in general, are relatively poor inhibitors of acetylcholinesterase but are converted by the cytochrome P450-containing mixed-function oxidase enzyme systems in insects and mammals to the corresponding phosphate triesters which are potent inhibitors of this enzyme.

In this chapter, we will discuss what is known currently about the mixed-function oxidase catalyzed metabolism of parathion. In addition, evidence is presented indicating that reactive intermediates formed in the mixed-function oxidase catalyzed metabolism of parathion *in vitro* covalently bind to the endoplasmic reticulum and more particularly to cytochrome P450, resulting in an inhibition of the reactions catalyzed by this enzyme.

METHODS

Cytochrome P450 from the livers of phenobarbital pretreated rabbits and rats was purified to apparent homogeneity using essentially the method of Imai and Sato (1). The enzyme NADPH-cytochrome *c* reductase from the livers of phenobarbital-pretreated rabbits and rats was purified by a method developed in this laboratory (2). The metabolites of parathion present in the *in vitro* incubations were isolated and quantitated as described previously (3). The binding of ^{35}S from [^{35}S]parathion, ^{14}C from [ethyl-^{14}C]parathion, and ^{32}P from [^{32}P]parathion was determined as described previously (3). Benzphetamine metabolism was determined by the method of Cochin and Axelrod (4). Cytochrome P450 was determined by the method of Omura and Sato (5). Thiocyanate was determined by the method of Sörbo (6).

RESULTS AND DISCUSSION

What is known concerning the metabolism of parathion by the rat and other experimental animals is shown in Fig. 1. There are five major products of parathion metabolism. One of these is the corresponding phosphate triester, paraoxon, which is formed in a mixed-function oxidase catalyzed reaction in which the sulfur atom of parathion is replaced by an oxygen atom (7). The other product of this reaction is atomic sulfur (3). The water formed in this reaction is common to all mixed-function oxidase catalyzed reactions. The paraoxon is subject to hydrolysis by esterases present in various tissues to form diethyl phosphate and *p*-nitrophenol (8). Parathion is not a substrate for these esterases, presumably because the phosphorus is not as electrophilic as that in paraoxon. The difference in the electrophilicity of the phosphorus in these two compounds is most likely due to the greater electronegativity of oxygen as compared to

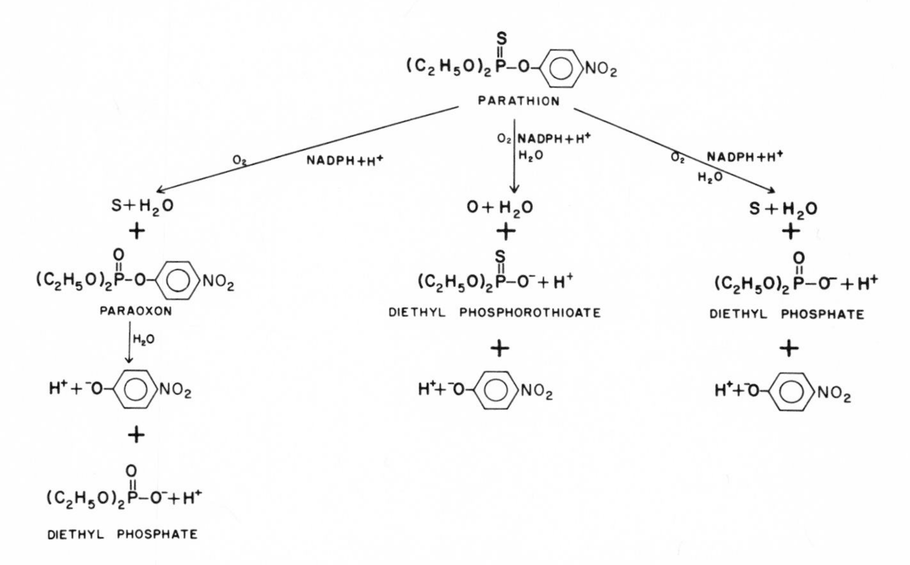

Fig. 1. Metabolic scheme for the metabolism of parathion by mammalian hepatic microsomes. From T. Kamataki, M. C. M. L. Lin, D. H. Belcher, and R. A. Neal, *Drug Metab. Dispos.* **4**, 180–189 (1976). Copyright by American Society for Pharmacology and Experimental Therapeutics.

sulfur. The lesser electrophilicity of the phosphorus in parathion is most probably the reason for its being a weaker inhibitor of acetylcholinesterase than paraoxon.

Parathion is also metabolized to diethyl phosphorothioic acid and *p*-nitrophenol in a reaction requiring a cytochrome P450-containing mixed-function oxidase enzyme system (9, 10). Studies with $H_2\,^{18}O$ have indicated that water in addition to molecular oxygen and NADPH is required in this reaction (11). The compound diethyl phosphate plus *p*-nitrophenol can also be formed from parathion in a mixed-function oxidase catalyzed reaction (12).

For the purpose of this chapter, we will concentrate our attention on the chemical mechanism of the mixed-function oxidase catalyzed metabolism of parathion to paraoxon.

In our studies of the mechanism of this reaction, we were concerned with two principal questions. These were, first, what was the site of attack of the mixed-function oxidase activated oxygen on parathion, and second, was the oxygen which reacts with parathion retained in the product paraoxon? The answer to the first question was approached in a number of ways. However, the finding which provided the most informative answer was that paraoxon was not a substrate for the mixed-function oxidase enzyme system. This and other data indicated that the site of attack of the oxygen on parathion was the sulfur atom. This conclusion was supported by experiments using the chemical model system for the mixed-function oxidase enzyme system, peroxytrifluoroacetic acid (13, 14). Parathion reacts exothermically with the peracid to form paraoxon and diethyl phosphorothioic acid (13). However, paraoxon did not react even on prolonged incubation. The question concerning whether the attacking oxygen is retained in the product paraoxon was answered by incubating parathion with rabbit liver microsomes and NADPH in an incubation medium enriched with $H_2\,^{18}O$ and in a separate experiment in an atmosphere enriched with $^{18}O_2$ (11). The paraoxon formed in these reactions was isolated and examined for an increased content of ^{18}O using gas chromatography–mass spectrometry. The results of these experiments are shown in Table 1. As can be seen, when the reaction was carried out in a medium enriched with $H_2\,^{18}O$ to the extent of 20

Table 1. ^{18}O Enrichment in Paraoxon Following Incubation of Parathion with Rabbit Liver Microsomes[a]

Expt.	Source of ^{18}O (atoms % excess)	Observed atom % excess ^{18}O in paraoxon
1	$H_2\,^{18}O$ (20)	0.0
2	$^{18}O_2$ (46)	42.0

[a]The details of this experiment are described in Ptashne *et al.* (11).

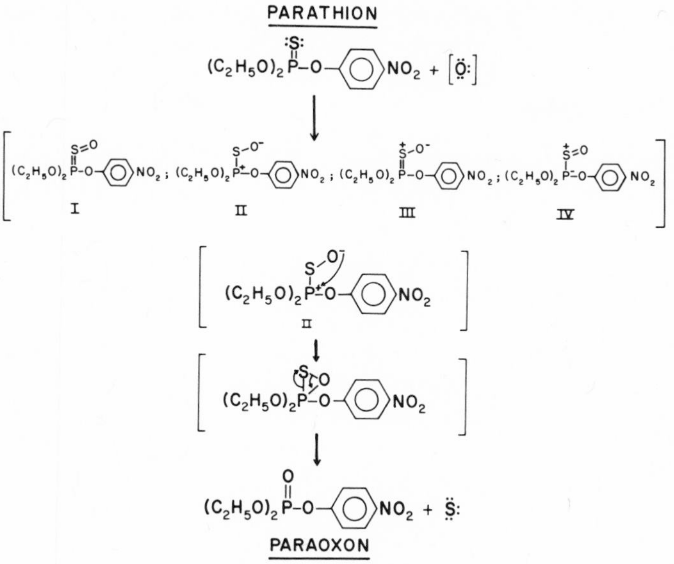

PARATHION

$(C_2H_5O)_2P(=S)-O-C_6H_4-NO_2 + [\ddot{O}:]$

$\downarrow$

$\left[(C_2H_5O)_2P(=S{=}O)-O-C_6H_4-NO_2 \; (I);\; (C_2H_5O)_2P^{+}(-S-O^{-})-O-C_6H_4-NO_2 \; (II);\; (C_2H_5O)_2P(=S^{+}-O^{-})-O-C_6H_4-NO_2 \; (III);\; (C_2H_5O)_2P^{-}(-S^{+}{=}O)-O-C_6H_4-NO_2 \; (IV) \right]$

$\left[(C_2H_5O)_2P^{+}(-S-O^{-})-O-C_6H_4-NO_2 \right]$ II

$\downarrow$

$\left[(C_2H_5O)_2P(SO)-O-C_6H_4-NO_2 \right]$

$\downarrow$

$(C_2H_5O)_2P(=O)-O-C_6H_4-NO_2 + \ddot{S}:$

PARAOXON

Fig. 2. Chemical mechanism for the metabolism of parathion to paraoxon by the mammalian hepatic mixed-function oxidase enzyme system.

atom-percent excess, no ^{18}O was found to be present in the paraoxon formed during the incubation. However, when the reaction was carried out in an atmosphere enriched with $^{18}O_2$, the resultant paraoxon contained ^{18}O to about the same atom-percent excess as the atmosphere in which the incubation took place. These data indicated that the oxygen atom attached to the phosphorus of paraoxon was that which was donated to the sulfur atom of parathion by the mixed-function oxidase enzyme system.

From these data, a chemical mechanism for the formation of paraoxon from parathion was proposed. This mechanism is shown in Fig. 2. It is postulated that a singlet oxygen atom generated in the mixed-function oxidase reaction is donated to the sulfur atom of parathion to yield a compound analogous to the sulfine formed in the reaction of peracids with thioketones (15, 16). The attacking singlet oxygen atom is shown in brackets by way of suggesting that it is more likely transferred from cytochrome P450 to the phosphorothionate sulfur in a concerted reaction. As shown in Fig. 2, there are four different structures that may contribute to the resonance stabilization of the resultant "sulfine." It is proposed that one of these resonance forms (perhaps form II) reacts internally to form a cyclic phosphorus–sulfur–oxygen intermediate analogous to the oxthirane which has been proposed by numerous investigators to be an intermediate in the reaction pathways of various sulfines (17). This resultant phosphooxythirane then undergoes a cyclic electron shift with the loss of sulfur, forming paraoxon. It is proposed that the sulfur atom that is released is in its singlet form. Although there is no evidence to support this hypothesis, it is probable that if the attacking oxygen atom is in its singlet state the departing sulfur atom may also be in its singlet state.

If the sulfur atom is released in its singlet state, it would be a highly reactive electrophil which would bind readily to nucleophils near the site of its release. The thiono-sulfur group of parathion has been found to be covalently bound to tissue macromolecules following administration of [^{35}S] parathion *in vivo* (18) and on incubation with hepatic microsomes *in vitro* (3). Table 2 shows the results of an experiment in which doubly labeled parathion (^{32}P, ^{35}S) was incubated with microsomes isolated from the livers of phenobarbital-pretreated rats. In the absence of NADPH, only a trace of radioactivity could be found bound to the microsomes. However, as shown in Table 2, when the incubation was carried out in the presence of NADPH, a substantial amount of sulfur became covalently bound to the microsomes. A small but significant amount of the phosphorus-containing portion of the parathion molecule was also covalently bound to the microsomes. The results of this experiment clearly indicate that the majority of the sulfur bound to the microsomes is free of the phosphorus-containing portion of the molecule and thus must be atomic sulfur released in the metabolism of parathion to paraoxon. This is further substantiated by the finding that the amount of sulfur bound under these conditions is equivalent to the amount of paraoxon formed in the incubation (3). Also shown in Table 2 are the results of an experiment in which an amount of paraoxon, which was approximately 5 times the amount that would be expected to be formed in the incubation containing doubly labeled parathion, was incubated with an equal aliquot of the same preparation of microsomes. As can be seen, a much smaller amount of ^{32}P was bound using [^{32}P] paraoxon that was bound using the doubly labeled parathion. Thus the binding of ^{32}P in a reaction of paraoxon with nucleophilic sites on the endoplasmic reticulum is responsible for only a small portion of the total binding of ^{32}P using doubly labeled parathion. It

Table 2. ^{35}S and ^{32}P Bound to Microsomes Following Incubation with Parathion and Its Metabolites[a]

Substrate	nmol ^{35}S bound/mg protein/15 min	nmol ^{32}P bound/mg protein/15 min
[^{35}S,^{32}P] Parathion (2.5×10^{-4} M)	12.93 ± 0.37	1.12 ± 0.21
[^{32}P] Paraoxon (2.5×10^{-5} M)	–	0.25 ± 0.03
[^{32}P] Diethylphosphate (4×10^{-5} M)	–	None
[^{35}S,^{32}P] Diethyl phosphorothioic acid (4×10^{-5} M)	None	None

[a]The details of this experiment are described in Poor and Neal (18). This experiment was carried out using hepatic microsomes from phenobarbital-pretreated rats.

appears that the greater portion of the ^{32}P binding in the incubation using doubly labeled parathion is the result of the reaction of one or more of the intermediate "sulfines" shown in Fig. 2 with nucleophils on the endoplasmic reticulum. In examining the remainder of the data in Table 2, it can be seen that the incubation of the other phosphorus-containing metabolites of parathion with microsomes in the presence of NADPH does not lead to the binding of any radioactivity.

The binding of sulfur and/or an activated intermediate of the phosphorus-containing portion of the parathion molecule to the endoplasmic reticulum leads to a decrease in the amount of cytochrome P450 detectable as its carbon monoxide complex and to a decrease in the rate of metabolism of substrates such as benzphetamine (3). Paraoxon, or any other of the isolatable metabolites of parathion, does not decrease the amount of cytochrome P450 or inhibit the ability of microsomes to metabolize substrates such as benzphetamine (3).

We have also examined the metabolism of parathion by reconstituted mixed-function oxidase enzyme systems isolated from the livers of phenobarbital-pretreated rabbits and rats. Cytochrome P450 was purified to apparent homogeneity from both species using essentially the method of Imai and Sato (1). The NADPH-cytochrome *c* reductase enzymes from both species that were used in these experiments were approximately 60% pure. They were, however, free of cytochrome P450, cytochrome b_5, and cytochrome b_5 reductase. Shown in Table 3 are the results of an experiment examining the metabolism of parathion by a reconstituted system from rabbit liver. The cytochrome P450 used in this experiment had a specific activity of 18.5 nmol/mg protein. As can be seen, there is a requirement for both cytochrome P450 and the reductase. The activity

Table 3. Parathion Metabolism by a Reconstituted Mixed-Function Oxidase System from Rabbit Liver[a,b]

	Product formation (nmol/nmol P450/5 min)		
Conditions	Paraoxon	Diethyl phosphorothioic acid	Diethyl phosphate
Complete	4.510 ± 0.424	2.420 ± 0.156	0.470 ± 0.107
Minus P450	0.048 ± 0.002	0.040 ± 0.002	0.016 ± 0.002
Minus reductase	0.018 ± 0.001	0.070 ± 0.003	0
Minus lipid	1.790 ± 0.021	0.816 ± 0.002	0.206 ± 0.049
Minus deoxycholate	3.860 ± 0.679	1.990 ± 0.347	0.381 ± 0.100
Minus both lipid and deoxycholate	1.690 ± 0.120	0.756 ± 0.008	0.245 ± 0.005

[a] From T. Kamataki, M. C. M. L. Lin, D. H. Belcher, and R. A. Neal, *Drug Metab. Dispos.* **4**, 180–189 (1976). Copyright by American Society for Pharmacology and Experimental Therapeutics.
[b] The incubations were carried out for a period of 5 min. The composition of the reaction mixture is described in Fig. 3.

was also stimulated by addition of dilauroyl phosphatidylcholine. Deoxycholate also had a slight stimulating effect. Another important aspect of the data shown in Table 3 is that all three of the major phosphorus-containing metabolites of parathion are formed by what appears to be a single species of cytochrome P450. These data and those from the studies using a peroxy acid model system (13) suggest that the mixed-function oxidase enzyme system is only involved in the addition of an oxygen atom to the phosphorothionate sulfur (Fig. 2) and that the various products are formed nonenzymatically from a common intermediate. It is believed that this intermediate is one or more of the resonance forms of the intermediate "sulfine" shown in Fig. 2.

The linearity of the formation of these various parathion metabolites and the linearity of the metabolism of benzphetamine by the reconstituted mixed-function oxidase enzyme system from rabbit liver are shown in Fig. 3. In these experiments, a reductase-to-P450 ratio of 3 units to 1 nmol was used. Using this ratio, cytochrome P450 is the rate-limiting enzyme. Using this enzyme system, the rate of formation of the various parathion metabolites departs rapidly from linearity. In contrast, the rate of metabolism of benzphetamine (formaldehyde) remained linear throughout the incubation period. As with the data using intact microsomes (3), these data suggest that the binding of some reactive product of parathion to one or both of the enzymes of this reconstituted system is responsible for the departure of the rate of metabolism of parathion from linearity. The inhibition was not due to inactivation of NADPH since an excess of this cofactor did not improve linearity. It is also not due to product inhibition by one of the isolatable parathion metabolites since, as noted previously (3, 19), none of these products has an inhibitory effect on parathion metabolism. The addition of glutathione (GSH) to these incubations did not improve the linearity of the reaction at the 3:1 reductase-to-P450 ratio. However, GSH and dithiothretol did improve the linearity of the formation of

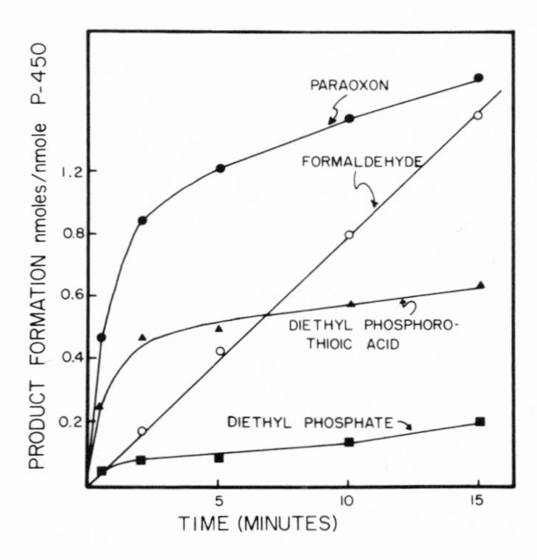

Fig. 3. Linearity of the metabolism of parathion and benzphetamine by a reconstituted mixed-function oxidase enzyme system from rabbit liver. The 0.5 ml reaction mixture contained 50 μg of sodium deoxycholate, 15 μg of dilauroyl L-3-phosphatidylcholine, 1.5 units of NADPH-cytochrome *c* reductase, 0.5 nmol of cytochrome P450, 0.05 M Hepes buffer (pH 7.8), 0.015 M $MgCl_2$, 0.1 mM EDTA, and 5×10^{-5} [ethyl-^{14}C]parathion or 1×10^{-3} M benzphetamine.

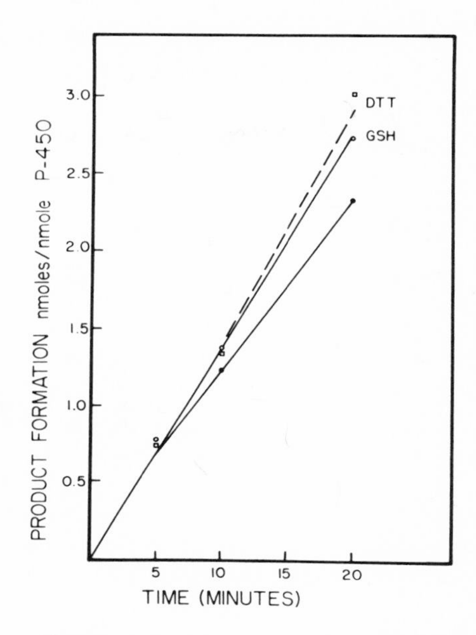

Fig. 4. Linearity of paraoxon formation in the presence and absence of glutathione and dithiothreitol using a reconstituted mixed-function oxidase enzyme system from rabbit liver. The 0.5 ml reaction mixture is described in Fig. 3 except that 0.05 unit of NADPH-cytochrome *c* reductase and 0.5 nmol of cytochrome P450 were used in these experiments.

paraoxon when an excess of cytochrome P450 was present, in other words, under conditions where the reductase was limiting. The linearity of paraoxon formation at a reductase-to-P450 ratio of 0.1 unit to 1 nmol, in the presence and absence of GSH and dithiothreitol, is shown in Fig. 4. At this reductase-to-P450 ratio, the rate of paraoxon formation departs from linearity to a lesser degree than that at a reductase-to-P450 ratio of 3:1 (Fig. 3). In the presence of GSH and dithiothreitol, the rate of formation of paraoxon is essentially linear for the time period examined. These data suggest that at this reductase-to-P450 ratio (reductase limiting) and in the absence of GSH or dithiothreitol the inactivation of cytochrome P450 as a result of the binding of some reactive products of parathion is still occurring but that sufficient active enzyme is available to maintain linearity for a longer period of time. It appears that in the presence of GSH and dithiothreitol the rate of inactivation of cytochrome P450 is decreased and the rate of formation of paraoxon is linear for a longer period of time. Table 4 shows the results of an experiment in which the effect of GSH on the decrease in cytochrome P450 seen on incubation of the rabbit liver reconstituted system with parathion is compared with the decrease seen in the absence of GSH. In addition, the effect of GSH on the binding of ^{14}C from [ethyl-^{14}C]parathion and of ^{35}S from [^{35}S]parathion and on the rate of paraoxon formation is also compared to the control. GSH partially prevents the decrease in cytochrome P450. Whereas approximately 35% of the P450 was no longer detectable in the control incubation after a 20-min incubation, only 15% was lost in the incubation containing GSH. A decrease in the amount of ^{14}C and ^{35}S binding was also seen in the presence of GSH as compared to control and the rate of formation of

Table 4. Examination of ^{14}C and ^{35}S Binding to Reconstituted System Proteins Following Incubation with ^{35}S or ^{14}C Parathion[a]

Incubation conditions	Time (min)	Decrease in P450 (nmol)	^{14}C binding (nmol/ nmol P450)	^{35}S binding (nmol/ nmol P450)	Paraoxon formation (nmol/nmol P450)
Complete	10	0.062	0.024	0.163	1.24
	20	0.178	0.027	0.403	2.32
Complete + GSH	10	0.012	0.003	0.091	1.38
	20	0.074	0.013	0.118	2.90

[a]The experimental conditions are described in Fig. 4.

paraoxon was increased. These data suggest that in the presence of GSH less of a sulfur- and/or [^{14}C]-containing metabolite or metabolites of parathion is bound to the proteins of the reconstituted system, resulting in an increase in the amount of active cytochrome P450.

We next examined whether the sulfur which was bound to the proteins of a reconstituted system from the liver of phenobarbital-treated rats was bound to both the reductase and cytochrome P450. In this experiment, the reconstituted system was incubated with [^{35}S]parathion. The reaction mixture was dialyzed and applied to a Sephadex G25 column to remove the last traces of unreacted parathion and its noncovalently bound metabolites. The protein fraction from the Sephadex column was reduced in volume and subjected to SDS-poly-

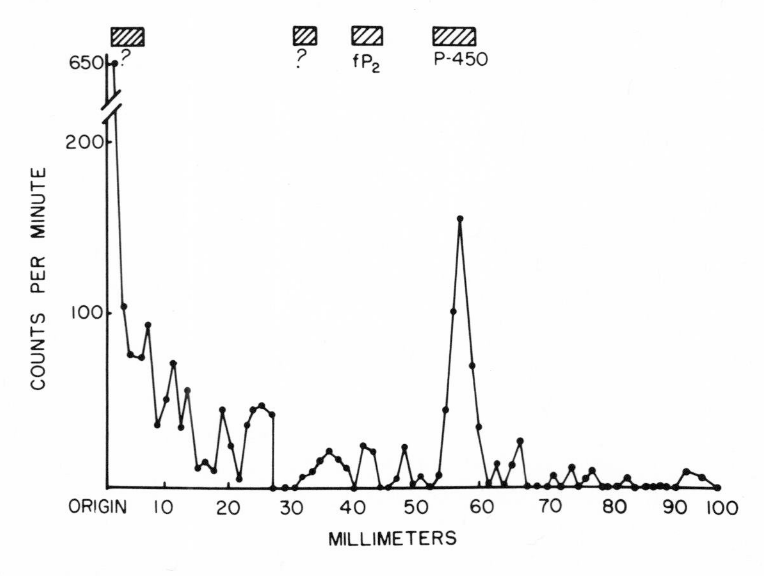

Fig. 5. SDS-polyacrylamide gel electrophoresis of a reconstituted mixed-function oxidase system from rat liver which had been labeled with ^{35}S by incubation with [^{35}S]parathion. The incubation procedures were essentially as described in Fig. 3.

acrylamide gel electrophoresis in the absence of either dithiothreitol or mercaptoethanol. The results are shown in Fig. 5. There was considerable protein and radioactivity at the origin. This material at the origin may represent an aggregate of reductase and P450 molecules since pretreatment of the concentrated eluant from the Sephadex column with 100 mM CN^- for 3 h at room temperature prior to gel electrophoresis prevented the accumulation of both protein and radioactivity at the origin. Little radioactivity was associated with the position of reductase (fP_2) on the gel, whereas a significant amount was found in the area of the gel corresponding to P450. This was the case in both the gel of the sample which had been incubated with CN^- prior to gel electrophoresis and that which had not been incubated with CN^- (Fig. 5). These data indicate the sulfur is bound predominantly or exclusively to cytochrome P450.

Previous work in our laboratory had shown that approximately 50% of the sulfur bound to rat liver microsomes incubated with [^{35}S]CS_2 was in the form of a hydrodisulfide (20). This form of bound sulfur could be released on treatment of the microsomes with CN^-. The chemical form of the sulfur released was thiocyanate (SCN^-). We therefore examined the ability of CN^- to release the sulfur bound to the proteins of the reconstituted system (cytochrome P450) as SCN^-. The results are shown in Figs. 6 and 7. In these experiments, a rat liver reconstituted system was incubated with [^{35}S]parathion, dialyzed, and applied

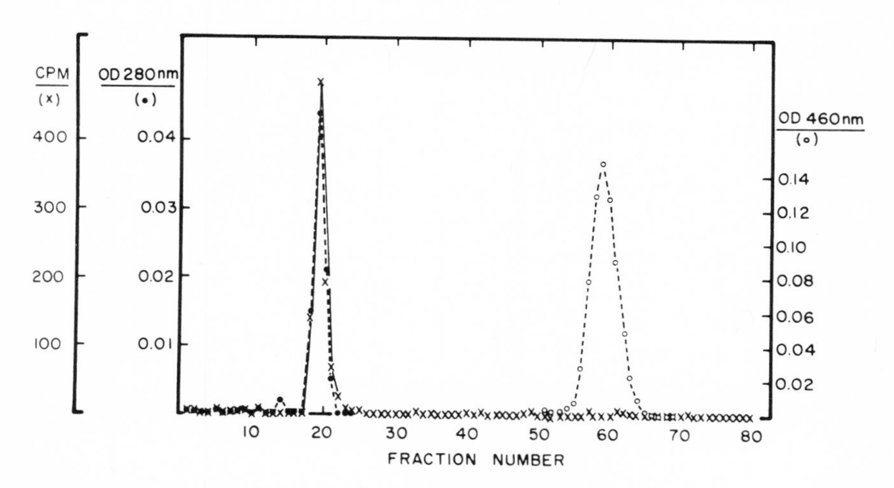

Fig. 6. Elution profile of protein, radioactivity, and thiocyanate from a Sephadex G25 column of a reconstituted mixed-function oxidase enzyme system from rat liver which had been incubated with [^{35}S]parathion. The 5 ml incubation mixture contained 20 nmol cytochrome P450 (specific activity 16.4 nmol/mg protein), 5 units NADPH-cytochrome *c* reductase, 600 μg dilauroyl L-3-phosphatidylcholine, 600 μg sodium deoxycholate, and 1×10^{-4} M [^{35}S]parathion. The remainder of the incubation mixture is described in Fig. 3. The incubation time was 5 min. One-milliliter fractions were collected. The radioactivity (X) represents cpm/0.1 ml. The OD_{280} (•) was measured on each 1-ml fraction. Thiocyanate (○) was detected as described previously (6).

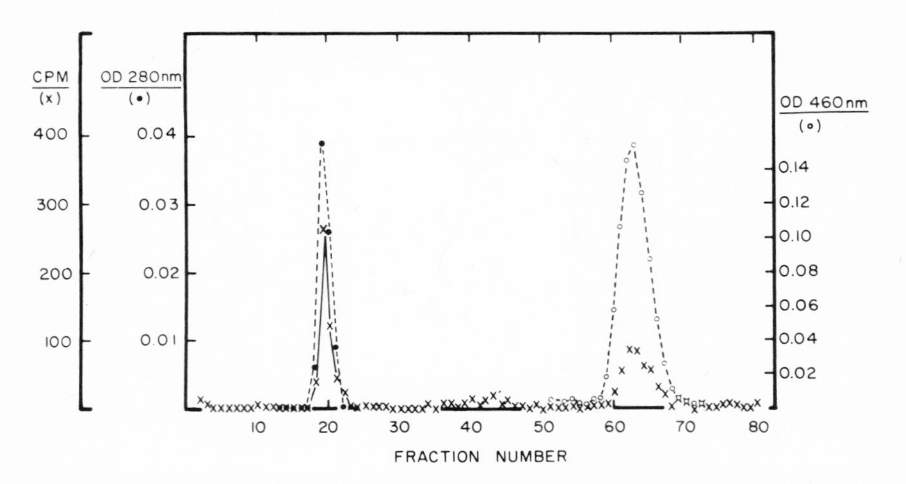

Fig. 7. Elution profile of protein, radioactivity, and thiocyanate from a Sephadex G25 column of a reconstituted mixed-function oxidase enzyme system from rat liver which had been incubated with [^{35}S]parathion followed by incubation with cyanide. The incubation conditions and analytical procedures were as described in Fig. 6 except that the labeled protein was incubated with 10 mM sodium cyanide for 3 h at room temperature prior to application to the Sephadex column.

to a Sephadex G25 column to remove the last traces of unmetabolized parathion and its noncovalently bound metabolites. The protein fraction from the Sephadex column was reduced in volume and divided into two portions. One portion was incubated with 10 mM CN^- at room temperature for 3 h and the other fraction was incubated for 3 h in the absence of CN^-. Figure 6 shows the elution profile of protein, radioactivity, and unlabeled SCN^- from a Sephadex G25 column of the sample not incubated with CN^-. The unlabeled SCN^- was added just before application to the column. The only peak of radioactivity was that associated with the protein peak. Figure 7 shows the elution profile of the sample incubated with CN^-. As can be seen, there was a decrease in the radioactivity associated with the protein and the appearance of a peak which exactly coincided with the colorimetric peak of exogenously added SCN^-. Parathion elutes from this Sephadex column in fractions 78–105 and the other sulfur-containing metabolite of parathion, diethyl phosphorothioic acid, elutes in fractions 35–50. The similarity of these data and those obtained with microsomes labeled with ^{35}S from [^{35}S]-CS_2, in which the form of the radioactivity was unequivocally identified as SCN^-, indicates that a portion of the sulfur bound to proteins of the reconstituted system and more likely to cytochrome P450 is present as a hydrodisulfide. Figure 8 shows the way in which it is proposed that the hydrodisulfide linkage is formed. A cysteine side chain of the cytochrome P450 molecule (designated R) reacts with atomic sulfur to form a hydrodisulfide. On treatment with CN^- there is a preferential attack of this nucleophile on the sulfhydryl sulfur, releasing it as SCN^-. Organic hydrodisulfides

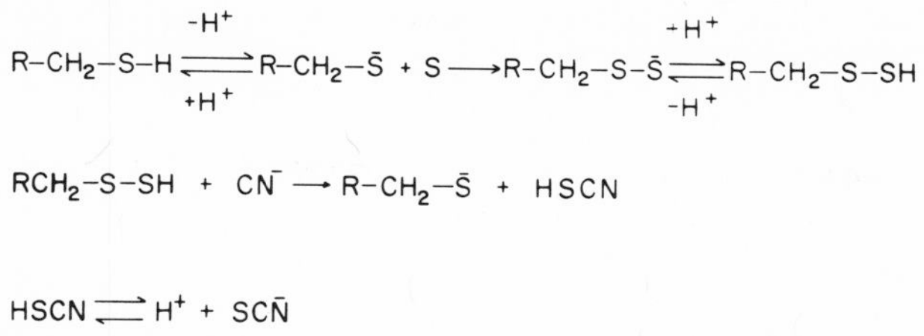

$$R{-}CH_2{-}S{-}H \underset{+H^+}{\overset{-H^+}{\rightleftharpoons}} R{-}CH_2{-}\bar{S} + S \longrightarrow R{-}CH_2{-}S{-}\bar{S} \underset{-H^+}{\overset{+H^+}{\rightleftharpoons}} R{-}CH_2{-}S{-}SH$$

$$RCH_2{-}S{-}SH + CN^- \longrightarrow R{-}CH_2{-}\bar{S} + HSCN$$

$$HSCN \rightleftharpoons H^+ + SC\bar{N}$$

Fig. 8. Scheme for formation of the hydrodisulfide linkage in a reaction for atomic sulfur with the side chain of a cysteine in the cytochrome P450 molecule. Reaction of cyanide ions with the hydrodisulfide.

have been synthesized and are relatively stable (21). When benzyl hydrodisulfide is reacted with CN^-, a significant portion of the terminal sulfur is released as SCN^- (21).

Whether the formation of the hydrodisulfide is the reason for all or a portion of the inactivation of cytochrome P450 is yet to be determined. In addition, the nature of the bound sulfur not released by CN^- has not yet been determined. The singlet sulfur atom participates in carbon–hydrogen insertion reactions in a manner analogous to that of carbenes and nitrenes (22, 23). It is possible that a portion of the bound sulfur is the result of carbon–hydrogen insertion reactions. If this is the case, the electronic form of the sulfur released from parathion and perhaps other thiono-sulfur containing compounds would be singlet since the triplet state of atomic sulfur does not participate in carbon–hydrogen insertion reactions.

ACKNOWLEDGMENTS

Support by USPHS Grants ES 00267, ES 00075, and ES 00112 is gratefully acknowledged.

REFERENCES

1. Y. Imai and R. Sato, A gel-electrophoretically homogeneous preparation of cytochrome P-450 from liver microsomes of phenobarbital-treated rabbits, *Biochem. Biophys. Res. Commun.* **60**, 8–14 (1974).
2. T. Kamataki and R. A. Neal, unpublished.
3. B. J. Norman, R. E. Poore, and R. A. Neal, Studies of the binding of sulfur released in the mixed-function oxidase catalyzed metabolism of diethyl-*p*-nitrophenyl phosphorothionate (parathion) to diethyl *p*-nitrophenyl phosphate (paraoxon), *Biochem. Pharmacol.* **23**, 1733–1744 (1974).
4. J. Cochin and J. Axelrod, Biochemical and pharmacological changes in the rat following chronic administration of morphine, nalorphine and normorphine, *J. Pharmacol. Exp. Ther.* **125**, 105–110 (1959).

5. T. Omura and R. Sato, The carbon monoxide-binding pigment of liver microsomes. I. Evidence for its hemoprotein nature, *J. Biol. Chem.* **239**, 2370–2378 (1964).
6. B. Sörbo, Enzymic transfer of sulfur from mercaptopyruvate to sulfite or sulfinates, *Biochim. Biophys. Acta* **24**, 324–329 (1957).
7. J. C. Gage, A cholinesterase inhibitor derived from *O,O*-diethyl *O-p*-nitrophenyl thiophosphate *in vivo, Biochem. J.* **54**, 426–430 (1953).
8. W. N. Aldridge, Serum esterases. 2. An enzyme hydrolyzing diethyl *p*-nitrophenyl phosphate (E600) and its identity with the A-esterase of mammalian sera, *Biochem. J.* **53**, 117–124 (1953).
9. R. A. Neal, Studies of the metabolism of diethyl 4-nitrophenyl phosphorothionate (parathion) *in vitro, Biochem. J.* **103**, 183–191 (1964).
10. T. Naksatsugawa and P. Dohm, Microsomal metabolism of parathion, *Biochem. Pharmacol.* **16**, 25–38 (1967).
11. K. A. Ptashne, R. M. Wolcott, and R. A. Neal, Oxygen-18 studies on the chemical mechanisms of the mixed-function oxidase catalyzed desulfuration of dearylation reactions or parathion, *J. Pharmacol. Exp. Ther.* **179**, 380–385 (1971).
12. M. Lin, T. Kamataki, and R. A. Neal, unpublished.
13. K. A. Ptashne and R. A. Neal, Reaction of parathion and malathion with peroxytrifluoroacetic acid, a model system for the mixed-function oxidases, *Biochemistry* **11**, 3224–3228 (1972).
14. A. W. Herriott, Peroxy acid oxidation of phosphinothioates a reversal of stereochemistry, *J. Am. Chem. Soc.* **93**, 3304–3350 (1971).
15. A. Battaglia, A. Dondoni, P. Giorgianni, G. Maccagnani, and G. Mazzanti, Sulfines. Part III. Kinetics of oxidation of thiobenzophenones with peroxybenzoic acid, *J. Chem. Soc.* 1547–1550 (1971).
16. J. Strating, L. Thijs, and B. Zwanenburg, Sulfines by oxidation of thioketones, *Tetrahedron Lett.* 65–67 (1966).
17. J. P. Snyder, Oxthiranes: Differential orbital correlation effects in the electrocyclic formation of sulfur-containing three-membered rings, *J. Am. Chem. Soc.* **96**, 5005–5007 (1974).
18. R. E. Poore and R. A. Neal, Evidence for extrahepatic metabolism of parathion, *Toxicol. Appl. Pharmacol.* **23**, 759–768 (1972).
19. R. A. Neal, Studies of the enzymic mechanism of the metabolism of diethyl-4-nitrophenyl (parathion) by rat liver microsomes, *Biochem. J.* **105**, 289–297 (1967).
20. G. L. Catignani and R. A. Neal, Evidence for the formation of a protein bound hydrodisulfide resulting from the microsomal mixed function oxidase catalyzed desulfuration of carbon disulfide, *Biochem. Biophys. Res. Commun.* **65**, 629–636 (1975).
21. S. Kawamura, Y. Otsuji, T. Nakabayashi, T. Kitao, and J. Tsurugi, Aralkyl hydrodisulfides. IV. The reaction of benzyl hydrodisulfide with several nucleophiles, *J. Org. Chem.* **30**, 2711–2714 (1965).
22. K. Gollnick and E. Leppin, Direct photolysis of carbonyl sulfide in solution. Mechanism of singlet D and triplet P sulfur atom formation, *J. Am. Chem. Soc.* **92**, 2217–2220 (1970).
23. E. Leppin and K. Gollnick, Direct photolysis of carbonyl sulfide in solution. Reactions of singlet D sulfur atoms in the liquid phase, *J. Am. Chem. Soc.* **92**, 2221–2227 (1970).

36

Discussion

De Matteis, when asked whether some of the sulfur which appeared to be covalently bound to microsomal protein might be merely dissolved in fat, replied that all noncovalently bound sulfur is extracted from the protein by extensive washing procedures. The fact that sulfur is then released as thiocyanate after treatment with cyanide indicates that it was bound as a sulfide. It is logical to assume that it existed as the hydrodisulfide.

De Matteis further explained that the time course of the effects of CS_2 on P450 differs from that of other toxic agents. In the first few hours of metabolism of CS_2, when there was a pronounced loss of P450 as measured by a decrease in CO binding, there was no corresponding loss of total heme. This contrasts to the effect of AIA treatment, which was followed by a loss of heme which stoichiometrically matched the P450 loss. It is possible that there is a steric effect which is responsible for the diminished P450–CO spectrum, and that there is not an actual loss of P450. He quoted a recent paper of Neal, in which it was suggested that either the reduction of the iron to Fe^{2+} or the accessibility of the iron for the CO has been impaired by the CS_2 treatment.

De Matteis also stated that the cyclic mechanism (see text of Neal's chapter) proposed for paraoxon formation from parathion is the mechanism most consistent with the retention of ^{18}O in carbon dioxide as well as the concept that the unshared pair of electrons of sulfur serves as the point of oxidative attack. Support for this mechanism includes the facts that oxidation of parathion and other thions by peracids results in the same products as those produced by the mixed-function oxidase, and that these peracid oxidations are cyclic.

In response to a question by Rentsch, De Matteis responded that the trichloroacetic acid washings of the microsomal protein contained the greatest amount of sulfur released from carbon. Any strong nucleophil such as reduced glutathione, dithiothreitol, mercaptoethanol, or cyanide can release the sulfur by displacement after the noncovalently bound portion has been removed.

Jerina suggested that it might not be necessary to postulate a role for free elemental sulfur in the parathion reactions but rather that a variety of sulfur–

oxygen–phosphorus complexes could be described which would provide a satisfactory explanation of the data. Neal responded that two reasons compel the postulation of a role for free sulfur: (1) elemental sulfur is a product of peracid oxidation of these compounds, and (2) free sulfur forms a hydrodisulfide linkage with microsomal protein. Jerina disagreed with the latter reason and suggested that sulfur–oxygen–phosphorus complexes could produce an attacking polysulfur species which could function as well as free sulfur. Neal countered with the argument that Jerina's postulated product would not form thiocyanate upon reaction with cyanide.

Ullrich suggested that a sulfine arising from CS_2 may bind to cytochrome P450 in a manner similar to that of the well-established cytochrome P450–carbene complex, but Neal thought that if such a complex were formed iron sulfide would be released under acid washing conditions. Ullrich went on to suggest that sulfine binding could occur in a different manner, i.e., by binding to the SH group of the axial ligand rather than to another SH group of the protein. This would account for limited accessibility of the Fe to CO and the resulting apparent low levels of P450 observed when measurements are made spectrophotometrically.

De Matteis commented that the disappearance of P450 is not accompanied by a proportional increase in cytochrome P420. Although some P420 is formed, he has not yet determined the amounts, nor was the absolute spectrum of the P450 studied. The half-life of P450 heme has not been determined during the metabolism of CS_2 or other sulfur compounds.

37

Metabolism and Biological Activity of Benzo[a]pyrene and Its Metabolic Products

A. H. Conney, A. W. Wood, W. Levin, A. Y. H. Lu, R. L. Chang, P. G. Wislocki, and R. L. Goode
Department of Biochemistry and Drug Metabolism
and
Technical Development Department
Hoffmann-LaRoche Inc.
Nutley, New Jersey 07110 U.S.A.

G. M. Holder, P. M. Dansette, H. Yagi, and D. M. Jerina

Section on Oxidation Mechanisms
Laboratory of Chemistry
National Institute of Arthritis, Metabolism, and Digestive Diseases
National Institutes of Health
Bethesda, Maryland 20014 U.S.A.

The carcinogenicity of benzo[*a*]pyrene (BP) and its widespread occurrence as an environmental pollutant have prompted numerous studies on the metabolic fate and biological activity of this hydrocarbon. Our laboratories in Nutley and Bethesda have jointly undertaken a systematic study of the metabolic fate and biological activity of BP and its many metabolites. The goal of this research program is to elucidate the molecular events that result in the metabolic activation of BP and to identify the proximate and ultimate carcinogenic forms of this environmental carcinogen. This chapter is a progress report that describes (1) a high-pressure liquid chromatography system for the separation of BP dihydrodiols, BP phenols, and BP quinones; (2) the metabolism of BP by a solubilized and reconstituted cytochrome P448 monooxygenase system in the presence and absence of purified epoxide hydrase; (3) the metabolic activation of BP to mutagenic metabolites by a cytochrome P448 or cytochrome P450 reconstituted monooxygenase system; (4) the effect of epoxide hydrase on the cytochrome P448-mediated activation of BP to mutagenic metabolites; (5) the mutagenic activity of 28 synthetic derivatives of BP; and (6) the carcinogenicity of three BP oxides on mouse skin.

SYNTHESIS OF BP DERIVATIVES

Reference compounds for high-pressure liquid chromatography (HPLC), as well as the materials which have been tested for mutagenicity and carcinogenicity, were obtained by direct chemical synthesis, with the exception of BP-9,10-dihydrodiol, which was formed by the action of epoxide hydrase on BP-9,10-oxide (1). For the 12 isomeric phenols of BP, the previously reported synthesis of seven of these isomers has been repeated and in some cases improved upon (2). The unknown 1-, 2-, 4-, 10-, and 12-HOBP isomers were prepared by unequivocal synthetic procedures in which the ring system was constructed from simpler molecules (2), thereby providing isomerically homogeneous samples in each case. The non-K-region BP-7,8- and -9,10-oxides were prepared by the halohydrin ester route (1), while the K-region BP-4,5- and -11,12-oxides were obtained from the corresponding dihydrodiols (3). Synthesis of BP-7,8-dihydrodiol has also been described (4). Earlier procedures for synthesis of BP-1,3-, -3,6-, and -6,12-quinones were repeated (5,6), while improved routes to BP-4,5- and -11,12-quinones were developed (3). Synthesis and properties of the stereoisomerically pure (±)-7β,8α-dihydroxy-9β,10β-epoxy-7,8,9,10-tetrahydro-BP, hereafter abbreviated as BP-7,8-diol-9,10-epoxide, in which the 7-hydroxyl group and the epoxide group are on the same face of the molecule, are described elsewhere in this volume (7). All of the materials used in these studies are of high purity and contain only occasional trace amounts of contaminants as detected by chromatography and NMR spectroscopy.

CHROMATOGRAPHIC PROPERTIES OF BP METABOLITES

High-pressure liquid chromatography is highly efficient for the separation and quantitation of phenols, quinones, and dihydrodiols formed from BP (Fig. 1) (8). Since separations could not be achieved for many of the reference compounds, rigorous identification of individual compounds within each group will require further study. Structural assignment to the metabolic dihydrodiols cannot be made with certainty until the unavailable 1,2- and 2,3-dihydrodiols are obtained, or until further spectroscopic criteria can be applied. Three of the five available quinones (1,6-, 3,6-, and 4,5-) emerge from the column as a single badly tailing peak. The tail of this peak contains BP-6,12- and -11,12-quinones, as well as BP-4,5-oxide. Mobility of the many other possible quinones is unknown. The 12 synthetic BP-phenols are well separated from the quinones.

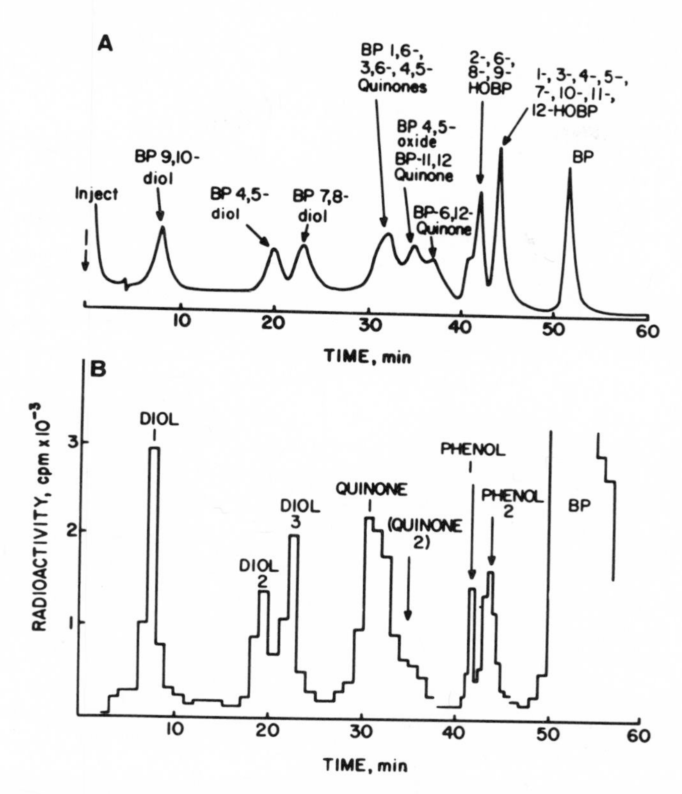

Fig. 1. Chromatographic profile of benzo[a]pyrene (BP) metabolites. (A) Composite high-pressure liquid chromatography profile obtained with available reference compounds. (B) Chromatographic distribution of metabolites obtained after incubation of 0.2 nmol of cytochrome P448 from 3-methylcholanthrene-pretreated rats, 216 units of epoxide hydrase, 120 units of NADPH–cytochrome *c* reductase, 0.1 mg of lipid, 0.5 μmol of NADPH, 3 μmol of $MgCl_2$, 100 μmol of potassium phosphate buffer (pH 6.8), and 95 nmol of [^{14}C] BP per milliliter. After incubation at 37°C for 5 min, the samples were extracted with acetone–ethyl acetate, dried, and redissolved in dioxane as previously described (8). Two coupled 1-m by 2.1-mm analytical DuPont ODS columns were used, with a linear gradient of 35–85% methanol in water over 50 min, an inlet pressure of 2500 psig, and a flow rate of 0.7–1.1 ml/min. Taken from studies by Holder *et al.* (8).

However, they separate into only two overlapping phenol peaks, the first containing 2-, 6-, 8-, and 9-HOBP, and the second containing 1-, 3-, 4-, 5-, 7-, 10-, 11-, and 12-HOBP. A composite plot of the separation of the available standards and the results of a study on the *in vitro* metabolism of [^{14}C] BP are shown in Fig. 1. Gelboin and his associates have independently described a similar HPLC system for the separation of BP metabolites, and they have used HPLC for metabolic experiments (9).

EFFECT OF EPOXIDE HYDRASE ON THE METABOLISM OF BP BY A PARTIALLY PURIFIED CYTOCHROME P448-CONTAINING SYSTEM

Hepatic epoxide hydrase is an important enzyme for the metabolism of aromatic compounds, since it converts intermediate arene oxides formed by the mixed-function oxidase system to the corresponding dihydrodiols. Thus the separation, purification, and reconstitution of the mixed-function oxidase system, and the removal of epoxide hydrase from this system (10,11), have allowed us to study the role of the oxidase and hydrase in controlling the levels of cytotoxic and mutagenic metabolites of polycyclic hydrocarbons. Studies with the solubilized and reconstituted monooxygenase system showed that BP was readily metabolized to quinones and phenols, but little or no dihydrodiol was formed from BP in the absence of added epoxide hydrase (Fig. 2). This is consistent with the absence or very low level of epoxide hydrase in our purified cytochrome P448 (10). On addition of epoxide hydrase to the system, diols 1, 2, and 3 appeared, while the radioactive peak in the region of BP-4,5-oxide and BP-6,12- and -11,12-quinones (quinone 2 fraction) was lost. Formation of diols 2 and 3 was maximized with the addition of 72 units of partially purified epoxide hydrase, while further addition of this enzyme resulted in continued increase of diol 1.

A good stoichiometric relationship was found between the increase in diol 2 (BP-4,5-dihydrodiol) and the decrease in quinone 2 (BP-4,5-oxide). The stoichiometry was not as good between diol 1 and phenol 1 or between diol 3 and phenol 2. The results demonstrate that the dihydrodiols and phenolic metabolites of BP share arene oxides as common precursors. Phenol fractions 1 and 2 could be reduced only to the extent of 85% and 45%, respectively, after addition of the maximum amount of epoxide hydrase. Diol 1 (BP-9,10-dihydrodiol) is formed mostly at the expense of phenol 1, which contains the rearrangement product from BP-9,10-oxide, predominantly 9-HOBP. Similarly, diol 3 (BP-7,8-dihydrodiol) is formed at the expense of phenol 2, which contains the preponderant isomerization product of BP-7,8-oxide, mainly 7-HOBP. Clearly, phenol 2 must, at the very least, contain 7-HOBP in addition to the known 3-HOBP. Similarly, phenol 1 probably contains other phenols in addition to 9-HOBP. On addition of epoxide hydrase, diols 1 and 3 increased more rapidly than phenols 1 and 2 decreased. The inability to markedly decrease the amount of quinone 1 by addition of epoxide hydrase suggests either that the arene oxides that ultimately result in quinone production are not good substrates for hydrase or that there is a non-arene oxide pathway (12,13) leading to the metabolites in quinone 1 fraction.

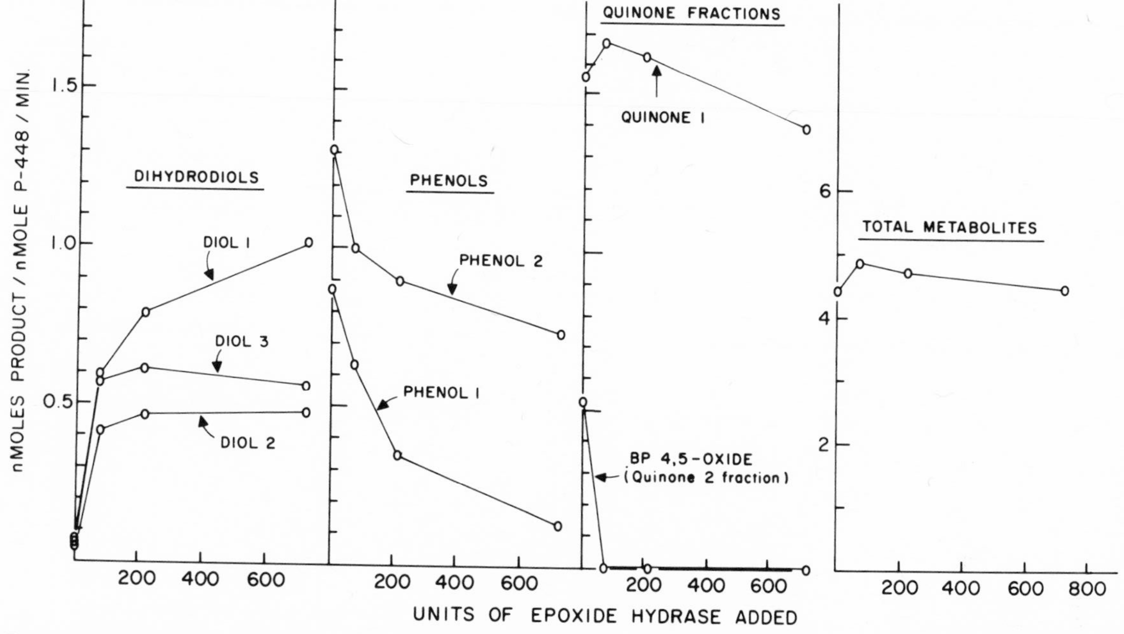

Fig. 2. Effect of epoxide hydrase on the metabolism of benzo[a]pyrene (BP) by a purified cytochrome P448-dependent monooxygenase system. The methodology for studies on the metabolism of [^{14}C]BP by a reconstituted liver microsomal cytochrome P448 system from 3-methylcholanthrene-treated rats is described in the caption of Fig. 1. Taken from studies by Holder *et al.* (8).

METABOLIC ACTIVATION OF BP TO MUTAGENIC METABOLITES BY A SOLUBILIZED, RECONSTITUTED CYTOCHROME P450 OR P448 MONOOXYGENASE SYSTEM

B. N. Ames and his associates have developed a series of mutants of *Salmonella typhimurium* which can be specifically reverted from a histidine requirement to histidine independence by a wide variety of mutagens (14–16). Malling (17) used these bacteria to demonstrate that liver microsomes metabolized dimethylnitrosamine to mutagenic intermediates, and Garner and his associates used the bacteria to demonstrate that liver microsomes metabolized aflatoxin B_1 to highly cytotoxic compounds (18,19). Subsequent investigations by Ames and his associates showed that incubations of selected strains of *Salmonella typhimurium* with rat liver fractions and appropriate cofactors resulted in the metabolism of aflatoxin B_1, benzo[*a*]pyrene, and a wide variety of other carcinogens to potent mutagens (16,20).

We have studied the metabolic activation of BP by a reconstituted system containing purified liver microsomal cytochrome P450 from phenobarbital-treated rats or liver microsomal cytochrome P448 from 3-methylcholanthrene-

Table 1. Requirements for the Metabolic Activation of Benzo[*a*]pyrene to Compounds That Are Mutagenic in *Salmonella typhimurium* TA 1538[a]

Addition	His^+ revertants per plate
None	10 ± 2
BP (50 μM) without hydroxylase system	9 ± 2
Complete hydroxylase system[a]	145 ± 10
–BP	17 ± 7
–NADPH-cytochrome *c* reductase	13 ± 6
–Cytochrome P448	14 ± 5
–Phosphatidylcholine	42 ± 4
–NADPH	13 ± 4

[a]The complete hydroxylase system consisted of 50 μg phosphatidylcholine, 150 units of purified NADPH-cytochrome *c* reductase (10), 0.1 nmol of purified cytochrome P448 (12.8 nmol/mg protein) from 3-methylcholanthrene-treated rats, 25 nmol BP, and 200 nmol NADPH incubated in 0.5 ml of 0.005 M sodium phosphate containing 0.9% NaCl. The final pH was 6.8. Approximately 2×10^8 *Salmonella typhimurium* (strain TA 1538) was added to each reaction mixture, and the samples were incubated at 37°C for 5 min. Menadione (15 μM) was then added to the incubation mixture, and 2 ml of top agar was immediately added. The mixture was poured onto a petri plate and incubated for 2 days with a histidine-deficient medium as described elsewhere (16,20). The addition of menadione effectively blocked the metabolic activation of BP during the 2-day incubation and did not interfere with the expression of mutations that resulted from the 5-min incubation. Each value is the mean ± SD from three plates.

treated rats. Studies on the requirements for the enzymatic activation of BP during a 5-min incubation with strain TA 1538 of *Salmonella typhimurium* and the reconstituted cytochrome P448 system indicated that optimal metabolic activation to mutagenic metabolites occurred in the presence of BP, NADPH-cytochrome *c* reductase, cytochrome P448, phosphatidylcholine, and NADPH (Table 1). An absolute requirement was observed for all components of the system except phosphatidylcholine. The cytochrome P448-mediated metabolism of BP during a 5-min incubation resulted in considerably more mutations than when cytochrome P450 was used as the source of monooxygenase (Fig. 3).

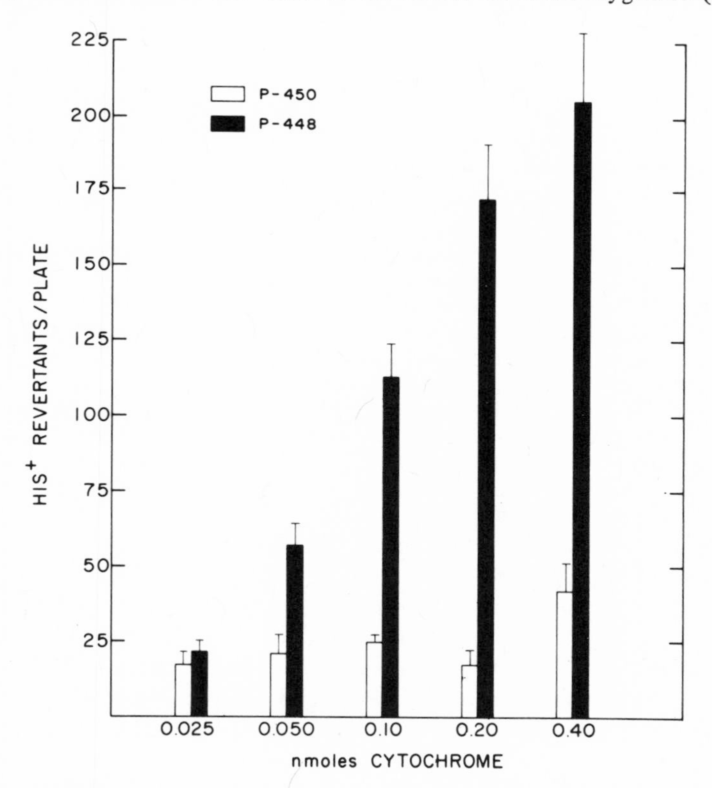

Fig. 3. Activation of benzo[a]pyrene to mutagenic metabolites by purified cytochrome P450 and cytochrome P448 from liver microsomes of rats treated with phenobarbital or 3-methylcholanthrene. The incubation conditions are described in the footnote to Table 1. Cytochrome P448 was purified from liver microsomes of 3-methylcholanthrene-treated rats to a specific content of 12.8 nmol/mg protein, and cytochrome P450 was purified from liver microsomes of phenobarbital-treated rats to a specific content of 8.3 nmol/mg protein (10). Reaction mixtures which contained no hemoprotein but were otherwise complete resulted in 19 reversions per petri plate. This blank value was subtracted from the data presented. Each value represents the mean ± SD obtained from three replicate incubation mixtures.

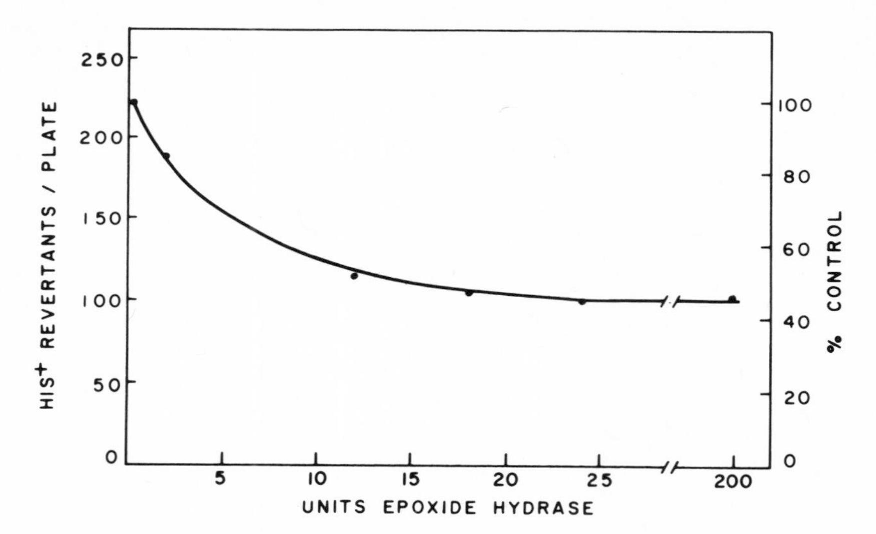

Fig. 4. Effect of epoxide hydrase on the activation of benzo[a]pyrene to mutagenic metabolites by purified cytochrome P448 from liver microsomes of 3-methylcholanthrene-treated rats. The incubation conditions are described in the footnote to Table 1. Liver microsomal epoxide hydrase was purified from phenobarbital-treated rats (22), and a preparation with a specific activity of 5044 units/mg ptotein was used. One unit of epoxide hydrase catalyzes the hydration of 1 nmol of styrene oxide in 15 min at 37°C. Reaction mixtures which lacked NADPH but were otherwise complete produced <20 revertants per plate and were subtracted from the plotted values. Epoxide hydrase alone did not alter the spontaneous reversion rate. The values presented represent the average of two experiments.

These results are in accord with other studies indicating that the purified cytochrome P448 is much more active than cytochrome P450 in metabolizing BP (21).

The effect of purified epoxide hydrase (22) on the cytochrome P448-mediated activation of BP to metabolites with mutagenic activity in *Salmonella typhimurium* strain TA 1538 is shown in Fig. 4. Addition of 1–12 units of epoxide hydrase gradually reduced the metabolic activation of BP to mutagenic metabolites.[1] The addition of 12 units of epoxide hydrase decreased mutagenesis by 50–60%, while continued addition of epoxide hydrase (total of 200 units) did not further influence the number of His^+ revertants observed. These results suggest that at least 50–60% of the cytochrome P448-mediated activation of BP to mutagenic metabolites is via arene oxides. The inability of epoxide hydrase to

[1] Observations by Oesch (34) and by Ames and his associates (16) also indicate mutagenic activity for BP-4,5-oxide. Observations by Oesch have also shown that epoxide hydrase inhibited the metabolic activation of BP to mutagenic metabolites (34). Recent studies by Malaveille and his associates suggested that BP-7,8-diol-9,10-epoxide has the same mutagenic activity as BP-4,5-oxide (35). The stereochemistry of their diol epoxide was not investigated.

completely block the cytochrome P448-mediated mutagenesis suggests that nonarene oxide metabolites of BP may also have mutagenic activity in *Salmonella typhimurium* strain TA 1538 or that some mutagenic arene oxides are poor substrates for the purified epoxide hydrase.

MUTAGENIC ACTIVITY OF BP DERIVATIVES IN *Salmonella typhimurium*

Since BP and other polycyclic hydrocarbons undergo enzymatic oxidation at multiple sites, we have initiated studies on the biological activity of many of these possible BP metabolites. We have studied the mutagenic activity of BP derivatives as an initial test of biological activity because the mutation test systems are generally well characterized, relatively rapid, and provide an index of an interaction of the BP derivatives with genetic material under conditions that result in an altered phenotype.

For studies on the mutagenic effects of BP derivatives in bacteria, we have utilized several histidine-dependent strains of *Salmonella typhimurium* developed by Ames[2] (14,15). As a result of a different base sequence at or near the site of histidine mutation, four of the *Salmonella* tester strains have different sensitivity profiles to a series of mutagens. Strain TA 1535 detects base-pair mutagens, while strains TA 1536, TA 1537, and TA 1538 detect frameshift mutagens. Strain TA 98 derived from TA 1538 and strain TA 100 derived from TA 1535 contain a plasmid factor that enhances error-prone recombinational repair and increases the sensitivity of these strains to many mutagens. Growth and maintenance of the bacteria and the pour plate procedure for mutagen testing were essentially as has been described in detail elsewhere (14). The mutagenic activity of a compound was quantified by counting the number of bacterial colonies that grew on a histidine-deficient medium.

The number of reversions to histidine independence that were induced in four strains of *Salmonella typhimurium* by BP, BP-4,5-oxide (K region), BP-7,8-oxide, and BP-9,10-oxide are shown in Table 2. BP-4,5-oxide was highly mutagenic in strains TA 1537 and TA 1538. As little as 0.25 μg of this K-region arene oxide added to the bacteria in 2 ml of top agar induced a greater than fortyfold increase in the mutation rate in strain TA 1538. In contrast to the results with BP-4,5-oxide, we found that BP-7,8- and -9,10-oxides were only weakly mutagenic, even at high concentrations. None of the arene oxides was capable of inducing histidine independence in strains TA 1535 and TA 1536. It was also found that BP and the six BP phenols (4-, 5-, 7-, 8-, 9-, and 10-HOBP)

[2] Six strains of *Salmonella typhimurium* were obtained from B. Ames, University of California at Berkeley. The Chinese hamster cell line V79-6 was generously provided by E. H. Y. Chu, University of Michigan, Ann Arbor.

Table 2. Reversion of Histidine Auxotrophs of *Salmonella typhimurium* by BP-Oxides[a]

Compound	Amount added (μg)	His$^+$ revertants per plate			
		1535	1536	1537	1538
–	–	27 ± 3	0	7 ± 2	15 ± 4
BP	1.0	28 ± 3	0	8 ± 2	11 ± 5
	5.0	31 ± 6	0	8 ± 2	11 ± 3
BP-4,5-oxide	0.25	29 ± 3	0	63 ± 7	606 ± 56
	1.0	39 ± 4	0	229 ± 23	3800
	5.0	40 ± 5	0	1235 ± 200	–
BP-7,8-oxide	0.25	31 ± 5	0	8 ± 2	28 ± 3
	1.0	30 ± 6	0	8 ± 4	43 ± 4
	5.0	28 ± 6	0	7 ± 1	73 ± 8
BP-9,10-oxide	0.25	33 ± 6	0	7 ± 2	22 ± 4
	1.0	30 ± 5	0	11 ± 1	31 ± 3
	5.0	28 ± 5	0	11 ± 3	61 ± 8

[a]BP and the BP-oxides were added to 2 ml of top agar containing 2 × 10^8 bacteria and a histidine-deficient medium. The petri plates were incubated at 37°C for 48 h. Each value represents the average ± SD from five plates. Taken from studies by Wood *et al.* (36)

which could form during the spontaneous isomerization of BP-4,5-, -7,8-, and -9,10-oxides in an aqueous environment were inactive in the above four *Salmonella* tester strains, except for 7-HOBP, which had very weak mutagenic activity in TA 1538 (2–3 times the control reversion rate) when 1–10 μg of this phenol was added per plate.

The low mutagenic activity of BP-7,8-oxide and BP-9,10-oxide could have been due to poor intrinsic activity of these arene oxides with the tester strains or to the instability (23–25) of these two non-K-region arene oxides, which undergo relatively rapid spontaneous isomerization to phenols. Similarly, the phenols could be inactive because of rapid decomposition. We therefore examined the half-lives of the arene oxides and phenols in nutrient broth at 37°C. All six phenols, BP, and BP-4,5-oxide showed no significant decomposition in this aqueous environment at 37°C for at least 90 min. Additional studies with BP-4,5-oxide indicated that it was stable for at least 8 h. BP-7,8- and -9,10-oxides, however, had half-lives of about 2 h and 0.5 h, respectively.[3] Since the pour plate procedure gives no indication of the minimum time necessary to induce mutations, an additional set of experiments was initiated to compensate for the differences in arene oxide stability (Table 3). BP-7,8- and -9,10-oxides (2.5 or 12.5 μg) were added every 20 min for 2 h to 5 ml of nutrient broth

[3]Although the $t_{1/2}$ of BP-7,8-oxide and BP-9,10-oxide in Difco nutrient broth at 37°C is 120 and 30 min, respectively, the $t_{1/2}$ in 1–100 mM phosphate buffer (pH 7.4) at 37°C is 30 and 2–5 min, respectively. BP-4,5-oxide was stable when incubated in either medium.

Table 3. Reversion of TA 1538 after Repeated Additions of BP-7,8-Oxide and BP-9,10-Oxide[a]

Addition	Amount added during 2-h incubation	Total amount added (μg)	His$^+$ revertants per plate
Acetone/NH_4OH	–	–	14 ± 6
BP-4,5-oxide	0.625 μg once	0.625	534 ± 31
BP-7,8-oxide	2.5 μg every 20 min	15	55 ± 6
	12.5 μg every 20 min	75	99 ± 10
BP-9,10-oxide	2.5 μg every 20 min	15	61 ± 6
	12.5 μg every 20 min	75	164 ± 17

[a]The incubation mixture consisted of 5.0 ml nutrient broth with an initial cell density of 1 × 10^8 bacteria/ml. Cells were incubated for 2 h at 37°C. At zero time or at the end of 2 h, 0.1 ml of the incubation mixture was added to 2.0 ml top agar and incubated for 2 days. The number of revertants observed in the zero time samples was never significantly above the background. Taken from studies by Wood *et al.* (36).

which contained 5 × 10^8 *Salmonella typhimurium* of strain TA 1538. BP-4,5-oxide (0.625 μg) was added once at zero time. Immediately after the first addition, and at the end of 2 h, a 0.1-ml sample of each incubation mixture was added to 2.0 ml of top agar, plated, and incubated for 2 days according to the usual procedure. As shown in Table 3, both the BP-7,8- and -9,10-oxides had low but significant mutagenic activity at both concentrations. However, BP-4,5-oxide was still more than 100 times as mutagenic as BP-9,10- or -7,8-oxide on a per microgram basis when compensated for differences in stability of the three arene oxides. The zero-time sample of BP-4,5-oxide gave the same reversion frequency as was obtained from an untreated control incubation. This result indicates that the colonies reverted by the 2-h exposure to BP-4,5-oxide did not result from the subsequent 2-day exposure to the small amount (0.0125 μg) of arene oxide in the 0.1-ml sample that was plated.

More detailed comparative studies on the mutagenic effects of several concentrations of BP-4,5-, -7,8-, and -9,10-oxides were initiated in the TA 1538, TA 98, and TA 100 strains of *Salmonella typhimurium.* In addition, the mutagenic activities of BP-11,12-oxide and BP-7,8-diol-9,10-epoxide were also evaluated. BP-4,5-oxide was highly mutagenic in all three *Salmonella* strains, and the number of revertant colonies was roughly proportional to the amount of oxide added, even at the highest concentration of 4 μg of BP-4,5-oxide per plate (Fig. 5). The data in Fig. 5 indicate that BP-7,8-, -9,10-, and -11,12-oxides were only weakly mutagenic in the three tester strains. The most interesting compound studied was BP-7,8-diol-9,10-epoxide. This compound, at low concentrations (0.07–0.28 μg per plate), was equally or more active than BP 4,5-oxide in all three tester strains (Fig. 5). In strain TA 100, BP-7,8-diol-9,10-epoxide was 10- to 25-fold more mutagenic than BP-4,5-oxide. BP-7,8-diol-9,10-epoxide was

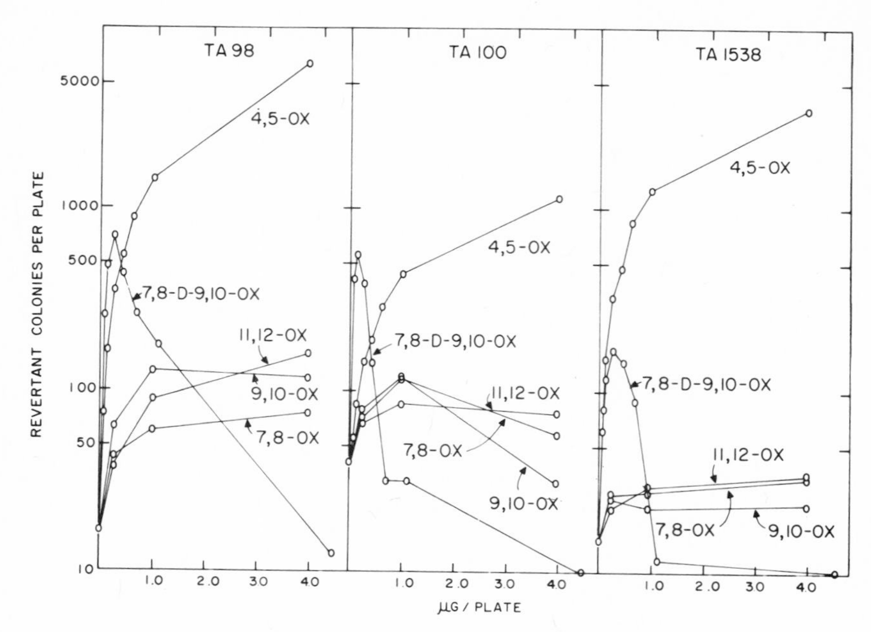

Fig. 5. Mutagenic activity of BP-7,8-diol-9,10-epoxide and BP-4,5-, -7,8-, -9,10-, and -11,12-oxides in the TA 1538, TA 98, and TA 100 strains of Salmonella typhimurium. The abbreviations 4,5-OX, 7,8-OX, 9,10-OX, and 11,12-OX refer to BP-4,5-, -7,8-, -9,10-, and -11,12-oxides, respectively. 7,8-D-9,10-OX refers to BP-7,8-diol-9,10-epoxide. The four simple oxides were added to 2 ml of top agar in 100 μl of DMSO–NH_4OH (1000:1). The diol-epoxide was added in 100 μl of anhydrous DMSO. Approximately 2×10^8 bacteria were present in the top agar.

also highly cytotoxic, and concentrations above 0.15 μg per plate resulted in appreciable cytotoxicity in all three strains of *Salmonella.* A poor lawn of cells[4] was observed at high concentrations of the diol-epoxide, and the number of revertant colonies per plate was decreased as the concentration of BP-7,8-diol-9,10-epoxide increased. The mutagenic BP-4,5-oxide was not cytotoxic, even when 4 μg of this compound was added per plate.

The results of testing 23 nonarene oxide derivatives of BP for mutagenicity in the TA 1538 strain of *Salmonella typhimurium* are summarized in Table 4. None of the BP dihydrodiols or BP quinones was mutagenic. Among the 12 BP phenols tested, 1-HOBP, 3-HOBP, 6-HOBP, 7-HOBP, and 12-HOBP were weakly mutagenic. The 6-HOBP and 12-HOBP were somewhat more mutagenic than the

[4] A lawn of bacterial cells is formed on the petri plates because of the few cell divisions that occur by the *Salmonella* tester strains grown on a medium with a low histidine content. When the histidine is consumed by the bacterial cells, growth stops, except for those revertant cells that can grow in the absence of histidine. A diminished or abnormal lawn of cells is an indication of cytotoxicity and cell death.

Table 4. Activity of Benzo[a]pyrene Derivatives in *Salmonella* Strain TA 1538[a]

Weakly mutagenic phenols[b]	1-HOBP, 3-HOBP, 6-HOBP, 7-HOBP, 12-HOBP
Nonmutangenic phenols	2-HOBP, 4-HOBP, 5-HOBP, 8-HOBP, 9-HOBP, 10-HOBP, 11-HOBP
Nonmutagenic dihydrodiols	4,5 (*cis*), 4,5 (*trans*), 7,8 (*trans*), 9,10 (*cis*), 11,12 (*cis*), 11,12 (*trans*)
Nonmutagenic quinones	1,6; 3,6; 4,5; 6,12; 11,12

[a] 1 or 10 μg of each compound per plate.
[b] Mutations that were 1–4 times above the background were observed for 1-HOBP, 3-HOBP, and 7-HOBP. Somewhat higher mutagenic activity was observed for 6-HOBP and 12-HOBP. The last two compounds were moderately active mutagens in *Salmonella typhimurium* strain TA 98.

other BP phenols in the TA 1538 strain, and these two phenols were moderately active mutagens in strain TA 98.

MUTAGENIC ACTIVITY OF BP DERIVATIVES IN CHINESE HAMSTER V79 CELLS

For mutagenesis studies with eukaryotic mammalian cells, we used an established line of V79 Chinese hamster cells developed by Chu and Malling (26,27). Resistance to the lethal effects of the purine analogue 8-azaguanine (8-AG) was used as the mutagenic marker. For the purposes of this work, we, like others (28), define a mutation as a stable and heritable change in phenotype. The cells were grown in Eagle's minimum essential medium (Gibco), which contained a 1.5-fold higher-than-normal concentration of glutamine and nonessential amino acids and dialyzed and heat-inactivated fetal calf serum (Reheis). Medium for stock cultures contained no antibiotics but contained hypoxanthine, thymidine, glycine, and aminopterin to select against spontaneously formed 8-AG-resistant variants (27). Four days prior to an experiment, the medium was removed and replaced with medium free of aminopterin. Medium used for experiments was not supplemented with thymidine, hypoxathine, aminopterin, or glycine but contained penicillin and streptomycin.

The cytotoxicity of BP and its derivatives was determined by comparing the plating efficiency of treated cultures to cultures receiving solvent alone. The procedure used was essentially as described by Huberman *et al.* (28), except that 100 cells were added to 60-mm culture dishes (Falcon). Seven days after treatment, the cells were fixed and stained with Giemsa, and colonies of more than 50 cells were counted. Induction of 8-AG-resistant variants of Chinese hamster cells was done according to the procedure developed by Chu and Malling (26), as modified by Huberman *et al.* (28). Cells (1×10^4) were added to 60- by 15-mm culture dishes in 5 ml medium 18 h prior to treatment with the test compounds. Addition of the compounds to be tested and all subsequent treatments of the cells were done in subdued light. After 1 h at 37°C, the medium was removed and the cells were washed once with phosphate-buffered saline at pH 7.2 and then cultured in fresh medium. After 2 days, drug-resistant variants were selected by adding 8-AG (10 μg/ml medium) to the cultures. Preliminary experiments indicated that when dialyzed fetal calf serum was used in the medium, a 10 μg/ml concentration of 8-AG was 4–5 times the minimum

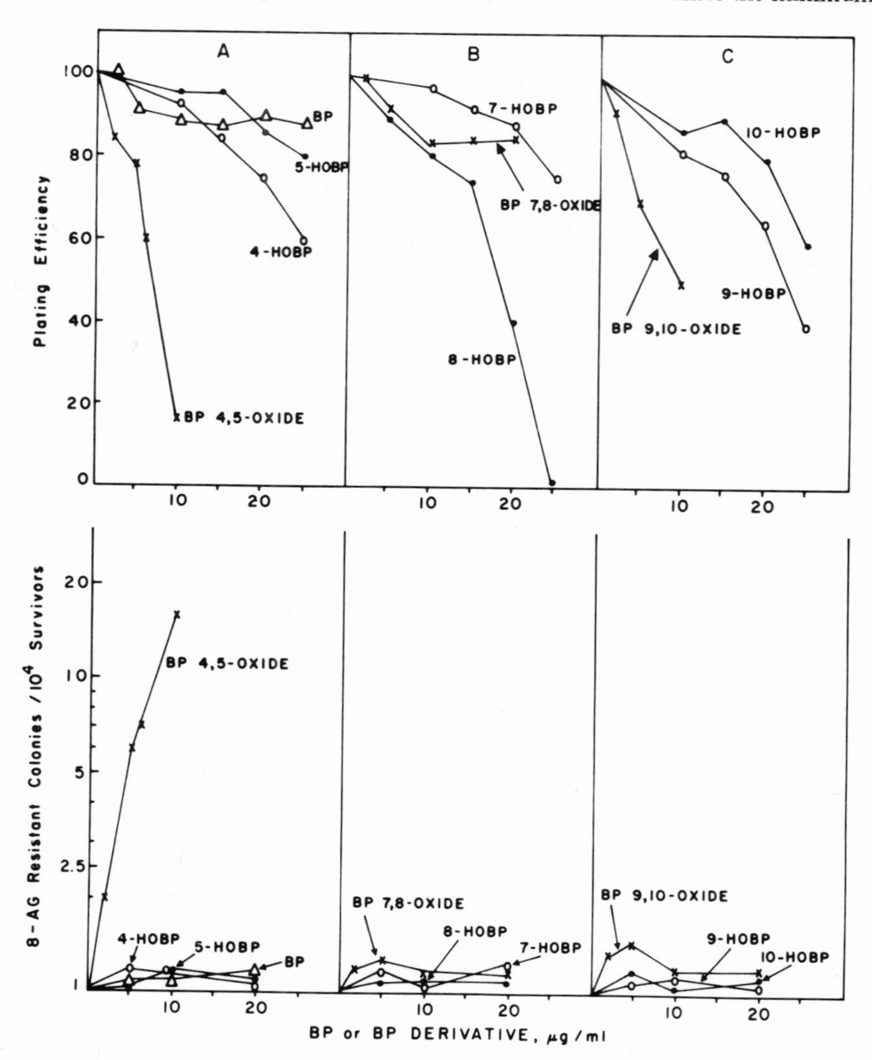

Fig. 6. Cytotoxic and mutagenic effects of BP-oxides and BP-phenols on Chinese hamster V79 cells. BP-oxides were added to 5 ml of medium in 20 μl of acetone–NH_4OH (1000:1). BP-phenols were added in 20 μl of acetone. Taken from studies by Wood *et al.* (36).

amount necessary to kill all 8-AG-sensitive cells. Culture medium was replaced on day 7 with 5 ml of fresh medium containing the same concentration of 8-AG. This procedure, using one medium change after selection, was employed to minimize the chance that secondary colonies would arise from the reattachment and growth of cells dislodged from a primary colony during manipulation (26). Drug-resistant colonies were fixed with methanol, stained with Giemsa, and counted on day 14.

Figure 6 illustrates the cytotoxic and mutagenic activity of BP-4,5-, -7,8-, and -9,10-oxides, the six corresponding phenols, and BP, as a function of concentration in cultures of Chinese hamster V79 cells. BP, as previously reported (29), had little cytotoxicity and failed to alter the spontaneous mutation frequency in the V79 cells. BP-4,5-oxide was the most cytotoxic of the three arene oxides and BP-7,8-oxide was the least cytotoxic. As was found with the bacterial studies, BP-4,5-oxide was the most active of the derivatives in inducing mutations. A 1-h exposure to BP-4,5-oxide (5 μg/ml) resulted in a sixfold increase in the number of 8-AG-resistant colonies seen over those in a control culture. In contrast, neither BP-7,8- nor BP-9,10-oxide at any concentration tested induced more than a 1.5-fold change in the mutation frequency. Of the six phenols tested, none changed the spontaneous mutation frequency, but large differences in their cytotoxicity were observed. 8-HOBP was the most toxic of the phenols, while 7-HOBP was the least. Both BP-4,5- and -9,10-oxides were more toxic than their corresponding phenols.

BP-11,12-oxide and BP-7,8-diol-9,10-epoxide were also tested for mutagenicity in Chinese hamster V79 cells (Fig. 7). In contrast to studies with *Salmonella typhimurium,* BP-11,12-oxide possessed the same high mutagenic activity as BP-4,5-oxide in the mammalian cells. BP-7,8-diol-9,10-epoxide was the most highly mutagenic and cytotoxic compound studied. The results indicate that BP-7,8-diol-9,10-epoxide is at least thirty- to fortyfold more mutagenic than BP-4,5-oxide in V79 cells. Since the diol-epoxide is relatively unstable compared to the 4,5-oxide, the mutagenicity of the diol-epoxide as described here may be an underestimate of its intrinsic activity. Exposure of the cells to 0.25 μg of BP-7,8-diol-9,10-epoxide per milliliter for 1 h resulted in 63% plating efficiency and 81 mutant colonies per 10^5 surviving cells, whereas exposure of the cells to 2 μg/ml killed all the cells. Exposure of the V79 cells to 5 μg of BP-4,5-oxide per milliliter for 1 h resulted in 86% plating efficiency and 48 mutant colonies per 10^5 surviving cells.

CARCINOGENICITY STUDIES WITH BP-4,5-, -7,8-, AND -9,10-OXIDES

Female C57BL/6J mice were treated topically in subdued light with 0.1 or 0.4 μmol of BP, BP-4,5-oxide, BP-7,8-oxide, or BP-9,10-oxide in 25 μl of acetone–NH_4OH (1000:1) once every 2 weeks for 60 weeks, and the animals

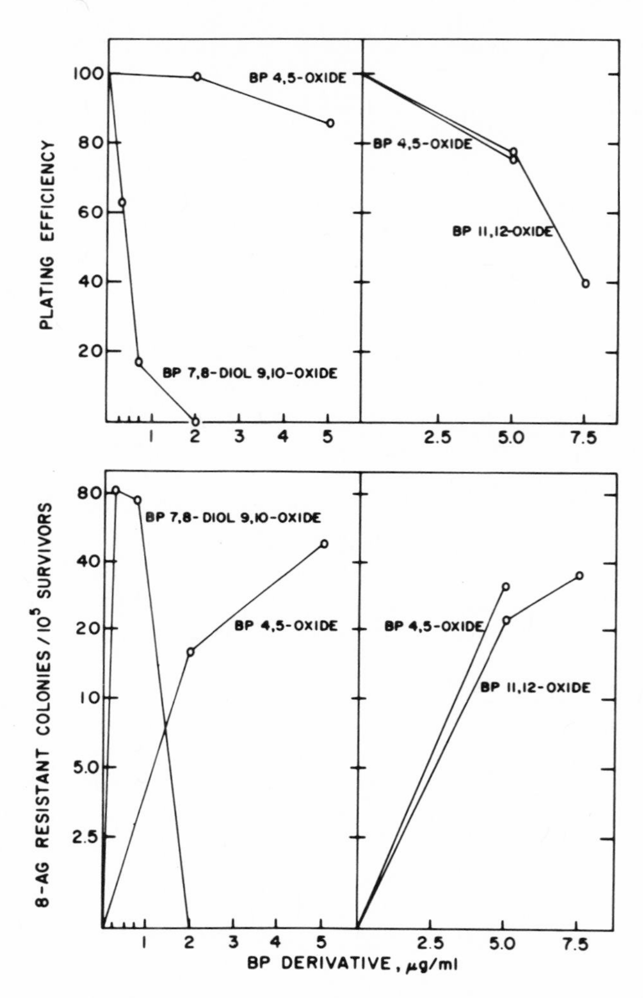

Fig. 7. Cytotoxic and mutagenic effects of BP-4,5-oxide, BP-11,12-oxide, and BP-7,8-diol-9,10-epoxide on Chinese hamster V79 cells. Recloning of the cells used in Fig. 6 resulted in a subclone of cells with a lower spontaneous mutation frequency, and these cells were used. BP-4,5-oxide and BP-11,12-oxide were added to 5 ml of medium in 20 μl of DMSO–NH_4OH (1000:1), and BP-7,8-diol-9,10-epoxide was added in 20 μl of anhydrous dimethylsulfoxide.

were placed in a dark room for 24 h after each application of hydrocarbon. The results of this study are summarized in Table 5. BP was considerably more carcinogenic than the three BP oxides tested. In the low-dose groups, 95% of the BP-treated animals had skin tumors at 60 weeks after the start of the experiment, whereas less than 10% of the animals treated with the BP oxides had skin tumors. Increasing the dose of BP and the BP-oxides fourfold did not increase

Table 5. Skin Tumors in Mice Treated Topically with BP-Oxides[a]

Group	Dose (μmol)	Percent animals with tumors at 60 weeks
Control	–	0
BP	0.1	95
BP-7,8-oxide	0.1	9
BP-4,5-oxide	0.1	6
BP-9,10-oxide	0.1	0
BP	0.4	100
BP-7,8-oxide	0.4	94
BP-4,5-oxide	0.4	4
BP-9,10-oxide	0.4	0

[a]Female C57BL/6J mice at 9 weeks of age were shaved with electric clippers while under light ether anesthesia. BP and its three oxides (0.1 or 0.4 μmol) dissolved in 25 μl of acetone–NH_4OH (1000:1) were applied topically to the backs of mice once every 2 weeks for 60 weeks. Control mice were treated with the solvent. Each group consisted of 30–39 animals. The percent of animals with tumors was calculated from tumor-bearing animals that died during the 60-week experiment, surviving tumor-bearing animals, and surviving non-tumor-bearing animals. Only 4–5 deaths occurred per group in non-tumor-bearing animals during the 60-week experiment.

the carcinogenicity of BP-4,5-oxide or BP-9,10-oxide. However, the fourfold increase in dose markedly increased the carcinogenicity of BP-7,8-oxide. Ninety-four percent of the mice treated with 0.4 μmol of BP-7,8-oxide every 2 weeks for 60 weeks had skin tumors, whereas less than 5% of the animals treated with 0.4 μmol of BP-4,5-oxide or BP-9,10-oxide had skin tumors. Histological studies revealed that the majority of tumors, after 60 weeks of treatment, were squamous cell carcinomas in both the BP-treated and BP-7,8-oxide-treated groups.

DISCUSSION

A solubilized, reconstituted cytochrome P448 monooxygenase system that has little or no epoxide hydrase activity has been prepared from rat liver microsomes. This monooxygenase system metabolizes BP primarily to phenols and quinones, and little if any metabolism to dihydrodiols occurs. Addition of purified epoxide hydrase to the monooxygenase system results in the formation of large amounts of dihydrodiols and decreased formation of phenolic metabolites. The results indicate that BP is metabolized to at least three arene oxides (BP-4,5-, -7,8-, and -9,10-oxides) that either isomerize to phenols or undergo metabolism by epoxide hydrase to dihydrodiols. Similar studies on the metabolism of naphthalene revealed that a partially purified cytochrome P450 preparation from rat liver microsomes metabolizes this hydrocarbon primarily to 1-naphthol, whereas addition of partially purified epoxide hydrase to the cytochrome P450 system results in the formation of large amounts of 1,2-dihydroxy-1,2-dihydronaphthalene and decreased amounts of 1-naphthol (11). The availability of purified cytochrome P450-dependent monooxygenases and epoxide hydrase thus facilitates the demonstration of intermediate arene oxides in the metabolism of aromatic hydrocarbons. The use of these purified enzymes should also be helpful in evaluating the role of intermediate epoxides and arene oxides in the mutations or malignant transformations that can be induced in living cells when organic chemicals undergo metabolic activation. The results of our studies with purified enzymes demonstrate that the metabolic activation of BP to mutagenic metabolites by the cytochrome P448-reconstituted monooxygenase is only partially inhibited by purified epoxide hydrase. Most of the protection which occurs on addition of epoxide hydrase can be attributed to the conversion of the mutagenic BP-4,5-oxide to the corresponding nonmutagenic dihydrodiol. It was of interest that the metabolic activation of BP could not be completely abolished by high levels of epoxide hydrase. This result indicates that a metabolite is formed which either is not inactivated by epoxide hydrase or reacts with critical cellular macromolecules (30) at a rate substantially faster than its enzymatic hydration rate.

Studies on the mutagenic activity of 28 BP metabolites and closely related derivatives revealed that BP-4,5-oxide and BP-7,8-diol-9,10-epoxide were highly mutagenic in several strains of *Salmonella typhimurium* and in Chinese hamster V79 cells. It was of considerable interest that our synthetic BP-7,8-diol-9,10-epoxide was 30- to 40-fold more mutagenic than BP-4,5-oxide in Chinese hamster V79 cells and 10- to 25-fold more mutagenic than BP-4,5-oxide in the TA 100 strain of *Salmonella typhimurium.* BP-7,8- and -9,10-oxides were weakly mutagenic in several strains of *Salmonella* and in Chinese hamster V79 cells. Interestingly, BP-11,12-oxide was only weakly active in several strains of *Salmonella typhimurium* but was highly active (equal to BP-4,5-oxide) in Chinese hamster V79 cells. These results with BP-11,12-oxide point out the importance of using several model systems for mutagenicity studies. Studies with *Salmonella* strain TA 1538 revealed that 1-HOBP, 3-HOBP, 6-HOBP, 7-HOBP, and 12-HOBP were weakly active mutagens. Seven other BP-phenols, six dihydrodiols, and five quinones were nonmutagenic.

Primary metabolites of BP can undergo further oxidative metabolism at other sites on the hydrocarbon molecule, and recent studies indicate that such secondary metabolites of BP can interact with DNA. Borgen *et al.* (31) have shown that BP-7,8-dihydrodiol, in the presence of hamster liver microsomes and NADPH, binds to DNA to a tenfold greater extent than does BP under the same conditions. These investigators concluded that BP-7,8-dihydrodiol, formed by the enzymatic hydration of BP-7,8-oxide, was further metabolized to reactive metabolites. One of these metabolites may be a 7,8-dihydrodiol-9,10-epoxide.

Sims *et al.* (32) have recently reported evidence for the *in vitro* formation of a BP-7,8-diol-9,10-epoxide by rat liver microsomes, and they indicated that their epoxide bound to DNA. The stereochemistry or purity of this diol epoxide was not provided. Although the carcinogenicity of such secondary metabolites has not been reported, our studies with BP-7,8-diol-9,10-epoxide of known stereochemistry indicate that this compound is a strong mutagen. It should be noted that numerous other secondary metabolites of BP which may bind to DNA are also possible.

Mutagenesis screens may reflect the extent to which compounds are capable of binding to DNA. In this system, compounds which are highly susceptible to nucleophilic attack by DNA relative to other types of reactions could be the most potent mutagens. Thus the relatively stable K-region arene oxides might be expected to be better mutagens than non-K-region arene oxides, which undergo facile isomerization to phenols (33). Furthermore, a diol-epoxide (which is not an arene oxide) could be expected to be an even better mutagen since the rate of nucleophilic attack relative to other reactions could be quite high.

It was of considerable interest that the highly mutagenic and stable BP-4,5-oxide (K-region oxide) has only weak carcinogenic activity for mouse skin, whereas the weakly mutagenic and unstable BP-7,8-oxide was strongly carcinogenic for mouse skin. The weakly mutagenic and unstable BP-9,10-oxide was noncarcinogenic. Although our data are consistent with the metabolic activation of BP to BP-7,8-oxide, BP-7,8-dihydrodiol, and BP-7,8-diol-9,10-epoxide, additional studies are needed on the relative carcinogenic activity of these and other reference compounds.

SUMMARY

In a solubilized, reconstituted cytochrome P448 monooxygenase system that is devoid of epoxide hydrase, benzo[*a*]pyrene (BP) is metabolized primarily to phenols and quinones but is not appreciably metabolized to dihydrodiols. Addition of purified epoxide hydrase results in the formation of large amounts of dihydrodiols and decreased formation of phenolic metabolites. Metabolic activation of BP to mutagenic metabolites by the cytochrome P448-reconstituted monooxygenase is inhibited by purified epoxide hydrase.

BP-4,5-oxide, BP-11,12-oxide, and the 7,8-diol-9,10-epoxide of BP are highly active mutagens in *Salmonella typhimurium* and/or Chinese hamster V79 cells. The diol-epoxide was the most mutagenic BP derivative studied. BP-7,8-oxide, BP-9,10-oxide, 1-HOBP, 3-HOBP, 6-HOBP, 7-HOBP, and 12-HOBP are weakly active mutagens in strain TA 1538 of *Salmonella typhimurium.* Seven other BP phenols, six dihydrodiols, and five quinones are nonmutagenic.

The weakly mutagenic and unstable BP-7,8-oxide is an active skin carcinogen in mice, whereas the weakly mutagenic and unstable BP-9,10-oxide and the highly mutagenic and stable BP-4,5-oxide have little or no carcinogenic activity. These results indicate that the mutagenicities or stabilities of these three BP oxides are not correlated with their carcinogenic activities.

Note Added in Proof

Since the occurrence of this symposium, several papers have appeared on the unusually high mutagenic activity of the two stereoisomeric BP 7,8-diol-9,10-epoxides [*Biochem. Biophys. Res. Comm.* **68**, 1006 (1976); *Proc. Nat. Acad. Sci. USA* **73**, 607 (1976); *Nature* **261**, 52, (1976); *Cancer Res.* **36**, 3358 (1976)], the mutagenic activities of 29 BP derivatives [*Cancer Res.* **36**, 3350 (1976)], the mutagenic activity of many of these compounds after metabolic activation [*J. Biol. Chem.* **251**, 4882 (1976)], and the carcinogenicity of BP 7,8-dihydrodiol and several BP oxides and phenols [*Proc. Nat. Acad. Sci. USA* **73**, 243 (1976); *Cancer Res.* **36**, 3625 (1976); *Proc. Nat. Acad. Sci. USA* **73** (Nov. 1976)].

Acknowledgments

We thank Drs. R. Marenpot and J. Weisburger (Naylor Dana Institute, American Health Foundation, Valhalla, New York) for their help with the histopathological examination of tissue. We thank Mrs. Mary Ann Augustin for her excellent help in the preparation of the manuscript.

REFERENCES

1. H. Yagi and D. M. Jerina, A general synthetic method for non-K-region arene oxides, *J. Am. Chem. Soc.* **97**, 3185–3192, (1975).
2. H. Yagi, G. M. Holder, P. Dansette, O. Hernandez, H. J. C. Yeh, R. A. LeMahieu, and D. M. Jerina, Synthesis and spectral properties of the isomeric hydroxybenzo[*a*]pyrenes, *J. Org. Chem.* **41**, 977–985 (1976).
3. P. M. Dansette, O. Hernandez, D. M. Jerina, H. D. Mah, and H. Yagi, manuscript in preparation.
4. D. T. Gibson, V. Mahadevan, D. M. Jerina, H. Yagi, and H. J. Yeh, Oxidation of the carcinogens benzo[*a*]pyrene and benzo[*a*]anthracene to dihydrodiols by bacterium, *Science* **189**, 295–297 (1975).
5. H. Vollman, H. Becker, M. Corell, and H. Streeck, Beitrage zur Kenntnis der Pyrens und seiner Derivate, *Ann. Chem.* **531**, 48–54 (1937).
6. H. E. Schroeder, F. B. Stilmar, and F. S. Palmer, Condensation of phthalideneacetic acid with naphthalene to form benzopyrenequinones, *J. Am. Chem. Soc.* **78**, 446–450 (1956).
7. D. M. Jerina, H. Yagi, and O. Hernandez, Stereoselective synthesis and reactions of a diol-epoxide derived from benzo[*a*]pyrene, Chap. 40, this volume.
8. G. Holder, H. Yagi, P. Dansette, D. M. Jerina, W. Levin, A. Y. H. Lu, and A. H. Conney,

Effects of inducers and epoxide hydrase on the metabolism of benzo[*a*]pyrene by liver microsomes and a reconstituted system: Analysis by high-pressure liquid chromatography, *Proc. Natl. Acad. Sci. USA* **71,** 4356–4360 (1974).

9. J. K. Selkirk, R. G. Croy, P. P. Roller, and H. V. Gelboin, High-pressure liquid chromatographic analysis of benzo[*a*]pyrene metabolism and covalent binding and the mechanism of action of 7,8-benzoflavone and 1,2-epoxy-3,3,3-tri-chloropropane, *Cancer Res.* **34,** 3474–3480 (1974).
10. W. Levin, D. Ryan, S. West, and A. Y. H. Lu, Preparation of partially purified, lipid-depleted cytochrome P-450 and reduced nicotinamide adenine dinucleotide phosphate–cytochrome *c* reductase from rat liver microsomes, *J. Biol. Chem.* **249,** 1747–1754 (1974).
11. P. M. Dansette, D. M. Jerina, J. W. Daly, W. Levin, A. Y. H. Lu, R. Kuntzman, and A. H. Conney, Assay and partial purification of epoxide hydrase from rat liver microsomes, *Arch. Biochem. Biophys.* **164,** 511–517 (1974).
12. H. G. Selander, D. M. Jerina, and J. W. Daly, Metabolism of chlorobenzene with hepatic microsomes and solubilized cytochrome P-450 systems, *Arch. Biochem. Biophys.* **168,** 309–321 (1975).
13. J. E. Tomaszewski, D. M. Jerina, and J. W. Daly, Metabolism of aromatic substrates to phenols by animal monooxygenases: Evidence for a direct oxidative pathway not involving arene oxide intermediates, *Biochemistry* **14,** 2024–2031 (1975).
14. B. N. Ames, F. D. Lee, and W. E. Durston, An improved bacterial test system for the detection and classification of mutagens and carcinogens, *Proc. Natl. Acad. Sci. USA* **70,** 782–786 (1973).
15. J. McCann, N. E. Spingarn, J. Kobori, and B. N. Ames, Detection of carcinogens as mutagens: Bacterial tester strains with R factor plasmids, *Proc. Natl. Acad. Sci. USA* **72,** 979–983 (1975).
16. B. N. Ames, J. McCann, and E. Yamasaki, Methods for detecting carcinogens and mutagens with the *Salmonella*/mammalian-microsome mutagenicity test, *Mutat. Res.* **31,** 347–364 (1975).
17. H. V. Malling, Dimethylnitrosamine: Formation of mutagenic compounds by interaction with mouse liver microsomes, *Mutat. Res.* **13,** 425–429 (1971).
18. R. C. Garner, E. C. Miller, J. A. Miller, J. V. Garner, and R. S. Hanson, Formation of a factor lethal for *S. typhimurium* TA 1530 and TA 1531 on incubation of aflatoxin B_1 with rat liver microsomes, *Biochem. Biophys. Res. Commun.* **45,** 774–780 (1971).
19. R. C. Garner, E. C. Miller, and J. A. Miller, Liver microsomal metabolism of aflatoxin B_1 to a reactive derivative toxic to *Salmonella typhimurium* TA 1530, *Cancer Res.* **32,** 2058–2066 (1972).
20. B. N. Ames, W. E. Durston, E. Yamasaki, and F. D. Lee, Carcinogens are mutagens: A simple system combining liver homogenates for activation and bacteria for detection, *Proc. Natl. Acad. Sci. USA* **70,** 2281–2285 (1973).
21. D. Ryan, A. Y. H. Lu, J. Kawalek, S. B. West, and W. Levin, Highly purified cytochrome P-448 and P-450 from rat liver microsomes, *Biochem. Biophys. Res. Commun.* **64,** 1134–1141 (1975).
22. A. Y. H. Lu, D. Ryan, D. M. Jerina, J. W. Daly, and W. Levin, Liver microsomal epoxide hydrase: Solubilization, purification and characterization, *J. Biol. Chem.* **250,** 8283–8288 (1975).
23. D. M. Jerina, H. Yagi, and J. W. Daly, Arene oxides-oxepins, *Heterocycles* **1,** 267–326 (1973).
24. P. Y. Bruice, T. C. Bruice, H. G. Selander, H. Yagi, and D. M. Jerina, Comparative mechanisms of reaction of K-region and non-K-region arene oxides of phenanthrene, *J. Am. Chem. Soc.* **96,** 6814–6815 (1974).
25. P. Y. Bruice, T. C. Bruice, P. M. Dansette, H. G. Selander, H. Yagi, and D. M. Jerina, A

comparison of the mechanism of solvolysis and rearrangement of K-region vs. non-K-region arene oxides of phenanthrene. Comparative solvolytic rate constants of K-region and non-K-region arene oxides, *J. Am. Chem. Soc.* 98, 2965–2973 (1976).
26. E. H. Y. Chu, and M. V. Malling, Mammalian cell genetics. II. Chemical induction of specific locus mutations in Chinese hamster cells *in vitro, Proc. Natl. Acad. Sci. USA* **61**, 1306–1312 (1968).
27. E. H. Y. Chu, Induction and analysis of gene mutations in mammalian cell cultures, in: *Chemical Mutagens: Principles and Methods for Their Detection* (A. Hollaender, ed.), pp. 411–444, Plenum Press, New York (1971).
28. E. Huberman, L. Aspiras, C. Heidelberger, P. L. Grover, and P. Sims, Mutagenicity to mammalian cells of epoxides and other derivatives of polycyclic hydrocarbons, *Proc. Natl. Acad. Sci. USA* **68**, 3195–3199 (1971).
29. E. Huberman and L. Sachs, Cell mediated mutagenesis of mammalian cells with chemical carcinogens, *Int. J. Cancer* **13**, 326–333 (1974).
30. E. C. Miller and J. A. Miller, Biochemical mechanisms of chemical carcinogenesis, in: *The Molecular Biology of Cancer* (H. Busch, ed.), pp. 377–402, Academic Press, New York (1974).
31. A. Borgen, H. Darvey, N. Castagnoli, T. T. Crocker, R. E. Rasmussen, and I. Y. Wang, Metabolic conversion of benzo[*a*]pyrene by Syrian hamster liver microsomes and binding of metabolites to DNA, *J. Med. Chem.* **16**, 502–504 (1973).
32. P. Sims, P. L. Grover, A. Swaisland, K. Pal, and A. Hewer, Metabolic activation of benzo[*a*]pyrene proceeds by a diol-epoxide, *Nature (London)* **252**, 326–328 (1974).
33. P. Y. Bruice, T. C. Bruice, H. Yagi, and D. M. Jerina, Nucleophilic displacements on the arene oxides of phenanthrene, *J. Am. Chem. Soc.* **98**, 2973–2981 (1976).
34. F. Oesch, Evaluation of the importance of enzymes involved in the control of mutagenic metabolites, in: *Screening Tests in Chemical Carcinogenesis* (R. Montesano, H. Bartsch, and L. Tomatis, eds.), IARC Scientific Publication No. 12, pp. 255–274, Lyon, France (1975).
35. C. Malaveille, H. Bartsch, P. L. Grover, and P. Sims, Mutagenicity of non-K-region diols and diol-epoxides of benz[*a*]anthracene and benzo[*a*]pyrene in *S. typhimurium* TA 100, *Biochem. Biophys. Res. Commun.* **66**, 693–700 (1975).
36. A. W. Wood, R. L. Goode, R. L. Chang, W. Levin, A. H. Conney, H. Yagi, P. M. Dansette, and D. M. Jerina, Mutagenic and cytotoxic activity of benzo[*a*]pyrene 4,5-, 7,8-, and 9,10-oxides and the six corresponding phenols, *Proc. Natl. Acad. Sci. USA* **72**, 3176–3180 (1975).

38

Discussion

Nebert commented that the carcinogenic index in C57BL and DBA2 mice did not correlate with benzo[*a*]pyrene-induced mutagenesis of *Salmonella typhimurium* tester strains TA 1537, TA 1538, TA 98, and TA 100.

Schenkman questioned whether because of the limitations in epoxide hydratase activity, some of the positions in benzo[*a*]pyrene might be hydroxylated by direct insertion. Conney responded that the metabolites were a complex mixture of phenols and indeed it was possible that some may be formed by direct insertion mechanisms.

Oesch commented that Conney's observation that low doses of benzo[*a*]pyrene painted on the skin were more effective in tumorigenesis than the epoxide agrees with similar results observed by Bresnick using methylcholanthrene and its epoxide. Bresnick has also observed that when doses of methylcholanthrene too low to cause tumors were applied to the skin, the methylcholanthrene acted as a promoter and increased the carcinogenicity of the epoxide. Schulman also raised the question of promoters and pointed out that a promoter had been isolated from croton oils. To both comments, Conney responded that he had not yet investigated the effects of promoters.

In response to a question, Conney indicated that there is a latency period of about 2 weeks after giving the benzo[*a*]pyrene before tumors are initiated; the time of onset was about the same whether the dose was small or large. Only large doses, however, produced tumors when the epoxide was given, but the latency period remained the same. In response to another question, he indicated that the carcinogens were painted on in subdued light and the animals were kept in the dark for the ensuing 24 h. All mutagenicity studies were performed in subdued light.

Lenk questioned the stability of the primary oxygenation products during high-pressure liquid chromatography. Conney responded that some of the metabolites may well be unstable and further work will be required to identify them.

39

Polycyclic Hydrocarbon Epoxides as Active Metabolic Intermediates

Peter Sims
Chester Beatty Research Institute
Institute of Cancer Research
Royal Cancer Hospital
London SW3 6JB, United Kingdom

Many chemical carcinogens, including the polycyclic hydrocarbons, require metabolic activation before they can react with cellular macromolecules and probably before they can exert their biological effects (1). With the polycyclic hydrocarbons, this activation is carried out by both human and animal tissues, and the intermediates thus formed are believed to be epoxides (2). In this chapter, the overall metabolic pathways by which different types of epoxides are formed from polycyclic hydrocarbons, the routes by which these epoxides are further metabolized, and the structures of the epoxides that react with glutathione and probably with DNA in cells are considered.

FORMATION OF EPOXIDES IN THE METABOLISM OF POLYCYCLIC HYDROCARBONS

Simple Epoxides

Studies on the metabolism of polycyclic hydrocarbons have been carried out in a variety of systems including whole animals (3), cells in culture (4, 5), tissues in short-term organ culture (6), and tissue preparations obtained from both human and animal sources (7–9). Although there are some variations between the products obtained in the different types of experiments, the overall pattern of metabolites is very similar and, because of this, a great deal of the published work on hydrocarbon metabolism has been carried out using readily available tissues such as rat liver, even though the hydrocarbons are usually considered to be biologically inactive in these tissues.

The initial process in the metabolism of the hydrocarbons is carried out by the monooxygenases present on the endoplasmic reticulum of cells, and the enzymes require NADPH and molecular oxygen. The primary products are simple epoxides which can be formed on either K-region or non-K-region bonds. The formation of K-region epoxides from a number of polycyclic hydrocarbons by microsomal fractions of rat liver (9) and lung (8) and of human lung (7) has been demonstrated directly, but with the exception of naphthalene, which is metabolized to naphthalene 1,2-oxide by rat liver microsomal fractions (10), direct evidence for the formation of simple non-K-region epoxides has not been obtained. However, since aromatic hydrocarbons are metabolized by tissue preparations to non-K-region dihydrodiols, it is presumed that these arise through the intermediate formation of non-K-region epoxides. Some tissues that probably metabolize polycyclic hydrocarbons to epoxides are listed in Table 1.

The simple epoxides are metabolized by one of the three metabolic routes shown in Fig. 1. The first route involves conjugation with glutathione and this is carried out by glutathione *S*-epoxide transferases that are present in the cytosol of cells of many tissues including liver and lung (11). K-region epoxides are much better substrates than non-K-region epoxides for these enzymes (12), and glutathione conjugates formed from non-K-region epoxides are not detected as metabolites when hydrocarbons such as benz[*a*]anthracene[1] are metabolized by rat liver preparations. The glutathione conjugates are usually considered to be detoxication products since they are water soluble and are unlikely to reenter the microsomal lipid membrane. Moreover, there is no evidence to suggest that they are further metabolized to active species by microsomal enzymes, although in whole animals the simpler hydrocarbons are excreted as mercapturic acids that probably arise by the metabolism of glutathione conjugates at the peptide side chains (13).

The simple epoxides, both K-region and non-K-region, are also metabolized by a route that involves the hydration of these compounds with the formation of dihydrodiols, usually of *trans* configuration. The metabolism is carried out by epoxide hydrases (or hydratases) present on the endoplasmic reticulum of cells of many tissues (14). The dihydrodiols were once thought to be true detoxica-

[1] Numbering system for benz[*a*]anthracene:

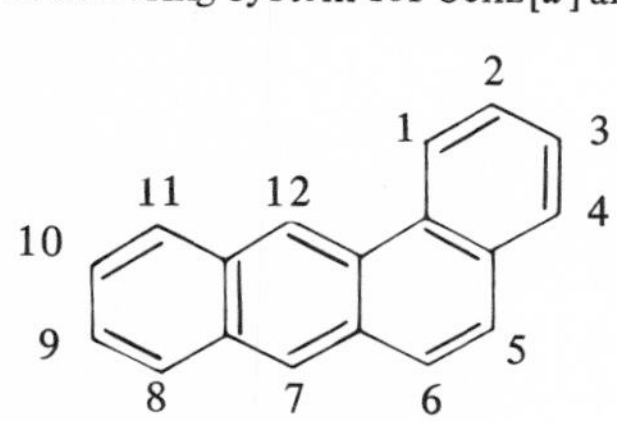

Table 1. Tissues That Convert Polycyclic Hydrocarbons into Epoxides

Tissue	Species	Evidence for epoxide formation
Liver	Rat	Epoxide detected
	Mouse	Dihydrodiols formed
	Hamster	Dihydrodiols formed
	Guinea pig	Dihydrodiols formed
Whole lung	Rat	Epoxides detected
	Man	Epoxides detected
Bronchial epithelium	Man	Dihydrodiols formed
Trachea	Hamster	Dihydrodiols formed
	Rat	Dihydrodiols formed
Skin	Mouse	Dihydrodiols formed
	Man	Dihydrodiols formed
Stomach	Mouse	Dihydrodiols formed
Small intestine	Mouse	Dihydrodiols formed
Lymphocytes	Man	Dihydrodiols formed
Embryo cells	Mouse	Dihydrodiols formed
	Hamster	Dihydrodiols formed

tion products and those from the simple hydrocarbons are converted into sulfuric and glucuronic acid conjugates that are excreted in the urine of treated animals (15). However, as is discussed below, recent work has shown that some dihydrodiols are further metabolized into products that may be biologically active.

A third route by which the simple epoxides, both K-region and non-K-region, are metabolized involves their enzymatic reduction to the parent hydrocarbons by a process that requires NADPH and microsomal fractions and is best demonstrated by carrying out the incubations under anaerobic conditions (16). Although this reduction has so far been shown to be brought about only by rat liver microsomal fractions, it is probable that other tissues could carry out similar reactions. The significance of these reactions in the overall metabolism of the hydrocarbons is not yet clear.

The K-region epoxides will also react with cellular proteins and nucleic acids (17), both chemically and in cells treated with the epoxides, but the evidence presented below suggests that these epoxides are not involved in the reactions with nucleic acids that occur when cells are treated with polycyclic hydrocarbons.

Diol-Epoxides

Over the past few years, it has become apparent that dihydrodiols are further metabolized by microsomal monooxygenases. For example, when 7,12-dimethylbenz[*a*]anthracene was metabolized by rat liver homogenates, the 8,9-dihydrodiol formed as a metabolite was itself further metabolized by enzymes present in the reaction mixture (18). At the same time, products were formed that were not extracted into ethyl acetate and were probably glutathione conjugates. Further investigation showed that the 8,9-dihydrodiol of benz[*a*]-anthracene was converted by rat liver microsomal fractions into the 8,9-dihydro-diol-10,11-epoxide by metabolism on the 10,11-olefinic double bond (19). The diol-epoxide was isolated from the reaction products by thin-layer chromatography and characterized by its UV and mass spectra. The diol-epoxide was also synthesized from the parent dihydrodiol using *m*-chloroperoxybenzoic acid, and its UV and mass spectra showed it to be identical with the metabolite. As shown in Fig. 2, both the dihydrodiol and the diol-epoxide are converted enzymically into a tetrahydrotetrol and a glutathione conjugate by rat liver preparations.

Polycyclic hydrocarbons are metabolized to a number of isomeric dihydrodiols, formed on various double bonds throughout the molecule. Each of these dihydrodiols can, in theory, be metabolized to several diol-epoxides, by metabolism either at adjacent olefinic double bonds or at aromatic bonds in other rings on the molecule, but only a few of these possible diol-epoxides derived from any one hydrocarbon are presently available for study. Methylated hydrocarbons such as 7,12-dimethylbenz[*a*]anthracene are metabolized at the methyl groups by tissue preparation to form hydroxymethyl compounds (20), but there is no evidence that hydroxymethyl groups are directly involved in hydrocarbon–nucleic acid reactions. Some hydroxymethyl derivatives of poly-

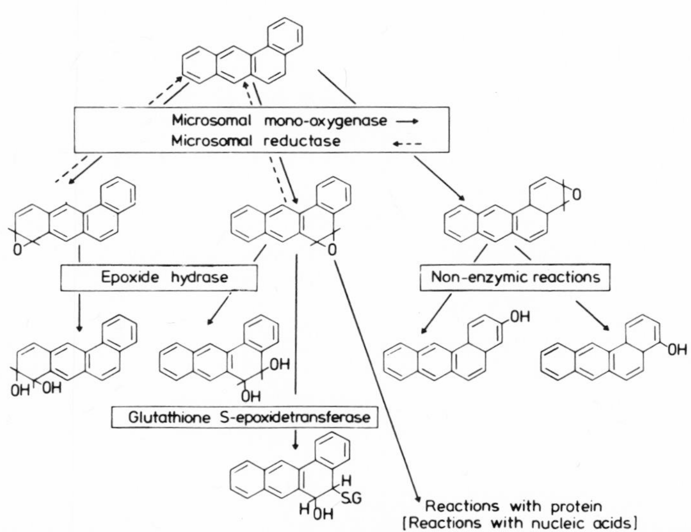

Fig. 1. Metabolism of benz[a]anthracene. From R. Montesano; H. Bartsch, and L. Tomatis (eds.), *Screening Tests in Chemical Carcinogenesis,* IARC Scientific Publication No. 12, Lyon, 1976.

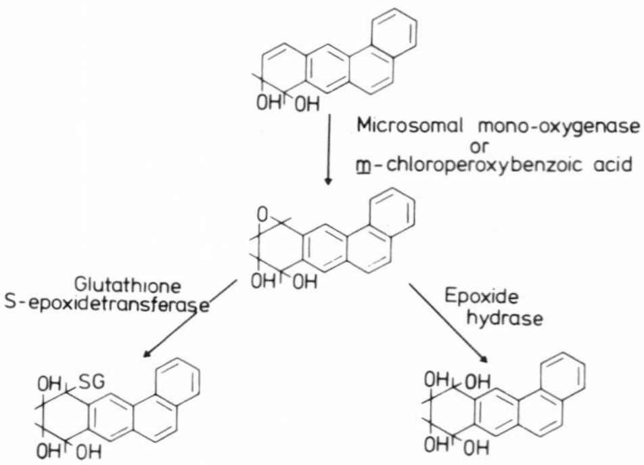

Fig. 2. Metabolism of 8,9-dihydro-8,9-dihydroxybenz[a]anthracene. From *Biochem. Soc. Trans.* **3**, 59–62 (1975).

cyclic hydrocarbons are carcinogenic, but, like their parent hydrocarbons, they are metabolized on the aromatic rings by routes similar to those shown in Fig. 1 to yield dihydrodiols that are probably further metabolized to diol-epoxides. With methylated hydrocarbons, the number of possible diol-epoxides is increased because hydroxymethyl derivatives of diol-epoxides may also be formed; very few of these intermediates have so far been synthesized.

INTERACTIONS OF DIOL-EPOXIDES WITH GLUTATHIONE

It is not yet known just how many diol-epoxides are actually formed from any one hydrocarbon by metabolism. However, experiments with the 8,9-dihydrodiol 10,11-epoxide of benz[*a*]anthracene suggest that they are probably further metabolized to tetrahydrotetrols by epoxide hydrase and to glutathione conjugates by glutathione transferase. Other experiments have shown that some dihydrodiols are further metabolized to more than one glutathione conjugate by rat liver preparations through the intermediate formation of diol-epoxides so that studies on the structures of glutathione conjugates derived from polycyclic hydrocarbons may provide evidence of the types of diol-epoxides initially formed in hydrocarbon metabolism. For example, when the glutathione conjugates obtained from the metabolism of ^{3}H-labeled benz[*a*]anthracene were chromatographed on Sephadex G25 columns in 0.5 M acetic acid, four peaks of radioactivity were obtained (21). One of these contained residual protein carrying covalently bound hydrocarbon and another peak contained a compound that was identified as the K-region glutathione conjugate since it was identical in its chromatographic and spectra properties with the conjugate obtained when the K-region epoxide was allowed to react with glutathione in the presence of glutathione transferase. Studies on the structure of the conjugates obtained when the ^{3}H-labeled 5,6- and 8,9-dihydrodiols of benz[*a*]anthracene nature of the products in the other two peaks of radioactivity. Chromatography on Sephadex G25 columns of the glutathione conjugates obtained from these two dihydrodiols showed that the 5,6-dihydrodiol gave only one conjugate, whereas the 8,9-dihydrodiol yielded two. The conjugate formed from the 5,6-dihydrodiol and one of those formed from the 8,9-dihydrodiol had chromatographic characteristics similar to those of the products forming the third of the radioactive peaks obtained in the experiments with benz[*a*]anthracene, which indicates that one or both of these conjugates are probably formed in the metabolism of the hydrocarbon. Since the UV absorption spectra of the conjugates were similar, it is probable that the 5,6-dihydrodiol is metabolized on the 8,9-bond and the 8,9-dihydrodiol is metabolized on the 5,6-bond since, as shown in Fig. 3, metabolism on these bonds would lead to conjugates with the same type of partially reduced benz[*a*]anthracene nucleus. The second conjugate obtained in the metabolism of the 8,9-dihydrodiol was identical both with the fourth conjugate formed in the metabolism of benz[*a*]anthracene and with the conjugate obtained when synthetic benz[*a*]anthracene 8,9-dihydrodiol-10,11-oxide was allowed to react with glutathione in the presence of glutathione transferase. Thus, as indicated in Fig. 3, at least three diol-epoxides are probably formed in the metabolism of benz[*a*]anthracene. Evidence that 7-methylbenz[*a*]anthracene and 7,12-dimethylbenz[*a*]anthracene are also metabolized to diol-epoxides of these types has been obtained in similar experiments.

As shown in Fig. 4, the 7,8- and 9,10-dihydrodiols derived from benzo[*a*]pyrene are both metabolized by rat liver preparations to products with the chromatographic and UV spectral properties of glutathione conjugates and a

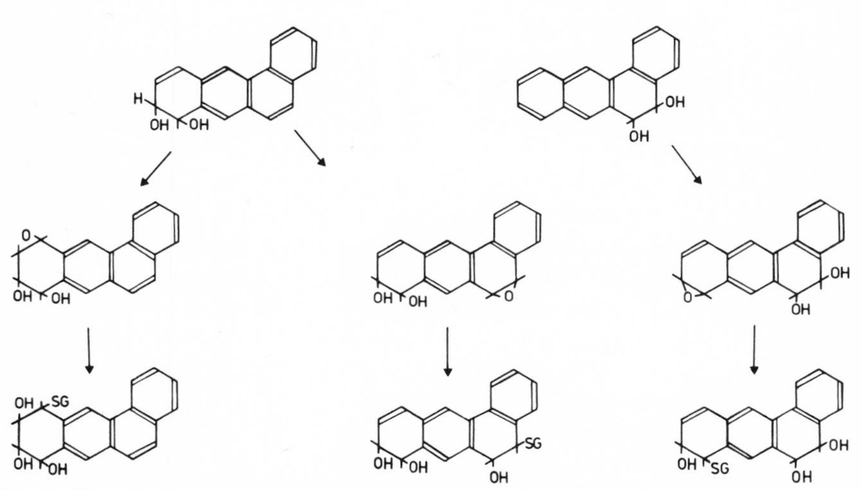

Fig. 3. Glutathione conjugates formed from benz[a]anthracene.—SG, glutathione. From R. Montesano, H. Bartsch, and L. Tomatis (eds.), *Screening Tests in Chemical Carcinogenesis,* IARC Scientific Publication No. 12, Lyon (1976).

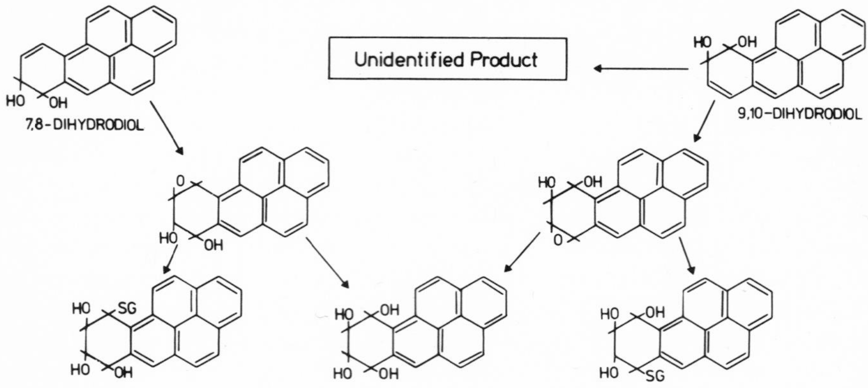

Fig. 4. Metabolism of some benzo[a]pyrene dihydrodiols.

tetrahydrotetrol, and both types of metabolites have probably arisen through the intermediate formation of diol-epoxides (J. Booth and P. Sims, unpublished observations). The mass spectrum of the tetrahydrotetrol obtained from each dihydrodiol was consistent with the proposed structure. In addition to being metabolized at the olefinic double bond, the 9,10-dihydrodiol is metabolized to a large extent to a second product, the formation of which does not appear to involve a diol-epoxide. The metabolite has not yet been identified, but it may be a catechol arising from the dehydrogenation of dihydrodiol, since it is formed from the dihydrodiol in the presence of NADPH by either rat liver soluble fractions (which contain dehydrogenases) or microsomal fractions. The 7,8-dihydrodiol does not form a comparable metabolite with either of these fractions.

The results described above thus suggest that polycyclic hydrocarbons are first metabolized to simple epoxides by microsomal monooxygenases and then are converted into dihydrodiols by epoxide hydrases. The dihydrodiols are further metabolized by the oxygenases to diol-epoxides, the non-K-region compounds being metabolized both on the adjacent olefinic double bonds and on other aromatic double bonds. These diol-epoxides may then be metabolized to tetrahydrotetrols and glutathione conjugates and it is also possible that some of them may react with cellular macromolecules.

INTERACTION OF K-REGION EPOXIDES AND DIOL-EPOXIDES WITH NUCLEIC ACIDS

Investigations that have been carried out over the past few years into the biochemical and biological properties of K-region epoxides have been reviewed (2). These epoxides have been obtained in crystalline form, and studies on some of their properties show that they are able to alkylate nucleic acids and proteins and to induce mutations in several test systems and malignant transformations in rodent cells in culture. These properties suggest that they might be involved in hydrocarbon carcinogenesis.

However, more recent work has made it clear that, whatever their role in carcinogenesis, the K-region epoxides are not the active hydrocarbon metabolites that react with nucleic acids in cells in culture and in tissues treated with polycyclic hydrocarbons. This was first demonstrated in experiments with 7-methylbenz[*a*]anthracene (22), where the hydrocarbon-deoxyribonucleoside products obtained from the enzymatic hydrolysis of DNA from mouse embryo cells that had been treated with the hydrocarbon were compared on columns of Sephadex LH20 with epoxide-deoxyribonucleoside products obtained by the hydrolysis of DNA that was allowed to react chemically with the K-region epoxide, 7-methylbenz[*a*]anthracene-5,6-oxide. The results showed that the two sets of products had different chromatographic properties and were therefore not the same. These experiments were later extended to benz[*a*]anthracene (23) and benzo[*a*]pyrene (24, 25), and similar conclusions were drawn.

Since hydrocarbon-nucleoside products are always eluted from the Sephadex column by methanol–water mixtures earlier than the K-region epoxide-nucleoside products, it is probable that the former products are more polar than the latter. The possibility that products such as diol-epoxides, arising from the further metabolism of dihydrodiols, are involved in the reactions of polycyclic hydrocarbons with cellular DNA has therefore been examined in experiments using either hamster embryo cells or mice. In these experiments, cells or mouse skin was treated with a radioactively labeled compound, either a hydrocarbon or one of its metabolites, and, after 24 h, the DNA was isolated by published procedures. The nucleic acid was hydrolyzed successively with DNAse, phosphodiesterase, and alkaline phosphatase and the deoxyribonucleoside products thus obtained were chromatographed on Sephadex LH20 columns in methanol–water gradient. Fractions were collected and the radioactivity and UV absorption of each fraction measured. Four types of experiments were carried out in which aromatic hydrocarbons labeled either with ^{3}H or with ^{14}C were either incubated with hamster embryo cells or, in a few experiments, painted on the backs of mice. ^{3}H-Labeled dihydrodiols were similarly used to treat embryo cells or mouse skin, ^{3}H-labeled dihydrodiols were incubated with rat liver microsomal fractions in the presence of DNA, or unlabeled or ^{3}H-labeled synthetic diol-epoxides were allowed to react chemically with DNA. In the experiments using unlabeled diol-epoxides, enough reactants were used to allow the hydrocarbon-deoxyribonucleoside products present in fractions from the column to be detected by their UV absorption. As far as possible, comparisons among products from the various types of experiments were made by co-chromatographing hydrolysates from two sources on the same column and measuring either ^{14}C and ^{3}H or, alternatively, ^{3}H and UV absorption in the same fraction. Whenever possible, products present in fractions from the columns were also compared by thin-layer chromatography.

Experiments with benz[*a*]anthracene have been carried out only with hamster embryo cells (23). The results showed that the hydrocarbon-deoxyribonucleoside products obtained from DNA treated with the 8,9-dihydrodiol-10,11-oxide of benz[*a*]anthracene possessed chromatographic characteristics similar to those formed when DNA was incubated with benz[*a*]anthracene-8,9-dihydrodiol in a microsomal incubation system. As already described, this system is known to convert the 8,9-dihydrodiol into the 8,9-dihydrodiol-10,11-epoxide. The epoxide-deoxyribonucleoside products obtained when the diol-epoxide was allowed to react chemically with DNA were also chromatographically similar to the products obtained from the DNA of cells treated with the parent hydrocarbon. No reaction was observed when the 5,6-dihydrodiol was incubated with DNA in a microsomal system: presumably the diol-epoxide that is probably formed from this dihydrodiol either is metabolized before it can react with the nucleic acid or is not an efficient alkylating agent. Although the above results indicate that the 8,9-dihydrodiol-10,11-epoxide may be the species involved in the binding of benz[*a*]anthracene to cellular DNA, the possibility that it is another product such as one of the other possible diol-epoxides shown in Fig. 3 cannot be excluded since the chromatographic properties of epoxide-deoxyribonucleoside products derived from other diol-epoxides of benz[*a*]-anthracene are not yet known.

Benzo[*a*]pyrene yields three major dihydrodiols on metabolism in rat liver microsomal systems, the K-region 4,5-dihydrodiol and the non-K-region 7,8- and 9,10-dihydrodiols (8); the non-K-region dihydrodiols are probably metabolized to the diol-epoxides shown in Fig. 4. Experiments with the hydrocarbon and its derivatives carried out both in hamster embryo cells (24) and on mouse skin (P. L. Grover, A. Hewer, and P. Sims, unpublished observations) gave essentially the same results. Benzo[*a*]pyrene and its 7,8-dihydrodiol each yielded deoxyribonucleoside products that were chromatographically identical with each other. The product obtained when the dihydrodiol was incubated in a rat liver microsomal system in the presence of DNA was chromatographically identical to the product obtained from cells treated with the hydrocarbon. Neither the 4,5-dihydrodiol nor the 9,10-dihydrodiol yielded deoxyribonucleoside products corresponding to those obtained when benzo[*a*]pyrene was used as substrate, indicating that these metabolites may not be involved in the reactions with cellular DNA produced by the hydrocarbon. However, when painted on mouse skin, the 9,10-dihydrodiol yielded a deoxyribonucleoside product, the chromatographic characteristics of which differed from those of the products derived from either the 7,8-dihydrodiol or benzo[*a*]pyrene itself, showing that the metabolic products formed from the 9,10-dihydrodiol are not involved in the interactions with DNA that occur in cells treated with the hydrocarbon.

The epoxide-deoxyribonucleoside products obtained when DNA was allowed to react chemically with preparations thought to contain the 7,8-dihydrodiol-

9,10-epoxide of benzo[*a*]pyrene were chromatographically very similar to the hydrocarbon-deoxyribonucleoside products obtained from the DNA of cells treated with the hydrocarbon. Moreover, evidence that at least some of the hydrocarbon moieties bound to the DNA of mouse skin treated with benzo[*a*]-pyrene retain intact pyrene nuclei has been obtained in experiments using a highly sensitive spectrophotofluorometer (P. Daudel, M. Duquesne, P. Vigny, P. L. Grover, and P. Sims, unpublished observations). These experiments showed that this DNA exhibits a fluorescence spectrum that contains two peaks at wavelengths very similar to those present in the fluorescence spectrum of DNA that was allowed to react with either the 7,8-dihydrodiol-9,10-epoxide or the 9,10-dihydrodiol-7,8-epoxide of benzo[*a*]pyrene. Other peaks in the spectrum of the DNA from mouse skin indicated that two other as yet unidentified products may also be present in this nucleic acid.

Hydrocarbon-deoxyribonucleoside products present in hydrolysates of the DNA from mouse embryo cells treated with 7-methylbenz[*a*]anthracene have been described and similar experiments with hamster embryo cells yielded products with the same chromatographic characteristics. However, the deoxyribonucleoside products obtained when either the 8,9-dihydrodiol-10,11-epoxide of 7-methylbenz[*a*]anthracene or a preparation that probably contains the 3,4-dihydrodiol-1,2-epoxide of the hydrocarbon was allowed to react chemically with DNA were chromatographically similar to, but not identical with, the deoxyribonucleoside products obtained from the DNA of cells treated with the hydrocarbon. Thus it will not be possible to establish if a diol-epoxide is the active metabolite formed from 7-methylbenz[*a*]anthracene until other compounds of this type are synthesized.

The presently available evidence supports the hypothesis that the active intermediates formed from polycyclic hydrocarbons by metabolism arise through the further metabolism of non-K-region dihydrodiols. Although diol-epoxides are formed from the dihydrodiols by metabolism, the role played by this type of compound in the reactions that occur with the DNA of cells treated with hydrocarbons is not yet clear, partly because of difficulties in obtaining sufficient amounts of the relevant dihydrodiols and in characterizing completely some of the diol-epoxides so far synthesized from them. The evidence so far available also suggests that of all the dihydrodiols that may be generated from any one hydrocarbon within cells perhaps only one gives rise to a product that is concerned in the reactions that occur with the DNA of cells treated with a hydrocarbon. The process thus appears to be selective, and this selection may be due to a number of factors. For example, some epoxides formed in the metabolism of hydrocarbons in cells may be destroyed before they can react with nuclear DNA, either by reacting with other macromolecules such as protein or by metabolism to tetrahydrotetrols and glutathione conjugates. It seems probable that in cells treated with hydrocarbons the simple K-region epoxides are metabolized by these reactions since, although the epoxides are formed in the metabolism of the hydrocarbons and will react with DNA in model systems, they do not appear to survive long enough to react with nuclear DNA in cells treated with hydrocarbons. It is also likely that some dihydrodiols formed from hydrocarbons are metabolized by routes that do not lead to epoxides. Thus, as indicated above, the 7,8- and 9,10-dihydrodiols of benzo[*a*]pyrene are metabolized in rat liver preparations to tetrahydrotetrols and glutathione conjugates, but the 9,10-dihydrodiol is also metabolized to a large extent to another product that may not arise through an epoxide intermediate.

Another factor that may contribute to the selective process lies in differences between various diol-epoxides in their effectiveness as alkylating agents toward nucleic acids. However, because only a few diol-epoxides derived from any one hydrocarbon are as yet available for study, little detailed work on these comparisons has been carried out, although all the diol-epoxides so far studied readily react with nucleic acids. In studies on the extent of reaction of some K-region epoxides with RNA and with polyribonucleotides (26), it was shown that reactions were higher with poly(G) than with poly(A) and were low with pyrimidine polynucleotides, although the actual sites of reaction in each case have not been identified. Preliminary results of similar experiments with the benzo[*a*]pyrene diol-epoxides indicate that these diol-epoxides also react to a greater extent with poly(G) and poly(A) than with the pyrimidine polymers (A. W. Murray, P. L. Grover, and P. Sims, unpublished observations).

CONCLUSIONS

Although the evidence presently available is based mainly on chromatographic comparisons, it suggests that non-K-region dihydrodiols give rise to the species involved in the reaction with DNA that occurs when cells or tissues are treated with polycyclic hydrocarbons. If this is correct, then the activation must take place in three stages: first, an initial oxidation by microsomal monooxygenases to form simple non-K-region epoxides; second, hydration of these epoxides by epoxide hydrase to form non-K-region dihydrodiols; third, metabolic oxidation of the dihydrodiols to the active species. The available evidence also suggests that dihydrodiols are metabolized to diol-epoxides, which makes it seem likely that the active species are compounds of this type. In non-K-region dihydrodiols, the double bonds adjacent to the hydroxyl groups are olefinic in type and may therefore be particularly susceptible to metabolism. The diol-epoxides thus formed may well differ in their chemical and biochemical properties from the simple K-region and non-K-region epoxides. Although diol-epoxides themselves have not yet been tested for biological activity, some non-K-region dihydrodiols appear to be more active than their parent hydrocarbons in mouse fibroblast transformation systems (27) and in the *Salmonella typhimurium* TA 100 mutagenicity system in the presence of rat liver homogenates (C. Malaveille, H. Bartsch, P. L. Grover, and P. Sims, unpublished observations). Some of the dihydrodiols so far tested are not necessarily those involved in the formation of active species from the parent hydrocarbons, but they are known to be metabolized on adjacent olefinic bonds by liver preparations to give diol-epoxides and are presumably similarly metabolized by mouse fibroblasts. It may be significant that the related K-region dihydrodiols, which do not possess olefinic-type bonds, are inactive in these transformation systems.

The relevance to hydrocarbon carcinogenesis of the reaction with DNA described above is not yet known. However, active species that react with DNA are also likely to react with RNA and protein, so that although the work described here does not necessarily implicate one specific type of macromolecule as a target in the initiation of carcinogenesis, it does give an indication of the types of reactive species that may be formed from hydrocarbons in cells. Studies of this kind may also give indications as to why some hydrocarbons are carcinogenic whereas others are not.

ACKNOWLEDGMENTS

I gratefully acknowledge the contributions made to this work by Professor A. W. Murray, Drs. J. Booth, P. L. Grover, and A. J. Swaisland, Mrs. Kalyani Pal, and Messrs. A. Hewer and G. R. Keysell. The investigations were supported by grants from the Medical Research Council and the Cancer Research Campaign.

REFERENCES

1. J. A. Miller, Carcinogenesis by chemicals: An overview, G. H. A. Clowes Memorial Lecture, *Cancer Res.* **30,** 559–576 (1970).
2. P. Sims and P. L. Grover, Epoxides in aromatic hydrocarbon metabolism and carcinogenesis, *Adv. Cancer Res.* **20,** 165–274 (1974).
3. E. Boyland and P. Sims, The metabolism of benz[*a*]anthracene, *Biochem. J.* **91,** 493–506 (1964).
4. P. Sims, The metabolism of some aromatic hydrocarbons by mouse embryo cell cultures, *Biochem. Pharmacol.* **19,** 285–297 (1970).
5. P. Sims, P. L. Grover, T. Kuroki, E. Huberman, H. Marquardt, J. K. Selkirk, and C. Heidelberger, The metabolism of benz[*a*]anthracene and dibenz[*a,h*]anthracene and their related "K-region" epoxides, *cis*-dihydrodiols and phenols by hamster embryo cells, *Biochem. Pharmacol.* **22,** 1–8 (1973).
6. K. Pal, P. L. Grover, and P. Sims, The metabolism of carcinogenic polycyclic hydrocarbons by tissues of the respiratory tract, *Biochem. Soc. Trans.* **3,** 174–175 (1974).
7. P. L. Grover, A. Hewer, and P. Sims, K-Region epoxides of polycyclic hydrocarbons: Formation and further metabolism of benz[*a*]anthracene 5,6-oxide by human lung preparations, *FEBS Lett.* **34,** 63–68 (1973).
8. P. L. Grover, A. Hewer, and P. Sims, Metabolism of polycyclic hydrocarbons by rat-liver preparations, *Biochem. Pharmacol.* **23,** 323–332 (1974).

9. P. Sims, Qualitative and quantitative studies on the metabolism of a series of aromatic hydrocarbons by rat-liver preparations, *Biochem. Pharmacol.* **19,** 795–818 (1970).
10. D. M. Jerina, J. W. Daly, B. Witkop, P. Zaltzman-Nireberg, and S. Udenfriend, 1,2-Naphthalene oxide as an intermediate in the microsomal hydroxylation of naphthalene, *Biochemistry,* **9,** 147–155 (1970).
11. P. L. Grover, K-Region epoxides of polycyclic hydrocarbons: Formation and further metabolism by rat-liver preparations, *Biochem. Pharmacol.* **23,** 333–343 (1974).
12. J. Booth and P. Sims, Metabolism of benz[*a*] anthracene epoxides by rat-liver, *Biochem. Pharmacol.* **23,** 2547–2555 (1974).
13. J. Booth, E. Boyland, and P. Sims, The conversion of naphthalene into a derivative of glutathione by rat-liver slices, *Biochem. J.* **74,** 117–122 (1960).
14. F. Oesch, Mammalian epoxide hydrases: Inducible enzymes catalysing the inactivation of carcinogenic and cytotoxic metabolites derived from aromatic and olefinic compounds, *Xenobiotica,* **3,** 305–340 (1973).
15. P. Sims, The conversion of naphthalene into compounds related to *trans*-1,2-dihydro-1,2-dihydroxynaphthalene by rabbits, *Biochem. J.* **73,** 389–395 (1959).
16. J. Booth, A. Hewer, G. R. Keysell, and P. Sims, Enzymic reduction of aromatic hydrocarbon epoxides by the microsomal fraction of rat liver, *Xenobiotica* **5,** 197–203 (1975).
17. P. L. Grover and P. Sims, Interactions of the K-region epoxides of phenanthrene and dibenz[*a,h*] anthracene with nucleic acids and histone, *Biochem. Pharmacol.* **19,** 2251–2259 (1970).
18. J. Booth, G. R. Keysell, and P. Sims, Formation of glutathione conjugates as metabolites of 7,12-dimethylbenz[*a*] anthracene by rat-liver homogenates, *Biochem. Pharmacol.* **22,** 1781–1791 (1973).
19. J. Booth and P. Sims, 8,9-Dihydro-8,9-dihydroxybenz[*a*] anthracene 10,11-oxide: A new type of polycyclic aromatic hydrocarbon metabolite, *FEBS Lett.* **47,** 30–33 (1974).
20. E. Boyland and P. Sims, The metabolism of 7,12-dimethylbenz[*a*] anthracene by rat-liver homogenates, *Biochem. J.* **95,** 780–787 (1965).
21. G. R. Keysell, J. Booth, and P. Sims, Glutathione conjugates as metabolites of benz[*a*] anthracene, *Xenobiotica* **5,** 439–448 (1975).
22. W. M. Baird, A. Dipple, P. L. Grover, and P. Brookes, Studies on the formation of hydrocarbon-deoxyribonucleoside products by the binding of derivatives of 7-methylbenz[*a*] anthracene to DNA in aqueous solution and in mouse embryo cells in culture, *Cancer Res.* **33,** 2386–2392 (1973).
23. A. J. Swaisland, A. Hewer, K. Pal, G. R. Keysell, J. Booth, P. L. Grover, and P. Sims, Polycyclic hydrocarbon epoxides: The involvement of 8,9-dihydro-8,9-dihydrobenz[*a*] anthracene 10,11-oxide in reactions with the DNA of benz[*a*] anthracene-treated hamster embryo cells, *FEBS Lett.* **47,** 34–38 (1974).
24. P. Sims, P. L. Grover, A. Swaisland, K. Pal, and A. Hewer, Metabolic activation of benzo[*a*] pyrene proceeds by a diol-epoxide, *Nature (London)* **252,** 326–328 (1974).
25. W. M. Baird, R. G. Harvey, and P. Brookes, Comparison of the cellular DNA-bound products of benzo[*c*] pyrene with the products formed by the reaction of benzo[*c*]-pyrene-4,5-oxide with DNA, *Cancer Res.* **35,** 54–57 (1975).
26. A. J. Swaisland, P. L. Grover, and P. Sims, Reactions of polycyclic hydrocarbon epoxides with RNA and polyribonucleotides, *Chem.-Biol. Interactions* **9,** 317–326 (1974).
27. H. Marquardt, P. L. Grover, and P. Sims, Transformation of mouse fibroblasts by 8,9-dihydrodiols of 7-methylbenz[*a*] anthracene and 7,12-dimethylbenz[*a*] anthracene, *Proc. Am. Assoc. Cancer Res.* **16,** 17 (1975).

40

Stereoselective Synthesis and Reactions of a Diol-Epoxide Derived from Benzo[a]pyrene

Donald M. Jerina, Haruhiko Yagi, and Oscar Hernandez
National Institute of Arthritis, Metabolism, and Digestive Diseases
National Institutes of Health
Bethesda, Maryland 20014, U.S.A.

In the course of examining the metabolism-induced binding of benzo[*a*]pyrene (BP) to DNA in the presence of active liver microsomes, Borgen *et al.* (1) made the intriguing observation that further metabolism of *trans*-7,8-dihydroxy-7,8-dihydro-BP resulted in much more extensive binding to DNA than that observed with two other metabolic dihydrodiols or with BP itself. Subsequently, Sims *et al.* (2) suggested that the facile binding of this dihydrodiol was mediated by formation of an oxide at the 9,10-double bond. Synthesis of the suspected binding agent, 7,8-dihydroxy-9,10-epoxy,7,8,9,10-tetrahydro-BP, was claimed to result from the action of *m*-chloroperoxybenzoic acid on the parent dihydrodiol (2). We have considered the possibility that the suspected high chemical reactivity of this diol-epoxide from BP might well be due to anchimeric assistance (3,4) by the neighboring hydroxyl group at C-7.

Reaction rates of epoxides with simple nucleophiles are well known to be enhanced by proximate hydroxyl groups. For example, a sterol epoxide (*1*) reacts 18 times faster with azide when a proximate hydroxyl group resides on the same face of the molecule as the oxirane ring when compared to the isomer in which these two groups are *trans* (5). Similarly, the antileukemic agent triptolide (*2*), in which the 14-OH and the 9,11-epoxide are on the same side of a cyclohexane ring, reacts 20 times faster with propanethiol than does the isomer in which these two groups are on opposite faces of the molecule (Fig. 1). Reaction of triptolide with macromolecular thiols has been suggested (6) as the basis of the high biological activity of this triepoxide. Based on the enhanced reactivity of triptolide (6) and the sterol-epoxide (5), 7β,8α-dihydroxy-9β,10β-epoxy-7,8,9,10-tetrahydro-BP would be the ideal isomer of the diol-epoxide to

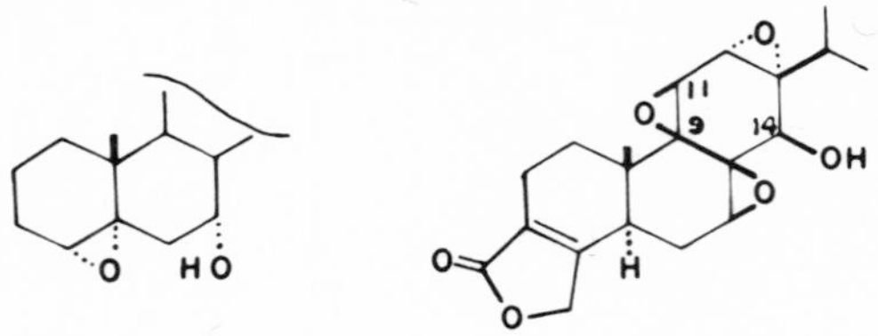

4α,5α-epoxy-7α-hydroxycholestane (1)   triptolide (2)

Fig. 1. Examples of epoxides which show enhanced rates of nucleophilic addition due to the presence of a proximate hydroxyl group (5,6).

employ in studies of mutagenicity, carcinogenicity, and binding to biopolymers since the 7-OH and the epoxide ring are *cis.*

The chemical literature contains ample precedent to indicate that the direct reaction of peroxyacid with the 7,8-dihydrodiol of BP should produce the diol-epoxide in which the 7-OH and the 9,10-epoxide are *trans,* the isomer for which accelerated rates of attack by nucleophiles due to anchimeric assistance cannot be expected. Although the 8-OH and the 9,10-epoxide are on the same face of the molecule, anchimeric assistance has not been observed for this geometric situation. The conformation of the 7,8-dihydrodiol in solution is such that both hydroxyl groups occupy pseudoequatorial positions (7), the ideal conformation for the 8-OH to direct attack by peroxyacid to the same face of the molecule (8,9). In addition, the 7-OH can be expected to direct attack of peroxyacid mainly to the opposite face of the molecule (8,9). Thus both hydroxyl groups can be expected to act in concert to produce the undesired stereoisomer of the diol-epoxide. These predictions on the direction of attack by peroxyacid on *trans*-dihydrodiols were proved correct by the observation that 1β,2α-dihydroxy-3α,4α-epoxy-1,2,3,4-tetrahydronaphthalene (*4*) was the sole diastereomer isolated on reaction of *trans*-1,2-dihydroxy-1,2-dihydronaphthalene (*3*) (10) with *m*-chloroperoxybenzoic acid (Fig. 2).

The initial attempt (2) to prepare the diol-epoxide of BP by direct epoxidation of the 7,8-dihydrodiol was run on such a small scale that adequate

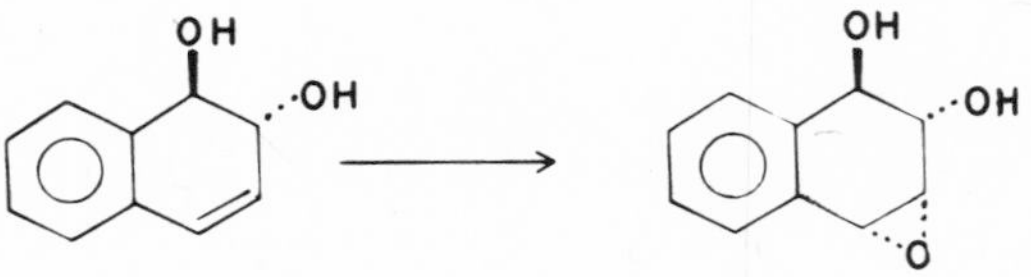

3   4 (5, BP series)

Fig. 2. Direct epoxidation of trans-1,2-dihydroxy-1,2-dihydronaphthalene to form 1β,2α-dihydroxy-3α,4α-eposy-1,2,3,4-tetrahydronaphthalene.

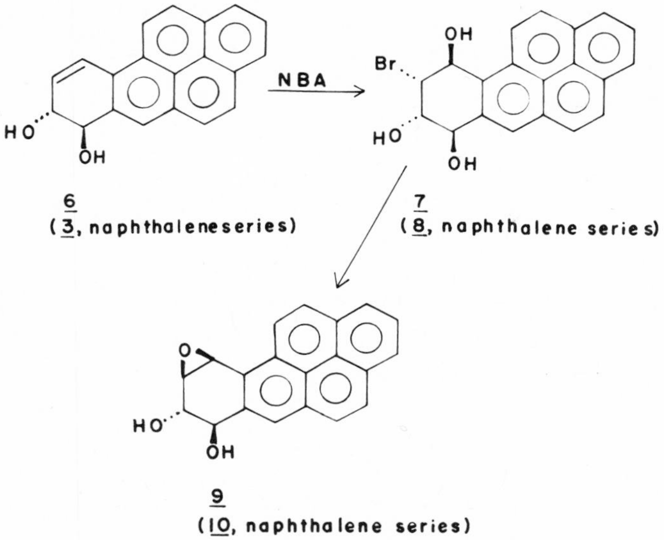

Fig. 3. Stereoselective synthesis of 7β,8α-dihydroxy-9β,10β-epoxy-7,8,9,10-tetrahydro-BP and the corresponding isomer in the naphthalene series via halohydrin pathways.

structural characterization of the TLC-purified reaction product was not possible. Reexamination of this reaction exactly as described but on a much larger scale with synthetic 7,8-dihydrodiol (7) has not allowed isolation of a diol-epoxide. Although the desired compound (presumably *5*, Fig. 2) is present as evidenced by the mass spectrum after silylation of the crude reaction mixture, only the *m*-chlorobenzoic acid adduct at C-10 could be isolated under a variety of workup conditions. Thus attention was directed toward synthesis of the desired diol-epoxide (*9*, Fig. 3) in which the 7-OH and the oxirane ring are *cis.*

The synthetic problem thus becomes one of finding a means of introducing the epoxide with relative stereochemistry opposite to that observed in the peroxyacid reaction. A common means of achieving this, when steric factors alone must be considered, is the formation of an intermediate halohydrin (11,12). Initial attack on the olefin by either peroxyacid or *N*-bromoacetamide occurs predominantly from the least-hindered face of the molecule, and cyclization of the halohydrin produces the epoxide stereoisomer opposite to that formed by peroxyacid. Although it is clear that the kinetic and steric course of epoxidation of dihydrodiols is subject to control by factors other than steric bulk, these factors can be anticipated to cause both *N*-bromoacetamide and peroxyacid to attack the same face of the molecule. This was indeed found to be the case. The dihydrodiols from naphthalene (*3*) and from BP (*6*) were stereoselectively converted to the desired halohydrins (*7* and *8*, Fig. 3) in which the benzylic hydroxyl groups are *cis.* Cyclization with base produced the diol-epoxides (*9* and *10*, Fig. 3) with stereochemistry similar to that in triptolide (*2*).

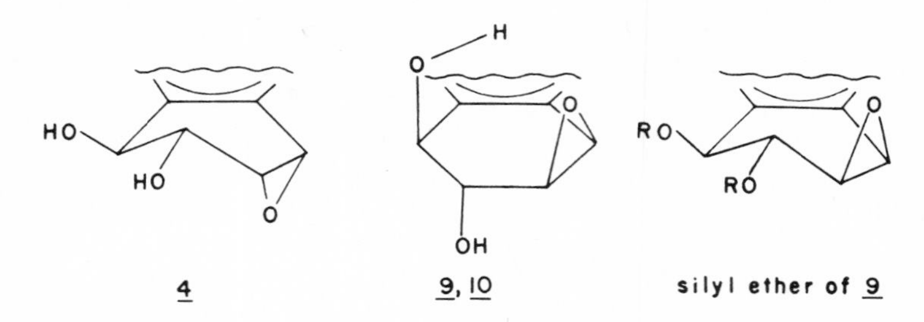

Fig. 4. Preferred conformation of the diol-epoxides 4, 9, *and* 10 *as well as the bis-silyl ether of* 9.

Detailed analysis of the nuclear magnetic resonance spectra of the diol-epoxides *4, 9,* and *10* confirmed that the relative stereochemistry in these molecules was as assigned. Magnitudes of the coupling constants (J) were also indicative of the most highly occupied conformations. Thus, for the diol-epoxide *4* from naphthalene in which the benzylic hydroxyl group and the epoxide are on opposite sides of the ring and cannot undergo intramolecular hydrogen bonding, the hydroxyl groups occupy mainly pseudoequatorial positions ($J_{1,2}$ = 9.0 Hz) as shown in Fig. 4. In contrast, the hydroxyl groups in both *9* and *10* are mainly pseudoaxial ($J_{1,2}$ = 3.0 Hz and $J_{7,8}$ = 6.0 Hz). The benzylic hydroxyl groups appear to be intramolecularly hydrogen-bonded to the epoxides (Fig. 4) since silylation of the hydroxyl groups causes a change in conformation ($J_{7,8}$ = 9.0 Hz) in the silyl ether of *9.* The change in conformation which puts the bulkier silyl ether groups into a more pseudoequatorial conformation argues for the importance of intramolecular hydrogen bonding in causing *9* and *10* to preferentially assume the conformation in which the free hydroxyl groups are pseudoaxial. In the absence of intramolecular hydrogen bonding to the epoxide, both the diols and the diol-epoxide produced by peroxyacid prefer the conformation with the hydroxyl groups pseudoequatorial. The complete synthetic and spectral details of these studies have been published (13).

In order to establish relative reactivity for the diol-epoxides *4, 9,* and *10,* a convenient spectrophotometric assay was sought in which the chromophore of the attacking nucleophile would change on reaction with the epoxide. 4-Nitrothiophenol with a pK_a~4.5 (14) proved ideal for this purpose. The anion form has a strong absorption in aqueous solution at 415 nm, while the acid form as well as reaction products resulting from alkylation by an epoxide would not absorb in this region. Second-order nucleophilic rate constants were measured by monitoring the disappearance of 4-nitrothiophenolate in the presence of a large excess of the epoxides at pH 7.4 (μ 0.1, phosphate) in water–ethanol (3:1). Comparison of the rate constants for the diol-epoxides from naphthalene (*4* and *10*) established that anchimeric assistance by the benzylic hydroxyl group at C-1 did not accelerate the rate of *10* (0.10 $M^{-1}s^{-1}$) relative to *4* (0.22 $M^{-1}s^{-1}$) as might have been anticipated. In fact, *4* reacted slightly faster. For purposes of comparison, phenanthrene-9,10-oxide reacts with a second-order rate constant of 2.1 $M^{-1}s^{-1}$ under these conditions. Measurement of the rate constant for the diol-epoxide *9* from BP was complicated by its exceptionally fast reaction rate with the thiolate as well as its instability in the solvent. Presumably solvolysis occurs to form a tetraol. The second-order rate constant for *9* is estimated to be 2 or possibly 3 orders of magnitude faster than that for *4* and *10*.

At present, a definitive explanation of the remarkably high reactivity of *9* is not possible. Although anchimeric assistance to opening of the oxirane ring by the benzylic hydroxyl group was not evident for the diol-epoxides (*4* and *10*) from naphthalene, slightly altered geometry in the diol-epoxide *9* from BP could lead to some degree of anchimeric assistance. An alteration in ground state and transition state geometry can be expected because of steric hindrance due to H_{11} in the "bay region" of the hydrocarbon. The size and nature of the aromatic portion of the hydrocarbon must also play a role in determining relative rates. Finally, changes in mechanism from direct displacement (S_N2), to tight ion pair trapping, to a free carbonium ion (S_N1) are possible. Further study of *9* may provide insight into the origin of its high mutagenic activity toward bacteria and mammalian cells in culture (15). Additionally chemical reaction of *9* with DNA produces a nucleoside product which cochromatographs with one of the nucleoside peaks obtained when BP and DNA are incubated in the presence of liver microsomes (2,16). Preliminary studies on the binding of *9* to poly(G) indicate that the bound product is particularly sensitive to cleavage by acid which may be indicative of alkylation at the oxygen on C-6 of guanine. Arguments have been presented which suggest that such alkylations may be important in the induction of cancer (17,18).

ADDENDUM

After preparation of the manuscript, an interesting article by Hulbert appeared (19) in which arguments are presented for the importance of the diol-epoxide stereoisomer *9* due to its potential for enhanced reactivity compared to *5*. In addition, McCaustland and Engel have described improved methodology (20) for the preparation of a diol-epoxide from *6* by treatment with peroxyacid (2). Although we were still unable to obtain a pure product by this modification (20), the use of tetrahydrofuran as solvent for the epoxidation did produce the desired compound. We thank McCaustland for this suggestion.

The diol-epoxide produced from *6* by reaction with *m*-chloroperoxybenzoic acid in tetrahydrofuran has the relative stereochemistry shown in structure *5* (Fig. 2). Examination of the nuclear magnetic resonance coupling constants ($J_{7,8}$

= 9.0 Hz) indicates that the preferred conformation is that in which the hydroxyl groups occupy pseudoequatorial positions (*cf.* structure *4,* Fig. 4) and thus do not participate in intramolecular hydrogen bonding to the oxirane ring.

Because of the high solvolytic rate for *9* in water–alcohol, a solvent was sought which would readily dissolve the compound and have low nucleophilicity. *tert*-Butyl alcohol met both of these requirements, and reaction rates between sodium *p*-nitrothiophenolate and large excesses of the diol-epoxides were determined at 30°C. The following rate constants were observed: 0.01 $M^{-1}s^{-1}$ for *4* and 3.3 $M^{-1}s^{-1}$ for *10* in the naphthalene series, 0.43 $M^{-1}s^{-1}$ for *5* and 70 $M^{-1}s^{-1}$ for *9* in the BP series, and 0.15M^{-1} for 7,8,9,10-tetrahydrobenzo[*a*]-pyrene 9,10-epoxide (21). Comparison of the reaction rates of the diol-epoxides prepared by peroxyacid (*4* and *5*) with those obtained via halohydrins (*10* and *9*) indicates that the isomers with stereochemistry similar to that in triptolide (*9* and *10*) are more than a hundredfold more reactive, a convincing argument for participation by the benzylic hydroxyl group in anchimeric assistance. The enhanced reactivity of *9* is even more dramatic when compared to 7,8,9,10-tetrahydrobenzo[*a*]pyrene-9,10-epoxide, which lacks the hydroxyl groups. Origin of the enhanced reactivity in the BP series relative to the naphthalene series is under study.

Note Added in Proof

The subject of diol epoxides has received wide attention and significant advances have occurred since the present Symposium was held. The current status of the testing of the 7,8-diol-9,10-epoxides of BP for mutagenic activity has been discussed elsewhere in this volume (15). Of particular interest to the present chapter is the observation that 7,8,9,10-tetrahydrobenzo[*a*]pyrene 9,10-epoxide, an analog of diol epoxides *5* and *9* which lacks the 7- and 8-hydroxyl groups, is more mutagenic toward *S. typhimurium* strain TA 98 than either of the diol epoxides (22). Rapid hydrolysis of the diol epoxides *9* and *10* by *cis* and *trans* addition of water at the 10-position (23) determines their half-lives in aqueous media. Half-life must, therefore, be considered when comparing relative mutagenic activity. Major covalent adducts which form when the diol epoxides react with polyguanylic acid consist of *trans* opening of the epoxide of *5* by the 2-amino group of guanosine (23) and *cis* and *trans* opening of the epoxide *9* by the 2-amino group of guanosine as well as alkylation of the phosphate backbone (24). A theory has been presented, based on existing carcinogenicity data for substituted hydrocarbons (25) and quantum mechanical calculations (26), which predicts that "bay region" epoxides of non-K-region dihydrodiols will have the highest biological activity. Confirmation of this theory has been obtained through study of the dihydrodiols of benzo[*a*]anthracene (27).

REFERENCES

1. A. Borgen, H. Darvey, N. Castagnoli, T. T. Crocker, R. E. Rasmussen, and I. Y. Wang, Metabolic conversion of benzo[a]pyrene by Syrian hamster liver microsomes and binding of metabolites to deoxyribonucleic Acid, *J. Med. Chem.* **16,** 502–506 (1973).
2. P. Sims, P. L. Grover, A. Swaisland, K. Pal, and A. Hewer, Metabolic activation of benzo[a]pyrene proceeds by a diol-epoxide, *Nature (London)* **252,** 326–328 (1974). P. Sims, Polycyclic hydrocarbon epoxides as active metabolic intermediates, Chap. 39, this volume.
3. B. Capon, Neighbouring group participation, *Q. Rev.* **18,** 45–111 (1964).
4. A. J. Kirby and A. R. Fersht, Intramolecular catalysis, in: *Progress in Bioorganic Chemistry,* Vol. 1 (E. T. Kaiser and F. J. Kezdy, eds.), pp. 1–82, Wiley-Interscience, New York (1971).
5. D. H. R. Barton and Y. Houminer, Neighbouring hydroxy-group participation in the opening of epoxides by nucleophiles, *J. Chem. Soc. Chem. Commun.* **1973,** 839–840.
6. S. M. Kupchan and R. M. Schubert, Selective aklylation: A biomimetic reaction of the antileukemic triptolides, *Science* **185,** 791–793 (1974).
7. D. T. Gibson, V. Mahadevan, D. M. Jerina, H. Yagi, and H. J. C. Yeh, Oxidation of the carcinogens benzo[a]pyrene and benzo[a]anthracene to dihydrodiols by a bacterium, *Science* **189,** 295–297 (1975).
8. H. B. Henbest, Stereoselectivity in the reactions of cyclic compounds, *Proc. Chem. Soc. (London)* **1963,** 159–165.
9. P. Chamberlain, M. L. Roberts, and G. H. Whitham, Epoxidation of allylic alcohols with peroxy-acids: Attempts to define transition state geometry, *J. Chem. Soc. (B)* **170,** 1374–1381 (1970).
10. J. Booth, E. Boyland, and E. E. Turner, The reduction of *o*-quinones with lithium aluminum hydride, *J. Chem. Soc.* **1950,** 1188–1190.
11. G. Berti, Stereochemical aspects of the synthesis of 1,2-epoxides, in: *Topics in Stereochemistry,* Vol. 7 (N. L. Allinger and E. L. Eliel, eds), pp. 93–251, Wiley-Interscience, New York (1973).
12. W. Cocker and D. H. Grayson, A convenient preparation of (–)-β-3,4-epoxycarane, *Tetrahedron Lett.* **1969,** 4451–4452.
13. H. Yagi, O. Hernandez, and D. M. Jerina, Synthesis of (±)-7β,8α-dihydroxy-9β,10β-epoxy-7,8,9,10-tetrahydrobenzo[a]pyrene, a potential metabolite of the carcinogen benzo[a]pyrene with stereochemistry related to the antileukemic tripolides, *J. Am. Chem. Soc.* **97,** 6881–6883 (1975).
14. W. P. Jencks and K. Salvesen, Equilibrium deuterium isotope effects on the ionization of thiol acids, *J. Am. Chem. Soc.* **93,** 4433–4436 (1971).
15. A. H. Conney, A. W. Wood, W. Levin, A. Y. H. Lu, R. L. Chang, P. G. Wislocki, R. L. Goode, G. M. Holder, P. M. Dansette, H. Yagi, and D. M. Jerina, Metabolism and biological activity of benzo[a]pyrene and its metabolic products, Chap. 37, this volume.
16. D. W. Nebert, A. R. Boobis, H. Yagi, D. M. Jerina, and R. E. Kouri, Genetic differences in benzo[a]pyrene carcinogenic index *in vivo* and in mouse cytochrome $P_1$450-mediated benzo[a] pyrene metabolite binding to DNA *in vitro,* Chap. 12, this volume.
17. P. Brookes, Role of covalent binding in carcinogenicity, Chap. 54, this volume.
18. P. D. Lawley, D. J. Orr, and M. Jarman, Isolation and identification of products from alkylation of nucleic acids: Ethyl- and isopropyl-purines, *Biochem. J.* **145,** 73–84 (1975).

19. P. Hulbert, Carbonium ion as ultimate carcinogen of polycyclic aromatic hydrocarbons, *Nature (London)* **256,** 146–148 (1975).
20. D. J. McCaustland and J. P. Engel, Metabolites of polycyclic aromatic hydrocarbons. II. Synthesis of 7,8-dihydrobenzo[*a*]pyrene-7,8-diol and 7,8-dihydrobenzo[*a*]pyrene-7,8-epoxide, *Tetrahedron Lett.* **1975,** 2549–2552.
21. J. F. Waterfall and P. Sims, Epoxy derivatives of aromatic polycyclic hydrocarbons: The preparation and metabolism of epoxides related to benzo[*a*]pyrene and to 7,8- and 9,10-dihydrobenzo[*a*]pyrene, *Biochem. J.* **128,** 265–277 (1972).
22. A. W. Wood, P. G. Wislocki, R. L. Chang, W. Levin, A. Y. H. Lu, H. Yagi, O. Hernandez, D. M. Jerina, and A. H. Conney, Mutagenicity and cytotoxicity of benzo[*a*]pyrene benzo-ring epoxides, *Cancer Res.* **36,** 3358–3366 (1976).
23. A. M. Jeffrey, K. W. Jennette, S. H. Blobstein, I. B. Weinstein, F. A. Beland, R. G. Harvey, H. Kasai, I. Miura, and K. Nakanishi, Benzo[*a*]pyrene-nucleic acid derivative found *in vivo:* Structure of a benzo[*a*]pyrene tetrahydrodiol, epoxide-guanosine adduct, *J. Am. Chem. Soc.* **98,** 5714–5715 (1976).
24. M. Koreeda, P. D. Moore, H. Yagi, H. J. C. Yeh, and D. M. Jerina, Alkylation of polyguanylic acid at the 2-amino group and phosphate by the potent mutagen (±)-7β-8α-dihydroxy-9β,10β-epoxy-7,8,9,10-tetrahydrobenzo[*a*]pyrene, *J. Am. Chem. Soc.* **98,** 6720–6722 (1976).
25. D. M. Jerina and J. W. Daly, Oxidation at carbon, in: *Drug Metabolism* (D. V. Parke and R. L. Smith, eds.), pp. 15–33, Taylor and Francis, Ltd., London (1976).
26. D. M. Jerina, R. E. Lehr, H. Yagi, O. Hernandez, P. M. Dansette, P. G. Wislocki, A. W. Wood, R. L. Chang, W. Levin, and A. H. Conney, Mutagenicity of benzo[*a*]pyrene derivatives and the description of a quantum mechanical model which predicts the ease of carbonium ion formation from diol epoxides, in: *In Vitro Metabolic Activation in Mutagenesis Testing* (F. J. de Serres, J. R. Bend, and R. M. Philpot, eds.), in press, Elsevier (1976).
27. D. M. Jerina, R. Lehr, M. Schaefer-Ridder, H. Yagi, J. M. Karle, D. R. Thakker, A. W. Wood, A. Y. H. Lu, D. Ryan, S. West, W. Levin, and A. H. Conney, Bay region epoxides of dihydrodiols: A concept which explains the mutagenic and carcinogenic activity of benzo[*a*]pyrene and benzo[*a*]anthracene, in: *Origins of Human Cancer* (J. D. Watson and H. Hiatt, eds.), in press, Cold Spring Harbor Laboratory (1977).

41

Discussion

Sims was asked if anything is known about the initial attack of benzpyrene epoxide on nucleic acid. He replied that the reaction sequence was not known but that a likely reaction sequence would be the intercalation of the larger planar aromatic system with the DNA, followed by the covalent binding reaction. Sims compared this reaction sequence with affinity labeling of an enzyme and suggested that it would have the interesting feature of being much like an S_N1 reaction. As a consequence, a poor nucleophil such as DNA could be attacked even in the presence of strong nucleophils such as sulfhydryl groups. If the interaction of the epoxide with DNA were a S_N2 reaction, preferential reaction with sulfhydryl groups would be expected.

Lijinski asked if anyone could tell him why benzanthracene is several orders of magnitude weaker as a carcinogen than benzpyrene and, since the epoxidation reaction takes place in the liver, why these compounds do not produce liver tumors. Jerina replied that the reason benzanthracene is a weaker carcinogen is not clear, but pointed out that it might be significant that the diol-epoxide group in benzanthracene diol-epoxide, which Sims has shown probably to be the reactive species, is not the bay-region. He suggested that the bay-region might impose conformation restraints on a diol-epoxide group which might greatly enhance its reactivity and possibly its specificity.

Ullrich asked whether the diol-epoxide is a substrate for epoxide hydratase. Sims replied that all diol-epoxides tested so far are substrates. Ullrich also asked if the reaction sequence in the formation of diol-epoxides is known: hydrocarbon → epoxide → diol → diol-epoxide, or hydrocarbon → epoxide → di(epoxide) → diol-epoxide. Sims replied that both situations seemed possible; starting with the hydrocarbon as substrate, one would get the epoxide product and possibly the second epoxide, whereas the dihydrodiol as substrate leads to the diol-epoxide.

42

Additional Routes in the Metabolism of Phenacetin

Tilman Fischbach, Werner Lenk, and Dieter Sackerer
Pharmakologisches Institut der Universität München
D-8 München, Germany

In a study on the biogenesis and metabolism of *N*-hydroxyphenacetin in rats and rabbits, *N*-hydroxyphenacetin and *N*-[4-(2-hydroxyethoxy)phenyl] acetamide were detected as new primary oxygenation products of phenacetin in addition to *N*-(4-ethoxyphenyl)glycolamide and *N*-(4-hydroxyphenyl)acetamide (NAPAP) (Fig. 1) in hepatic microsomal suspensions from untreated or phenobarbital (PB) or 3-methylcholanthrene (3-MC) treated rabbits.

METHODS

Formation *in Vitro*

Incubates (100 ml total volume) contained $NADP^+$ (1.2 mM), glucose-6-phosphate (10 mM), 350 IU of glucose-6-phosphate-dehydrogenase per liter, $MgCl_2$ (6 mM), NaF (0.1 M), microsomal protein (3 mg/ml), and phenacetin (1 mM), which was added in methanol solution (0.5 ml) to the incubates. The flasks were shaken at 37°C for 30 min at high frequency.

Isolation and Determination

In the early experiments, the incubation was stopped by addition of an equal amount of ether. After the addition of 3 ml of 2 M HCl, a second extraction followed. The combined extracts were shaken 2 times with 15 ml of 2 M NaOH each to separate the acidic from the neutral and basic metabolites. The ether extracts were evaporated at reduced pressures and the metabolites were separ-

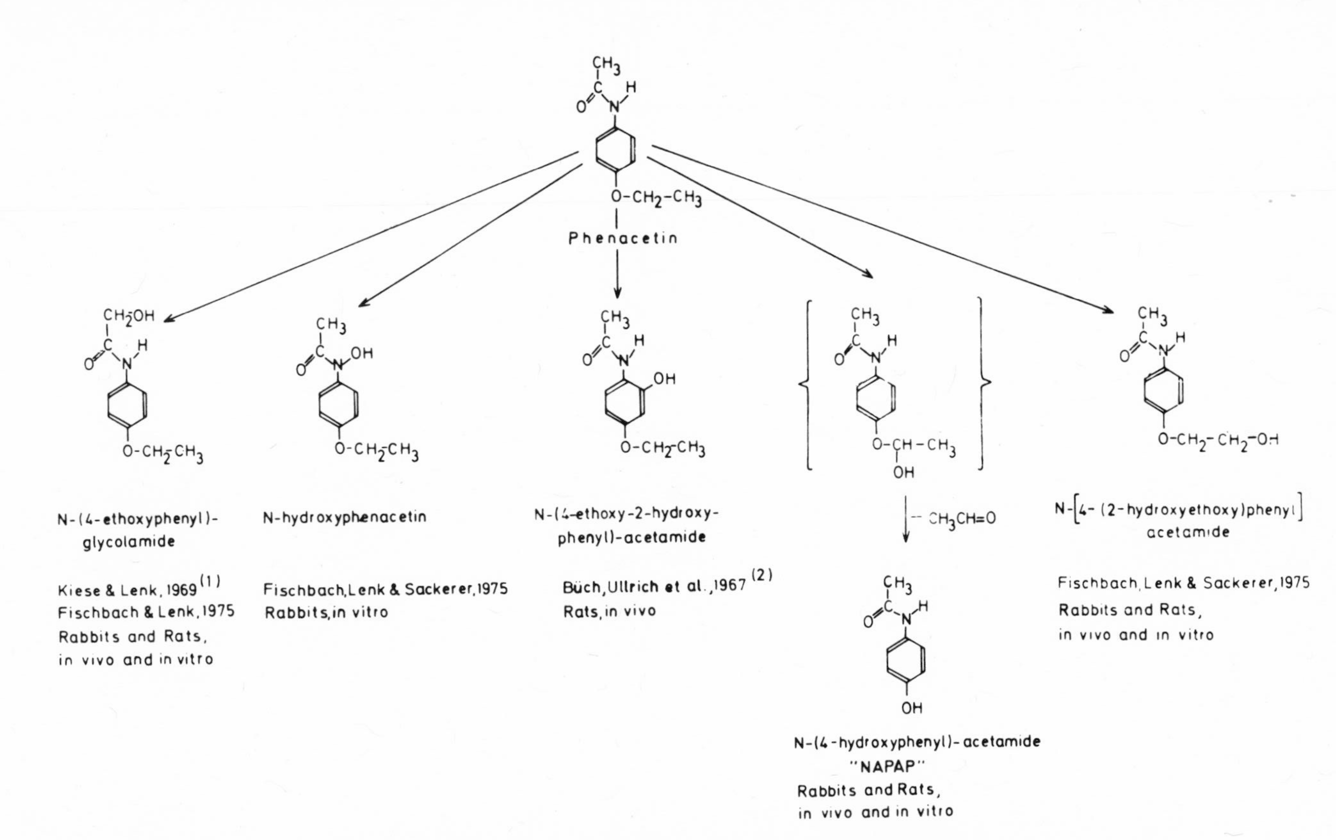

Fig. 1. Primary oxygenation products of phenacetin in rats and rabbits.

ated on thin-layer plates. The amount of metabolite was determined by reading the extinction at 247 nm of the methanol eluates of the zones scraped off the plates. The alkaline layer was acidified to pH 6.0 and extracted twice with an equal volume of ether. The solvent was removed at reduced pressures and the acidic metabolites were first applied to cellulose thin-layer plates to separate *N*-hydroxyphenacetin and then to Kieselgel 60 PF_{254} (Merck) to purify the others. The amounts of metabolites were determined by reading the extinction of the methanol eluates of the zones scraped off the plates at 246 and 250 nm. The shifts in λ_{max} values which occurred on adding 3 drops of 2 M NaOH to the methanol eluates in the cuvettes further contributed to the identification of the metabolites (Fig. 2). Methanol eluates from a blank cellulose layer were employed as reference for the optical determination of *N*-hydroxyphenacetin. In later experiments, ether was replaced by ethylacetate for extracting the metabolites.

TLC of Phenacetin and Some of Its Primary Oxygenation Products

N-Hydroxphenacetin

Repeated chromatography was carried out on thin layers of cellulose F (Merck AG) with the solvent system cyclohexane–acetic acid (92.5:7.5). R_f values after eight developments were 0.43 for *N*-hydroxyphenacetin and 0.35 for phenacetin.

Phenacetin, N-(4-Ethoxyphenyl)glycolamide, and N-[4-(2-Hydroxyethoxy)phenyl] acetamide

Repeated chromatography was carried out on thin layers of Kieselgel 60 PF_{254} (Merck) with the solvent system chloroform–methanol (95:5) or benzene–dioxane–acetic acid 90:25:4). R_f values after three developments with chloroform–methanol (95:5) were 0.51 for phenacetin, 0.28 for *N*-(4-ethoxyphenyl)glycolamide, and 0.13 for *N*[4-(2-hydroxyethoxy)phenyl] acetamide.

RESULTS

Preliminary results of the *in vitro* experiments are shown in Table 1. They show that *N*-hydroxylation and hydroxylation of the acetic acid moiety are slow and that, at least with PB-stimulated hepatic microsomes, ω-hydroxylation of the *O*-ethyl group is as rapid as is the (ω-1)-hydroxylation which gives NAPAP. Results from other experiments have shown that *N*-hydroxyphenacetin (I) as well as *N*-[4-(2-hydroxyethoxy)phenyl] acetamide (II) or *N*-(4-ethoxyphenyl)-

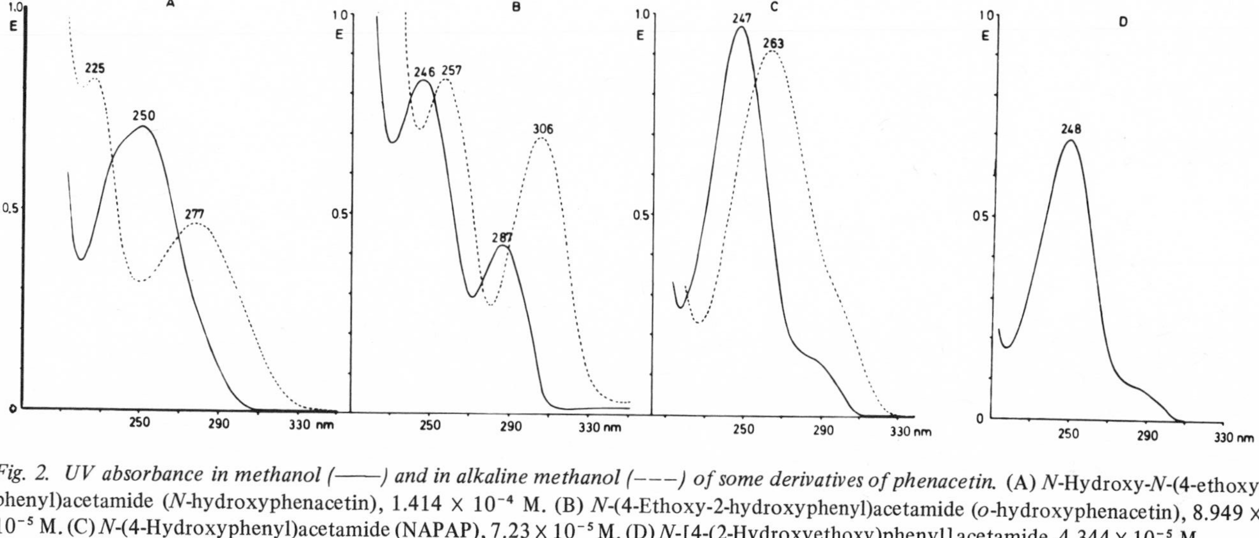

Fig. 2. UV absorbance in methanol (——) and in alkaline methanol (– – –) of some derivatives of phenacetin. (A) *N*-Hydroxy-*N*-(4-ethoxyphenyl)acetamide (*N*-hydroxyphenacetin), 1.414×10^{-4} M. (B) *N*-(4-Ethoxy-2-hydroxyphenyl)acetamide (*o*-hydroxyphenacetin), 8.949×10^{-5} M. (C) *N*-(4-Hydroxyphenyl)acetamide (NAPAP), 7.23×10^{-5} M. (D) *N*-[4-(2-Hydroxyethoxy)phenyl] acetamide, 4.344×10^{-5} M.

Table 1. Oxygenation Rates for Specific Pathways of Phenacetin Metabolism[a]

Enzyme source	*N*-(4-Ethoxyphenyl)-glycolamide	*N*-Hydroxyphenacetin	*N*-[4-(2-Hydroxy-ethoxy)phenyl] acetamide	*N*-(4-Hydroxy-phenyl)acetamide
Microsomes from untreated rabbits	0.77	0.9		
	0.38	n.d.		
	1.64	1.2		
Microsomes from PB-treated rabbits	0.95	1.4	12.0	22.0
	1.2	0.4	18.3	19.2
	0.52		17.0	22.6
Microsomes from 3-MC-treated rabbits	1.1	0.8		
	1.9	0.61		
	1.4	n.d.		

[a] All data are expressed on the basis of μmol/30 min/300 mg microsomal protein.

Table 2. Urinary Excretion of Phenacetin and Metabolites by Rabbits and Rats After Injection of Phenacetin or *N*-Hydroxyphenacetin[a]

				Phenacetin recovered[b]		N-(4-Ethoxyphenyl)-glycolamide		*N*-[4-(2-Hydroxyethoxy)-phenyl] acetamide	
Number of animals	Substance administered	Total dose (g)	Total urine (ml)	mg	Percent of dose	mg	Percent of dose	mg	Percent of dose
3 rabbits	Phenacetin	2.4	680	4.1	0.17	1.3	0.05	0.3	0.01
20 rats (♀)	Phenacetin	2.7	670	2.2	0.08	0.1	0.003	0.1	0.002
9 rats (3 ♂,6 ♀)	*N*-Hydroxy-phenacetin	1.29	240	23.1	1.95	Not detected	–	0.1	0.008

[a]The animals were repeatedly injected i.p. with 100 mg/kg each of the compounds indicated, and the urines excreted within 24 h were pooled for the determination of metabolites.

[b]Recoveries of phenacetin and metabolites have been rounded to the nearest 0.1 mg while the percent recovery was not corrected [Editors].

glycolamide (III) are intermediates in the metabolism of phenacetin, because (I) can be reduced or transformed into *N*-(4-ethoxy-2-hydroxyphenyl)acetamide, whereas (II) and (III) can be further oxidized.

Experiments *in vivo* have shown that *N*-[4-(2-hydroxyethoxy)phenyl]-acetamide is also excreted in small amounts in the urine of rats and rabbits injected i.p. with phenacetin or *N*-hydroxyphenacetin, and that *N*-(4-ethoxyphenyl)glycolamide also is a minor urinary metabolite of phenacetin in rats and rabbits (Table 2).

REFERENCES

1. M. Kiese and W. Lenk, Oxidation of acetanilides to glycoanilides and oxanilic acids in rabbits, *Biochem. Pharmacol.* **18,** 1325–1333 (1969).
2. H. Büch, K. Pfleger, W. Rummel, V. Ullrich, D. Hey, and H. Staudinger, Untersuchungen über den oxydativen Stoffwechsel des Phenacetins bei der Ratte, *Biochem. Pharmacol.* **16,** 2247–2256 (1967).

43

Biliary Metabolites of Paracetamol in Relation to Dose Level and Various Pretreatments

A. Malnoë* and M. Strolin Benedetti*
Battelle Research Centre
Geneva, Switzerland

R. L. Smith
St. Mary's Hospital Medical School
University of London, England

A. Frigerio
Istituto di Ricerche Farmacologiche "Mario Negri"
Milano, Italy

Large doses of paracetamol are hepatotoxic in laboratory animals and man (1–3). Recent mechanistic studies have indicated that paracetamol hepatotoxicity is related to the metabolic conversion of the drug to a chemically reactive arylating agent which binds covalently to tissue macromolecules (2,4,5) and that the extent of hepatic damage is directly related to the degree of covalent binding that occurs. Extensive binding of the reactive metabolite and tissue damage do not occur, however, until after the liver glutathione has been depleted, suggesting that the reactive intermediate is normally detoxified by conjugation with glutathione (6).

The formation of a glutathione conjugate of paracetamol has not been previously described, although a mercapturic acid, which is probably derived from this conjugate, has been identified as a urinary metabolite (7). Glutathione conjugates, probably because of their relatively high molecular weight and polarity, are commonly preferentially excreted in the bile and not in the urine. We have therefore examined the bile of rats dosed with [^{14}C]paracetamol for

*Present address: Centre de Recherche Delalande, 10, Rue des Carrieres, 92500 Rueil-Malmaison, France.

the presence of a glutathione conjugate. As conjugation with glutathione may be an important pathway for the detoxication of the active metabolite of paracetamol, the amount of this conjugate found in bile might be expected to reflect the activity of the metabolic pathway leading to the formation of the putative active metabolite. In this chapter, we report the identification in the bile of rats dosed with [^{14}C]paracetamol of a metabolite which is absent from the urine and which appears to be a glutathione conjugate. We also report the effect of dose level and various pretreatments, which are known to affect the toxicity of paracetamol, on the elimination of this conjugate in bile.

Male Sprague-Dawley rats (200–300 g body weight) were anesthetized with sodium pentobarbital (60 mg/kg i.p.) and bile-duct-cannulated using polyethylene No. 10 tubing. Body temperature was maintained using a heating lamp and bile was collected for up to 5 h over dry ice. [^{14}C]-Paracetamol was administered either orally (50–5000 mg/kg) as a suspension in 1% carboxymethylcellulose or intravenously (50 mg/kg, 3 μCi) dissolved in 0.9% saline. In separate experiments, urine was collected for 24 h from rats dosed orally with [^{14}C]paracetamol (50 mg/kg). ^{14}C in bile was determined by liquid scintillation counting using Instagel as the scintillation fluid. Aliquots of bile and urine were chromatographed on silica gel thin-layer plates (Merck, 0.25 mm thick) and developed with *n*-propanol–0.4 M ammonium hydroxide (8:2) or *n*-butanol–acetic acid–water (4:1:2) and then scanned to reveal ^{14}C areas corresponding

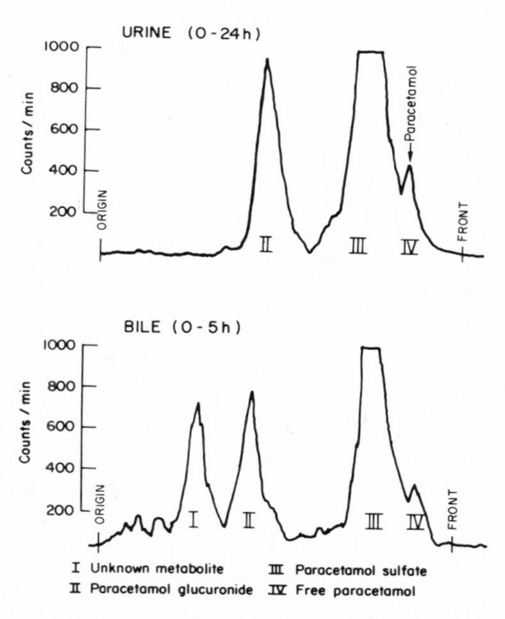

Fig. 1. Radiochromatogram of an aliquot of bile and of urine from rats receiving [^{14}C]paracetamol orally (50 mg/kg). TLC. Solvent system: *n*-propanol–0.4M ammonium hydroxide (8:2).

to the metabolites of paracetamol. For quantification, the radioactive areas were scraped directly into vials containing the scintillation fluid.

Figure 1 shows representative radiochromatogram scans for bile and urine obtained from rats dosed with [^{14}C] paracetamol (50 mg/kg orally). Bile showed four major ^{14}C peaks and urine showed three. Metabolites II, III, and IV were present in bile and urine and were shown by the effect of incubation with β-glucuronidase and sulfatase and by comparison of chromatographic properties with those given by authentic samples to correspond to paracetamol *O*-glucuronide, paracetamol *O*-sulfate, and free paracetamol, respectively. Metabolite I, which was found only in bile, gave a positive test for divalent sulfur with the potassium dichromate–silver nitrate spray reagent (8) and was ninhydrin positive. On acid hydrolysis (6 N HCl at 105°C for 16 h), it afforded glutamic acid and glycine (Unichrom amino acid analyzer, Beckman) in a molar ratio of 1:1.

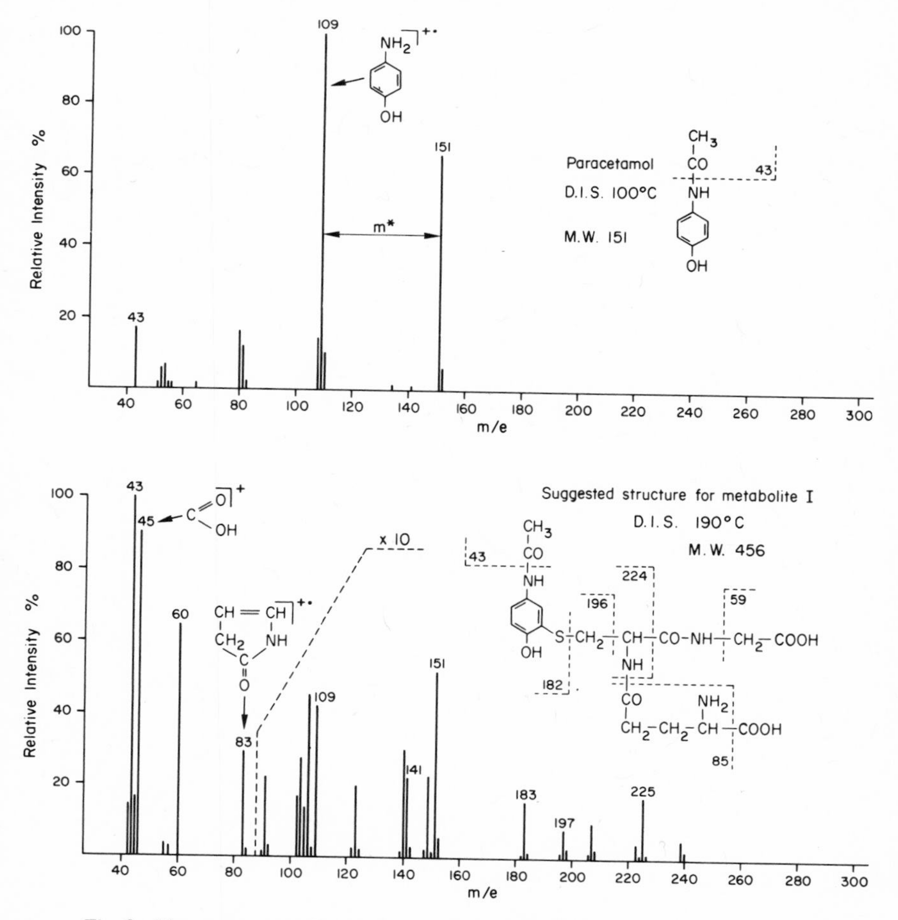

Fig. 2. Mass spectrum of paracetamol and metabolite I isolated from rat bile.

Table 1. Biliary Elimination of Paracetamol Metabolites after Oral and Intravenous Administration of [^{14}C]Paracetamol in the Rat[a]

Dose (mg/kg)	Route of adminis-tration	Percent of dose recovered in bile	Paracetamol metabolites			
			Paracetamol glutathione	Paracetamol glucuronide	Paracetamol sulfate	Free paracetamol
50	p.o.	7.2 ± 0.5	20.2 ± 0.6	25.9 ± 3.9	51.3 ± 2.6	2.6 ± 0.2
50	i.v.	12.0 ± 0.4[b]	38.4 ± 4.4[b]	27.9 ± 2.8	29.5 ± 3.7[b]	4.2 ± 0.9

[a]Each value represents the mean ± SEM of four rats. Figures show percentage of bile radioactivity.
[b]Significantly different ($p < 0.05$) from the group receiving paracetamol orally.

The metabolite was further analyzed by mass spectrometry (LKB 9000, direct inlet system, probe temperature 190°C, ionization current 60 μA, ionization voltage 70 eV). The mass spectrum (Fig. 2) of metabolite I is consistent with the view that it is a glutathione conjugate of paracetamol.

The qualitative and quantitative aspects of the biliary excretion of [^{14}C]paracetamol in the rat are shown in Table 1. Bile-duct-cannulated rats excrete 7% of an oral dose and 12% of an intravenous dose of the drug in the bile within 5 h. The biliary material consists largely of the glutathione, glucuronic acid, and sulfate conjugates of paracetamol together with small amounts of the unchanged drug. The glutathione conjugate accounts for 20% of the biliary material when the drug is given orally but this increases to 38% when it is injected intravenously. The latter is associated with a concomitant fall in the amount of the sulfate conjugate in the bile. This may be due to rapid depletion of the sulfate pool, thereby leading to an increase in the amount of drug undergoing glutathione conjugation.

Table 2 shows the effect of dose level on the biliary metabolites of paracetamol. As previously reported for urine (9), there is a decrease in the elimina-

Table 2. Effect of Dose Level on the Biliary Elimination of Paracetamol Metabolites[a]

Dose of paracetamol (mg/kg, p.o.)	Percent of dose recovered in bile (0–5 h)	Paracetamol metabolites			
		Paracetamol glutathione	Paracetamol glucuronide	Parcetamol sulfate	Free paracetamol
50	7.2 ± 0.5	20.2 ± 0.6	25.9 ± 3.9	51.3 ± 2.6	2.6 ± 0.2
500	7.6 ± 1.0	21.2 ± 2.1	47.4 ± 4.5[b]	26.5 ± 3.8[b]	4.9 ± 0.2[b]
1000	7.5 ± 0.6	34.6 ± 4.7[b]	48.6 ± 5.1[b]	8.7 ± 1.1[b]	8.1 ± 0.7[b]
5000	2.7 ± 0.4[b]	25.6 ± 2.4	56.0 ± 1.4[b]	7.0 b 0.7[b]	11.4 ± 2.2[b]

[a]Each value represents the mean ± SEM of three to five rats. Figures show percentage of bile radioactivity.
[b]Significantly different ($p < 0.05$) from the group receiving 50 mg/kg.

Table 3. Effect of Various Pretreatments on the Biliary Elimination of Paracetamol Metabolites[a]

Pretreatment	Dose of paracetamol (mg/kg, p.o.)	Percent of dose recovered in bile (0–5 h)	Paracetamol metabolites			
			Paracetamol glutathione	Paracetamol glucuronide	Paracetamol sulfate	Free paracetamol
–	50	7.2 ± 0.5	20.2 ± 0.6	25.9 ± 3.9	51.3 ± 2.6	2.6 ± 0.2
Piperonyl butoxide[b]	50	6.1 ± 0.7	11.8 ± 1.3[g]	21.8 ± 2.1	63.3 ± 2.4[g]	3.0 ± 0.4
Phenobarbital[c]	50	6.6 ± 0.3	26.1 ± 0.9[g]	25.8 ± 2.0	45.0 ± 2.6[g]	3.0 ± 0.4
–	1000	7.5 ± 0.6	34.6 ± 4.7	48.6 ± 5.1	8.7 ± 1.1	8.1 ± 0.7
Phenobarbital[c]	1000	7.9 ± 0.7	59.0 ± 2.8[h]	22.6 ± 1.5[h]	12.4 ± 1.9	5.6 ± 1.0
Diethyl maleate[d]	1000	6.7 ± 0.3	18.0 ± 1.8[h]	56.4 ± 2.5	19.8 ± 0.9[h]	5.8 ± 0.4
Diethyl maleate + L-glutathione[e]	1000	7.1 ± 0.7	28.4 ± 2.2[i]	42.2 ± 6.7	21.6 ± 4.6[h]	8.2 ± 1.2
L-cysteine[f]	1000	8.8 ± 1.5	25.6 ± 1.4	43.7 ± 6.6	23.8 ± 5.4[h]	6.8 ± 0.3

[a]Each value is the mean ± SEM of three to five rats. Figures show percentage of bile radioactivity.
[b]800 mg/kg, i.p. 30 min before paracetamol.
[c]Pretreatment for 4 days (90 mg/kg/day, i.p.).
[d]0.7 ml/kg i.p. 30 min before paracetamol.
[e]Diethyl maleate was administered as above and L-glutathione (500 mg/kg) was given i.v. 15 min after paracetamol.
[f]500 mg/kg i.v. 15 min after paracetamol.
[g]Significantly different ($p < 0.05$) from the group receiving 50 mg/kg without pretreatment.
[h]Significantly different ($p < 0.05$) from the group receiving 1000 mg/kg without pretreatment.
[i]Significantly different ($p < 0.05$) from the group receiving diethyl maleate alone.

tion of the sulfate conjugate when the dose is increased, probably due to sulfate depletion. In the bile, this decrease is compensated by relatively higher amounts of the drug being excreted unchanged or as the glucuronic acid and glutathione conjugates. The percentage of the glutathione conjugate excreted into the bile increased up to the dose of 1 g/kg. However, at 5 g/kg (approximate LD_{50} for paracetamol) (10), it diminished, probably because of a depletion of liver glutathione.

Table 3 shows the effect of various pretreatments on the biliary excretion of the glutathione and other conjugates of paracetamol in the rat. Pretreatment with phenobarbital (90 mg/kg i.p. for 4 days), which both induces the cytochrome P450 microsomal enzymes and enhances the toxicity of paracetamol (2), causes a significant increase in the amount of glutathione conjugate eliminated in the bile, presumably by enhancing the formation of the reactive intermediate of paracetamol. Conversely, pretreatment with piperonyl butoxide, which both inhibits cytochrome P450 microsomal enzymes and reduces the toxicity of the drug (2), also markedly reduces the amount of glutathione conjugate in bile. Pretreatment with diethyl maleate, which depletes liver glutathione (11), also markedly reduces the amount of drug excreted in the bile as a glutathione conjugate. This effect of the diethyl maleate was, however, reversed by the intravenous administration of L-glutathione, suggesting that exogenously administered glutathione can be available to the hepatic cells for conjugation with the reactive intermediate. This view is supported by the recent finding that exogenously administered glutathione protects mice against paracetamol-induced liver necrosis (12). The intravenous administration of L-cysteine (500 mg/kg i.v. 15 min after the paracetamol), which also protects against paracetamol-induced liver necrosis (6), had little effect on the amount of glutathione conjugate excreted in bile but it did cause a marked increase in the proportion of sulfate conjugate. L-Cysteine is a precursor of both glutathione and active sulfate (13,14) and the protective effect of the amino acid against paracetamol hepatotoxicity may be due to an increase in the amount of active sulfate available for conjugation, thereby decreasing the amount of drug metabolized through the toxic pathway.

The findings indicate that rat bile contains a new conjugate of paracetamol which is absent from urine, and which appears to be a glutathione conjugate. Furthermore, the amount of the glutathione conjugate present in rat bile varies in direct relationship with those factors known to affect the hepatotoxicity of the drug.

REFERENCES

1. E. H. Boyd and G. M. Bereczky, Liver necrosis from paracetamol, *Br. J. Pharmacol.* **26**, 606–614 (1966).

2. J. R. Mitchell, D. J. Jollow, W. Z. Potter, D. C. Davis, J. R. Gillette, and B. B. Brodie, Acetaminophen-induced hepatic necrosis. I. Role of drug metabolism, *J. Pharmacol. Exp. Ther.* **187**, 185–194 (1973).
3. L. F. Prescott, N. Wright, P. Roscoe, and S. S. Brown, Plasma-paracetamol half-life and hepatic necrosis in patients with paracetamol overdosage, *Lancet* **i**, 519–522 (1971).
4. D. J. Jollow, J. R. Mitchell, W. Z. Potter, D. C. Davis, J. R. Gillette, and B. B. Brodie, Acetaminophen-induced hepatic necrosis. II. Role of covalent binding *in vivo, J. Pharmacol. Exp. Ther.* **187**, 195–202 (1973).
5. W. Z. Potter, D. C. Davis, J. R. Mitchell, D. J. Jollow, J. R. Gillette, and B. B. Brodie, Acetaminophen-induced hepatic necrosis. III. Cytochrome P-450 mediated covalent binding *in vitro, J. Pharmacol. Exp. Ther.* **187**, 203–210 (1973).
6. J. R. Mitchell, D. J. Jollow, W. Z. Potter, J. R. Gillette, and B. B. Brodie, Acetaminophen-induced hepatic necrosis. IV. Protective role of glutathione, *J. Pharmacol. Exp. Ther.* **187**, 211–217 (1973).
7. J. R. Mitchell, S. S. Thorgeirsson, W. Z. Potter, D. J. Jollow, and H. Keiser, Acetaminophen-induced hepatic injury: Protective role of glutathione in man and rationale for therapy, *Clin. Pharmacol. Ther.* **16**, 676–684 (1974).
8. R. H. Knight and L. Young, Biochemical studies of toxic agents. II. The occurrence of premercapturic acids, *Biochem. J.* **70**, 111–119 (1958).
9. D. J. Jollow, S. S. Thorgeirsson, W. Z. Potter, M. Hashimoto, and J. R. Mitchell, Acetaminophen-induced hepatic necrosis. VI. Metabolic disposition of toxic and non-toxic doses of acetaminophen, *Pharmacology* **12**, 251–271 (1974).
10. A. E. M. McLean and A. P. Day, The effect of diet on the toxicity of paracetamol and the safety of paracetamol-methionine mixtures, *Biochem. Pharmacol.* **24**, 37–42 (1975).
11. E. Boyland and L. F. Chasseaud, The effect of some carbonyl compounds on rat liver glutathione levels, *Biochem. Pharmacol.* **19**, 1526–1528 (1970).
12. M. Strolin-Benedetti, A. Louis, A. Malnoë, M. Schneider, R. Lam, L. Kreber, and R. L. Smith, Prevention of paracetamol-induced liver damage in mice with glutathione, *J. Pharm. Pharmacol.* **27**, 629–632 (1975).
13. E. Boyland and L. F. Chasseaud, The role of glutathione and glutathione *S*-transferases in mercapturic acid biosynthesis, *Adv. Enzymol.* **32**, 173–217 (1969).
14. A. B. Roy, Sulfate conjugation enzymes, in: *Concepts in Biochemical Pharmacology,* Vol. XXVIII, Part 2 (B. B. Brodie and J. R. Gillette, eds.), pp. 536–563, Springer, New York (1971).

44

Studies on Pharmacokinetics of *N*-Demethyldiazepam, the Active Main Metabolite of Diazepam

Lauri Kangas, Jussi Kanto, Raija Sellman, and Aimo Pekkarinen
Department of Pharmacology
University of Turku
Turku 52, Finland

The main metabolite of diazepam (D) in man is *N*-demethyldiazepam (ND), which, after hydroxylation to oxazepam, is excreted mainly in conjugated form in the urine. The aim of these studies was to get more information about the pharmacokinetics of D and ND, especially in humans, because of wide use of D and scanty knowledge of its metabolism and pharmacokinetics in patients.

CHEMICAL METHODS

The concentrations of D and its metabolites in plasma and in cerebrospinal fluid as well as in bile and in urine were measured by gas chromatography using a 63-Ni-EC detector according to Kangas *et al.* (1). The tissue concentrations of D and ND were measured by the same method with some alterations. The binding of D and ND to plasma proteins was determined using a technique involving ultracentrifugation according to van der Kleijn (2).

RESULTS AND DISCUSSION

Effect of Administration Route of Diazepam on the Plasma Concentrations of Diazepam and *N*-Demethyldiazepam

In adult patients or volunteers, the route of a single administration of 5 mg or 10 mg D p.o., i.m., i.v., or rectally had no significant effect on the plasma concentrations of ND. Small concentrations of ND in the plasma could be detected 10–15 min after the administration of D (2–23 ng/ml), and they were increased over a 24-h period (12–48 ng/ml). After i.v. administration of D, the formation of ND began somewhat sooner than after p.o. or i.m. administration, indicating faster entry of D to the sites of metabolism in the liver. In contrast to ND, the route of D administration had a marked effect on the plasma concentrations of D. After i.m. injection and rectal administration with a suppository containing a lipophilic basal component, the absorption of D was delayed and the maximum concentrations were 2–5 times lower than after peroral administration (3).

Effect of Chronic Alcohol Intake on the Plasma Concentrations of *N*-Demethyldiazepam

After 10 mg i.v. injection of D, the alcoholics had in the beginning of their alcohol-free period significantly smaller concentrations of ND in the plasma than healthy volunteers (4). In the healthy volunteers, the plasma concentration of ND increased from the mean value of 8 ng/ml at 15 min to the mean value of 29 ng/ml at 3 h. The same values in the alcoholics were 7 and 12 ng/ml, respectively. The alcoholics had significantly lower plasma concentrations of D than the volunteers. The lower concentrations of ND in the alcoholics may indicate decreased demethylation of D in the liver, or increased volume of distribution of D and ND in alcoholics.

Diazepam as Inducer of Its Own Metabolism—Long-Term Treatment

The i.v. injection of 10 mg D in seven psychiatric patients with continuous D therapy caused significantly lower plasma concentrations of D (from 15 min to 24 h) than in 12 healthy volunteers or in 12 psychiatric patients without D, and respectively significantly higher increases in the concentrations of ND (5). Chronic therapy with D (15 mg/day) was continued until a steady-state level was reached (1–6 weeks) (6). There was a six- to sevenfold interindividual variation in the plasma concentrations of ND (200–1400 ng/ml) and D (100–700 ng/ml). After this period, the plasma concentrations began to decrease significantly, which indicated enzyme induction. The lowest values in chronic therapy with the same dose were only 1/5–1/10 of those resulting after 2 weeks of therapy. After discontinuation of D therapy, one patient had measurable amounts of D for 4 days and ND for 7 days in plasma. These results show the tendency of D and especially of ND to accumulate early during chronic D administration. After an i.v. dose of D, the concentration of ND in the plasma of chronic D users was 10–20 times higher than in controls. D was administered i.v. to one patient before and after an abstinence period. The patient was not able to metabolize D to ND as well after the abstinence as before. All these phenomena indicate that D can induce its own metabolism during chronic use.

Placental Transfer of *N*-Demethyldiazepam

After a single i.m. 10-mg D dose to the pregnant women 40–120 min before a legal abortion, only small traces of ND were found in maternal plasma. In contrast, D was rapidly transported across the placenta and the fetomaternal ratio of D concentrations in the plasma was 1.2 (7).

After subchronic p.o. administration of D to pregnant women at a dose of 15 mg daily before legal abortions, the fetomaternal ratios of both D and ND in the plasma were 0.4, and significant accumulation of these compounds in fetal tissues and placenta was observed (8). The highest ND concentration (mean 310 ng/ml) was found in fetal liver. It was significantly higher than the concentration of ND in the fetal plasma (mean 110 ng/ml). This may be indirect evidence for metabolism of D in the fetal liver. During the abortion, after a single 10 mg i.m. or 0.2 mg/kg i.v. dose of D, the concentrations of both D and ND were significantly higher in both umbilical vessels than in the pregnant woman. The fetomaternal ratio for D was 1.3–2.0, and for ND, 2.0–2.5.

After subchronic p.o. therapy of D (10–15 mg/day) to the mother before the labor, the fetomaternal ratios of D and ND were 0.8 and 0.9, respectively, and the postnatal disappearance of these compounds from the plasmas of the newborns lasted for 10 days or more (9).

After intraamniotic application of D early in human pregnancy, rapid disappearance of D from the amniotic fluid was found. Fetal tissue concentrations of D and ND were comparable to those found after a single administration of D to the mother (10).

Concentrations of *N*-Demethyldiazepam in Bile

After a 10-mg i.v. dose of D, patients with a T-tube in the common bile duct had significantly lower concentrations of D in plasma than control patients, indicating an enterohepatic cycling of D. The main conjugated metabolite in the bile was ND (about 74% in the conjugated form). Diazepam and oxazepam may undergo significant enterohepatic circulation, but *N*-demethyldiazepam probably does not, because its concentrations in the plasma of the patients with a T-tube were not lower than those of controls (11).

Binding of *N*-Demethyldiazepam to Plasma Proteins

In healthy volunteers, in neurological patients with normal renal function, and in patients with renal disease, 98, 97, and 95% of ND was bound to plasma proteins, respectively. The same values for D were 98, 98, and 92%, respectively (12). The protein binding of both D and ND was significantly lower in patients with renal disease. Free D and ND concentrations in the plasma were in equilibrium with their concentrations in cerebrospinal fluid in neurological patients (13). The patients with renal disease had significantly higher concentrations of ND in plasma after a single p.o. 10-mg D dose than healthy volunteers or patients with normal renal function (ref. 12 and unpublished results). This may depend on the increased free form of D at sites of metabolism and/or decreased renal excretion of ND in the patients with renal disease.

REFERENCES

1. L. Kangas, A. Pekkarinen, C. Sourander, and E. Raijola, A comparative gas chromatographic study on absorption of diazepam tablets in man, *Ann. Clin. Res.* **6,** Suppl. 11, 12–20 (1974).
2. E. van der Kleijn, Protein binding and lipophilic nature of ataractics of the meprobamate- and diazepin-group, *Arch. Int. Pharmacodyn. Ther.* **179,** 225–250 (1969).
3. J. Kanto, L. Kangas, R. Sellman, and E. Syvälahti, Tensopamin ja sen metaboliittien pitoisuus plasmassa erilaisten annosten ja antotapojen jälkeen, *Lääkeuutiset* **1/74,** 17–20 (1974).
4. R. Sellman, J. Kanto, E. Raijola, and A. Pekkarinen, Human and animal study on elimination from plasma and metabolism of diazepam after chronic alcohol intake, *Acta Pharmacol. Toxicol.* **36,** 33–38 (1975).
5. R. Sellman, J. Kanto, E. Raijola, and A. Pekkarinen, Induction effect of diazepam on its own metabolism, *Acta Pharmacol. Toxicol.,* **37,** 345–351 (1975).
6. J. Kanto, E. Iisalo, V. Lehtinen, and J. Salminen, The concentrations of diazepam and its metabolites in the plasma after an acute and chronic administration, *Psychopharmacologia (Berl.)* **36,** 123–131 (1974).
7. J. Kanto and R. Erkkola, The feto-maternal distribution of diazepam in early human pregnancy, *Ann. Chir. Gynecol. Fenn.* **63,** 489–491 (1974).
8. R. Erkkola, J. Kanto, and R. Sellman, Diazepam in early human pregnancy, *Acta Obstet. Gynecol. Scand.* **53,** 135–138 (1974).
9. J. Kanto, R. Erkkola, and R. Sellman, Perinatal metabolism of diazepam, *Brit. Med. J.* **1,** 641–642 (1974).
10. J. Kanto, R. Erkkola, L. Kangas, and R. Sellman, Pharmacokinetics of diazepam in early human pregnancy, in: *VI International Congress of Pharmacology,* Helsinki, Finland, July 20–25, 1232 (1975).
11. R. Sellman, J. Kanto, and J. Pekkarinen, Biliary excretion of diazepam and its metabolites in man, *Acta Pharmacol. Toxicol.* **37,** 242–249 (1975).
12. L. Kangas, J. Kanto, and R. Sellman, Binding of diazepam and *N*-demethyldiazepam to plasma proteins, in: *VI International Congress of Pharmacology,* Helsinki, Finland, July 20–25, p. 123 (1975).
13. J. Kanto, L. Kangas, and T. Siirtola, Cerebrospinal-fluid concentrations of diazepam and its metabolites in man, *Acta Pharmacol. Toxicol.* **36,** 328–334 (1975).

V

Reactive Intermediates in Lipid Peroxidation

45

Regulation and Effects of Lipid Peroxidation in Isolated Hepatocytes

Johan Högberg
Department of Forensic Medicine
Karolinska Institutet
Stockholm 60, Sweden

Lipid peroxidation is a free radical chain reaction in biological structures (1). Deleterious effects of free radical intermediates on many subcellular structures have been suggested, and a relationship between this type of lipid oxidation and pathological conditions was recognized early (*cf.* ref. 2). It was suggested by Barber and Bernheim (3) that lipid peroxidation *in vivo* is catalyzed by iron and ascorbate, but enzyme-catalyzed lipid peroxidation in the presence of iron has been described and carefully documented in microsomes (4–7).

Detailed information on the effects of lipid peroxidation on isolated subcellular structures and especially on microsomes has been reported. It was shown (8) that during such peroxidation there is an extensive loss of polyunsaturated fatty acids from the membrane, and membrane protein is also solubilized (7,9). Ultrastructural alterations in the microsomal fraction have been observed after peroxidation (10,11), and membrane-bound enzymes such as glucose-6-phosphatase (11,12), the microsomal drug-hydroxylating system (11,13,14), and mitochondrial isocitrate dehydrogenases (15) have been found to be inactivated. Reports on NADPH-cytochrome *c* reductase are contradictory (9,11). The UDPglucuronyltransferase activity of microsomes can be increased by lipid peroxidation (11), and latent lysosomal enzymes can be activated (6). Membrane destruction as a result of lipid peroxidation has long been suspected to be the cause of these effects (4), and the effects of lipid peroxidation have been compared to those of detergents (7,11,12). SH groups (16) and heme groups (14,17) seem to be susceptible to lipid peroxide attack, and a cross-linking effect has been ascribed to malonaldehyde, one of the cleavage products (2) of lipid peroxidation.

In order to examine the significance of lipid peroxidation in pathological states, animals have been treated in various ways. Vitamin E deficiency has been known for many years to be correlated with peroxide formation and serious pathological alterations (*cf.* ref. 3). Choline deficiency also may cause lipid peroxidation (18,19). Iron toxicity has been claimed to involve lipid peroxidation (29), as have ozone (21) and oxygen (22) toxicity. The toxicity of several foreign compounds has been ascribed to their prooxidant effect on endogenous lipids, and the formation of free radical intermediates, which initiate lipid peroxidation, has been suggested. The most studied effect is liver necrosis induced by carbon tetrachloride (*cf.* ref. 23). Recknagel and Glende (24,25) suggest the involvement of cytochrome P450 in such necrosis, while Slater (23) proposes a free radical formation via cytochrome P450 reductase.

Even if a direct correlation between lipid peroxidation and pathological effects *in vivo* is difficult to establish, it can be concluded that tissues have a "lipid peroxidation potential" (3) which under normal conditions is balanced by endogenous defense mechanisms. The effects of vitamin E defiency might thus be explained by a loss of endogenous free radical scavengers (2). This lipid-soluble antioxidant is present in relatively small amounts in membranes (2), but several others, usually cytoplasmic scavengers, have been discussed (*cf.* refs. 23,26). *In vitro* experiments (6,27,28) indicate a defense role for superoxide dismutase. Metal chelators are used extensively to inhibit lipid peroxidation *in vitro,* and are also used clinically to counteract the effects of an overdose of iron. Worth noting is the finding by Wills (29) that the naturally occurring iron chelate, ferritin, inhibits enzymatic lipid peroxidation. Malonaldehyde may be toxic (2), but it has been found to be further metabolized (30). Of principal interest is the fate of the lipid peroxides formed, as it has been suggested that the interaction of reduced iron with these species results in propagation of lipid peroxidation (31). Cristophersen (32) and O'Brien and Little (33) have suggested two routes for lipid peroxide decomposition, one via radical formation in the presence of iron and another via formation of hydroxy lipids. Hydroxy fatty acid formation from lipid peroxides catalyzed by glutathione peroxidase (32,34) or semidehydroascorbate reductase (35) was described and induction of glutathione-metabolizing enzymes *in vivo* has been documented (21,36).

In summary, lipid peroxidation, which is readily inducible in isolated organelles, has been fairly well characterized *in vitro,* and its pathological significance is suggested by numerous *in vivo* studies. However, the relationship between *in vitro* findings and the events leading to cell necrosis is not well understood and information about the interaction between lipid peroxidation and defense mechanisms is lacking. It was felt that isolated hepatocytes would provide further possibilities for characterizing these important aspects of lipid peroxidation, since lipid peroxidation can be readily induced and monitored in this system with a full set of targets and defense mechanisms present.

METHODS

The cell preparation technique has been described in detail previously (37,38). Male rats of the Sprague-Dawley strain were used throughout. The liver was perfused with Hanks' buffer containing collagenase and hyaluronidase to achieve dispersion of the liver cells. The dispersed cells were washed so that mainly hepatocytes with a uniform morphological appearance were harvested. More than 95% of these cells excluded trypan blue. The cell preparation technique is based on the method described by Berry and Friend (39), but several modifications [to a large extent based on the work of Seglen and co-workers (40,41)] have been introduced. Lipid peroxidation was induced and studied in a buffered cell suspension containing about 3×10^6 cells/ml by the addition of ADP-complexed iron (50 mM ADP plus 1.87 mM $FeCl_3$) or an aqueous solution of cumene hydroperoxide. The reaction was monitored by the withdrawal of 0.2-ml aliquots of the cell suspension at various times for the determination of substances reacting with thiobarbituric acid (malonaldehyde) (42).

RESULTS

Most systems used for studying lipid peroxidation in isolated organelles involve iron. The rate of peroxidation can be increased by the addition of reducing agents such as ascorbate, thiol compounds, or (when working with microsomes) NADPH. The isolated hepatocyte has been shown (43) to contain amounts of NADPH sufficient to possibly sustain iron-induced lipid peroxidation and the plasma membrane to constitute a permeability barrier for NADH (44). The level of $NADP^+$ is low (43). As could be expected from studies on iron uptake in liver (45,46), the plasma membrane seemed to constitute a permeability barrier for iron also, and the response in intact cells was thus largely dependent on the type of iron complex used. ADP-complexed iron was the most effective of those tested (37), much more effective, for example, than pyrophosphate-complexed iron [which is more effective in a microsomal preparation than the ADP complex (37)]. However, it could be shown (Fig. 1) that with incubation times exceeding 1 h the response to the different complexes became more uniform, indicating plasma membrane damage. If the cells were preincubated up to 50 min (Fig. 2), the response to the iron complexes gradually increased. In this situation, however, the ADP complex and the pyrophosphate

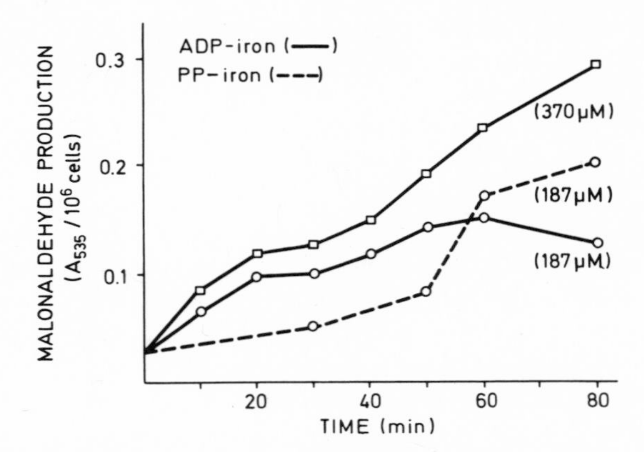

Fig. 1. Malonaldehyde production in isolated hepatocytes. About 12 × 10⁶ cells were incubated in a final volume of 4 ml. Lipid peroxidation was initiated by the addition of the iron complexes. At times indicated, 0.2 ml was taken for malonaldehyde measurements.

complex gave responses indicating that the permeability barrier of the plasma membrane was preserved; i.e., the effect of preincubation was similar with both complexes. In earlier studies (37) it was found that diethyl maleate, which is known to form conjugates with glutathione (GSH) (47), could, when added to isolated cells, induce an increase in the response to the iron complex similar to that caused by preincubation. It was thus suspected that cells lost GSH during preincubation and that GSH in some way affected iron-induced lipid peroxidation. Sies *et al.* (48) found a release of oxidized glutathione (GSSG) from perfused livers upon exposure to stable organic peroxides and interpreted this finding as a result of glutathione peroxidase activity. Consistent with these observations, it was found here (Table 1) that fresh isolated cells contained no GSSG while a 20-min incubation resulted in a substantial conversion of GSH to

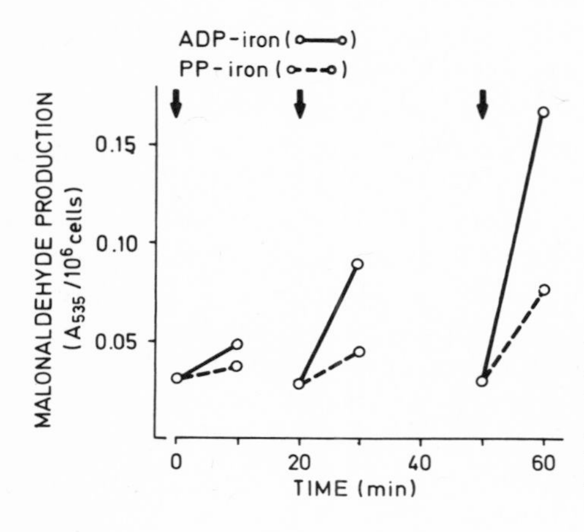

Fig. 2. Malonaldehyde production in isolated hepatocytes. About 6 × 10⁶ cells were incubated in a final volume of 2 ml. After preincubation (0, 20, and 50 min), peroxidation was started by addition of the iron complexes (indicated by arrows).

Table 1. Effect of ADP·Iron or Cumene Hydroperoxide on Glutathione in Isolated Hepatocytes[a]

Treatment	Time (min)	GSH		GSSG		GSH/GSSG
		$\mu g/10^6$ cells	Intracellular/extracellular	$\mu g/10^6$ cells	Intracellular/extracellular	
None (fresh cells)	0	6.7	6.1	0	–	–
Control incubation	20	5.2	3.2	1.7	0.5	3.1
ADP-complexed Fe^{3+} (187 μM)	5	6.0	–	0.3	–	20.0
	10	5.2	–	1.0	–	5.2
	20	3.9	–	2.8	–	1.4
Cumene-OOH (75 μM)	5	2.6	–	3.5	–	0.7
	10	3.5	–	1.5	–	2.3
	20	3.7	–	1.5	–	2.5

[a]Glutathione content was measured in 1-ml aliquots of the cell suspension (3×10^6 cells/ml) using a fluorometric method described by Cohn and Lyle (49). GSH and GSSG were measured in the total cell suspension, in reharvested cells (by centrifugation), and in cell-free incubation medium.

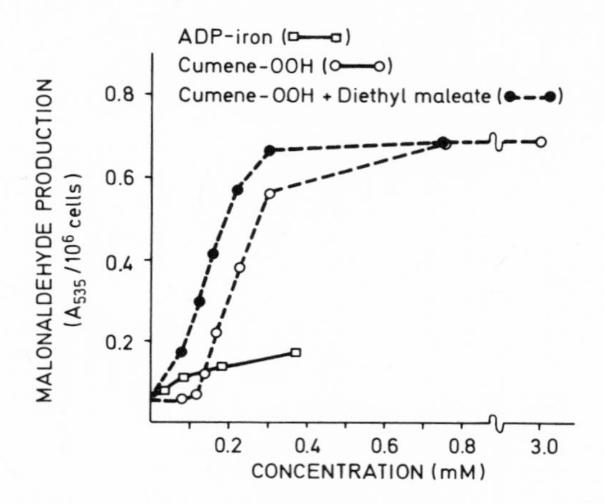

Fig. 3. Malonaldehyde production in isolated hepatocytes. About 5×10^6 cells were incubated in a final volume of 2 ml. Peroxidation was initiated by addition of either ADP-complexed iron, cumene hydroperoxide, or cumene hydroperoxide plus diethylmaleate (2 μl/ml). After 20 min, 0.2 ml was taken for malonaldehyde measurements.

GSSG and that GSSG to a large extent was found outside the cells in the incubation medium. ADP-iron addition was accompanied by more rapid oxidation of GSH. Cumene hydroperoxide in concentrations below 150 μM induced a rapid oxidation but no increase in malonaldehyde production. It has, however, been shown (50) that cumene hydroperoxide induces lipid peroxidation in microsomes and, by increasing the concentration of this organic hydroperoxide above 150 μM, malonaldehyde production in isolated cells could be demonstrated (Fig. 3). The lag phase in malonaldehyde production below 150 μM disappeared in the presence of diethyl maleate.

The well-documented difference between iron-induced lipid peroxidation (which involves NADPH-cytochrome P450 reductase) and cumene hydroperoxide-induced lipid peroxidation (involving cytochrome P450) in microsomes (50) may explain some differences in the effect of inhibitors on the two reactions. It was found (38) that menadione and *p*-benzoquinone markedly inhibit iron-induced but not cumene hydroperoxide-induced lipid peroxidation. *N,N,N,'N'*-Tetramethyl-*p*-phenylenediamine (TMPD), on the other hand, inhibited the hydroperoxide-induced but not the iron-induced reaction at low concentrations (20 μM). TMPD in high concentration (100 μM) inhibited both reactions and also decreased the low levels of malonaldehyde normally present in fresh cells. These results thus supported the thesis that the iron-induced reaction in isolated hepatocytes was dependent on the microsomal NADPH-oxidizing flavoprotein and that the cumene hydroperoxide reaction was dependent on

Table 2. Effect of ADP·Iron or Cumene Hydroperoxide on Microsomal Enzymes[a]

Treatment	Time (min)	Cytochrome P450 (nmol/10^6 cells)	NADPH-cytochrome *c* reductase (nmol cytochrome *c* reduced/min/10^6 cells)	Alprenolol metabolism (nmol/15 min/10^6 cells)
Control incubation	20	0.20	24.0	11.1
ADP-complexed Fe^{3+}				
187 μM	20	0.21	19.7	9.0
187 μM	60	0.20	–	–
370 μM	20	0.19	–	5.9
Cumene-OOH				
150 μM	20	0.20	20.6	8.7
750 μM	20	0.02	11.8	0.9

[a]The amount of cytochrome P450 was estimated from the CO-difference spectrum (51). The incubation volume of 2 ml NADPH-cytochrome *c* reductase activity was measured in an incubation medium which had been diluted tenfold to lyse the cells. Incubation volume was 2 ml. Alprenolol metabolism was measured as described by Moldéus *et al.* (43). Incubation volume was 4 ml and the concentration of [^{14}C]alprenolol was 100 μM.

Table 3. Effect of ADP·Iron or Cumene Hydroperoxide on Microsomal Activities[a]

Treatment	Time (min)	Glucose-6-phosphatase (nmol P_i/20 min/10^6 cells)	*p*-Nitrophenol metabolism (nmol/10 min/10^6 cells)
Control incubation	20	77	9.5
ADP-complexed Fe^{3+}			
47 μM	20	–	6.1
187 μM	20	–	4.0
187 μM	30	42	–
Cumene-OOH (300 μM)	30	41	–

[a] Glucose-6-phosphatase activity was measured in the 105,000*g* pellet of lysed (by dilution in water) cells. The incubation volume was 1 ml *p*-nitrophenol metabolism was measured as described by Moldéus *et al.* (53). The concentration of *p*-nitrophenol was 200 μM and the incubation volume was 4 ml. (The *p*-nitrophenol experiment was done in cooperation with P. Moldéus.)

cytochrome P450. The results also suggested that in control cells there is a basal lipid peroxidation dependent on one or both of the two microsomal enzymes and, finally that the malonaldehyde formed is rapidly metabolized (38).

Cumene hydroperoxide-induced lipid peroxidation was linear only between 150 and 300 μM (Fig. 3). Above 800 μM, no further malonaldehyde production was seen with cumene hydroperoxide, while additional iron promoted further peroxidation (38). It was subsequently found that high concentrations of cumene hydroperoxide destroy cytochrome P450 but do not totally inhibit NADPH-cytochrome *c* reductase (Table 2). Addition of iron to liver cells, on the other hand, did not affect cytochrome P450 but decreased cytochrome P450-dependent drug oxidation (Table 2). This can be explained at least partially by the observation that the flavoprotein in intact cells, as in microsomes (7), was solubilized (44). Consistent with a primary involvement of the endoplasmic reticulum were the findings that glucose-6-phosphatase activity and the UDPglucuronyl-transferase-dependent *p*-nitrophenol metabolism were inhibited by iron-induced lipid peroxidation (Table 3).

It is well known that liver cells contain stores of iron in the form of ferritin and that excess iron is taken up by this protein (52). As it can be expected that iron in the ferritin-bound form does not participate in any reactions in the cell (29), it can also be suspected that ferritin can affect iron-induced lipid peroxidation. It was found that 30–40 min after iron addition there was an increased amount of iron bound to a ferritin fraction isolated from the incubated cells (Table 4). This finding might explain why the iron-induced malonaldehyde production ceased spontaneously in cells (Fig. 1). Furthermore, iron bound to ferritin can be expected to be in equilibrium with the surrounding medium, and its presence in control cells might thus explain the basal malonaldehyde production discussed above.

Table 4. Effect of ADP·Iron on Ferritin Iron Content in Isolated Hepatocytes[a]

Incubation time (min)	Iron bound to protein (ferritin)	
	Control	ADP·iron
10	1.4	1.5
20	1.6	1.6
30	1.6	1.8
40	1.7	2.4
50	1.8	2.5

[a]The ferritin fraction was isolated as previously described (38). About 6 × 10^6 cells were incubated in a final volume of 2 ml. ADP-complexed iron was added to a final concentration of 187 μM, which roughly corresponds to the amount of iron expected to be contained in the cells at the start of the incubation (52).

DISCUSSION

The accumulation of lipofuscin pigments (54) may be taken as evidence for a basal and usually concealed lipid peroxidation in tissues. In light of the lively transfer of iron into and out of liver cells, it can be postulated that iron in these cells can catalyze reactions such as lipid peroxidation on its way from the plasma membrane to the apoferritin binding site. Furthermore, it can be postulated that iron exists as a Fe^{2+} chelate in the cell for a short time period as it has been shown that iron transfer from transferritin to apoferritin is greatly facilitated by iron reduction (55). Several groups (5–7) have shown that enzymatic lipid peroxidation is catalyzed by NADPH-cytochrome *c* reductase. The presence of iron, as well as its reduction, seems to be a necessity, and the reduction may be mediated by the superoxide anion (5,6). Fong *et al.* and Zimmermann *et al.* further suggest that hydroxyl free radicals (6,28) or singlet oxygen (28), species that attack membrane lipids and thus induce lipid peroxidation, may be generated. Bidlack and Tappel (31), on the other hand, suggest that reduced iron interacts with lipid peroxides and in this way free radicals are formed. These free radicals can attack new membrane lipids and thus accelerate lipid peroxidation.

The cell preparations were constantly found to contain low amounts of malonaldehyde. The rapid decrease of this substance in the presence of an inhibitor of lipid peroxidation indicates a substantial formation and metabolism of malonaldehyde. Furthermore, the oxidation of GSH in control cells can be taken as evidence that malonaldehyde production was accompanied by peroxide production and that these peroxides produced were at least partially metabolized by GSH peroxidase.

Thus there were signs of a basal and probably harmless lipid peroxidation in liver cells, but several treatments have been found to bring the process out of balance. Malonaldehyde accumulated in the cells upon the addition of chloral hydrate (37). This was probably due to an inhibition of malonaldehyde oxidation and not to increased lipid peroxidation; however, such a process would not be unimportant, since the accumulation of malonaldehyde in cells may inhibit enzyme activities (44). It has also been found that when the level of GSH was decreased to 25% of control an accumulation of malonaldehyde occurred (38). Iron, like cumene hydroperoxide, induced increased lipid peroxidation, malonaldehyde accumulation, and toxic effects in the cell. With limited amounts of iron, the accumulation was transient and the toxic effects were limited, while high amounts of cumene hydroperoxide quickly destroyed the cells (44). Even rather high concentrations of iron did not cause any morphological changes (44) but biochemical alterations such as membrane leakage and enzyme inhibition were observed.

The findings in this study are consistent with a central role for iron reduction in lipid peroxidation in isolated hepatocytes. Whether reduced iron mainly attacks lipid peroxides or participates in the production of hydroxyl free

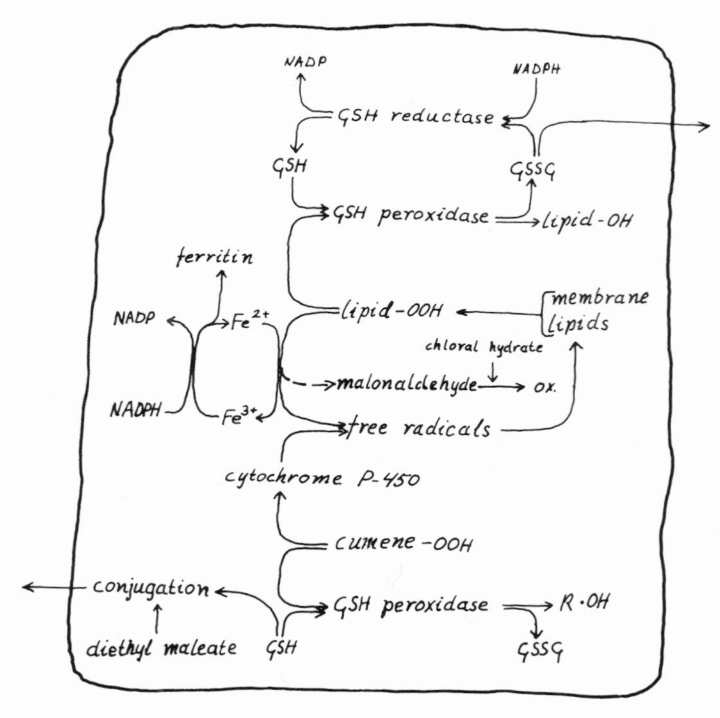

Fig. 4. Schematic presentation of suggested mechanism for lipid peroxidation in isolated hepatocytes. Borders indicate plasma membrane.

radicals is not clear; of course, both of these mechanisms may be significant. However, competition between routes involving radicals and other routes for the decomposition of lipid peroxides formed seems to be important for the propagation of lipid peroxidation in the cell. Thus this may be a point where basal lipid peroxide formation is held in check by endogenous defense mechanisms. The scheme we propose for lipid peroxidation in isolated hepatocytes is summarized in Fig. 4. A slow rate of lipid peroxide formation is kept at a steady state by GSH peroxidase in the presence of low concentrations of iron and can be accelerated by the addition of more iron. Iron is activated to a prooxidant and its binding to ferritin is facilitated by reduction by NADPH-cytochrome *c* reductase. The proposed effects of chloral hydrate, diethyl maleate, and cumene hydroperoxide are also indicated in the figure, as is the loss of GSSG.

A detergentlike effect on the microsomal membrane is observed during the initial stages of lipid peroxidation (7,11), such as an increased UDPglucuronyltransferase activity. That this may be an artifact due to compartmentation in microsomal vesicles was indicated by the opposite effect of lipid peroxidation on this enzyme in isolated hepatocytes. However, one noteworthy detergentlike effect that was demonstrable in isolated cells was the solubilization of NADPH-cytochrome *c* reductase. What consequences this might have for the spreading of lipid peroxidation in a cell is yet not clear. It may be that the radicals produced in the vicinity of the flavoprotein are transient species that will not travel far in the cell and that the solubilization of this enzyme allows these radicals to be produced throughout the cytoplasm. However, in light of the limited effects of lipid peroxidation induced by relatively high concentrations of iron, it seems likely that endogenous defense systems can also cope with this situation as long as ferritin can bind up the iron.

ACKNOWLEDGMENTS

The author wishes to thank Mrs. Annika Kristoferson for the technical assistance and Drs. S. Orrenius and J. DePierre for help with the preparation of the manuscript.

REFERENCES

1. W. A. Pryor, Free radical reactions and their importance in biochemical systems, *Fed. Proc.* **32**, 1862–1869 (1973).
2. A. L. Tappel, Lipid peroxidation damage to cell components, *Fed. Proc.* **32**, 1870–1874 (1973).
3. A. A. Barber and F. Bernheim, Lipid peroxidation: Its measurement, occurrence and significance in animal tissues, *Adv. Gerontol. Res.* **2**, 335–403 (1967).
4. P. Hochstein and L. Ernster, in: *Ciba Foundation Symposium on Cellular Injury* (A. V. S. de Reuck and J. Knight, eds.), pp. 123–124, Churchill, London (1964).

5. T. C. Pederson, J. A. Buege, and S. D. Aust, The role of NADPH cytochrome *c* reductase in liver microsomal lipid peroxidation, *J. Biol. Chem.* **248**, 7134–7141 (1973).
6. K. Fong, P. B. McCay, J. L. Poyer, B. B. Keele, and H. Misra, Evidence that peroxidation of lysosomal membranes is initiated by hydroxyl free radicals produced during flavin enzyme activity, *J. Biol. Chem.* **248**, 7792–7797 (1973).
7. J. Högberg, R. E. Larson, A. Kristoferson, and S. Orrenius, NADPH-dependent reductase solubilized from microsomes by peroxidation and its activity, *Biochem. Biophys. Res. Commun.* **56**, 836–842 (1974).
8. H. E. May and P. B. McCay, TPNH oxidase-catalyzed alterations of membrane phospholipids. I. Nature of the lipid alterations, *J. Biol. Chem.* **243**, 2288–2295 (1968).
9. W. R. Bidlack and A. L. Tappel, Damage to microsomal membrane by lipid peroxidation, *Lipids* **8**, 177–182 (1973).
10. A. U. Arstila, M. H. Smith, and B. F. Trump, Microsomal lipid peroxidation: Morphological characterization, *Science* **175**, 530–532 (1972).
11. J. Högberg, A. Bergstrand, and S. V. Jakobsson, Lipid peroxidation of rat-liver microsomes; its effect on the microsomal membrane and some membrane-bound microsomal enzymes, *Eur. J. Biochem.* **37**, 51–59 (1973).
12. E. D. Wills, Effects of lipid peroxidation on membrane-bound enzymes of the endoplasmic reticulum, *Biochem. J.* **123**, 983–991 (1971).
13. E. D. Wills, Lipid peroxide formation in microsomes; relationship of hydroxylation to lipid peroxide formation, *Biochem. J.* **113**, 333–341 (1969).
14. W. Levin, A. Y. H. Lu, M. Jacobson, R. Kuntzman, J. L. Poyer, and P. B. McCay, Lipid peroxidation and the degradation of cytochrome P-450 heme, *Arch. Biochem. Biophys.* **158**, 842–852 (1973).
15. R. C. McKnight and F. E. Hunter, Mitochondrial membrane ghosts produced by lipid peroxidation induced by ferrous ion. II. Composition and enzymatic activity, *J. Biol. Chem.* **241**, 2757–2765 (1966).
16. R. C. Green, C. Little, and P. J. O'Brien, The inactivation of isocitrate dehydrogenase by lipid peroxide, *Arch. Biochem. Biophys.* **142**, 598–605 (1971).
17. F. De Matteis and R. G. Sparks, Iron-dependent loss of liver cytochrome P-450 haem *in vivo* and *in vitro*, *FEBS Lett.* **29**, 141–144 (1973).
18. A. J. Monserrat, A. K. Ghoshal, W. S. Hartroft, and E. A. Porta, Lipoperoxidation in the pathogenesis of renal necrosis in choline-deficient rats, *Am. J. Pathol.* **55**, 163–190 (1969).
19. R. B. Wilson, N. S. Kula, P. M. Newberne, and M. W. Conner, Vascular damage and lipid peroxidation in choline-deficient rats, *Exp. Mol. Pathol.* **18**, 357–368 (1973).
20. L. Goldberg, L. E. Martin, and A. Batchelor, Biochemical changes in the tissues of animals injected with iron. III. Lipid peroxidation, *Biochem. J.* **83**, 291–298 (1962).
21. C. K. Chow and A. L. Tappel, An enzymatic protective mechanism against lipid peroxidation damage to lungs of ozone exposed rats, *Lipids* **7**, 518–524 (1972).
22. J. E. Allen, D. B. P. Goodman, A. Besarab, and H. Rasmussen, Studies on the biochemical basis of oxygen toxicity, *Biochim. Biophys. Acta* **320**, 708–728 (1973).
23. T. F. Slater, *Free Radical Mechanisms in Tissue Injury*, Pion, London (1972).
24. E. A. Glende, Jr., and R. O. Recknagel, Microsomal electron transport and *in vitro* CCl_4-induced lipoperoxidation, *Fed. Proc.* **29**, 755 (1970).
25. E. A. Glende, Jr., Carbon tetrachloride-induced protection against carbon tetrachloride toxicity; the role of the liver microsomal drug-metabolizing system, *Biochem. Pharmacol.* **21**, 1697–1702 (1972).
26. N. R. Di Luzio, Antioxidants, lipid peroxidation and chemical-induced liver injury, *Fed. Proc.* **32**, 1875–1881 (1973).

27. T. C. Pederson and S. D. Aust, NADPH-dependent lipid peroxidation catalyzed by purified NADPH cytochrome *c* reductase from rat liver microsomes, *Biochem. Biophys. Res. Commun.* **48**, 789–795 (1972).
28. R. Zimmerman, L. Flohé, U. Weser, and H. Hartman, Inhibition of lipid peroxidation in isolated inner membrane of rat liver mitochondria by superoxide dismutase, *FEBS Lett.* **29**, 117–120 (1973).
29. E. D. Wills, Mechanisms of lipid peroxide formation in animal tissues, *Biochem. J.* **99**, 667–676 (1966).
30. R. O. Recknagel and A. K. Ghoshal, Lipoperoxidation as a vector in carbon tetrachloride hepatotoxicity, *Lab Invest.* **15**, 132–146 (1966).
31. W. R. Bidlack and A. L. Tappel, A proposed mechanism for the TPNH enzymatic lipid peroxidizing system of rat liver microsomes, *Lipids* **7**, 564–565 (1972).
32. B. O. Christophersen, Formation of monohydroxypolyenoic fatty acids from lipid peroxides by a glutathione peroxidase, *Biochim. Biophys. Acta* **164**, 35–46 (1968).
33. P. J. O'Brien and C. Little, Intracellular mechanisms for the decomposition of a lipid peroxide. II. Decomposition of a lipid peroxide by subcellular fractions, *Can. J. Biochem* **47**, 493–499 (1969).
34. C. Little and P. J. O'Brien, An intracellular GSH-peroxidase with a lipid peroxide substrate, *Biochem. Biophys. Res. Commun.* **31**, 145–150 (1968).
35. R. C. Green and P. J. O'Brien, The involvement of semidehydroascorbate reductase in the oxidation of NADH by lipid peroxide in mitochondria and microsomes, *Biochim. Biophys. Acta* **293**, 334–342 (1973).
36. C. M. MacDonald, The effects of ethanol on hepatic lipid peroxidation and on the activities of glutathione reductase and perioxidase, *FEBS Lett.* **35**, 227–230 (1973).
37. J. Högberg, S. Orrenius, and R. E. Larson, Lipid peroxidation in isolated hepatocytes, *Eur. J. Biochem.* **50**, 595–602 (1975).
38. J. Högberg, S. Orrenius, and P. J. O'Brien, Further studies on lipid peroxide formation in isolated hepatocytes, *Eur. J. Biochem.* **59**, 449–455 (1975).
39. M. N. Berry and D. S. Friend, High-yield preparation of isolated rat liver parenchymal cells; a biochemical and fine structural study, *J. Cell Biol.* **43**, 506–520 (1969).
40. T. Berg, D. Boman, and P. O. Seglen, Induction of tryptophan oxygenase in primary rat liver cell suspensions by glucocorticoid hormone, *Exp. Cell Res.* **72**, 571–574 (1972).
41. P. O. Seglen, Preparation of rat liver cells. I. Effect of Ca^{2+} on enzymatic dispersion of isolated, perfused liver, *Exp. Cell Res.* **74**, 450–454 (1972).
42. F. Bernheim, M. L. C. Bernheim, and K. M. Wilbur, The reaction between thiobarbituric acid and the oxidation products of certain lipids, *J. Biol. Chem.* **174**, 257–264 (1948).
43. P. Moldéus, R. Grundin, H. Vadi, and S. Orrenius, A study of drug metabolism linked to cytochrome P-450 in isolated rat-liver cells, *Eur. J. Biochem.* **46**, 351–360 (1974).
44. J. Högeberg, P. Moldéus, B. Arborgh, P. O'Brien, and S. Orrenius, The consequences of lipid peroxidation in isolated hepatocytes, *Eur. J. Biochem.* **59**, 457–462 (1975).
45. A. Mazur, S. Green, and A. Carleton, Mechanism of plasma iron incorporation into hepatic ferritin, *J. Biol. Chem.* **235**, 595–603 (1960).
46. P. Charley, M. Rosenstein, E. Shore, and P. Saltman, The role of chelation and binding equilibria in iron metabolism, *Arch. Biochem. Biophys.* **88**, 222–226 (1960).
47. C. F. Chasseaud, in: *Glutathione* (L. Flohé *et al.*, eds.), pp. 90–109, Georg Thieme, Stuttgart (1974).
48. H. Sies, C. Gerstenecker, H. Menzel, and L. Flohé, Oxidation in the NADP system and release of GSSG from hemoglobin-free perfused rat liver during peroxidatic oxidation of glutathione by hydroperoxides, *FEBS Lett.* **27**, 171–175 (1972).
49. V. H. Cohn and J. Lyle, A fluorometric assay for glutathione, *Anal. Biochem.* **14**, 434–440 (1966).

50. P. J. O'Brien and A. Rahimtula, Involvement of cytochrome P-450 in the intracellular formation of lipid peroxides, *J. Agric. Food Chem.* **23**, 154–158 (1975).
51. D. Kupfer and S. Orrenius, Characteristics of guinea pig liver and adrenal monooxygenase systems, *Mol. Pharmacol.* **6**, 221–230 (1970).
52. C. P. Van Wyk, M. Linder-Horowitz, and H. N. Munro, Effect of iron loading on non-heme iron compounds in different liver cell populations, *J. Biol. Chem.* **246**, 1025–1031 (1971).
53. P. Moldéus, H. Vadi, and M. Berggren, Oxidative and conjugative metabolism of *p*-nitroanisole and *p*-nitrophenol in isolated rat liver cells, *Acta Pharmacol. Toxicol.* **39**, 17–32 (1976).
54. K. S. Chio, U. Reiss, B. Fletcher, and A. L. Tappel, Peroxidation of subcellular organelles: Formation of lipofuscinlike fluorescent pigments, *Science* **166**, 1535–1536 (1969).
55. J. P. G. Miller and D. J. Perkins, Model experiments for the study of iron transfer from transferrin to ferritin, *Eur. J. Biochem.* **10**, 146–151 (1969).

46

Discussion

In response to Schulman's question regarding morphological changes in the cells, Högberg stated that in the case of cumene hydroperoxide-induced lipid peroxidation the observed high concentrations of malonaldehyde are accompanied by cellular changes, including preliminary indications of condensation of endoplasmic reticulum.

Schenkman inquired about the relative contributions of mitochondria and endoplasmic reticulum in the accumulation of malonaldehyde, since both are known to play a role in lipid peroxidation. Högberg felt that the malonaldehyde production is primarily associated with the endoplasmic reticulum, based on experiments in which malonaldehyde production was blocked with inhibitors of NADPH-cytochrome *c* reductase. He also pointed out that lipid peroxidation in mitochondria is nonenzymatic, further complicating the picture.

Högberg did not find any decrease in P450 levels when iron was used as an inducer of lipid peroxidation. This was surprising since there is a decrease when cumene hydroperoxide is used. A decrease is seen in microsomes.

Anders inquired about release of enzymes into the medium when either one or the other of the inducers of lipid peroxidation was employed. Högberg stated that while he has some indications that leakage occurs he has not made any direct measurements of, for example, lactic dehydrogenase levels in the medium.

Zimmerman commented that he had shown earlier in another model that carbon tetrachloride and several other agents cause rapid release of enzymes into the medium. Anders had obtained similar results in the same system, using chlorpromazine.

Hildebrandt was interested in the effect of prolonged incubation on lipid peroxidation, in the absence of ADP- or pyrophosphate-complexed iron. Högberg found that the level of peroxidation remains constant, having measured it for up to 2 h. The assay requires added ADP.

Kocsis wished to know whether the increased levels of malonaldehyde are due to an effect on the formation or the inactivation of the compound. Högberg had not studied this aspect. At the end of the typical incubation time there is a decrease in malonaldehyde levels, which suggested to him that continuous decomposition takes place. He felt that the added substances must also have a stimulatory effect on malonaldehyde formation.

Vainio was interested in the underlying mechanisms for the observed changes in the rate of *p*-nitrophenol metabolism. He referred to the known association of increased UDPglucuronyltransferase activity with lipid peroxidation. Högberg noted that this aspect had not been part of the study. He suggested that the increased UDPglucuronyltransferase activity in microsomal preparations may be due to the detergentlike effect of lipid peroxidation. Apparently this effect is not demonstrable in intact liver cells.

47

New Data Supporting an Obligatory Role for Lipid Peroxidation in Carbon Tetrachloride-Induced Loss of Aminopyrine Demethylase, Cytochrome P450, and Glucose-6-phosphatase

Richard O. Recknagel, Eric A. Glende, Jr., and Andrew M. Hruszkewycz
Department of Physiology
School of Medicine
Case Western Reserve University
Cleveland, Ohio 44106, U.S.A.

An attractive theme running through the work of many authors is the idea that drug-induced cellular injury may involve covalent binding of drug metabolites to functionally important structures of the cells that are injured. This idea has been widely discussed, especially rather recently, and it has been the central conceptual frame of reference for a considerable body of work (1–5). No attempt to review this idea in its entirety is intended here. Rather, this chapter presents new data which bear directly on part of the problem. An outline of the main idea, as applied to acute toxic hepatocellular necrosis, is given in Fig. 1, which was constructed from the text as given by Gillette *et al.* (2). It is postulated that a potentially toxigenic therapeutic agent or foreign chemical is first converted to an active metabolite. The latter reacts with a so-called action site on a tissue macromolecule, and is covalently bound there. The covalent binding of the active metabolite then, in some way, leads to hepatocellular necrosis. It was pointed out (2) that various factors would be expected to affect the severity of the ultimate lesion. One set of such factors could be thought of as controlling the rate of formation of and reactivity of active metabolites, e.g., more or less absorption, more or less excretion, more or less monooxygenase activity. An-

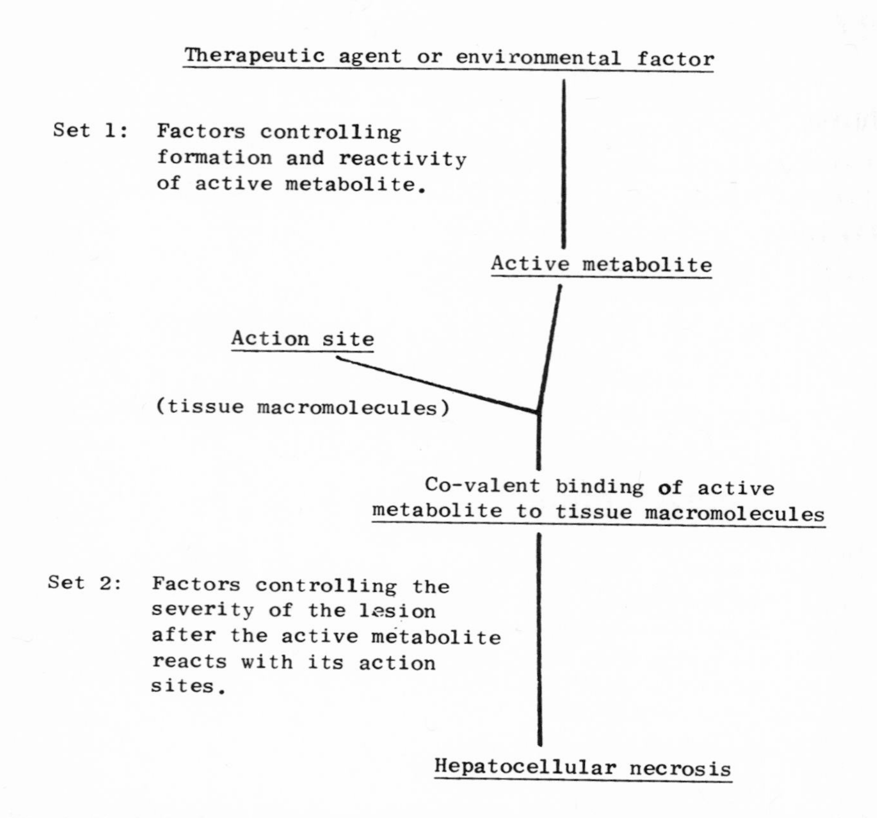

Fig. 1. The Gillette, Mitchell, Brodie hypothesis of covalent binding of toxigenic metabolites as applied to the problem of hepatocellular necrosis.

other set of factors could be thought of as those which come into play after covalent binding of the active metabolite to tissue macromolecules has taken place. It was pointed out that little is known of the latter set of factors. Thus, although great gaps in our current knowledge are recognized, the cornerstone of the central conception is the idea that covalent binding of active metabolites is the key event leading eventually to cellular necrosis.

There is no doubt that this idea has stimulated much new work, especially with regard to factors controlling formation and reactivity of active metabolites (set 1 of Fig. 1). With regard to factors controlling the severity of the lesion (set 2 of Fig. 1) and especially with regard to what is usually regarded as the mechanism of necrosis, the situation is less satisfactory. When we speak of a mechanism of necrosis, we mean the cause of the necrosis. Uncovering the mechanism of a necrogenic process is equivalent to uncovering its cause, in the sense of precise delineation of the key structural and functional abnormalities, i.e., the necessary and sufficient pathological perturbations, which, once initiated, lead inexorably to death of the cell.

In our opinion, the highly seductive idea as given in Fig. 1 suffers to a certain extent in that it may lead to unwarranted suppositions regarding intimate chemical mechanisms of toxicity. For example, it has been stated (2) that for toxigenic therapeutic agents "for any particular compound there should be a relationship between the severity of the lesion and the amount of covalently bound active metabolite." As a specific example, in the case of bromobenzene, liver cell necrosis correlates with covalent binding of a metabolite, and not with the amount of bromobenzene itself, and the statement was made that "the necrosis is caused by the covalent binding of a metabolite and not by bromobenzene itself . . . ." Although admittedly provisional, the latter statement is a statement on mechanism. The difficulty here is that the mechanism of the necrosis is still so totally obscure that the assumption of a causal imperative leading from covalent binding to necrosis may not be justified. Furthermore, it has sometimes been taken for granted that even though a correlation of metabolite covalent binding with necrosis may not constitute a mechanism of necrosis, the fact of covalent binding may at least be regarded as a step in the right direction toward elucidation of underlying mechanisms. However, it has been pointed out (5) that "when toxins interact with a number of biochemical systems simultaneously, as chemical metabolites frequently do, it is difficult to determine whether changes in cell function result from a sequence of changes originating from a single initial biochemical alteration or from the concerted action of a number of different initial biochemical alterations. In any event, such studies frequently fail to provide any practical information about the nature of the active form of the toxin or any clues to methods for preventing the toxicity." In other words, localized and critical covalent binding of a toxic drug metabolite may be surrounded by unknown degrees of noncritical covalent binding. Search for critical cellular functions compromised by the toxic drug could thus be thwarted by an intractable maze of side reactions. In the special case of CCl_4 poisoning, the case for a "covalent binding to macromolecules" hypothesis is even more tenuous. For, despite data from *in vivo* experiments (6–8) which have indicated a correlation between the degree of covalent binding of ^{14}C from $^{14}CCl_4$ and severity of CCl_4-induced pathological changes, the covalent binding may be incidental to the key toxigenic events. This possibility emerges from data presented in this chapter which show conclusively that for three enzymes known to be severely depressed by CCl_4, covalent binding of metabolites of CCl_4 does not, in and of itself, constitute a mechanism of injury. Thus, because of inherent complexities resident in the cellular systems on which toxigenic drugs and foreign compounds exert their undesirable effects, the disarmingly simple notion of covalent binding (as shown in Fig. 1) may require modifications and elaborations before its potential usefulness as a guide toward unraveling mechanisms of toxicity can be realized.

EXPERIMENTAL PROCEDURE

The microsome fraction was isolated from rat liver homogenates by routine centrifugation procedures. A full description of all procedural details will be published elsewhere. The essential experimental design involved an initial incubation of the microsome fraction, after which the microsomes were recovered and assayed in a second incubation for residual enzyme activity. Conditions for the initial incubation were designed with two distinct purposes in mind. One set was designed to maximize microsomal metabolism of CCl_4 in the absence of lipid peroxidation. This was achieved by carrying out the initial incubation under anaerobic conditions, with added EDTA. Use of anaerobic conditions for study of CCl_3—Cl bond cleavage was first developed by Uehleke and co-workers (9–11). An alternate set of conditions allowed lipid peroxidation to proceed. This was accomplished by reproducing the experimental conditions for the anaerobic incubation, except that the system was aerobic and no EDTA was added. After the appropriate initial incubation, the microsomes were recovered and assayed for enzyme potency. Any residual enzyme activity was compared with corresponding activity of intact control microsomes which had neither supported CCl_4 metabolism nor undergone lipid peroxidation. The point of these experiments was to determine quantitatively any effects of the initial incubation conditions on selected enzyme activity of the isolated microsomes.

RESULTS

Rat liver microsomes catalyze the rapid disappearance of added CCl_4 (Fig. 2). Note that for the experiment shown in Fig. 2 the incubation was anaerobic, and lipid peroxidation was prevented by addition of EDTA. Unreacted CCl_4 and newly appearing $CHCl_3$ were extracted into heptane and quantitatively assayed with a gas chromatograph equipped with a ^{63}Ni electron capture detector. All added CCl_4 disappears within 90 min, and about one-third of added CCl_4 appears as $CHCl_3$. During the anaerobic conversion of CCl_4 to $CHCl_3$, if lipid peroxidation is prevented, there is no detectable diminution in microsomal cytochrome P450 content (Fig. 3, left panel). In an exactly equivalent incubation system which was made aerobic and with no added EDTA, there is a marked decrease in cytochrome P450 content associated with vigorous lipid peroxidation (Fig. 3, right panel; note MDA production). An even more dramatic finding occurred with respect to G6Pase. During 90 min of incubation under anaerobic conditions with added EDTA, although there was substantial conversion of CCl_4 to $CHCl_3$, G6Pase activity was essentially unaffected (Fig. 4). The rate of metabolism of CCl_4 in this anaerobic system is about 5 times greater than estimated rates of CCl_4 metabolism *in vivo.* In sharp contrast to its stability during anaerobic CCl_4 metabolism, microsomal G6Pase decreases

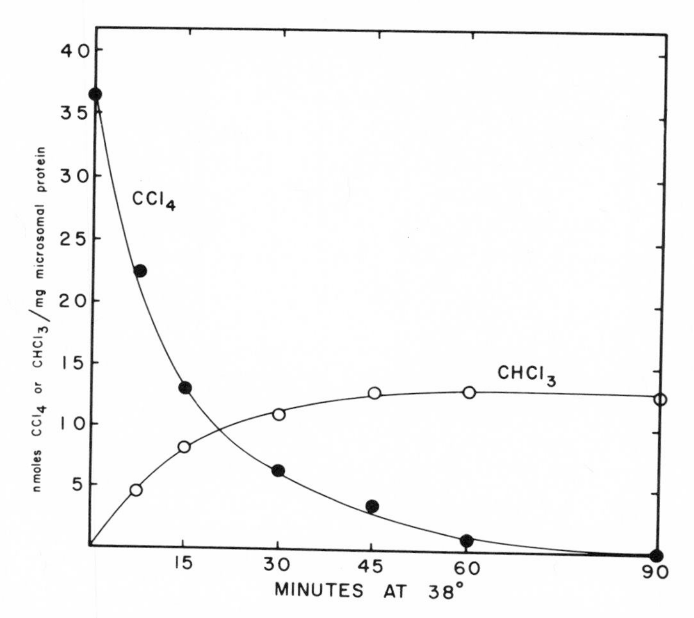

Fig. 2. Conversion of CCl_4 to $CHCl_3$ in vitro *catalyzed by rat liver microsomes.* Conditions: Final volume, 8 ml, containing 100 mg equivalents of rat liver microsomes per milliliter. Incubation medium was 0.9% NaCl containing 0.05 M sodium phosphate buffer at pH 6.5. NADPH was generated from NADP by isocitric dehydrogenase and isocitrate, in the presence of Mg^{2+} and nicotinamide. Total CCl_4 added, 0.081 μl. EDTA, 0.003 M. Anaerobic, 38°C.

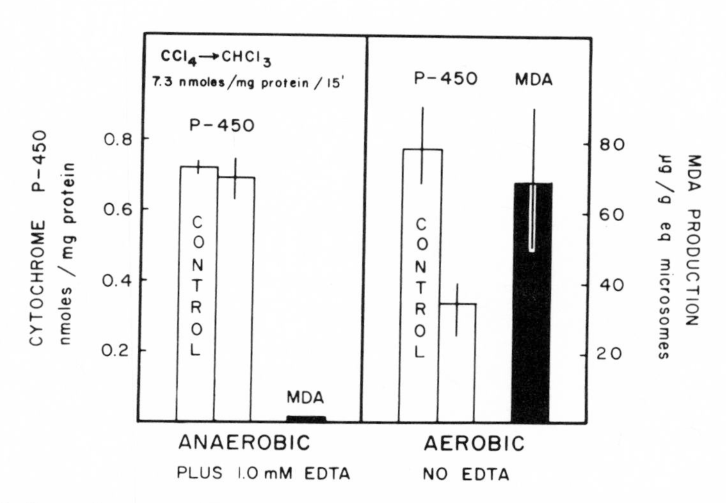

Fig. 3. Left panel: preservation of microsomal cytochrome P450 content during anaerobic conversion of CCl_4 to $CHCl_3$ in presence of added EDTA. Right panel: loss of cytochrome P450 associated with lipid peroxidation. Experimental conditions were essentially as in Fig. 2.

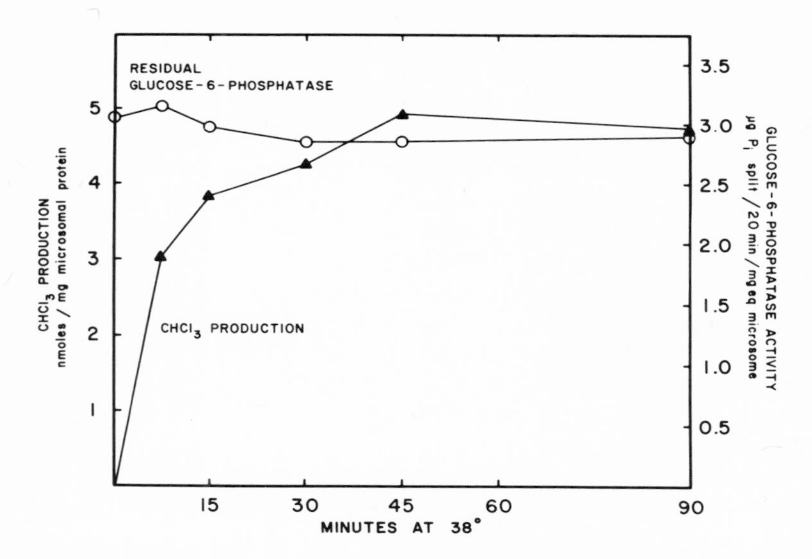

Fig. 4. Preservation of glucose-6-phosphatase activity during the anaerobic conversion of CCl_4 to $CHCl_3$ by rat liver microsomes. Note: lipid peroxidation was prevented by addition of EDTA. Experimental conditions were essentially as in Fig. 2, except that the incubation medium was tris-maleate buffer.

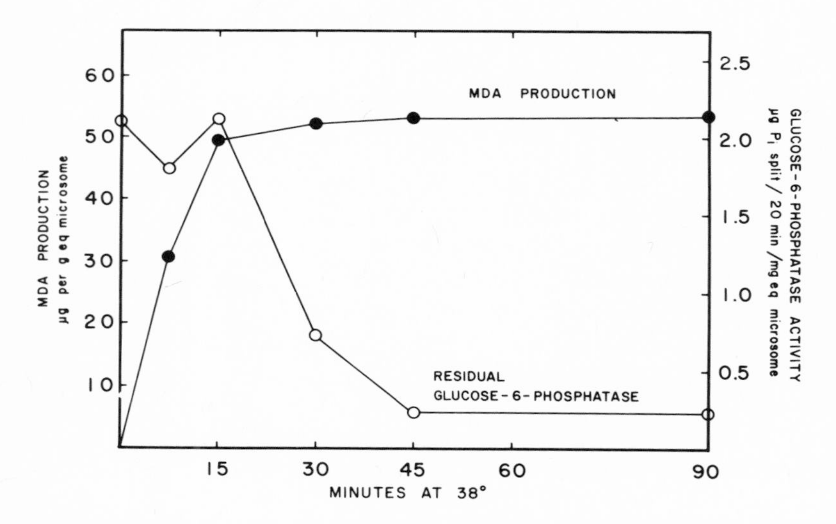

Fig. 5. Loss of glucose-6-phosphatase activity associated with rat liver microsomal lipid peroxidation. Essential conditions were the same as in Fig. 4 except that the system was aerobic and contained no added EDTA.

rapidly in an aerobic system in which lipid peroxidation is allowed to occur (Fig. 5).

Data shown in Figs. 3 and 4 clearly show that if lipid peroxidation is prevented, vigorous metabolism of CCl_4 is totally unable to produce any significant decrease in either cytochrome P450 content or G6Pase activity of rat liver microsomes. Both of these enzymes decrease markedly *in vivo* in the CCl_4-poisoned rat, and their disappearance requires cleavage of the CCl_3-Cl bond (12, 13). Since the bond cleavage in and of itself is not a sufficient cause for disappearance of these enzymes, it is manifestly clear that some process must intervene between initial CCl_3-Cl bond cleavage and loss of these enzymes. In effect, the "covalent binding to macromolecules" hypothesis, in what we may call its primitive form, holds that covalent binding of toxic metabolites is the decisive intervening event.

Actually this idea as a mechanism of CCl_4-induced cell injury is not new, since Butler (14) had suggested 14 years ago that covalent binding of free radical cleavage products onto protein SH groups might be the decisive event. In a critical review (13), no evidence could be found in support of Butler's suggestion. In the meantime, a number of authors have from time to time suggested that covalent binding of cleavage products of CCl_4 might be involved in CCl_4 liver cell injury, in the primitive sense of direct attack by such cleavage products on susceptible sites. Maling *et al.* (15) have offered a recent version of this "covalent binding to macromolecules" hypothesis which combines the idea of covalent binding with lipid peroxidation. These workers comment that free radical products of CCl_4 metabolism "may bind covalently to the proteins and lipids of the hepatocytes. The binding of free radical derivatives to the double bonds of polyunsaturated fatty acids may initiate processes leading to diene conjugation in liver microsomal phospholipids and lipoperoxidation. Thus, the covalent binding of the free radical derivatives of CCl_4 to liver lipids and proteins presumably initiates the processes responsible for CCl_4 hepatotoxicity." This formulation as a specific mechanism for CCl_4 liver injury emerges from the more general statement as given in Fig. 1. Although not a primitive version of the covalent binding hypothesis, it clearly places the emphasis on covalent binding of metabolic products of CCl_4 as the key toxigenic event. For example, these workers (15) in their discussion of effects of dibenamine pretreatment on CCl_4 hepatotoxicity state that "the hepatotoxicity is less as the result of reduced binding of free radical derivatives to liver phospholipids and proteins." Availability of the *in vitro* systems used to generate data of Figs. 2–5 provided an opportunity to subject the "covalent binding to macromolecules" hypothesis to experimental test. We monitored the fate of cytochrome P450, aminopyrine demethylase, and G6Pase during *in vitro* covalent binding of ^{14}C from metabolism of $^{14}CCl_4$.

For 120 min of incubation, during which there was extensive covalent binding of ^{14}C from added $^{14}CCl_4$, microsomal cytochrome P450 was essen-

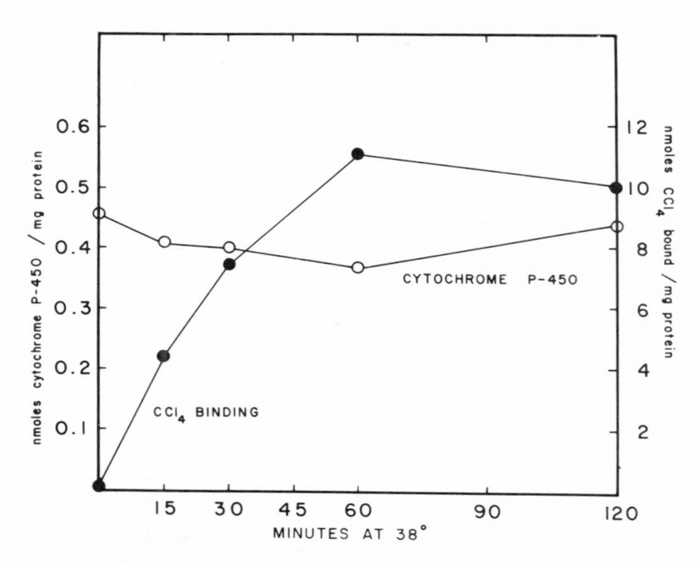

Fig. 6. Stability of rat liver microsomal cytochrome P450 during covalent binding of ^{14}C *from* $^{14}CCl_4$. Essential conditions: microsomes supplemented with NADPH, anaerobic, plus EDTA.

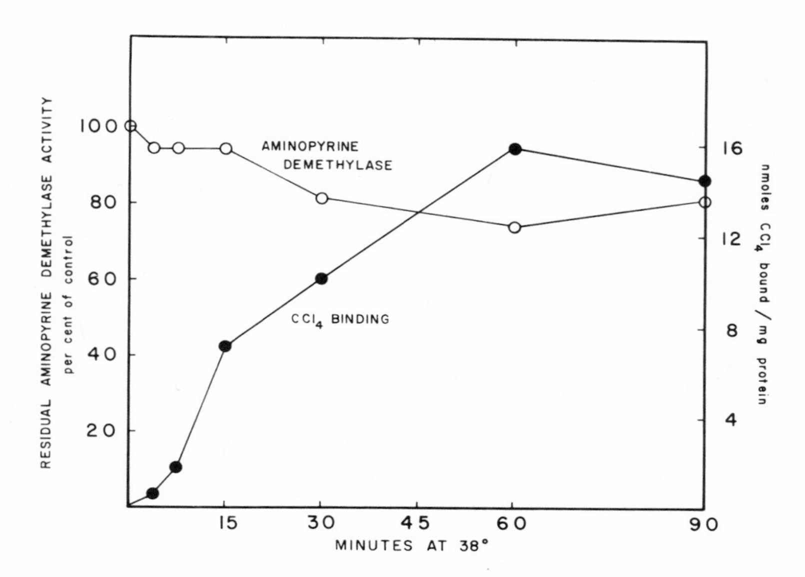

Fig. 7. Stability of rat liver microsomal aminopyrine demethylase during covalent binding of ^{14}C *from* $^{14}CCl_4$. Essential conditions were as in Fig. 6.

Table 1. Failure of CCl_4 Covalent Binding to Affect Rat Liver Microsomal Glucose-6-Phosphatase

Additions to rat liver microsomes[a]	Residual G6Pase μg P_i split/20 min/mg eq microsomes	CCl_4 bound nanomoles/mg protein
None	2.44	–
NADPH + $^{14}CCl_4$	2.28	18.9

[a]Incubation conditions: anaerobic, plus EDTA; 60 min of incubation at 38°

tially stable (Fig. 6). In a similar experiment, we found that during 90 min of incubation microsomal aminopyrine demethylase activity decreased only modestly, despite extensive covalent binding of ^{14}C from added $^{14}CCl_4$ (Fig. 7). When lipid peroxidation is allowed to occur, loss of aminopyrine demethylase activity is much greater than the modest decline shown in Fig. 7.

G6Pase is similarly indifferent to covalent binding of split products of CCl_4 metabolism (Table 1). We have estimated that in the anaerobic system employed in these studies, covalent binding of CCl_4 carbon was about 7 times greater than covalent binding of CCl_4 cleavage products *in vivo* at doses of CCl_4 which cause marked decreases of these enzymatic functions.

DISCUSSION

The work reported in this chapter has significance with regard to an important general problem of biochemical pathology, *viz.*, the problem of action at a distance. Specifically, for CCl_4 poisoning, this problem takes the following form. It is well known that CCl_4 toxicity depends on cleavage of the CCl_3–Cl bond (12, 13). Although it is not possible to decide the matter with certainty, nevertheless a number of lines of evidence suggest that this bond cleavage probably takes place uniquely at the cytochrome P450 locus (13). The fact that some ^{14}C from $^{14}CCl_4$ administered to the whole animal eventually appears at sites removed from the cytochrome P450 locus (e.g., mitochondrial lipids and proteins) can easily be understood in terms of lipid migrations (16, 17) or biosynthetic processes of one kind or another. If the CCl_3–Cl bond cleavage does indeed take place uniquely at the cytochrome P450 locus, the primitive form of the "covalent binding to macromolecules" hypothesis as applied to CCl_4 liver cell injury faces a theoretical difficulty. It seems highly probable to us that the cytochrome P450 locus could not coincide spatially with various other loci of the endoplasmic reticulum which house functions such as G6Pase, formation and intracellular movement of precursors of very low density lipoproteins, and especially protein synthesis, known to be associated with bulky polyribosomes. All of these functions decrease rapidly in the CCl_4-poisoned rat. Thus it would appear that the chemical pathological mechanism of CCl_4-induced loss of such functions must include the concept of some process, secondary to initial localized CCl_3–Cl bond cleavage, which can act at a distance. This creates a difficulty for the "covalent binding to macromolecules" hypothesis, which, in its primitive form, holds that the split products of CCl_3–Cl bond cleavage are the agents which migrate from their point of origin to induce functional derangement at a distance, i.e., by covalent binding to critical but unknown functional sites. The difficulty emerges from the likelihood that the primary products of cleavage of the CCl_3–Cl bond are most probably trichlormethyl and monatomic chlorine free radicals (13), and it is highly unlikely that these free radicals could survive long enough to act at sites removed from their immediate point of origin. The central significance of the data presented here is that they decisively remove at least part of the problem from the realm of speculation. The data presented in Figs. 6 and 7 and in Table 1 represent a formidable obstacle standing in the way of accepting the "covalent binding to macromolecules" hypothesis for CCl_4-induced liver cell injury, at least in its primitive form.

The data clearly show that for the enzymes studied, covalent binding of CCl_4 carbon is not the decisive event responsible for loss of these enzymes. Rather, some process, set into motion by CCl_3–Cl bond cleavage and dependent on intervention of molecular oxygen, appears to be involved. We believe that the evidence is very strong in favor of this process being none other than lipid peroxidation.

We know of no data which support the suggestion (15) that initiation of lipid peroxidation itself requires addition of free radical cleavage products of CCl_4 across double bonds of polyenoic fatty acids. It has been suggested (12, 18) that the key step is a hydrogen abstraction attack on methylene hydrogens of polyenoic fatty acids. If such is the case, any covalent binding would be incidental to the main event, which is initiation of lipid peroxidation. In any event, at least for G6Pase, cytochrome P450, and aminopyrine demethylase, it is quite clear that destructive processes intimately linked to lipid peroxidation are decisive for loss of these enzymes, not covalent binding of CCl_4 cleavage products.

The lipid peroxidation hypothesis does not suffer from difficulties inherent in the "covalent binding to macromolecules" hypothesis. The hypothesis holds that spreading peroxidative decomposition of lipids within the membrane of the endoplasmic reticulum coupled with emergence of toxic lipid peroxides or their breakdown products are the agencies which act at a distance. Such toxic entities, not being free radicals, would be expected to have half-lives much longer than those of the primary products of homolytic cleavage of the CCl_3–Cl bond. These chemical agents, as yet unknown, could induce pathological change at sites far removed from the discrete locus of CCl_3–Cl bond cleavage, and such pathological change would not be dependent in any way on covalent binding of metabolites of the primary toxic entity. That lipid peroxidation may indeed involve action at a distance is strongly supported by a large body of work on the toxicity of lipid hydroperoxides, as well as by the reports (19, 20) that microsomal lipid peroxidation *in vitro* induces hemolysis of red cells present in the incubation medium. Recent studies in our laboratory do not support the conclusion (20) that such red cell hemolysis is not due to a toxic entity arising from peroxidizing lipids.

The significance of this work in a more general sense consists of the following. Chemical reactions, by their nature, take place at discrete molecular loci. If toxicity of drugs and foreign compounds depends on their metabolic conversion of active metabolites, the loci of such conversions will in general be restricted. Yet, characteristically, the end results of toxic liver injury involve various cell structures, frequently all of them. In general, action at a distance is probably involved. In our view, the "covalent binding to macromolecules" hypothesis, at least in its various versions known to us, is too general and perhaps too simple, at least in its present form, to encompass the evident complexities which would be expected to be involved in the sequence of events leading from appearance of the toxigenic agent in the cell to eventual observable consequences. This work on the chemical pathology of CCl_4-induced liver cell injury may be taken by the skeptical investigator as a signal to the effect that the "covalent binding to macromolecules" hypothesis may be a reliable guide to fruitful work with regard to study of factors controlling formation and reactivity of active metabolites of drugs and foreign compounds, but it may not be a reliable guide with respect to unraveling the subsequent events, which, in effect, constitute in each individual case the mechanism of toxicity.

ACKNOWLEDGMENT

This work was supported by Grant AM-01489 from the National Institute of Arthritis, Metabolism, and Digestive Diseases, United States Public Health Service.

REFERENCES

1. B. B. Brodie, in: *Drug Responses in Man* (G. Wolstenholme and R. Porter, eds.), pp. 188–213, Little, Brown, Boston (1967).
2. J. R. Gillette, J. R. Mitchell, and B. B. Brodie, Biochemical mechanisms of drug toxicity, *Annu. Rev. Pharmacol.* **14**, 271–288 (1974).
3. J. R. Gillette, A perspective on the role of chemically reactive metabolites of foreign compounds in toxicity. I. Correlation of changes in covalent binding of reactive metabolites with changes in the incidence and severity of toxicity, *Biochem. Pharmacol.* **23**, 2785–2794 (1974).
4. J. R. Gillette, A perspective on the role of chemically reactive metabolites of foreign compounds in toxicity. II. Alterations in the kinetics of covalent binding, *Biochem. Pharmacol.* **23**, 2927–2938 (1974).

5. J. R. Mitchell and D. J. Jollow, Metabolic activation of drugs to toxic substances, *Gastroenterology* **68,** 392–410 (1975).
6. J. A. Castro and M. I. Diaz Gomez, Studies on the irreversible binding of $^{14}C-CCl_4$ to microsomal lipids in rats under varying experimental conditions, *Toxicol. Appl. Pharmacol.* **23,** 541–552 (1972).
7. J. A. Castro, C. R. deCastro, N. D'Acosta, M. I. Diaz Gomez, and E. C. deFerreyra, Carbon tetrachloride activation in liver microsomes from rats induced with 3-methylcholanthrene, *Biochem. Biophys. Res. Commun.* **50,** 273–279 (1973).
8. N. D'Acosta, J. A. Castro, M. I. Diaz Gomez, E. C. deFerreyra, C. R. deCastro, and O. M. deFenos, Role of cytochrome P-450 in carbon tetrachloride activation and CCl_4-induced necrosis. Effects of inhibitors of heme synthesis. I. 3-Amino-1,2,4-triazole, *Res. Commun. Chem. Pathol. Pharmacol.* **6,** 175–183 (1973).
9. O. Reiner and H. Uehleke, The effect of CCl_4 on lipid peroxidation and cytochrome P-450 in the endoplasmic reticulum, *Naunyn-Schmiedeberg's Arch. Pharmacol.* **270** (Suppl.) R111 (1970).
10. O. Reiner, S. Athanassopoulos, K. H. Hellmer, R. E. Murray, and H. Uehleke, Bildung von Chloroform aus Tetrachlorkohlenstoff in Lebermikrosomen, Lipid-peroxidation und Zerstorüng von Cytochrom P-450, *Arch. Toxikol.* **29,** 219–233 (1972).
11. H. Uehleke, K. H. Hellmer, and S. Tabarelli, Binding of $^{14}CCl_4$ to microsomal proteins *in vitro* and formation of $CHCl_3$ by reduced liver microsomes, *Xenobiotica* **3,** 1–11 (1973).
12. R. O. Recknagel, Carbon tetrachloride hepatotoxicity, *Pharmacol. Rev.* **19,** 145–208 (1967).
13. R. O. Recknagel and E. A. Glende, Jr., Carbon tetrachloride hepatotoxicity: An example of lethal cleavage, *CRC Crit. Rev. Toxicol.* **2,** 263–297 (1973).
14. T. C. Butler, Reduction of carbon tetrachloride *in vivo* and reduction of carbon tetrachloride and chloroform *in vitro* by tissues and tissue homogenates, *J. Pharmacol. Exp. Ther.* **134,** 311–319 (1961).
15. H. M. Maling, R. M. Eichelbaum, W. Saul, I. G. Sipes, E. A. B. Brown, and J. R. Gillette, Nature of the protection against carbon tetrachloride induced hepatotoxicity produced by pretreatment with dibenamine [*N*-(2-chloroethyl)-dibenzylamine], *Biochem. Pharmacol.* **23,** 1479–1491 (1974).
16. K. W. A. Wirtz and D. B. Zilversmit, The use of phenobarbital and carbon tetrachloride to examine liver phospholipid exchange in intact rats, *Biochim. Biophys. Acta* **187,** 468–476 (1969).
17. K. W. A. Wirtz and D. B. Zilversmit, Participation of soluble liver proteins in the exchange of membrane phospholipids, *Biochim. Biophys. Acta* **193,** 105–116 (1969).
18. R. O. Recknagel and A. K. Ghoshal, Lipoperoxidation as a vector in carbon tetrachloride hepatotoxicity, *Lab. Invest.* **15,** 132–148 (1966).
19. P. Hochstein, in: *Third International Conference on Hyperbaric Medicine* (I. W. Brown, Jr., and B. G. Cox, eds.), pp. 61–64, National Academy of Science, National Research Council, Washington, D.C. (1966).
20. P. M. Pfeifer and P. B. McCay, Reduced triphosphopyridine nucleotide oxidase-catalyzed alterations of membrane phospholipids. V. Use of erythrocytes to demonstrate enzyme-dependent production of a component with the properties of a free radical, *J. Biol. Chem.* **246,** 6401–6408 (1971).

48

Discussion

Anders referred to his findings which indicated that protein synthesis inhibitors (cycloheximide and actinomycin D) prevent CCl_4 toxicity when monitored by GOT release. These data can be seen as evidence either for or against the role of lipid peroxidation, but the interesting fact is that toxic manifestations can be prevented by inhibitors of protein synthesis. Recknagel agreed that the role of protein has yet to be examined. He reiterated the main thesis of his presentation, emphasizing that lipid peroxidation and not covalent binding is likely to be the key phenomenon in CCl_4 toxicity.

Mitchell's suggestion that the use of protein synthesis inhibitors might aid in isolating necrotizing effects from effects on the endoplasmic reticulum initiated an exchange among several participants. He inquired whether CCl_4 causes destruction of P450 and release of glucose-6-phosphatase in the presence of protein synthesis inhibitors. Recknagel quoted work by Farber stating that cycloheximide did not prevent membrane damage. Anders reported that GOT is not detected at increased levels in the presence of protein synthesis inhibitors, which would indicate a degree of protection; however, P450 destruction is not prevented by these agents.

Zimmerman noted that fatty degeneration *per se* leads to very little enzyme leakage, and inquired whether the above inhibitors prevent fatty degeneration or necrosis. Anders' data show that cycloheximide prevents mucolipid accumulation, but actinomycin D does not.

Snyder pointed out that in the anaerobic system one would expect a reaction to occur between the lipid and the free radical and inquired whether there is any evidence of lesion under these conditions, especially with regard to the arachidonic acid molecule. Recknagel stated that he is attempting to study this question and postulated that he expects to find shifts in double bonds. Such changes may not lead to any detectable pathology; only an analysis of the lipid structure may give the desired information. In response to Conney's question regarding the need to add EDTA to the anaerobic incubation mixture, Recknagel stated that EDTA was added to make certain that all lipid peroxidation is prevented. Whether making the system anaerobic would suffice was not investigated. When Mitchell observed that this experimental arrangement appeared to involve changing two variables simultaneously, Recknagel replied that preserving "the geometry of the experiment" was not important, but prevention of lipid peroxidation was.

49

Binding of Haloalkanes to Liver Microsomes

Hartmut Uehleke
Department of Toxicology
Bundesgesundheitsamt
D-1 Berlin 33, Germany

During the past two decades, many investigations have shown that numerous toxic compounds are inert *per se* but are converted in the body to chemically reactive metabolites (1–3). The reactive metabolites or metabolic intermediates can combine with cellular constituents and may initiate methemoglobin formation (4–8), sensitization (9, 10), tumors (1), mutation (11, 12), lipid peroxidation (13, 14), or cellular necrosis (3, 15).

The first suggestion that degradation of haloalkanes might be connected with their toxicity was proposed in 1883 by Zeller (16) and in 1887 by Kast (17). After application of chloroform narcosis, sodium chloride depleted dogs showed an increased urinary chloride excretion. In several dogs, 20–50% of the administered dose was probably degraded to chloride. Binz (18) in 1894 described the analogously increased bromide excretion in the urine of rabbits after administration of tetrabromomethane.

More than 60 years later, McCollister *et al.* (19) "rediscovered" halomethane metabolism when they observed that in monkeys a small percentage of labeled CCl_4 is expired as CO_2. In 1961 Butler (20) confirmed those results in dogs and found $CHCl_3$ among the exhaled compounds. Butler proposed the formation of radicals from CCl_4 as the aggressive species that might be responsible for the liver toxicity of CCl_4. This theory was expanded for other haloalkanes and solvents by Wirtschafter and Cronyn (21). Slater (13) and Recknagel (14) presented evidence to show that free-radical-generated lipid peroxidation is the most important factor in CCl_4-initiated liver damage.

In vitro, liver fractions and liver microsomes catalyze the conversion of CCl_4 and $CHCl_3$ to CO_2 (22). It was further demonstrated that this reaction increased both *in vivo* and in liver microsomes after pretreatment of the animals with phenobarbital or chlorophenothane (23, 24). Therefore, reductive dehalogenation at the microsomal cytochrome system was considered. Accordingly, acceleration or inhibition of drug metabolism should result in altered toxicity. This has been shown to be so for CCl_4 (23–25), halothane (26), and $CHCl_3$ (27).

Several investigators reported the *in vivo* irreversible binding of radioactivity from ^{14}C-labeled CCl_4 to liver protein and lipid (28–31); preferential binding was found in endoplasmic protein and lipid. Also, ^{36}Cl from $C^{36}Cl_4$ is irreversibly bound to liver constituents (29, 31). In similar studies, the covalent binding of labeled $CHCl_3$ (32) and halothane (33, 34) was elaborated. The connection between the binding of CCl_4 and halothane to reduced hepatic cytochromes, their microsomal activation and metabolism, and the irreversible binding to endoplasmic macromolecules (35–40) gave some insight into the basic mechanism of haloalkane activation. An understanding of the relationships between haloalkane activation and covalent binding to haloalkane toxicity and side effects was the goal of our investigations.

METHODS

Radiochemicals (38, 39, 41), animals (38, 41), tissue fractionation (42), *in vitro* incubates (38, 39, 41), estimation procedures for bound radioactivity (38–41), and isolation and purification of mitochondrial and microsomal lipid and protein (38–41, 43) are described in detail in the publications cited.

RESULTS AND DISCUSSION

In Vivo Binding

Several earlier experiments on the *in vivo* irreversible binding of radioactivity from individual haloalkanes are not comparable. Different doses were given and the bound radioactivity was indicated, for example, as "cpm per microsomal protein from liver of 100 g rat." Those values are not translatable when the yield of microsomes and other factors are not reported.

Consequently, we have compared the irreversible binding of CCl_4, $CHCl_3$, and halothane in mice. The solution of the ^{14}C-labeled haloalkanes in peanut oil was injected intraperitoneally; 50 μl contained 100 μmol of the haloalkane with 2 μCi ^{14}C per 10 g of mouse. The injection of halothane and chloroform produced light narcosis. A considerable amount of the radioactivity remains irreversibly bound to liver protein and lipid (Fig. 1). Chloroform binding to mitochondrial and microsomal protein and lipid reached a peak approximately 6 h after administration. In contrast, halothane binding to mitochondrial and microsomal protein plateaued early, but the binding to lipids reached a sharp

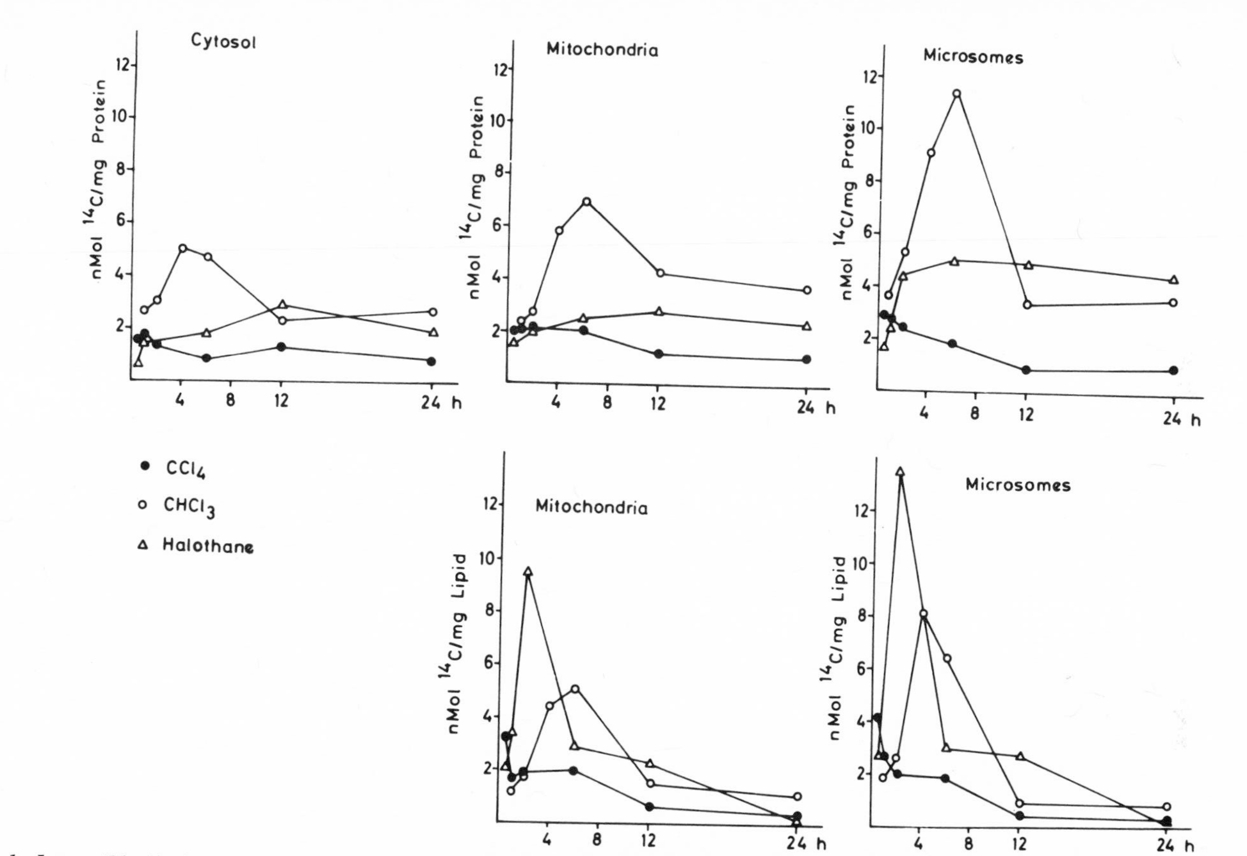

Fig. 1. Irreversible binding of radioactivity from ^{14}*C-labeled* CCl_4*,* $CHCl_3$*, and halothane to isolated and purified protein and lipid of liver fractions.* Mice were injected intraperitoneally with 100 μmol of each labeled haloalkane containing 2 μCi in 50 μl of peanut oil per 10 g of mouse.

peak within 2 h after administration. After maximal protein binding, a very slow decrease of bound radioactivity was observed, corresponding approximately to the half-life of liver protein. A remarkably high amount of radioactive material was still bound 24 h after application.

The irreversible binding of ^{14}C-labeled trichlorofluoromethane (not included in Fig. 1) to liver endoplasmic protein and lipid resembles the CCl_4 binding (41). Additional experiments with ^{14}C-labeled 1,1,2-trichloroethylene in mice resulted in protein and lipid binding curves with peaks 4–6 h after administration (44).

The magnitude of binding is defined more clearly by percentage of the dose of haloalkane bound to the liver endoplasmic protein and lipid. Assuming that the microsomes account for 5% of total liver weight and the liver contains 40 mg of microsomal protein per gram of liver, the maximal value for protein is approximately 0.23% of the $CHCl_3$ dose and 0.1% of the halothane and 0.06% of the CCl_4 doses given. The analogous estimation for endoplasmic lipids shows that 0.09% of the administered $CHCl_3$ dose, 0.14% of the halothane, and 0.06% of the CCl_4 are bound.

Thus the covalent binding of $CHCl_3$, halothane, and CCl_4 to liver microsomal and mitochondrial proteins and lipids reaches 0.4, 0.3, and 0.15% of the dose. It must be mentioned that also in extrahepatic tissues containing the cytochrome P450 system, such as lung, kidney, and adrenals, covalent binding of labeled haloalkanes was demonstrated (32, 39). Obviously, there is no relation between the liver toxicity of the individual haloalkanes and the amount of firmly bound material from the labeled halogenated alkanes in the liver.

A dose–response curve also exists for the $CHCl_3$ binding in mice with saturation at 3–5 mmol/kg (32). Howard *et al.* (34) administered only 1/25 of the narcotic halothane dose to mice. Correspondingly, the irreversible binding after 24 h was very low. Some comparable data from the publications cited can be calculated. Rao and Recknagel (30) gave oral doses of 2.5 ml/kg CCl_4 to rats (approximately 25 mmol/kg). Thirty minutes after administration, they found 9 μg (60 nmol) CCl_4 bound to "microsomal protein per 100 g body weight." Assuming 40 mg of microsomal protein per gram of liver and 4 g liver per 100 g body weight, a specific binding of 0.38 nmol/mg microsomal protein is the result. However, if Rao and Recknagel referred only to the microsomal protein which is isolated by the routine homogenization and centrifugation procedures, the actual binding would be approximately 3 times higher, i.e., about 1 nmol/mg protein. This value is still lower than our *in vivo* CCl_4 binding in mice. For similar reasons, a proper comparison of lipid binding is not possible. Reynolds (29) reported binding of 100 nmol ^{14}C per gram liver (obviously lipid and protein) after an oral dose of 100 μmol/100 g body weight to rats. This dose is only one-tenth of ours.

Van Dyke and Wood (45) added 100 μl (1 mmol) labeled halothane (10 μCi) to the perfusion medium (130 ml) of isolated perfused rat livers. The recovery of nonvolatile radioactivity in the separated microsomal fraction was 44% of the total nonvolatile activity in the liver. Total nonvolatile activity in the crude liver homogenate corresponded to about 160 nmol ^{14}C per gram of liver. Consequently, the binding to microsomal constituents was lower than 4 nmol/mg protein and lipid together, and appreciably lower than the binding in mice in our investigations. The question of whether the rate of total metabolism of CCl_4 or $CHCl_3$ to CO_2 in animals is somehow related to the irreversible binding cannot be answered accurately. Only one proper comparison of CCl_4 and $CHCl_3$ metabolism in rats was performed by Paul and Rubinstein (46). After oral administration of the ^{14}C-labeled haloalkanes, $CHCl_3$ was converted 3 times faster to CO_2 than CCl_4 was. Rubinstein and Kanics (22) reported that in rat liver homogenates or with isolated liver microsomes, $CHCl_3$ is metabolized to CO_2 3–5 times faster than CCl_4.

Van Dyke and Gandolfi (47) recently reported on the separation and purification of liver microsomal lipids 2 and 12 h after the intraperitoneal injection of ^{14}C-labeled halothane to rats (33 nmol/100 g body weight). They observed a maximum of phospholipid binding after 2 h, and the bound radioactivity had disappeared from the phospholipids after 12 h. In contrast, at this time, the irreversibly bound radioactivity from halothane in microsomal proteins was found to be higher than after 2 h. This observation corresponds well with our results. From the data given by Van Dyke and Gandolfi (47), it is not possible to calculate the specific binding rates (nmol/mg protein or lipid) of halothane.

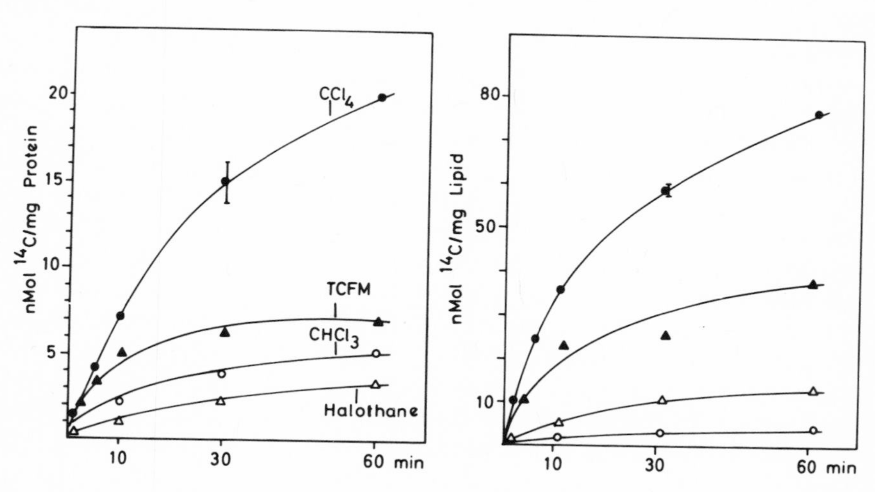

Fig. 2. Covalent binding of ^{14}C-labeled haloalkanes in anaerobic incubates of liver microsomes from phenobarbital-pretreated rabbits. TCFM, trichlorofluoromethane. The incubates contained, in 1 ml, 1 μmol haloalkane, 5 mg microsomal protein, 15 μmol glucose-6-phosphate, 0.6 μmol NADP, 9 μmol $MgCl_2$, and 0.75 U glucose-6-phosphate dehydrogenase, made to 1 ml with Krebs-Ringer phosphate solution, pH 7.4. Data represent means from seven experiments in the case of CCl_4 and four experiments in the case of each of the other haloalkanes. Only the greatest standard error (for CCl_4) is indicated.

The different kinetics of halothane binding to endoplasmic protein and lipid *in vivo* led Van Dyke and Gandolfi (47) to assume the possibility of two reactive intermediates during halothane metabolism. The rapid reaction with lipids should be mediated by the "radicals," whereas the slower protein binding agent could be trifluoroacetaldehyde. Our observations *in vitro* (Fig. 2) might in part agree, since the binding of halothane persists longer than the CCl_4 binding, but we found no dramatic differences between protein and lipid binding. It must be stated that the proposed trifluoroacetaldehyde intermediate (48) during degradation of halothane to the biological end product trifluoroacetic acid has never been substantiated.[1]

Irreversible Binding *in Vitro* with Liver Microsomes

In anaerobic suspensions of liver microsomes from rabbits pretreated with phenobarbital, CCl_4 is bound considerably faster to microsomal protein and lipid than halothane or $CHCl_3$ (Fig. 2). However, the binding rate of trichlorofluoromethane was unexpectedly high, and reached nearly 50% of the CCl_4 binding. The irreversible binding of labeled CCl_4, halothane, and trichlorofluoromethane to endoplasmic lipids proceeded approximately 5 times faster than the protein binding; the binding of radioactivity from $CHCl_3$ to protein and lipid was of similar magnitude. One could speculate that the binding rates of CCl_4 and $CHCl_3$ reflect the differences in their inhalation toxicities. The maximal allowable concentration (MAC) for occupational exposure to CCl_4 is 10 ppm, and for exposure to $CHCl_3$ it is 50 ppm.[2] However, halothane and trichlorofluoromethane, which display practically no acute liver toxicity, do not fit into the same classification.

After 60 min of incubation, about 10% of the total ^{14}C from $^{14}CCl_4$ in the incubates was irreversibly bound to the isolated microsomal protein, and more than 30% was found firmly associated with the purified lipid. Because the substrate saturation curve with CCl_4 is rather steep (38), more than 50% of the available CCl_4 is irreversibly bound to microsomal protein and lipid when 0.25 mM CCl_4 is added to the incubates. In anaerobic incubates, $CHCl_3$ formation from CCl_4 can be followed (38), and it accounts for the disappearance of an additional 20–25% of the CCl_4 substrate. Only small amounts of free CCl_4 remain after 60 min of incubation.

The covalent binding of radioactivity from labeled CCl_4 (38, 40) and halothane (39, 40, 49) to microsomal protein and lipid *in vitro* has been described in detail earlier, and a report on the comparative binding of CCl_4, $CHCl_3$, and halothane has been presented (50). In these reports, the incubation

[1] The demonstration of halothane defluoridation [Cohen *et al.*, *Anesthesiology* **43,** 392 (1975); Widger *et al.*, *Anesthesiology* **44,** 197 (1976)] and the possible radical formation during this pathway might explain the divergent experimental results.

[2] Recently, $CHCl_3$ has been shown to produce carcinoma in mice and rats after long term oral application. Consequently, MAC values were lowered in most countries.

Table 1. Irreversible Binding of ^{14}C to Microsomal Phospholipid Fractions after 60 min Anaerobic Incubation of Labeled CCl_4 or Halothane with Rabbit (Pretreated) Liver Microsomes and NADPH

Phospholipids	Percent phospholipid fraction	CCl_4[a]		Halothane[a]	
		PBR[b]	SR[c]	PBR[b]	SR[c]
Sphingomyelin	2.1	8.8	0.141	6.6	0.033
Lecithin	56.0	48.9	0.030	48.3	0.009
Phosphatidylinositol	11.2	10.7	0.033	10.3	0.010
Phosphatidylserine	3.8	5.6	0.050	5.1	0.014
Cephalin	28.3	26.0	0.032	29.7	0.011

[a]Total lipid binding: CCl_4, 76 nmol/mg lipid; halothane, 12 nmol/mg lipid.
[b]Percent of bound radioactivity in the individual phospholipid fractions, corrected for loss during TLC, related to the total phospholipid fraction.
[c]Specific radioactivity of the phospholipid fractions in nmol ^{14}C per nmol phospholipid-phosphate.

conditions and the influence of metabolic inhibitors, cytochrome blockers, and radical scavengers were described. Additional experiments on the covalent binding of CCl_4 and $CHCl_3$ to liver microsomes were summarized by Gillette *et al.* (3).

The labeled microsomal lipids from incubates with CCl_4 or halothane were further purified by TLC. Phospholipids account for approximately 80–85% of the microsomal lipid fraction. The radioactivity bound to the individual lipid components is indicated in Table 1. The largest amount of radioactivity was found in the lecithin fraction. However, the specific labeling was highest in the sphingomyelin.

Influence of Oxygen

The binding rates of ^{14}C from labeled CCl_4, halothane, and trichlorofluoromethane in anaerobic incubates with liver microsomes were greater than under aerobic conditions. However, the binding of $CHCl_3$ to protein and lipid increased in the presence of oxygen (Fig. 3). The metabolism of $CHCl_3$ to CO_2 in liver homogenates and liver microsomes has been shown to be a typical O_2-dependent mixed-function oxidation (22). In such a system, $CHCl_3$ was more rapidly converted to CO_2 than was CCl_4. However, in our experiments, the anaerobic covalent binding rate of CCl_4 (Fig. 2) was still 2 times greater than the binding rate of $CHCl_3$ in aerobic incubates (Fig. 3).

Species Differences

With CCl_4 and halothane there were no major differences in the initial velocities of protein and lipid binding for mouse, rat, and rabbit liver micro-

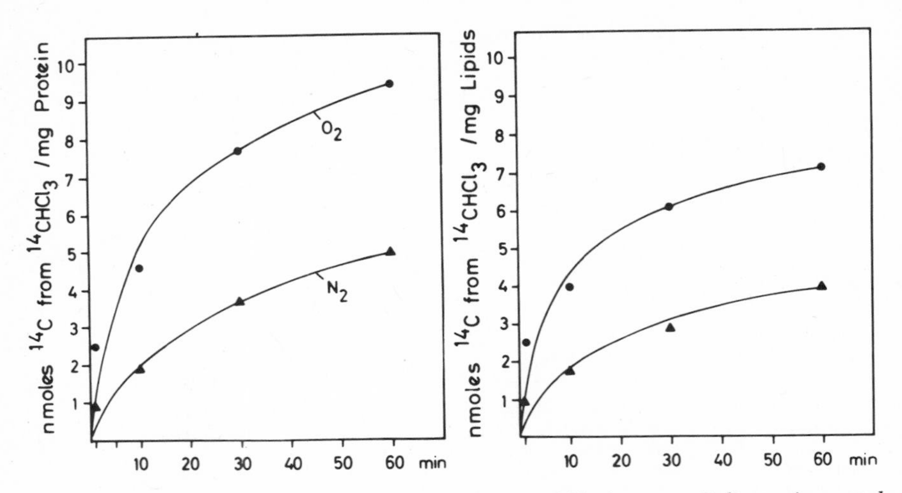

Fig. 3. Irreversible binding of ^{14}C *from* $CHCl_3$ *to rabbit (pretreated) liver microsomal protein and lipid during incubation in a gas phase of* N_2 *or* O_2. Results represent means from 4 experiments.

somes *in vitro*. Greater differences were observed with $CHCl_3$ in microsomes of the three species (Fig. 4). In one series of experiments, the binding of labeled CCl_4, $CHCl_3$, and halothane was studied in human liver microsomes. During the first 10 min the binding velocity of CCl_4 was similar to that in rat and rabbit liver microsomes but the binding curve plateaued much earlier. The initial velocity of $CHCl_3$ binding was slower than in rabbit liver microsomes and faster than in mice, but the irreversible binding reached the rabbit values after 60 min

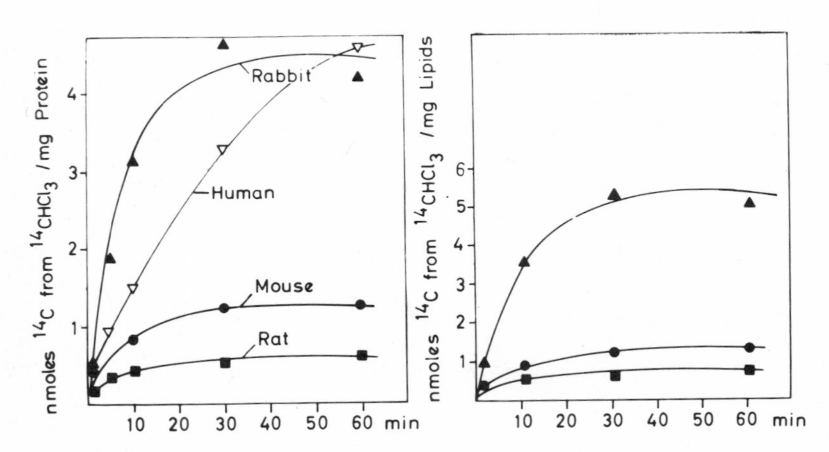

Fig. 4. Comparison of the irreversible binding of ^{14}C*-labeled* $CHCl_3$ in vitro *to protein and lipid of microsomes from normal rabbit, rat, mouse, and human liver.* Incubations were conducted in O_2 under the same conditions as given in Fig. 2.

(Fig. 4). The binding of halothane to human liver microsomal protein was only about 30% that of rat or rabbit liver microsomes. After oral administration of ^{14}C-labeled chloroform, mice excreted 80% of the dose in the form of CO_2 within 48 h (exhalation and urine), rats excreted 60%, and squirrel monkeys only 20% (51). Humans exhaled up to 50% of an oral dose of 500 mg ^{13}C-labeled chloroform as CO_2 (52).

Influence of Cytochrome P450 Concentrations

In microsomes from phenobarbital-pretreated animals the rate of covalent haloalkane binding was greater than in microsomes from control animals. The differences were smaller for lipid binding in rat and rabbit liver microsomes. A comparison of the protein binding rates during the incubation of the three labeled haloalkanes with microsomes from control and phenobarbital-pretreated rabbits is given in Fig. 5.

The greatest effect of phenobarbital stimulation on the rate of *in vitro* microsomal lipid and protein covalent binding was observed in mice with halothane (Fig. 6). Phenobarbital stimulation produced the least effects in rats (see Table 2).

Several reports have stated that pretreatment of rats with 3-methylcholanthrene (3-MC) reduced the toxicity of CCl_4 (3, 41, 53). It was postulated that 3-MC induces a distinct cytochrome variety which might differ in its ability to activate or to metabolize CCl_4. In liver microsomes from 3-MC-pretreated rats the covalent binding of ^{14}C from CCl_4 to protein increased, but proportionally less than the cytochrome P450 concentrations (Table 2). There was no significant change in lipid binding. Addition of CCl_4 to reduced liver microsomes from 3-MC-pretreated rats produced an analogous shift in the light absorption from 448 nm (CO) to 452 nm (CCl_4). As reported earlier (36), this shift in normal

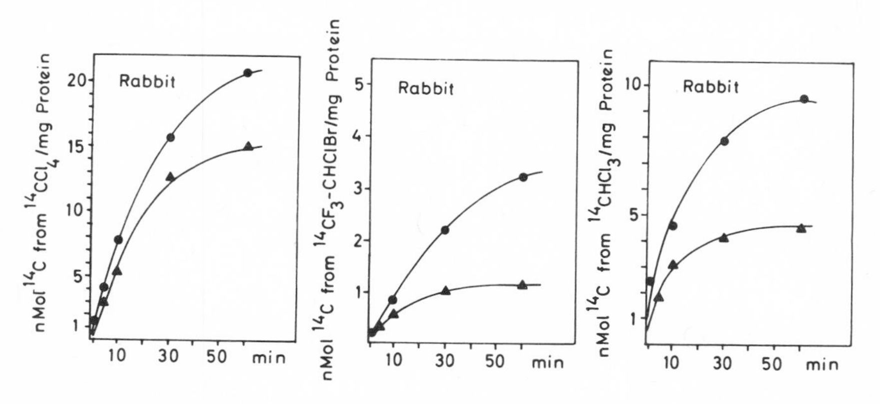

Fig. 5. Influence of increased cytochrome P450 concentrations after phenobarbital treatment on the irreversible in vitro *binding of labeled CCl_4, $CHCl_3$, and halothane to protein of rabbit liver microsomes.* ▲, Control; ●, stimulated.

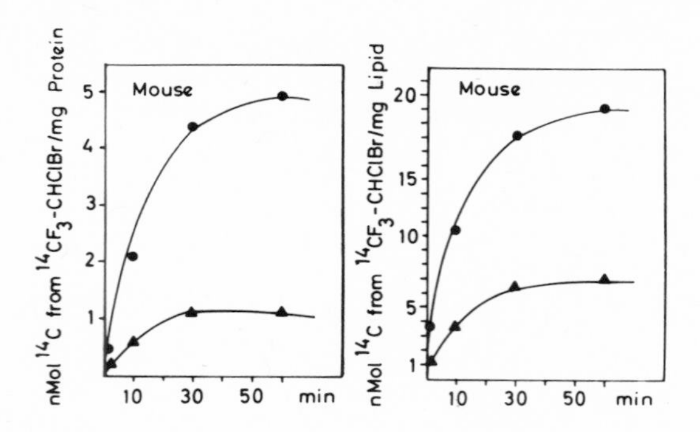

Fig. 6. Alteration of in vitro *^{14}C-binding from halothane to lipid and protein of liver microsomes from control and phenobarbital-treated mice.* ▲, Control; ●, stimulated.

liver microsomes was from 450 nm to 454 nm. From these results, one can conclude that the primary activation step during CCl_4 metabolism is not inhibited by 3-MC pretreatment. Additional experiments in our laboratory showed that the CCl_4-induced lipid peroxidation in microsomes from 3-MC-pretreated rats was not diminished.

After treatment of rats with cobalt chloride, liver microsomes with reduced cytochrome P450 concentrations had reduced covalent binding rates (Table 2). Again, the differences were smaller than the differences in cytochrome concentrations. Isopropanol given to rats prior to CCl_4 greatly increases the liver toxicity of CCl_4 (54). However, in suspensions of liver microsomes of rats pretreated with 2.5 ml isopropanol per kilogram the covalent binding rate of $^{14}CCl_4$ increased by a factor of only 1.2 (Table 2). With the same microsomes the *p-C*-hydroxylation of *N*-butylaniline was accelerated 1.6-fold, but the cytochrome P450 concentrations were not increased.

Covalent binding was also observed in microsomes from newborn rats which had low cytochrome P450 concentrations. The incorporation of radioactivity

Table 2. Influence of Cytochrome P450 Concentrations on the Irreversible Binding of CCl_4 in Incubates of Rat Liver Microsomes with NADPH

	Dose	*n*	P450[a]	Protein[b]	Lipid[b]
Control	–	7	1.05	10.2	50.2
Phb	0.1% water, 8 d	4	2.0	15.4	54.0
3-MC	20 mg/kg, 3 d	5	2.2	16.5	56.0
Isopropanol	2.5 ml/kg	2	1.1	12.8	52.0
$CoCl_2$	40 mg/kg, 2 d	3	0.64	7.4	40.1

[a]Cytochrome P450 expressed in nmol/mg protein.
[b] $^{14}CCl_4$ binding expressed as nmol/30 min/mg protein.

Table 3. Irreversible Protein Binding of Labeled Haloalkanes to Liver Microsomes of Newborn and Adult Rats *in Vitro*

Age	P450[a]	CCl_4[b]	$CHCl_3$[b]	Halothane[b]
18 h	0.32	1.5	0.16	0.13
32 d	1.05	6.7	0.39	0.44

[a]P450 in nmol/mg.
[b]Covalent binding in nmol/mg in 10 min.

into microsomal protein was less reduced than the cytochrome P450 concentrations. Binding to lipid was only slightly influenced (Table 3) (41, 49, 55).

The observation that the binding rates of haloalkanes do not correspond proportionally to the concentrations of cytochrome P450 in the individual microsomal preparation suggests that other activation mechanisms may exist in the microsomes. However, irreversible binding and lipid peroxidation are secondary reactions. The primary activation step produces reactive intermediate(s) which are probably different from those species which combine with reduced cytochrome P450. Reactive intermediates, e.g., free radicals, preferentially combine with cellular constituents close to the site at which they are formed. Binding to protein and lipid, recombination of radicals, and radical scavenging by oxygen and certain chemical groups occur simultaneously. Therefore, in our opinion, any acceleration of the primary step must not result in a proportional increase of the secondary events which we followed here. The strong influence of CO, O_2, and metyrapone on covalent binding and chloroform production from CCl_4 points to the participation of cytochrome P450 in the activation of the haloalkanes. The possible participation of microsomal flavines in radical formation and stabilization requires experimental proof.

Binding to Extraendoplasmic Molecules

CCl_4 is assumed to be a weak carcinogen, and halothane has been alleged to exert its occasional toxic liver effects on an allergic basis (41). From the work of Reynolds (29), it is known that, in contrast to liver lipid and protein, endoplasmic RNA is not labeled after oral doses of $^{14}CCl_4$.

In attempts to visualize the formation of reactive metabolites indirectly, model systems have been developed using the binding to protein and other acceptors in metabolizing *in vitro* systems (40, 56). In incubates of isolated microsomes, NADPH, and additional soluble protein or RNA, only rather low irreversible binding of ^{14}C from $^{14}CCl_4$ to the added albumin was observed (Table 4). No significant radioactivity was found associated with the isolated nonendoplasmic protein or RNA during incubation with labeled halothane. Chloroform binds only in aerobic incubates. Consequently, the low binding of

Table 4. Covalent Binding of ^{14}C from CCl_4 (N_2), $CHCl_3$ (O_2), and Halothane (N_2) in Incubates of Rabbit Liver Microsomes (5 mg/ml) to Added Soluble Bovine Albumin (5 mg/ml)

	Protein[a]	Lipid[a]	Added protein[b]
CCl_4	20	76	1.4
$CHCl_3$	5.1	4.1	1.3
Halothane	3.4	12	–0.2

[a]Binding to microsomal protein and lipid.
[b]Binding to added albumin in nmol/mg protein during 60 min. In anaerobic incubates the binding of $CHCl_3$ was 0.8.

radioactivity from $^{14}CCl_4$ in aerobic incubates could derive from the $^{14}CCl_3\cdot$intermediate which does not accumulate in the presence of oxygen. CCl_4, $CHCl_3$, halothane, and trichlorofluoromethane produced no mutagenic effects in metabolizing *in vitro* systems consisting of *Salmonella typhimurium* and microsomes to activate the potential mutagens (43). Obviously, the short-lived metabolic intermediates of the haloalkanes investigated do not reach or cannot penetrate the test organisms.

However, these observations do not account for the relatively high binding *in vivo* of haloalkanes to the cytosol and to mitochondrial constituents. Possibly, the intact cell structure facilitates the transport of reactive intermediates to the mitochondria (41).

ACKNOWLEDGMENT

Our investigations in this field were supported by the Deutsche Forschungsgemeinschaft.

REFERENCES

1. J. A. Miller, Carcinogenesis by chemicals: An overview, *Cancer Res.* **30,** 559–576 (1970).
2. H. Uehleke, Stoffwechsel von Arzneimitteln als Ursache von Wirkungen, Nebenwirkungen und Toxizitat, *Progr. Drug Res. (Basel)* **15,** 147–203 (1971).
3. J. R. Gillette, J. R. Mitchell, and B. B. Brodie, Biochemical mechanisms of drug toxicity, *Annu. Rev. Pharmacol.* **14,** 271–289 (1974).
4. W. Heubner, Methämoglobinbildende Gifte, *Ergeb. Physiol.* **43,** 8–56 (1940).
5. H. Uehleke, Biologische Oxydation und Reduktion am Stickstoff aromatischer Amino-

und Nitroderivate und ihre Folgen für den Organismus, *Progr. Drug. Res. (Basel)* **8**, 195–260 (1964).

6. A. Burger, J. Wagner, H. Uehleke, and E. Götz, Beeinflüssung von Pentosephosphatzyklus und Glykolyse in Erythrocyten während Methämoglobinbildung durch Phenylhydroxylamin, *Arch. Exp. Path. Pharmak.* **256**, 333–347 (1967).
7. H. Uehleke, Mechanisms of methemoglobin formation by therapeutic and environmental agents, in: *Toxicological Problems* (T. A. Loomis, ed.), Vol. 2, pp. 124–136, *Proceedings Fifth International Congress of Pharmacology,* Karger, Basel (1973).
8. M. Kiese, *Methemoglobinemia: A Comprehensive Treatise,* CRC Press, Cleveland (1974).
9. H. Uehleke, Biochemische Reaktionen als Ursache erworbener. Überempfindlichkeit gegen Fremdstoffe, *Z. Immunitätsforsch.* **123**, 447–457 (1962).
10. H. Uehleke, Metabolite von Arznei- und Fremdstoffen als Allergene, *Z. Immunitätsforsch. Suppl.* **1**, 22–36 (1974).
11. A. Hollaender, *Chemical Mutagens: Principles and Methods for Their Detection,* Plenum Press, New York (1971).
12. P. Czygan, H. Greim, A. J. Carro, F. Hutterer, F. Schaffner, H. Popper, O. Rosenthal, and D. Y. Cooper, Microsomal metabolism of dimethylnitrosamine and the cytochrome P-450 dependency of its activation to a mutagen, *Cancer Res.* **33**, 2983–2986 (1973).
13. T. F. Slater, *Free Radical Mechanisms in Tissue Injury,* Pion, London (1962).
14. R. O. Recknagel, Carbon tetrachloride hepatotoxicity, *Pharmacol. Rev.* **19**, 145–208 (1967).
15. B. B. Brodie, W. D. Reid, A. K. Cho, G. Sipes, G. Krishna, and J. R. Gillette, Possible mechanism of liver necrosis caused by aromatic organic compounds, *Proc. Natl. Acad. Sci. USA* **68**, 160–164 (1971).
16. A. Zeller, Über die Schicksale des Jodoforms und Chloroforms im Organismus, *Hoppe-Seyler's Z. Physiol. Chem.* **8**, 70–78 (1883).
17. A. Kast, Über die Schicksale einiger organischer Chlorverbindungen im Organismus, *Hoppe-Seyler's Z. Physiol. Chem.* **11**, 278–285 (1887).
18. C. Binz, Beiträge zur pharmakologischen Kenntnis der Halogene, *Naunyn-Schmiedeberg's Arch. Exp. Path. Pharmak.* **34**, 185–207 (1894).
19. D. D. McCollister, W. H. Beamer, G. J. Atchison, and M. C. Spencer, The absorption, distribution and elimination of radioactive carbon tetrachloride by monkeys upon exposure to low vapour concentrations, *J. Pharmacol. Exp. Ther.* **102**, 112–124 (1951).
20. T. C. Butler, Reduction of carbon tetrachloride *in vivo* and reduction of carbon tetrachloride and chloroform *in vitro* by tissues and tissue constituents, *J. Pharmacol. Exp. Ther.* **134**, 311–319 (1961).
21. Z. T. Wirtschafter and M. W. Cronyn, Free radical mechanism for solvent toxicity, *Arch. Environ. Health* **9**, 186–191 (1964).
22. D. Rubinstein and L. Kanics, The conversion of carbon tetrachloride and chloroform to carbon dioxide by rat liver homogenates, *Can. J. Biochem.* **42**, 1577–1585 (1964).
23. A. A. Seawright and A. E. M. McLean, The effect of diet on carbon tetrachloride metabolism, *Biochem. J.* **105**, 1055–1060 (1967).
24. R. C. Garner and A. E. M. McLean, Increased susceptibility to carbon tetrachloride poisoning in the rat after pretreatment with oral phenobarbitone, *Biochem. Pharmacol.* **18**, 645–650 (1969).
25. E. Cignoli and J. A. Castro, Effect of inhibitors of drug metabolizing enzymes on carbon tetrachloride hepatotoxicity, *Toxicol. Appl. Pharmacol.* **18**, 625–637 (1971).
26. R. J. Stenger and E. A. Johnson, Effects of phenobarbital pretreatment on the response of rat liver to halothane administration, *Proc. Soc. Exp. Biol. Med.* **140**, 1319–1324 (1972).

27. J. G. Lavigne and C. Marchand, The role of metabolism in chloroform hepatotoxicity, *Toxicol. Appl. Pharmacol.* **29**, 312–326 (1974).
28. C. Cessi, C. Colombini, and L. Mameli, The reaction of liver proteins with a metabolite of carbon tetrachloride, *Biochem. J.* **101**, 46–47c (1966).
29. E. Reynolds, Liver parenchymal cell injury. IV. Pattern of incorporation of carbon and chlorine from carbon tetrachloride into chemical constituents of liver *in vivo, J. Pharmacol. Exp. Ther.* **155**, 117–126 (1967).
30. K. S. Rao and R. O. Recknagel, Early incorporation of carbon-labeled carbon tetrachloride into rat liver particulate lipids and proteins, *Exp. Mol. Pathol.* **10**, 219–228 (1969).
31. E. Gordis, Lipid metabolites of carbon tetrachloride, *J. Clin. Invest.* **48**, 203–209 (1969).
32. K. F. Ilett, W. D. Reid, I. G. Sipes, and G. Krishna, Chloroform toxicity in mice: Correlation of renal and hepatic necrosis with covalent binding of metabolites to tissue macromolecules, *Exp. Mol. Pathol.* **19**, 215–229 (1973).
33. E. N. Cohen, Metabolism of halothane-2-^{14}C in the mouse, *Anesthesiology* **31**, 560–565 (1969).
34. L. C. Howard, D. R. Brown, and D. A. Blake, Subcellular binding of halothane-1-^{14}C in mouse liver and brain, *J. Pharm. Sci.* **62**, 1021–1023 (1973).
35. F. Schnitger and H. Uehleke, Der Einfluss von Dimethylnitrosamin, Tetrachlorkohlenstoff, Buttergelb und Cyclophosphamid auf den Aminosäureneinbau in Fraktionen von Leberhomogenaten nach metabolischer Aktivierung *in vitro, Arch. Toxikol.* **25**, 169–182 (1969).
36. O. Reiner and H. Uehleke, Bindung von Tetrachlorkohlenstoff an reduziertes mikrosomales Cytochrom P-450 und an Häm, *Hoppe-Seyler's Z. Physiol. Chem.* **352**, 1048–1052 (1971).
37. O. Reiner, S. Athanassopoulos, K. H. Hellmer, R. E. Murray, and H. Uehleke, Bildung von Chloroform aus Tetrachlorkohlenstoff in Lebermikrosomen, Lipidperoxidation und Zerstörung von Cytochrom P-450, *Arch Toxikol.* **29**, 219–233 (1972).
38. H. Uehleke, K. H. Hellmer, and S. Tabarelli, Binding of ^{14}C-carbon tetrachloride to microsomal proteins *in vitro* and formation of $CHCl_3$ by reduced microsomes, *Xenobiotica* **3**, 1–11 (1973).
39. H. Uehleke, K. H. Hellmer, and S. Tabarelli-Poplawski, Metabolic activation of halothane and its covalent binding to liver endoplasmic proteins *in vitro, Naunyn-Schmiedeberg's Arch. Pharmacol.* **279**, 39–52 (1973).
40. H. Uehleke, The model system of microsomal drug activation and covalent binding to endoplasmic proteins, in: *Experimental Model Systems in Toxicology and Their Significance in Man: Proceedings European Society for the Study of Drug Toxicity,* Vol. XV (H. Duncan, ed.), pp. 119–129, Excerpta Medica Foundation, Amsterdam (1974).
41. H. Uehleke and T. Werner, A comparative study on the irreversible binding of labelled halothane, trichlorofluoromethane, chloroform and carbon tetrachloride to hepatic protein and lipids *in vitro* and *in vivo, Arch. Toxikol.* **34**, 289–308 (1975).
42. H. Uehleke, F. Schnitger, and K. H. Hellmer, Verhalten verschiedener mikrosomaler Fremdstoff-Oxidationen nach Inaktivierung von Cytochrom P-450 durch UV-Bestrahlung oder durch Desoxycholatbehandlung, *Hoppe-Seyler's Z. Physiol. Chem.* **351**, 1475–1484 (1970).
43. H. Greim, H. Krämer, T. Werner, and H. Uehleke, Metabolic activation of haloalkanes and *in vitro* tests for mutagenicity, *Xenobiotica* in press (1976).
44. H. Uehleke, S. Poplawski, G. Bonse, and D. Henschler, Spectral evidence for 2,2,3-trichloro-oxirane formation during microsomal trichloroethylene oxidation, *Arch. Pharmacol.* **293**, Suppl. R 64 (1976).

45. R. A. Van Dyke and C. L. Wood, Binding of radioactivity from ^{14}C-labeled halothane in isolated perfused rat livers, *Anesthesiology* **38**, 328–332 (1973).
46. B. B. Paul and D. Rubinstein, Metabolism of carbon tetrachloride and chloroform by the rat, *J. Pharmacol. Exp. Ther.* **141**, 141–148 (1963).
47. R. A. Van Dyke and A. J. Gandolfi, Studies on irreversible binding of radioactivity from (^{14}C) halothane to rat hepatic microsomal lipids and protein, *Drug Metab. Dispos.* **2**, 469–476 (1974).
48. M. M. Airaksinen, P. H. Rosenberg, and T. Tammisto, A possible mechanism of toxicity of trifluoroethanol and other halothane metabolites, *Acta Pharmacol. Toxicol.* **28**, 299–304 (1970).
49. H. Uehleke and T. Werner, Postnatal development of halothane and other haloalkane metabolism and covalent binding in rat liver microsomes, in: *Basic and Therapeutic Aspects of Perinatal Pharmacology* (P. L. Morselli, S. Garattini, and F. Sereni, eds.), pp. 277–287, Raven Press, New York (1975).
50. T. Werner and H. Uehleke, Covalent binding of halothane and other haloalkanes to microsomal proteins and lipids *in vivo* and *in vitro, Naunyn-Schmiedeberg's Arch. Pharmacol.* **285**, Suppl. R90 (1974).
51. D. M. Brown, P. F. Langley, D. Smith, and D. C. Taylor, Metabolism of chloroform. I. The metabolism of ^{14}C-chloroform by different species, *Xenobiotica* **3**, 151–163 (1974).
52. B. J. Fry, T. Taylor, and D. E. Hathway, Pulmonary elimination of chloroform and its metabolite in man, *Arch. Int. Pharmacodyn. Ther.* **196**, 98–111 (1972).
53. R. O. Recknagel and E. A. Glende, Carbon tetrachloride hepatotoxicity: An example of lethal cleavage, *CRC Critical Reviews in Toxicology* **2**, 263–297 (1973).
54. G. J. Traiger and G. L. Plaa, Differences in the potential of carbon tetrachloride in rats by ethanol and isopropanol pretreatment, *Toxicol. Appl. Pharmacol.* **20**, 105–112 (1971).
55. H. Uehleke, Age-dependent role of biotransformation in toxic drug action, in: *Developmental and Genetic Aspects of Drug and Environmental Toxicity* (H. D. Duncan, ed.), pp. 11–21, Excerpta Medica Foundation, Amsterdam (1975).
56. H. Uehleke, The formation and kinetics of reactive drug metabolites in mammals, *Mutat. Res.* **25**, 159–167 (1974).

50

Discussion

In response to a question by Oesch, Uehleke estimated that about 50–60% of the covalently bound CCl_4 is metabolized in mitochondria. Oesch referred to the concept that the free radicals which are often thought of as intermediates may not be stable enough to survive translocation from the endoplasmic reticulum (ER) to mitochondria. He reported that there is evidence that ER can form complexes with mitochondria and suggested that the complexes play a role in heme protein synthesis, because inducing conditions increase the number of complexes. Thus it may be possible that the reactive species need not leave the lipid matrix of the ER. He inquired about the relative half-life of the radicals, and about the possibility that part of the metabolism would occur in mitochondria. Uehleke felt that the compounds are not metabolized by purified mitochondrial preparations, as these contain only about 0.1% P450. He agreed that there is electron microscopic evidence of tubular structures of the ER penetrating mitochondria. He pointed out that the half-life of the free radical in the gas phase is only one- to two-thousandths of a second. However, Steers (Göttingen) has obtained good evidence that radicals derived from aromatic amines, and possibly also the trichloromethyl radical, are stable in the lipid layer for minutes, some even for hours.

In response to questions by Arrhenius and Zimmerman, Uehleke stated that his mitochondrial preparations were prepared by ultracentrifugation and were checked for presence of P450. The preparation consists of about 67% mitochondria, the balance being lysosomal and other membranes. The P450 activity of this preparation was 0.1–0.2% that of microsomal preparations.

Ullrich referred to his data which show that a series of polyhalogenated compounds when interacting with P450 form carbenes by a two-electron reduction. He finds that the position of the P450 peak varies from 450 to 480 nm depending on the number of electronegative atoms in the carbene molecule. For example, in the absence of lipid peroxidation, $CHCl_3$ gives the same spectrum as $:CCl_2$ with a peak at 454 nm, indicating that the reduction by P450 would usually involve two electrons. He noted that Recknagel proposes that conversion of CCl_4 to the radical would involve a single electron. He then went on to express the point of view that the reductase may be the direct electron source for the formation of the $\cdot CCl_3$ free radical. Responding to a question by Uehleke, Ullrich felt that such a mechanism could operate in mitochondria; for instance, any reduced flavin could bring about the reaction.

Recknagel emphasized that the key question is the manner by which the chemical pathology in liver injury comes about, and, more specifically, whether covalent binding or lipid peroxidation is the key event in this process. He felt that the question of whether formation of a radical occurs immediately at the cytochrome P450 locus or whether another electron coming from a flavoprotein is involved is not important, as the enzymes which signal chemical injury, such as glucose-6-phosphatase, are not localized where the flavoprotein is. He agreed that in terms of the detailed reaction mechanism the latter question is of interest. He reemphasized that in his view the main question is whether the key reaction is covalent binding or lipid peroxidation.

Gillette pointed out that toxicity is blocked by CO, which is difficult to explain on the basis of covalent binding. In reply, Ullrich reiterated that his work indicates that the compound which binds covalently is a carbene, not the radical, and that the binding site is the iron (Fe^{2+}) of the P450. Therefore, CO would competitively inhibit the toxicity caused by this covalent binding.

51

Degradation of Cytochrome P450 Heme and Lipid Peroxidation in Lead-Poisoned Rats

W. Penning and P. Scoppa
Biology Group
Ispra G. D. Research
Science and Education, C. E. C.
EURATOM Joint Research Centre
21020 Ispra, Italy

Much of the research concerning the influence of lead on the metabolism of heme has been focused on hemoglobin synthesis. However, other heme moieties may be affected by this metal. We have observed that in acute lead poisoning there is a decrease of hepatic cytochrome P450 with a concomitant impairment of the oxidative metabolism of foreign compounds (1). In this chapter, some results obtained during an investigation on the turnover of cytochrome P450 in lead-poisoned rats are presented.

MATERIALS AND METHODS

Male rats (Sprague-Dawley derived, CD strain) weighing 160–180 g were used. Treated animals received lead nitrate (100 μmol/kg) by intravenous injection at 9 A.M. Twenty-four hours following lead treatment and after overnight fasting, 10 μCi 5-[3,5-^{3}H]aminolevulinic acid (New England Nuclear Corp., specific activity 2.63 Ci/mmol) in 0.9% NaCl was injected through a caudal vein. The animals were sacrificed at various times after injection. Breakdown of cytochrome P450 heme was followed by measuring the radioactivity remaining in the "CO-binding particles" as described by Lucier *et al.* (2). In experiments involving vitamin E, α-tocopheryl acetate (50 mg/kg) in olive oil was administered intraperitoneally 14 h after lead poisoning and the animals were sacrificed 10 h later, after overnight fasting.

Microsomal lipid peroxidation was measured by a slight modification of the method used by Levin *et al.* (3). Before stopping the reaction, an aliquot of the incubation mixture was removed for spectrophotometric determination of cytochrome P450 according to Schoene *et al.* (4). Drug-metabolizing enzyme activities were assayed as described in detail elsewhere (5). Protein content was estimated by the method of Lowry *et al.* (6) using bovine serum albumin as the standard.

RESULTS

Disappearance of radioactivity from both fast-phase and slow-phase components of CO-binding pigments is shown in Fig. 1. Corrected half-lives were obtained by the method of Levin and Kuntzman (7). The fast-phase component had a half life of 10 h in control compared to 2.5 h in lead-poisoned animals, whereas the half-lives of the slow-phase components were not significantly different. The fast-phase component represented 74% of the CO-binding pigment in control and 95% in lead-poisoned animals.

Breakdown of cytochrome P450 *in vitro*, as measured by incubation of liver

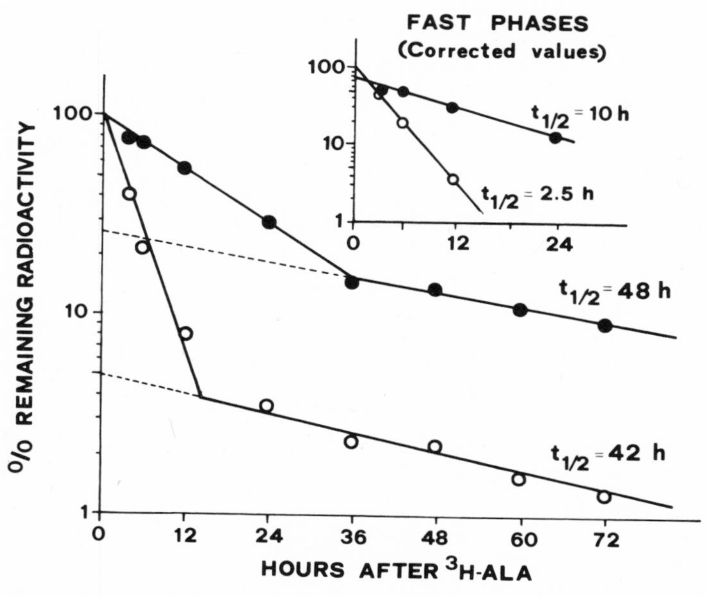

Fig. 1. Disappearance of labeled hemoprotein from CO-binding particles obtained from control (•) and lead-treated (○) rats. Each value represents the mean from eight rats. The zero intercept of the uncorrected fast phase represents the maximal incorporation into the CO-binding particles and was set equal to 100% for both control and lead-treated animals for comparative purposes.

Table 1. Breakdown of Cytochrome P450 and Lipid Peroxidation in Hepatic Microsomes of Normal and Lead-Treated Rats[a]

Treatment	Cytochrome P450 (% loss after 30 min)	Malondialdehyde (nmol/30 min/mg protein)
NaCl	21	18.75
Lead nitrate	51	27.70

[a]Liver microsomes from 500 mg wet wt of liver (containing 13.95 and 7.94 nmol cytochrome P450 in control and lead-treated rats, respectively) were incubated for 30 min at 37°C in a final volume of 9 ml containing 315 μmol Hepes buffer (pH 7.5), 12 μmol glucose-6-phosphate, 1.5 μmol $NADP^+$, and 5 IU glucose-6-phosphate dehydrogenase.

microsomes with an NADPH-generating system, occurred at a faster rate in lead-poisoned rats and was accompanied by an increase of lipid peroxidation (Table 1). When lipid peroxidation was blocked by 1 mM EDTA, $MnCl_2$, or $CoCl_2$, the concentration of cytochrome P450 remained unchanged at the end of the incubation time. Almost normal levels of lipid peroxidation, cytochrome P450 concentration, and drug-metabolizing enzyme activities were observed in hepatic microsomes from rats intoxicated with lead nitrate and treated with α-tocopheryl acetate (Table 2).

DISCUSSION

Studies carried out with ^{203}Pb show that little accumulation of lead in liver microsomes occurs in lead-poisoned rats (5). The resulting concentrations of lead are not sufficient to inhibit *in vitro* the activity of drug-metabolizing enzymes (1,5). Since it has been demonstrated that the biosynthesis of cytochrome P450 is slightly stimulated after lead poisoning (8), the data presented in this chapter show that the low concentration of hepatic cytochrome P450 observed in lead-poisoned rats (1) can be accounted for by a much faster catabolism of cytochrome P450 heme. About 95% of the CO-binding pigments present in liver microsomes of treated animals is in a very labile form having a half-life of 2.5 h.

In hepatic microsomes from lead-poisoned rats, both NADPH-dependent peroxidation of endogenous lipids and the breakdown of cytochrome P450 occurred at a faster rate.

The effects of lead poisoning on microsomal lipid peroxidation, cytochrome P450 concentration, and drug-metabolizing enzyme activities were almost abolished when the animals received α-tocopheryl acetate several hours after administration of lead nitrate.

Although the mechanism of loss of cytochrome P450 during stimulation of lipid peroxidation remains to be clarified, these studies suggest that in acute lead

Table 2. Effects of Vitamin E Treatment on Lipid Peroxidation, Cytochrome P450 Concentration, and Demethylating Enzyme Activities in Hepatic Microsomes of Lead-Poisoned Rats[a]

Treatment	Lipid peroxidation[b]	Cytochrome P450[c]	Aminopyrine demethylase[d]	p-Nitroanisole demethylase[e]
NaCl	0.617 ± 0.087	0.93 ± 0.05	6.79 ± 0.13	0.96 ± 0.08
Lead nitrate	0.909 ± 0.023	0.56 ± 0.04	3.46 ± 0.52	0.59 ± 0.14
NaCl + vitamin E	0.234 ± 0.023	0.91 ± 0.08	6.30 ± 1.48	1.09 ± 0.13
Lead nitrate + vitamin E	0.488 ± 0.083	0.81 ± 0.07	5.59 ± 1.72	0.93 ± 0.14

[a]The values represent the mean ± SE from determinations performed on hepatic microsomes of four animals.
[b]nmol malondialdehyde/min/mg protein.
[c]nmol/mg protein.
[d]nmol formaldehyde/min/mg protein.
[e]nmol *p*-nitrophenol/min/mg protein.

poisoning the breakdown of cytochrome P450 occurs by a mechanism involving its conversion into a very labile form.

Acknowledgments

This is contribution No. 1230 of the Biology Programme, Directorate General XII of the Commission of the European Communities.

REFERENCES

1. P. Scoppa, M. Roumengous, and W. Penning, Hepatic drug metabolizing activity in lead-poisoned rats, *Experientia* **29**, 970–972 (1973).
2. G. W. Lucier, R. Klein, H. B. Matthews, and O. S. McDaniel, Increased degradation of rat liver CO-binding pigment by methylmercury hydroxide, *Life Sci.* **11**, 597–604 (1972).
3. W. Levin, A. Y. H. Lu, M. Jacobson, R. Kuntzman, J. L. Poyer, and P. B. McCay, Lipid peroxidation and degradation of cytochrome P-450 heme, *Arch. Biochem. Biophys.* **158**, 842–852 (1973).
4. B. Schoene, R. A. Fleischmann, and H. Remmer, Determination of drug metabolizing enzymes in needle biopsies of human liver, *Eur. J. Clin. Pharmacol.* **4**, 65–73 (1972).
5. W. Penning, Invloed van de Akute Loodvergiftiging op het Geneesmiddelen Metaboliserend Systeem van Rattenlever, Ph.D. thesis, State University of Ghent, Belgium (1975).
6. O. H. Lowry, N. J. Rosebrough, A. L. Farr, and R. J. Randall, Protein measurement with Folin phenol reagent, *J. Biol. Chem.* **193**, 265–257 (1951).
7. W. Levin and R. Kuntzman, Biphasic decrease of radioactive hemoprotein from liver microsomal carbon monoxide-binding pigment particles: Effect of phenobarbital and chlordane, *Mol. Pharmacol.* **5**, 499–506 (1969).
8. P. Scoppa and W. Penning, Metabolismo dei farmaci nell'intossicazione acuta da piombo: il turnover del citrocromo P-450, *Riv. Tossicol. Sper. Clin.* **1–2**, 23–31 (1975).

VI

Reactive Intermediates in Carcinogenesis

52

The Effects of Microsomal Enzymes on Chemical Oncogenesis in Culture

Stephen Nesnow and Charles Heidelberger
Wisconsin Clinical Cancer Center and
McArdle Laboratory for Cancer Research
University of Wisconsin
Madison, Wisconsin 53706, U.S.A.

We have observed that fibroblastic cell lines derived from the same mouse strain can have different qualitative and quantitative responses to chemical oncogens with respect both to cytotoxicity and to oncogenic transformation. These variations may be the result of the inherent ability of each cell line to metabolize (and/or activate) these oncogens to more cytotoxic and oncogenic forms and/or detoxify these intermediates to less harmful xenobiotics. This chapter describes studies of the relationship between microsomal enzymes and chemical oncogenesis in cultured C3H mouse cells.

RESULTS AND DISCUSSION

10Tl/2CL8 cells are a cloned line of C3H mouse embryo fibroblasts which was developed in our laboratory (1) and is highly sensitive to postconfluence inhibition of cell division. Their plating efficiency is 20–30% and their saturation density per 60-mm dish is 7.5×10^5 cells. 10Tl/2CL8 cells can be transformed by various chemical oncogens (e.g., 3-methylcholanthrene (3-MC), 7,12-dimethylbenz[*a*]anthracene (DMBA), benzo[*a*]pyrene (BP), dibenz[*a,h*]anthracene, and *N*-methyl-*N'*-nitro-*N*-nitrosoguanidine (2,3)). These morphologically transformed cells, when injected into irradiated, syngeneic mice, gave fibrosarcomas, indicating that the cells are malignantly transformed as well. Recently, a new cell line was developed in our laboratory by Dr. Catherine Reznikoff from the ventral prostate gland of an adult C3H mouse. This line was developed using

Table 1. Comparison of the Biological Effects of Hydrocarbon Oncogens on C3H Mouse Fibroblastic Cells[a]

	Hydrocarbon								
	DMBA (3.9 μM)			BP (40 μM)			3-MC (37 μM)		
Cell line	%S[b]	CN[c]	T[d]	%S	CN	T	%S	CN	T
10Tl/2CL8	33	25	10	71	10	22	100	50	7
CVP3SC6	35	15	0	27	9	0	50	32	0

[a]Cell cultures were treated with the oncogen for 24 h. See Discussion for details of assay procedure.
[b]Percent survival as measured by plating efficienty (PE).
[c]Effect on the growth rate of cells as measured by cell number (CN) and expressed as percent of control.
[d]Transformation (T) is represented as foci per dish adjusted for PE.

the same procedures employed in the derivation of 10Tl/2CL8 cells. This new line,[1] named CVP3SC6, is an adult fibroblastic line of cells sensitive to postconfluence inhibition of division. CVP3SC6 cells have a plating efficiency of 15–25% and a saturation density per 60-mm dish of 8.1×10^5 cells. These two C3H mouse cell lines have very different biological responses to chemical oncogens (Table 1). Treatment of 10Tl/2CL8 cells with DMBA or BP caused significant cytotoxicity as measured by reduction in plating efficiency or cell number, and transformed these normal cells into malignant ones. 3-MC treatment of 10Tl/2CL8 cells also caused malignant transformation, but had no effect on toxicity as measured by plating efficiency. 3-MC, BP, and DMBA did not transform CVP3SC6 cells, but did cause considerable cytotoxicity.

The metabolism of 3-MC at the K-region is described in Fig. 1. The mixed-function oxidases form the intermediate 3-MC-11,12-oxide (epoxide), which can then undergo a number of reactions (4,5): nonenzymatic rearrangement to phenols, hydration to a dihydrodiol (3-MC-11,12-dihydrodiol) by a microsomal epoxide hydrase, conjugation with glutathione by a soluble enzyme, and covalent reaction with cellular macromolecules. 3-MC-11,12-oxide has been shown to be more oncogenic and cytotoxic than 3-MC in two other cell transformation systems (6,7). It is interesting to note that 10Tl/2CL8 or CVP3SC6 cells are not transformed with 3-MC-11,12-oxide, although severe cytotoxicity does occur in both cell lines (unpublished observations). *In vivo,* 3-MC-11,12-oxide was not tumorigenic when injected subcutaneously in newborn mice (8). Arene oxides are very labile species and it is possible that they might react with cellular nucleophils or rearrange to phenols before they are able to react with the critical

[1] This new prostate cell line is not to be confused with an earlier prostate line developed in our laboratory (see ref. 19).

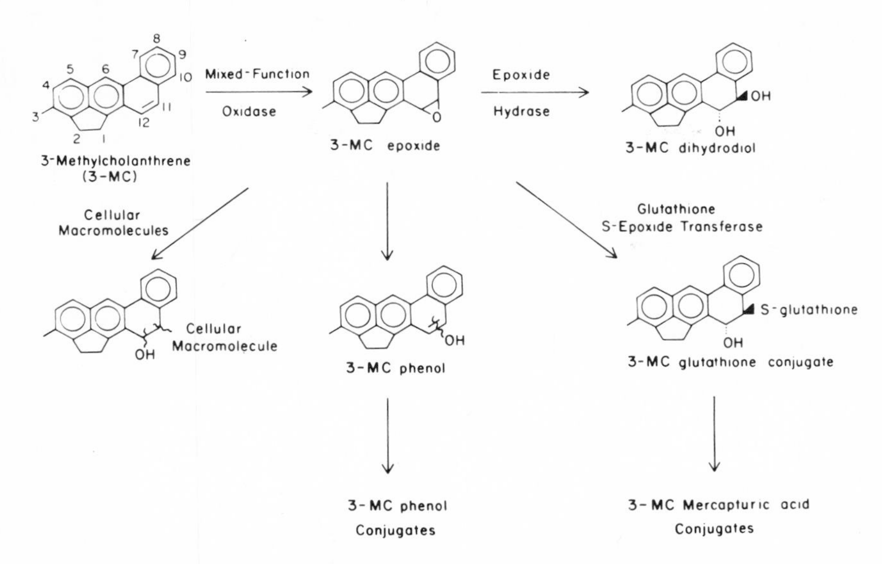

Fig. 1. Metabolites and enzymes involved in the K-region metabolism of 3-MC.

targets in the cell. The half-life of 3-MC-11,12-oxide in 10Tl/2CL8 cells has been found to be 38 min.[2] This means that 4½ h after treatment less than 1% of the original concentration is present.

Conversion of [^{3}H]3-MC to Water-Soluble Metabolites

One measure of the overall metabolizing ability of cells is their ability to convert aromatic hydrocarbons to water-soluble metabolites, which include phenol conjugates, cysteine and glutathione conjugates, and hydrocarbon-bound macromolecules. The procedure developed in our laboratory (9) was employed, which measures the radioactivity remaining in the aqueous layer after thorough extraction of SDS-treated cells (pretreated with [^{3}H]3-MC) with chloroform–methanol. The overall recovery of radioactivity is greater than 90% in this assay.

The saturating concentration of 3-MC was found to be 10^4 pmol per dish (2 μM) and there was a linear relationship between cell number (either at time of seeding or at time of assay) and activity per dish. It was also found that cells at confluence converted more 3-MC to water-soluble metabolites than cells in log phase growth (Fig. 2), a finding in agreement with our previously published results (9). After 96 h of incubation, 10Tl/2CL8 cells converted 3 times more 3-MC to water-soluble metabolites than did the CVP3SC6 cell line (Table 2).

[2] John Keller, McArdle Laboratory for Cancer Research, personal communication.

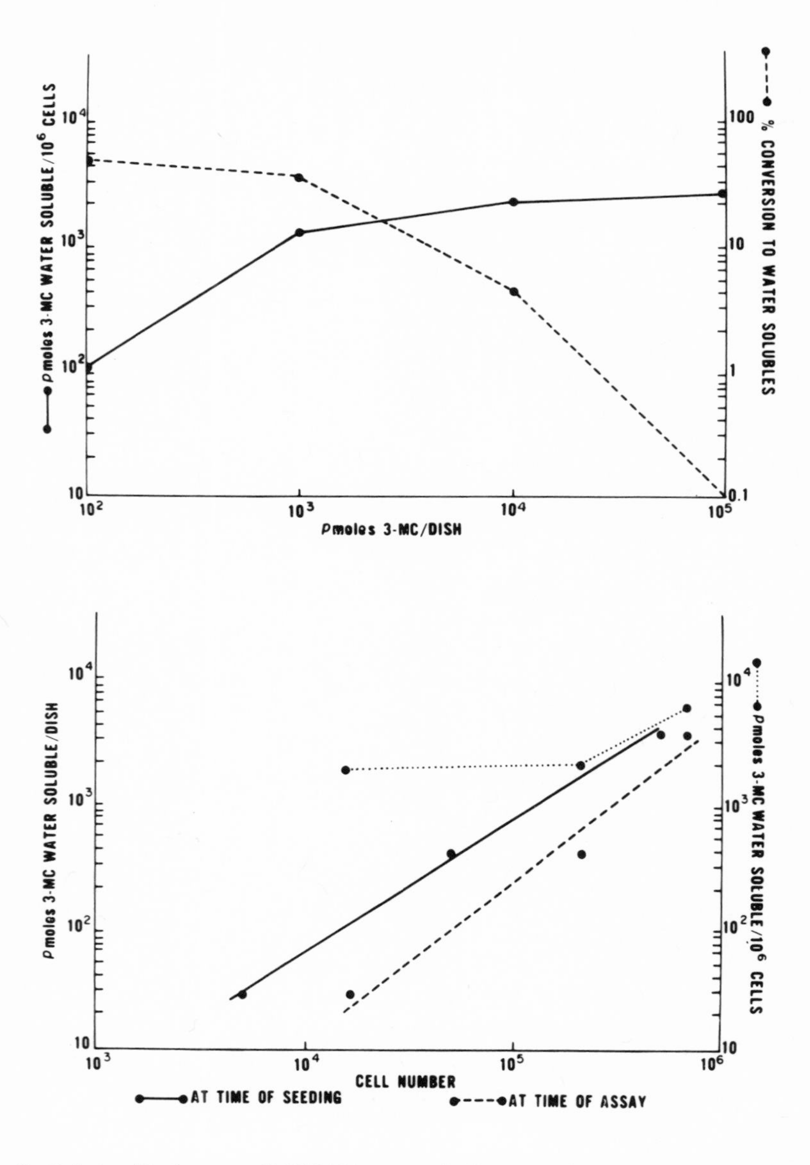

Fig. 2. Relationship between [³H] 3-MC water-soluble metabolites formed and the dose of [³H] 3-MC (top), and the number of 10Tl/2CL8 cells (bottom). In both cases, cells were seeded in 60-mm dishes and were treated with varying concentrations of [³H] 3-MC as noted (top graph) or with 2 μM [³H] 3-MC (bottom graph). The amount of water-soluble metabolites was determined (72 h after treatment) after subtraction of the background in medium without cells (approximately 0.05% of the radioactivity administered). The specific activity of [³H] 3-MC was 0.77 Ci/mmol. In the dose response experiment (top graph) 50,000 cells were seeded per dish.

Table 2. Comparison of Some Biochemical Properties of C3H Mouse Cell Lines

Cell line	$[^3H]$ 3-MC water-soluble metabolites[a]	AHH[b]				EH[d]		
		Treated[c]				Treated[e]		
		3-MC	BA	Control	T/C	3-MC	Control	T/C
10Tl/2CL8	3300	1.04	3.44	0.72	1.4;4.8	53.4±12	28.0±6	1.9
CVP3SC6	1150	0.32	4.20	0.31	1.0;13.5	82.8±12	41.6±8	2.0

[a] Values are expressed in pmol$[^3H]$3-MC water-soluble metabolites formed/10^6 cells. Cells were treated as previously described (9) with $[^3H]$3-MC (2 μM, specific activity 0.77 Ci/mmol) for 96 h. Each value is the average of duplicate determinations.
[b] Values are expressed in pmol hydroxylated BP formed (based on a fluorescence equivalent to that of 3-hydroxy-BP)/min/10^6 cells. The coefficient of variation within samples is 20%.
[c] Cell cultures were treated in roller bottles with 10 μM 3-MC or 13 μM BA and incubated for 48 h at 37°C.
[d] Values are expressed in pmol 3-MC-11,12-dihydrodiol formed/min/10^6 cells from 3-MC-11,12-oxide. Values are mean ± SD.
[e] Cell cultures were treated with 10 μM 3-MC for 48 h as described above.

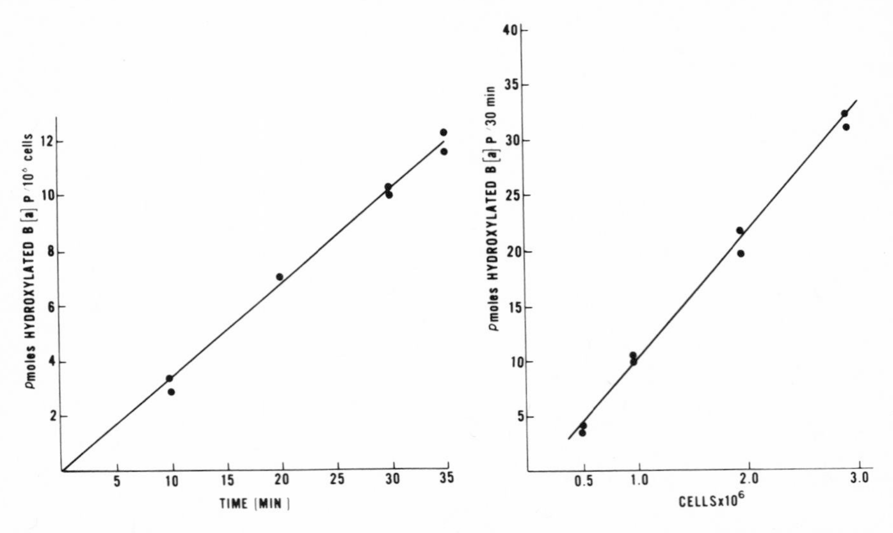

Fig. 3. AHH activity in 10Tl/2CL8 cells as a function of time (left) and cell number (right).

Aryl Hydrocarbon Hydroxylase

Aryl hydrocarbon hydroxylase (AHH) activity in these cell lines was determined using the fluorometric procedure of Nebert and Gelboin (10). In these experiments, cells from confluent monolayers were sparsely seeded into roller bottles in Eagle's basal medium supplemented with 10% heat-inactivated fetal calf serum. They were incubated at 37°C under an atmosphere of 5% CO_2 in air and rotated at 0.8 rpm. After incubation, the cells were washed, removed by scraping, counted, lysed by freeze-thawing, and assayed immediately. All cells assayed were in logarithmic growth. Initial experiments in 10Tl/2CL8 cells indicated that AHH activity was linear with time up to 35 min and with cell number up to 3×10^6 cells (Fig. 3).

The effects of benz[*a*]anthracene (BA) and 3-MC on the induction of AHH in these two cell lines were determined after 48 h of treatment (Table 2). Basal activity of the CVP3SC6 cells was one-half that of the mouse embryo line. With 10 μM 3-MC as the inducer, induction (T/C = 1.4) was observed at 48 h in 10Tl/2CL8 cells, similar to that reported in hamster embryo cells (11). Induction was not observed with 3-MC in the CVP3SC6 cells. BA (13 μM) was a potent inducer of AHH activity in CVP3SC6 cells (T/C = 13.5) and a better inducer than 3-MC in 10Tl/2CL8 cells (T/C = 4.8) after 48 h of treatment.

Epoxide Hydrase

Epoxide hydrase (EH) activity was determined in these C3H mouse cell lines by measuring the conversion of 3-MC-11,12-oxide to 3-MC-11,12-dihydrodiol

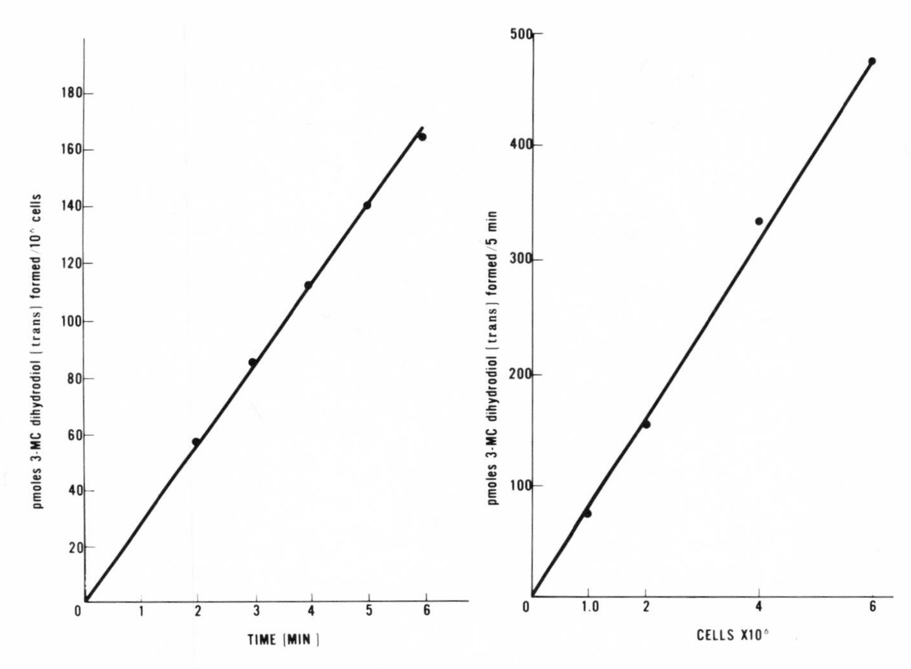

Fig. 4. EH activity in 10Tl/2CL8 cells as a function of time (left) and cell number (right).

(Fig. 1) using the high-pressure liquid chromatographic procedure of Nesnow and Heidelberger (12). Cells were grown in roller bottles and prepared for assay as previously described for AHH.

EH activity was linear with time (to 6 min) and cell number (to 6×10^6 cells) (Fig. 4). Basal levels in CVP3SC6 cells (Table 2) were consistently higher than those in 10Tl/2CL8 cells, and at 48 h, EH activity was induced to about the same extent ($T/C = 2$) in both cell lines by 10 μM 3-MC.

Chemical Oncogenesis in Culture

The ratios of the absolute activities (on a picomolar basis) of AHH to EH in these C3H mouse cell lines which have been treated with transforming doses of 3-MC are 0.0195 for 10Tl/2CL8 and 0.0039 for CVP3SC6. These ratios of induced AHH/EH activity were found to be consistent when the cells were grown in the same lot of fetal calf serum, but varied when different lots were used. In all cases, the CVP3SC6 ratio was always considerably lower than that of 10Tl/2CL8 cells. Thus 10Tl/2CL8 cells can metabolize hydrocarbons to a greater extent than CVP3SC6 cells (as seen from the water-soluble data) (Table 2), and can also produce and maintain higher steady state concentrations of oncogenic epoxides.

One method of altering the induced AHH/EH ratio is to treat the cells with

Table 3. The Effects of Modifiers of Microsomal Enzymes on Transformation and Cytotoxicity in 10Tl/2CL8 Cells[a]

Carcinogen (solvent)	Dose (μM)	Start treatment[b]	Length of treatment[c]	Modifier(s)	Dose (μM)	Start treatment	Length of treatment	PE[d] (% survival)[e]	Transformation[f]
(Acetone)	0.07	–48	120					30(100)	0
3-MC	3.7	0	72					30(100)	3.4
	0			BA	13	–48	48	27(90)	0
3-MC	3.7	0	72	BA	13	–48	48	25(83)	6.7
	0			7,8-BF	25	–24	96	21(70)	0
3-MC	3.7	0	72	7,8-BF	25	–24	96	20(67)	0
(Acetone)	0.07	–48	96					24(100)	0
3-MC	37	0	48					24(100)	4.2
	0			SO	50	0	48	22(92)	0
3-MC	37	0	48	SO	50	0	48	10(42)	9.4
	0			CYO	50	0	48	19(79)	0
3-MC	37	0	48	CYO	50	0	48	11(46)	15.1
	0			TNO	50	0	48	18(75)	0
3-MC	37	0	48	TNO	50	0	48	14(58)	11.5
(DMSO)	0.07	–24	72					21(100)	0

3-MC	37	0	48					20(95)	7.5
	0			SE	10	−24	72	13(65)	0
3-MC	37	0	48	SE	10	−24	72	12(60)	8.1
	0			CE	10	−24	72	17(85)	0
3-MC	37	0	48	CE	10	−24	72	16(80)	18.0
	0			DNE	50	−24	72	16(80)	0
3-MC	37	0	48	DNE	50	−24	72	8(40)	18.9
(Acetone)	0.07	−48	96					17(100)	0
3-MC	37	0	48					17(100)	6.0
	0			(BA+ [f]	13	−48	96		
	0			CE)	10	−24	72	11(65)	0
3-MC	37	0	48	(BA+	13	−48	96		
				CE)	10	−24	72	8(46)	47.1

[a]The abbreviations used are: 3-MC, 3-methylcholanthrene; DMSO, dimethylsulfoxide; BA, benz[*a*]anthracene; 7,8-BF, 7,8-benzoflavone; SO, styrene oxide; CYO, cyclohexene oxide; TNO, 1,2,3,4-tetrahydronaphthalene-1,2-oxide; SE, styrene; CE, cyclohexene; DNE, 1,2-dihydronaphthalene.

[b]Start treatment refers to the time (in hours) that the chemical was added to the cells relative to the time of oncogen addition (zero time); thus cells treated at −48 were treated 48 h prior to oncogen addition.

[c]Length of treatment is the total time (in hours) that the cells were in contact with the chemical.

[d]The PE is the percentage of the number of colony-forming cells relative to the number of cells seeded.

[e]The percent survival is the percent of the ratio of the treated to control PE.

[f]Transformation is expressed as the number of transformed foci per dish, adjusted for PE.

[g](+) means both chemicals were added in the same experiment.

Table 4. The Effects of Modifiers of Microsomal Enzymes on Transformation and Cytotoxicity in CVP3SC6 Cells[a]

Carcinogen (solvent)	Dose (μM)	Start treatment	Length of treatment	Modifier(s)	Dose (μM)	Start treatment	Length of treatment	PE (% survival)	Transformation
(Acetone)	0.07	–48	96					23(100)	0
3-MC	37	0	48					15(66)	0
	0			(BA+	13	–48	96		
	0			DNE)	10	–24	72	11(49)	0
3-MC	37	0	48	(BA+	13	–48	96		
				DNE)	10	–24	72	4(17)	18

[a]See Table 3 for explanation of column headings.

inducers and inhibitors of AHH. Low-passage cells (less than 12 passages) obtained from freshly confluent cultures were seeded in 60-mm dishes (1000 cells/5 ml medium) in Eagle's basal medium supplemented with 10% heat-inactivated fetal calf serum and incubated in an atmosphere of 5% CO_2 in air at 37°C (2). Twenty-four hours later, treatment was begun. Compounds were dissolved in acetone (0.5%) and either added directly to the cell cultures or added to fresh medium, and the cells were given this medium. In some experiments, cells were given multiple treatments of chemicals, and their times of addition and lengths of treatment are described in Tables 3 and 4. At the end of the last treatment, the cells were given fresh medium supplemented with penicillin (100 units/ml) and streptomycin (50 μg/ml). Medium was then changed twice weekly until the cells reached confluence, then once weekly. At 6 weeks, the cultures were fixed, stained, and scored for transformation (2). Cytotoxicity tests were performed for each experimental group using the same protocol as used for the transformation assay except that each dish was seeded with 200 cells and the dishes were fixed, stained, and scored for plating efficiency 7–10 days after seeding. Pretreatment of 10Tl/2CL8 cells with BA, a good inducer of AHH and a poor inducer of EH (data not shown), increased 3-MC-mediated transformation twofold (Table 3). These results are in agreement with similar experiments performed in hamster embryo cells (13) and mouse prostate cells (7). Pretreatment of 3-MC-treated 10Tl/2CL8 cells with 7,8-benzoflavone (7,8-BF), an inhibitor of AHH activity (14), completely inhibited oncogenic transformation. Similar results have been observed in mouse prostate cells (7) but not in hamster embryo cells (13). Cytotoxicity was not affected significantly by these treatments.

Treatment of cells with inhibitors of EH will also increase the AHH/EH ratio. 10Tl/2CL8 cells were treated with three epoxide hydrase inhibitors (15,16): styrene oxide (SO), cyclohexene oxide (CYO), and 1,2,3,4-tetrahydronaphthalene-1,2-oxide (17) (TNO); all increased 3-MC-mediated transformation two- to threefold when given concurrently with the oncogen. With these treatments, cytotoxicity due to the combination of oncogen and inhibitor was dramatically increased.

It was also of interest to see if these EH inhibitors could be formed *in situ* by the cells, and could increase 3-MC-mediated oncogenic transformation. Oesch and Daly (18) have shown that a tightly coupled mixed-function oxidase–EH system and a free EH enzyme exist in liver microsomes and that 3,3,3-trichloropropylene oxide, a potent inhibitor of free EH activity, does not inhibit the coupled EH enzymes(s). Possibly, that inhibitor has little or no effect on coupled EH activity because it cannot "enter" the coupled system. By forming the inhibitor *in situ* from the precursor alkene via the mixed-function oxidase, the coupled EH could be inhibited, and, consequently, the 3-MC-11,12-oxide formed by the mixed-function oxidases would not be enzymatically converted to the dihydrodiol. Transformation experiments revealed that pretreatment of

10Tl/2CL8 cells with cyclohexene (CE) and 1,2-dihydronaphthalene (DNE), precursors of CYO and TNO, respectively, increased 3-MC-mediated transformation about 2.5-fold, while pretreatment with styrene (SE) had no effect on transformation. Increased cytotoxicity was evident only with the DNE treatment. Combination pretreatment of an inducer of AHH, BA, and a precursor of an inhibitor of EH, CE, gave the largest increase in 3-MC-mediated transformation of 10Tl/2CL8 cells measured in these experiments, i.e., 7.8-fold.

CVP3SC6 cells which were not transformable with hydrocarbon oncogens were morphologically transformed by a combination of BA, DNE, and 3-MC. We believe that transformation of CVP3SC6 cells was accomplished by altering the activities of microsomal enzymes and thus increasing the steady-state cellular levels of epoxide(s) formed by the cells (20).

ACKNOWLEDGMENTS

This work has been supported in part by Grants CA-07175 and CA-14520 from the NCI, National Institute of Health, U.S. Public Health Service. Charles Heidelberger is an American Cancer Society Professor of Oncology.

REFERENCES

1. C. A. Reznikoff, D. W. Brankow, and C. Heidelberger, Establishment and characterization of a cloned line of C3H mouse embryo cells sensitive to postconfluence inhibition of division, *Cancer Res.* **33**, 3231–3238 (1973).
2. C. A. Reznikoff, J. S. Bertram, D. W. Brankow, and C. Heidelberger, Quantitative and qualitative studies of chemical transformation of cloned C3H mouse embryo cells sensitive to postconfluence inhibition of cell division, *Cancer Res.* **33**, 3239–3249 (1973).
3. J. S. Bertram and C. Heidelberger, Cell cycle dependence of oncogenic transformation induced by *N*-methyl-*N*′-nitro-*N*-nitrosoguanidine in culture, *Cancer Res.* **34**, 526–537 (1974).
4. F. Oesch, Mammalian epoxide hydrases: Inducible enzymes catalysing the inactivation of carcinogenic and cytotoxic metabolites derived from aromatic and olefinic compounds, *Xenobiotica* **3**, 305–340 (1973).
5. D. M. Jerina and J. W. Daly, Arene oxides: A new aspect of drug metabolism, *Science* **185**, 573–582 (1974).
6. E. Huberman, T. Kuroki, H. Marquardt, J. K. Selkirk, C. Heidelberger, P. L. Grover, and P. Sims, Transformation of hamster embryo cells by epoxides and other derivatives of polycyclic hydrocarbons, *Cancer Res.* **32**, 1391–1396 (1972).
7. H. Marquardt and C. Heidelberger, Influence of "feeder cells" and inducers and inhibitors of microsomal mixed-function oxidases on hydrocarbon induced malignant transformation of cells derived from C3H mouse prostate, *Cancer Res.* **32**, 721–725 (1972).
8. P. L. Grover, P. Sims, B. C. V. Mitchley, and F. J. C. Roe, The carcinogenicity of polycyclic hydrocarbon epoxides in newborn mice, *Br. J. Cancer* **31**, 182–188 (1975).

9. E. Huberman, J. K. Selkirk, and C. Heidelberger, Metabolism of polycyclic aromatic hydrocarbons in cell cultures, *Cancer Res.* **31**, 2161–2167 (1971).
10. D. W. Nebert and H. V. Gelboin, Substrate-inducible microsomal aryl hydroxylase in mammalian cell culture. I. Assay and properties of the induced enzyme, *J. Biol. Chem.* **243**, 6242–6249 (1968).
11. D. W. Nebert and H. V. Gelboin, Substrate inducible microsomal aryl hydroxylase in mammalian cell culture. II. Cellular responses during enzyme induction, *J. Biol. Chem.* **243**, 6250–6261 (1968).
12. S. Nesnow and C. Heidelberger, A rapid and sensitive liquid chromatographic assay for epoxide hydrase, *Anal. Biochem.* **67**, 525–530 (1975).
13. J. A. DiPaolo, P. J. Donovan, and R. L. Nelson, Transformation of hamster cells *in vitro* by polycyclic hydrocarbons without cytotoxicity, *Proc. Natl. Acad. Sci. USA* **68**, 2958–2961 (1971).
14. L. Diamond and H. V. Gelboin, Alpha-naphthoflavone: An inhibitor of hydrocarbon cytotoxicity and microsomal hydroxylase, *Science* **166**, 1023–1025 (1969).
15. F. Oesch, N. Kaubisch, D. M. Jerina, and J. W. Daly, Hepatic epoxide hydrase, structure–activity relationships for substrates and inhibitors, *Biochemistry* **10**, 4858–4866 (1971).
16. T. A. Stoming and E. Bresnick, Hepatic epoxide hydrase in neonatal and partially hepatectomized rats, *Cancer Res.* **34**, 2810–2813 (1974).
17. E. Straus and A. Rohrbacher, Überführung von Δ'-dihydro-naphthalin in alicyclische substitutionsprodukte des tetrahydro-naphthalins, *Chem. Ber.* **54**, 40–69, (1921).
18. F. Oesch and J. Daly, Conversion of naphthalene to *trans*-naphthalene dihydrodiol: Evidence for the presence of a coupled aryl monooxygenase-epoxide hydrase system in hepatic microsomes, *Biochem. Biophys. Res. Commun.* **46**, 1713–1720 (1972).
19. T. T. Chen and C. Heidelberger, Cultivation *in vitro* of cells derived from adult C3H mouse ventral prostate, *J. Natl. Cancer Inst.* **42**, 902–914 (1969).
20. S. Nesnow and C. Heidelberger, The effects of modifiers of microsomal enzymes on chemical oncogenesis in cultures of C3H mouse cell lines, *Cancer Res.* **36**, 1801–1808 (1976).

53

Discussion

Gelboin referred to earlier work by Diamond and himself showing that benzoflavone effectively blocked formation of water-soluble products and inhibited the hydroxylase. Kinoshita then showed that benzoflavone prevented carcinogen binding to macromolecules and inhibited tumorigenesis in mouse skin but that with benzpyrene as the carcinogen (or the initiator) benzoflavone did not inhibit and in some cases even stimulated tumorigenesis. This has been confirmed by others. Since methylcholanthrene metabolism is known to be inhibited by benzoflavone, Gelboin inquired whether the author had compared other carcinogens with methylcholanthrene as to their behavior in the presence of the inhibitor.

Nesnow noted that such experiments have been planned. He pointed out that the epoxide hydrase inhibitors used also react with glutathione. It is possible that some of the findings may have resulted from the combined effects of decreased glutathione levels and inhibition of epoxide hydrase. Experiments have been planned to distinguish between these two effects. Using benzpyrene, for example, with an experimental system similar to the one described to produce oncogenic transformation, various metabolic inhibitors and inducers can be used to study the detailed mechanisms responsible for tumor formation in mouse skin.

Jakobsson inquired whether the malignant transformed cells have been studied by means of electron microscopy. Nesnow noted that while the 10Tl/2CL8 cells have been examined rather thoroughly the relatively new CVP line has not yet been studied in detail. He pointed out that it is still to be determined whether the transformation with the CVP line is malignant. This may be done by repeatedly subcloning the cultures and injecting such cells into mice.

Snyder inquired whether cells obtained from cultures in which inhibitors have prevented extensive (metabolic or morphological) transformation would cause tumors when injected into animals. Nesnow agreed that the results would be instructive but indicated that the experiments have not yet been done.

Gelboin referred to a paper in press by Whitlock and himself showing that enzyme induction is a very stable phenomenon in liver cell cultures which have been in culture for a long time. Cell populations derived by subcloning such cultures, however, showed a high degree of heterogeneity in hydroxylase production. Further subcloning led to further heterogeneity in enzyme formation. The sum of these varying activities probably gives the average seen in the cultures before cloning. He wished to caution everyone working with cell cultures that cloning does not necessarily lead to a homogeneous cell population with regard to enzyme induction and inquired whether the author had examined this aspect. Nesnow replied that to some degree similar observations were made. Subclones of the original 10Tl/2 monolayer have exhibited such heterogeneity.

54

Role of Covalent Binding in Carcinogenicity

Peter Brookes
Chemical Carcinogenesis Division
Institute of Cancer Research
Pollards Wood Research Station
Chalfont St. Giles
Buckinghamshire HP8 4SP, England

To be asked to discuss the role of covalent binding in carcinogenicity at the present time and within the context of this symposium on active intermediates raises the problem of preaching to the converted. The work reported in the earlier sections indicates the importance which is now attached to an understanding of the nature of reactive metabolites. The comparatively recent change from the concept that metabolism was a detoxifying mechanism leading to the production of inactive readily excreted products has revolutionized thinking about chemical carcinogenesis. In 1964, the report by Brookes and Lawley (1, 2) that polycyclic hydrocarbons were covalently bound to the DNA and RNA of mouse skin under conditions of tumor induction was greeted with considerable scepticism. Since that time, elegant studies with a wide range of synthetic and naturally occurring carcinogens have indicated that covalent binding to cellular macromolecules is the rule rather than the exception. Several recent reviews (3–5) have detailed the evidence for this concept and I do not wish to reiterate them here. I will rather attempt to consider the questions which now arise:

1. Do tumors result as a direct consequence of covalent binding, and, if so, by what mechanism?
2. Which macromolecule is the critical target for the carcinogen?
3. Are all covalent binding reactions equally relevant in carcinogenesis, and, if not, can the significant reaction be identified?

Since these questions are clearly related, I will not deal with each separately but rather will discuss some data which it is to be hoped will provide the answers to at least some part of the problem.

BINDING STUDIES

The fact that covalent binding to cellular macromolecules has been found for virtually every carcinogen which has been adequately studied is in itself evidence that this binding is necessary for tumor induction. It becomes better evidence when one considers the enormous variation of chemical structures involved. I would suggest that the only common features of, for example, sulfur mustard, dimethyl nitrosamine, aflatoxin B_1, benzo[*a*]pyrene, and safrole is that they are carcinogenic and bind covalently to macromolecules in the cells in which they induce tumors.

Negative evidence is always less satisfactory, but it is worth recalling that compounds such as the acridine dyes (6) and ethidium bromide (7), which bind firmly but noncovalently to macromolecules, are not carcinogens. Furthermore, within a series of structurally related alkylating agents (8), azo dyes (9), and polycyclic hydrocarbons (1, 10, 11), a correlation was found between covalent binding and carcinogenicity. The later studies with the hydrocarbons used rodent embryo cells in culture in which it was possible to quantitatively relate binding to the amount of hydrocarbon metabolized by the cells (10, 11). Although all the hydrocarbons tested were metabolized, the carcinogens benzo[*a*]pyrene and dibenz[*a,h*]anthracene were distinguished from their closely related, weakly active analogues, benzo[*e*]pyrene and dibenz[*a,c*]anthracene, in that their binding to DNA and RNA was much greater than for the weak initiators.

Proteins as the Vital Cellular Receptor

In most cases the evidence that a particular carcinogen reacted with cellular macromolecules was obtained before the mechanism of metabolic activation was understood. The availability of highly radioactively labeled compounds was usually the limiting factor in such studies since the level of reaction with macromolecules following biologically meaningful doses is always low. An exception to this last statement was the early work with the azo dyes where the color of the compound and the availability of a large quantity of liver protein indicated binding to this macromolecule and led directly to the protein deletion theory of carcinogenesis (12).

The concept of the inactivation of a growth-controlling protein by the reaction of a carcinogen was very appealing and the finding that carcinogenic hydrocarbons also showed preferential binding to a particular protein fraction (13) added credence to the theory. Furthermore, Pitot and Heidelberger (14) used the Jacob–Monod model of bacterial gene suppression to overcome the objection that protein deletion would be a transient phenotypic event and not expected to confer the inherited character essential to the cancer cell. However, a number of difficulties are apparent in accepting protein as the critical cellular receptor for chemical carcinogens.

The use of radioactively labeled carcinogens and the development of methods for the isolation of pure macromolecules made it clear that under the conditions of tumor initiation the level of covalent binding for all carcinogens was very low. For example, chronic feeding of 4-dimethylaminoazobenzene to rats resulted in a maximum protein binding of about 0.5 μmol/g (15), which means that for proteins of 5×10^4 molecular weight only one in 40 would have a bound dye molecule. In the case of total mouse skin proteins, the binding of hydrocarbons was of the order of 0.1 μmol/g (1) and Heidelberger's data (13) for the reaction with the specific fraction of soluble protein to which hydrocarbons are bound indicated that only one in 170 protein molecules would be modified. At these levels of reaction, a very high degree of specificity of attack on a particularly significant protein would need to occur to explain carcinogenesis by this mechanism. Furthermore, the biological activity of protein molecules appears to be resistant to chemical modification, and those agents such as iodoacetamide which react readily with proteins (but not nucleic acids) (16) are not usually carcinogenic (8). Finally, it is significant that despite the considerable efforts of a number of laboratories the vital growth-regulating proteins have not been isolated, nor has the concept of their importance in cancer induction been substantiated or led to any theoretical or practical advances.

RNA as the Vital Cellular Receptor

The electrophilic nature of ultimate carcinogens leads to the prediction that the bases of nucleic acids would be targets for attack. It is also predictable that DNA and RNA would provide equally good nucleophilic centers, and this has been confirmed by experiment (17, 18).

A vital role for tRNA as a cellular receptor for carcinogens has been proposed by Weinstein (19). The idea was based on the suggestion that tRNAs play an important role in cell regulation and differentiation in higher organisms. In an elegant series of experiments, Weinstein and his colleagues showed that the *N*-acetoxy derivative of the liver carcinogen *N*-2-acetylaminofluorene reacted with certain tRNAs and inhibited their amino acid acceptor activity, ability to bind to ribosomes, and codon recognition function (20). It was also shown that the reaction of this bulky fluorene residue at the C-8 position of the guanine moiety of RNA resulted in the displacement of the guanine base from its normal coplanar relation with the adjacent base residues, its position being taken by the fluorene molecule (21).

Enthusiasm for ribosomal RNA as the critical cellular receptor (22) seems to have passed, and while modification of mRNA would no doubt affect its coding property, no theory of tumor initiation based on this reaction has received support.

The arguments aginst RNA being the vital cellular receptor again involve the question of target size. From Weinstein's data (23) on the level of reaction of a series of carcinogens with tRNAs, it may be calculated that only one in every

300–600 such molecules would be modified, and since there are many copies of any one particular tRNA and an active gene capable of producing many more, it requires a series of unlikely assumptions to make the hypothesis tenable.

DNA as the Vital Cellular Receptor

The concept that cancer was the result of a somatic mutation preceded any knowledge of what constituted the genetic material. Support for the theory formulated in 1928 by K. H. Bauer (24) reached its lowest point about 1960. Burdette in his review (25) of mutation and the origin of tumors decided that "the conclusion that a general correlation exists between mutagenicity and carcinogenicity of chemical compounds is not warranted." Most workers in the field of carcinogenesis seemed to agree, and when Lawley and I were encouraged by our data on hydrocarbon binding to mouse skin DNA to support the mutation theory at the Gatlinburg symposium in 1964 (2), Drs. Kaplan and Heidelberger pointed out that the available evidence was more in favor of an inverse correlation between carcinogenesis and mutagenesis. The work leading up to the present symposium has very largely eliminated these objections but has convinced very few researchers of the fundamental truth of the theory. I would like to summarize recent evidence which strongly implicates covalent binding to DNA and could lead to an understanding of tumor induction at the level of molecular biology.

Earlier I mentioned the correlation which exists between the level of nucleic acid reaction and carcinogenicity. In fact, it was the lack of such a correlation for a series of alkylating agents which gave a valuable clue to the mechanism of tumor initiation by compounds of this type. Swann and Magee (26) injected rats with a series of methylating agents and showed that while these compounds all produced about the same level of methylation of the DNA of the kidney, tumors in this organ were obtained with dimethylnitrosamine (DMN) and *N*-methylnitrosourea (MNUA) but not with dimethyl sulfate (DMS) or methyl methanesulfonate (MMS). A possible answer to this anomaly was provided by the work of Loveless (27), who was seeking an explanation of the fact that MNUA was mutagenic in bacteriophage whereas MMS was not, despite a comparable degree of phage DNA methylation. Loveless showed that while both methylating agents reacted predominantly with the N^7-position of guanine in DNA, MNUA also reacted to a significant extent with the O^6-position, which he postulated would result in mispairing on replication. Evidence for such a mispairing by an O^6-methylguanine-containing template has been reported by Gerchman and Ludlum (28).

These ideas on the mechanism of mutation induction in phage provide a possible explanation of the kidney tumor induction studies of Swann and Magee (26) if it is assumed that carcinogenesis by simple methylating agents is a consequence of O^6-guanine alkylation. Support for this idea comes from a series of recent experiments.

Frei (29) found that, as expected, MNUA was a potent inducer of thymoma in mice, while MMS was inactive. However, *N*-methyl-*N*′-nitro-*N*-nitrosoguanidine (MNNG), which had been shown to give a pattern of DNA methylation products similar to MNUA, did not induce thymoma (30). Frei and Lawley have now shown that following intraperitoneal injection of MNNG into mice extensive methylation of the DNA of the small bowel occurs, but very little reaction was found in the thymus and bone marrow (5), i.e., the target organs for thymoma induction. A number of laboratories have studied the excision by repair enzyme systems of alkylated bases from the DNA of animal tissues following the administration of nitrosamine carcinogens (31–36). It is apparent that, in the same animal, tissues may differ in their DNA repair function, particularly in relation to the excision of O^6-alkylguanine moieties. This work has suggested a positive correlation between susceptibility of rat tissues to tumor induction by MNUA and DMN, namely brain > kidney > liver, and the persistence of O^6-methylguanine in the DNA of the tissue (5).

Lawley (5), in an extensive review of the relationship between DNA alkylation products and tumor induction, reaches the tentative conclusion that if the level of O^6-methylguanine in the DNA of a tissue reaches about 20 μmol/mol DNA-P, then a 50% tumor yield will result. It will be interesting to see if this bold prediction is substantiated by further experimentation, but it must encourage those who strive to understand carcinogenesis as a problem in molecular biology.

Further evidence that the site of attachment to DNA of the same moiety may influence the carcinogenic response comes from studies with the polycyclic hydrocarbons. The highly reactive 7-bromomethylbenz[*a*]anthracene was considered as a possible model for the ultimate carcinogenic form of 7-methylbenz[*a*]-anthracene (7-MBA). The chemistry of its reaction with DNA was studied and found to involve attachment of the hydrocarbon via the methyl group to the extranuclear amino groups of guanine, adenine, and cytosine (37). An alternative possibility for the ultimate carcinogen derived from 7-MBA was the K-region epoxide. The chemistry of the reaction of this epoxide with DNA has yet to be established in detail, but it seems to involve attack on the purine bases, presumably via the K-region of the hydrocarbon (38). When the parent hydrocarbon, 7-MBA itself, is added to growing rodent embryo cells, it is metabolized and becomes bound to the cellular DNA (10). The chemistry of this latter reaction is least well understood at present, but enzymatic degradation of DNA with tritium-labeled 7-MBA bound in this way, and subsequent chromatography of the deoxyribonucleoside products on a column of LH20 Sephadex, allowed them to be characterized in terms of their elution volume (39). A similar degradation and fractionation of DNA after reaction with 7-bromomethylbenz-[*a*]anthracene or the K-epoxide of 7-MBA indicated that the products of reaction of each of these derivatives were different from each other as expected and different again from those resulting from the *in vivo* metabolism of 7-MBA (40). Although the substituted 7-MBA derivatives are carcinogenic (41–43), there seems no doubt that they are less potent than the parent 7-MBA, implying that there is some feature of the *in vivo* binding which makes it biologically more important. Whether this difference is due to a particular chemical reaction with DNA, to specificity of attack on a particular region of the genome, or to some reason quite unconnected with DNA reactions remains to be determined.

DNA Repair and Carcinogenesis

As discussed above, it appears that excision of alkylated bases from DNA may protect against tumor induction. The first indication that DNA repair (44) might operate to protect man from carcinogenic agents was the discovery that skin fibroblasts of xeroderma pigmentosum patients had a defective DNA repair ability (45). Such patients were known to be extremely sensitive to UV-induced skin cancers. Subsequent studies have shown that repair enzymes can excise damage induced by many chemical carcinogens, as discussed in reviews by Irving (4) and Roberts (46).

Much current research is concerned with the "postreplication repair" system which has been identified in mammalian cells (47). As with other molecular biological systems, postreplication repair was initially studied in bacteria (48). It was found that the operation of this repair mechanism resulted in mutation by a number of agents, including known carcinogens, which were more toxic but nonmutagenic to cells lacking this form of "error-prone" repair (49). It is premature but tempting to speculate that chemically induced lesions in DNA which have not been removed by excision repair are subject to "error-prone" repair, with consequent induction of mutation. In this way, the great diversity of damage introduced into DNA by different carcinogens might yield similar mutational events and so provide the basis of a unifying mechanism of carcinogenesis.

MUTAGENESIS AND CARCINOGENESIS

The nature of the relationship between mutagenesis and carcinogenesis must be at the center of any discussion on the role of covalent binding in tumor initiation. Having passed through an era when it was seriously suggested that an inverse relationship existed, it is now widely accepted that the "ultimate carcinogen" is a mutagen in some test system. The area of doubt concerns the correlation between different types of mutation and cancer.

The work with the simple methylating agents mentioned above (5) implicates base pair transitions as of prime importance *for this class of carcinogens.* Such mutations would arise during DNA replication without the intervention of any repair mechanism and might be equivalent to spontaneous mutations. It has been estimated (50) that in man the spontaneous mutation rate is about 10^{-6} per gene per cell division and the body must have developed protective mechanisms to control the effect of this genetic variation. Such mechanisms may break down if the mutation rate is increased very considerably. Lawley (5) has estimated that at a carcinogenic dose of a methylating carcinogen, with about 20 μmol O^6-methylguanine per mol P, the chance of miscoding would be about 10^5 per cell per division.

For the majority of carcinogens, the chemistry of their DNA reaction and its likely molecular biological effect are less clearly understood than is the case for the methylating agents. Ames (51) studied the mutagenic effect of a variety of carcinogens in a series of *Salmonella typhimurium* strains, using liver microsomal activation where necessary. He had previously concluded that a strain designed to respond to frameshift mutagens was most sensitive to mutation by a wide range of carcinogens (52). As mentioned above, the operation of a repair enzyme system may be responsible for converting a DNA lesion into a mutagenic event. Since the exact mechanism of this process is not known, it is premature to interpret mutation data which might involve this repair system. However, a frameshift mechanism with the implication of a large target size for a mutation leading to the loss of a gene function would offer an explanation of the greater potency of carcinogens such as hydrocarbons and aflatoxin compared to the simple methylating agents.

At present, not all known chemical carcinogens have been shown to be mutagenic, but fresh examples are being reported at an accelerating rate. Recent papers have indicated the mutagenicity of chromium salts (53), vinyl chloride (54), and chloroprene (55), each of which has been implicated as being a carcinogenic hazard to exposed production workers.

In relation to the present topic, it is the theoretical implications of the correlation between mutagenesis and carcinogenesis which is significant, but the practical aspect should not be overlooked. The introduction into the environment of large numbers of new chemicals is a fact of modern life. Animal carcinogenicity testing facilities throughout the world are totally inadequate to deal with the number of compounds involved. If microorganism mutation test systems can develop into reliable indicators of potential carcinogenic activity, the problem of screening new products becomes feasible. Considerable progress toward this objective is being made by an extensive program under the auspices of the National Cancer Institute, Bethesda, Maryland, and in a similar nationally organized study in Japan. Results so far available suggest that for ultimate carcinogens the detectability by mutation tests is near 100%; the lower success rate of about 60% for compounds requiring metabolic activation indicates the need for the type of study which is the subject of this symposium.

CONCLUSION

Looking back through the present chapter to the questions raised at the beginning, one is forced to conclude that no final answers have been found to any of the problems. However, I would suggest that a number of obvious leads to future research programs are indicated.

Personally, I have no regrets at having based my research for the past 10–12 years around the working hypothesis that reaction with DNA is an essential step in the process of cancer initiation (2). During this time, many experiments have been performed in many laboratories which could have made this hypothesis untenable, but I would suggest that the opposite has in fact happened. In these circumstances, it would seem more profitable to pursue this approach rather than to look for alternatives simply because they are novel. It is not the intention to suggest that future research should be directed to demonstrating covalent DNA reaction and mutagenic activity for more and more carcinogens. It seems possible that the answers to the questions posed may lie in an understanding of the structure and function of that large part of the DNA of the mammalian cell which is not concerned with coding for proteins or for functional RNA. This in turn probably involves a knowledge of the structure of the chromosome.

It is perhaps worth emphasizing that belief in the significance of DNA reaction does not necessarily imply acceptance of a mutational cause of cancer. The particular type of toxic effect produced in mammalian cells by carcinogen–DNA reaction has been proposed (J. J. Roberts, personal communication) as the event which could lead to tissue disorganization and regeneration (perhaps by a process akin to wound healing (56, 57), from which the cancer cell might arise. A similar viewpoint is contained in the recent discussion of the relation of mutation selection and cancer by Cairns (58).

ACKNOWLEDGMENTS

I would like to express my gratitude to my colleagues Drs. P. D. Lawley and J. J. Roberts for valuable discussions during the preparation of the manuscript. The author's research is supported by grants to the Institute of Cancer Research, from the Medical Research Council and Cancer Research Campaign, Great Britain, and by NIH (U.S.A.) contract number N01-CP-33367.

REFERENCES

1. P. Brookes and P. D. Lawley, Evidence for the binding of polynuclear aromatic hydrocarbons to the nucleic acid of mouse skin: Relation between carcinogenic power of hydrocarbons and their binding to DNA, *Nature (London)* **202**, 781–784 (1964).
2. P. Brookes and P. D. Lawley, The reaction of some mutagenic and carcinogenic compounds with nucleic acid, *J. Cell. Comp. Physiol.* **64**, Suppl. 1, 111–128 (1964).
3. E. C. Miller and J. A. Miller, Biochemical mechanisms of chemical carcinogenesis, in: *The Molecular Biology of Cancer* (H. Busch, ed.), pp. 377–402, Academic Press, New York (1974).

4. C. C. Irving, Interaction of chemical carcinogens with DNA, in: *Methods in Cancer Research,* Vol. 7 (H. Busch, ed.), pp. 189–244, Academic Press, New York (1973).
5. P. D. Lawley, Methylation of DNA by carcinogens: some applications of chemical analytical methods, in: *Screening Tests in Chemical Carcinogenesis* (R. Montesano, H. Bartsch and L. Tomatis eds.) I.A.R.C. Scientific Publications No. 12. International Agency for Research on Cancer, Lyon, pp. 181-210 (1976).
6. N. F. Gersch and D. O. Jordan, Interaction of DNA with aminoacridines, *J. Mol. Biol.* **13**, 138–156 (1965).
7. M. J. Waring, Complex formation between ethidium bromide and nucleic acids, *J. Mol. Biol.* **3**, 269–282 (1965).
8. N. H. Colburn and R. K. Boutwell, The binding of β-propiolactone and some related alkylating agents to DNA, RNA and protein of mouse skin, *Cancer Res.* **28**, 653–660 (1968).
9. C. W. Dingman and M. B. Sporn, The binding of metabolites of aminoazodyes to rat liver DNA *in vivo, Cancer Res.* **27**, 938–944 (1967).
10. M. Duncan, P. Brookes, and A. Dipple, Metabolism and binding to cellular macromolecules of a series of hydrocarbons by mouse embryo cells in culture, *Int. J. Cancer* **4**, 813–819 (1969).
11. M. Duncan and P. Brookes, Metabolism and macromolecular binding of dibenz [*a,h*] anthracene by mouse embryo cells in culture, *Int. J. Cancer* **9**, 349–352 (1972).
12. J. A. Miller and E. C. Miller, The carcinogenic aminoazo dyes, *Adv. Cancer Res.* **1**, 339–396 (1953).
13. C. Heidelberger, Studies on the molecular mechanism of hydrocarbon carcinogenesis, *J. Cell. Comp. Physiol.* **64**, Suppl. 1, 129–148 (1964).
14. H. C. Pitot and C. Heidelberger, Metabolic regulatory circuits and carcinogenesis, *Cancer Res.* **23**, 1694–1700 (1963).
15. J. J. Roberts, The binding of metabolites of 4-dimethylaminoazobenzene and 2-methyl-4-dimethylaminoazobenzene to hooded rat macromolecules during chronic feeding, in: *Physicochemical Mechanisms of Carcinogenesis* (E. D. Bergmann and P. Pullman, eds.), pp. 229–236, Israel Academy of Sciences and Humanities, Jerusalem (1969).
16. P. D. Lawley and P. Brookes, Cytotoxicity of alkylating agents towards sensitive and resistant strains of *E. coli* in relation to extent and mode of alkylation of cellular macromolecules and repair of alkylation lesions in DNA, *Biochem. J.* **109**, 433–447 (1968).
17. P. D. Lawley, Effects of some chemical mutagens and carcinogens on nucleic acids, in: *Progress in Nucleic Acid Research*, Vol. 5 (J. N. Davidson and W. E. Cohn, eds.), pp. 89–131, Academic Press, New York (1966).
18. P. N. Magee, V. M. Craddock, and P. F. Swann, The possible significance of alkylation of nucleic acids in carcinogenesis of the liver and other organs, in: *Carcinogenesis: A Broad Critique,* pp. 421–439, Williams and Wilkins, Baltimore (1967).
19. I. B. Weinstein, Modifications in transfer RNA during chemical carcinogenesis, in: *Symposium on Fundamental Cancer Research, 23rd, M.D. Anderson Hospital and Tumor Institute. Genetic Concepts and Neoplasia, A Collection of Papers,* pp. 380–409, Williams and Wilkins, Baltimore (1970).
20. L. M. Fink, S. Nishimura, and I. B. Weinstein, Modifications of RNA by chemical carcinogens, *Biochemistry* **9**, 496–502 (1970).
21. I. B. Weinstein and D. Grunberger, Structural and functional changes in nucleic acids modified by chemical carcinogens, in: *Chemical Carcinogenesis Part A* (P. O. P. Ts'o and J. A. DiPaolo, eds.), pp. 217–235, Dekker, New York (1974).
22. J. J. Roberts and G. P. Warwick, The covalent binding of metabolites of dimethyl

aminoazobenzene, β-naphthylamine and aniline to nucleic acids *in vivo, Int. J. Cancer* **1**, 179–196 (1966).

23. I. B. Weinstein, D. Grunberger, S. Fujimura, and L. M. Fink, Chemical carcinogens and RNA, *Cancer Res.* **31**, 651–655 (1971).
24. K. H. Bauer, *Mutationstheorie der Geschwulst-Entstehung*, Springer, Berlin (1928).
25. W. J. Burdette, The significance of mutations in relation to the origin of tumours: A review, *Cancer Res.* **15**, 201–226 (1955).
26. P. F. Swann and P. N. Magee, Nitrosamine-induced carcinogensis. Alkylation of nucleic acids of the rat by MNUA, DMN, DMS and MMS, *Biochem. J.* **110**, 39–47 (1968).
27. A. Loveless, Possible relevance of O^6-alkylation of deoxyguanosine to mutagenicity of nitrosamines and nitrosamides, *Nature (London)* **223**, 206–208 (1969).
28. L. L. Gerchman and D. B. Ludlum, Properties of O^6-methylguanine in template for RNA polymerase, *Biochim. Biophys. Acta* **308**, 310–316 (1973).
29. J. V. Frei, Tumour induction by low molecular weight alkylating agents, *Chem.-Biol. Interactions* **3**, 117–121 (1971).
30. J. V. Frei and V. V. Jashi, Lack of induction of thymomas and pulmonary adenomas in inbred Swiss mice by MNNG, *Chem.-Biol. Interactions* **8**, 131–133 (1974).
31. P. J. O'Connor, M. J. Capps, and A. W. Craig, Comparative studies of the hepatocarcinogen DMN *in vivo:* Reaction sites in rat liver DNA and the significance of their relative stabilities, *Br. J. Cancer* **27**, 153–166 (1973).
32. L. Den Engelse, Formation of methylated bases in DNA by DMN and its relation to differences in formation of tumours in liver of GR and C3Hf mice, *Chem-Biol. Interactions* **8**, 329–338 (1974).
33. R. Goth and M. F. Rajewsky, Persistence of O^6-ethylguanine in rat brain: Correlation with nervous system-specific carcinogenesis by ethylnitrosourea, *Proc. Natl. Acad. Sci. USA* **71**, 639–643 (1974).
34. P. Kleihues and G. P. Margison, Carcinogenicity of MNUA: Possible role of excision repair of O^6-methylguanine from DNA, *J. Natl. Cancer Inst.* **53**, 1839–1841 (1974).
35. J. V. Frei and P. D. Lawley, Methylation of DNA in various organs of C57Bl mice by a carcinogenic dose of MNUA and stability of some products up to 18 hours, *Chem.-Biol. Interactions* **10**, 413–427 (1975).
36. V. M. Craddock, Effect of a single treatment with the alkylating carcinogens DMN, DEN and MMS on liver regenerating after partial hepatectomy. II. Alkylation of DNA and inhibition of DNA replication, *Chem.-Biol. Interactions* **10**, 323–332 (1975).
37. A. Dipple, P. Brookes, D. S. Mackintosh, and M. P. Rayman, Reaction of 7-bromomethylbenz[*a*]anthracene with nucleic acids, polynucleotides and nucleosides, *Biochemistry* **10**, 4323–4330 (1971).
38. A. J. Swaisland, P. L. Grover, and P. Sims, Reactions of polycyclic hydrocarbon epoxides with RNA and polyribonucleotides, *Chem.-Biol. Interactions* **9**, 317–326 (1974).
39. W. M. Baird and P. Brookes, Isolation of the hydrocarbon–deoxyribonucleoside products from the DNA of mouse embryo cells treated in culture with 7-methylbenz[*a*]-anthracene-^{3}H, *Cancer Res.* **33**, 2378–2385 (1973).
40. W. M. Baird, A. Dipple, P. L. Grover, P. Sims, and P. Brookes, Studies on the formation of hydrocarbon-deoxyribonucleoside products by the binding of derivatives of 7-methylbenz[*a*]anthracene to DNA in aqueous solution and in mouse embryo cells in culture, *Cancer Res.* **33**, 2386–2392 (1973).
41. A. Dipple and T. A. Slade, Structure and activity in chemical carcinogenesis, *Eur. J. Cancer* **7**, 473–476 (1971).
42. E. Boyland and P. Sims, The carcinogenic activities in mice of compounds related to benz[*a*]anthracene, *Int. J. Cancer* **2**, 500–504 (1967).

43. E. C. Miller and J. A. Miller, Low carcinogenicity of the K-region epoxides of 7-methylbenz[*a*]anthracene and benz[*a*]anthracene in the mouse, *Proc. Soc. Exp. Biol. Med.* **124**, 915–919 (1967).
44. P. Howard-Flanders, DNA repair, *Annu. Rev. Biochem.* **37**, 175–200 (1968).
45. J. E. Cleaver, Defective repair replication of DNA in *Xeroderma pigmentosum, Nature (London)* **218**, 652–656 (1968).
46. J. J. Roberts, Repair of alkylated DNA in mammalian cells, in: *Molecular and Cellular Repair Processes* (R. F. Beers, R. M. Herriott, and R. C. Tilgham, eds.), pp. 226–238, Johns Hopkins University Press, Baltimore (1972).
47. A. R. Lehman, Post-replication repair of ultra-violet irradiated mammalian cells, *J. Mol. Biol.* **66**, 319–337 (1972).
48. W. D. Rupp, C. E. Wilde, D. L. Reno, and P. Howard-Flanders, Exchanges between DNA strands in ultra-violet irradiated *E. coli, J. Mol. Biol.* **61**, 25–44 (1971).
49. E. M. Witkin, Mutation-proof and mutation-prone modes of survival in derivatives of *E. coli* B differing in sensitivity to ultraviolet light, *Brookhaven Symp. Biol.* **20**, 17–55 (1967).
50. A. C. Stevenson and C. B. Kerr, On the distribution of frequencies of mutation to genes determining harmful traits in man, *Mutat. Res.* **4**, 339–352 (1967).
51. B. N. Ames, W. E. Durston, E. Yamasaki, and F. D. Lee, Carcinogens are mutagens. A simple test system combining liver homogenates for activation and bacteria for detection, *Proc. Natl. Acad. Sci. USA* **70**, 2281–2285 (1973).
52. B. N. Ames, E. G. Gurney, J. A. Miller, and H. Bartsch, Carcinogens as frameshift mutagens: Metabolites and derivatives of 2-acetylaminofluorene and other aromatic amine carcinogens, *Proc. Natl. Acad. Sci. USA* **69**, 3128–3132 (1972).
53. S. Venitt and L. S. Levy, Mutagenicity of chromates in bacteria and its relevance to chromate carcinogenesis, *Nature (London)* **250**, 493–495 (1974).
54. H. Bartsch, C. Malaveille, and R. Montesano, Human, rat and mouse liver-mediated mutagenicity of vinyl chloride in *S. typhimurium* strains, *Int. J. Cancer* **15**, 429–437 (1975).
55. H. Bartsch, C. Malaveille, R. Montesano, and L. Tomatis, Tissue mediated mutagenicity of vinylidene chloride and 2-chlorobutadiene in *S. typhimurium, Nature (London)* **255**, 641–643 (1975).
56. A Haddow, Molecular repair, wound healing, and carcinogenesis: Tumour production a possible overhealing? in: *Advances in Cancer Research*, Vol. 16 (G. Klein and S. Weinhouse, eds.), pp. 181–234, Academic Press, New York (1972).
57. A. Haddow, Addendum to "Molecular repair, wound healing, and carcinogenesis", in: *Advances in Cancer Research,* Vol. 20 (G. Klein and S. Weinhouse, eds.), pp. 343–366, Academic Press, New York (1974).
58. J. Cairns, Mutation selection and the natural history of cancer, *Nature (London)* **255**, 197–200 (1975).

55

Discussion

Jerina commented that he has observed binding of the diol-epoxide of benzo[*a*] pyrene to polyguanylic acid and found that it binds extensively. Miller suggested to Jerina that it was probably an example of *O*-6 alkylation. Brookes reported that binding with benzo[*a*] pyrene and 7-methylbenzanthracene *in vivo* involves a purine. He has observed hydrocarbon binding to purine but not pyrimidine moieties.

Gelboin felt that it was odd that one type of binding produced mutagenesis or interference with gene action while another type did not because it would be expected that molecules as large as hydrocarbons bound to nucleotides should interfere with transcription. Brookes agreed that he had at first come to the same conclusion but that later evidence appears to indicate that it is necessary to distinguish between the importance of binding at *O*-6 or *N*-7 versus binding directed by the nature of the reacting codon. He suggested that an alkylating agent may react indiscriminately, but if a hydrocarbon which is metabolized in a rather complex manner is inserted, perhaps then it is directed to a particular point on the genome by the nature of that reaction. Furthermore, since at any given time 90% of DNA is inactive, it would be of value to determine whether hydrocarbons reacted selectively with inactive DNA. He raised questions concerning the role that might be played by reiterated versus nonreiterated sequences in determining the specificity of the interaction. Finally, he stressed the role of repair enzymes by pointing out that if the hydrocarbon is put into certain positions on the DNA which involve its localization on the narrow groove the hydrocarbon may be ignored by repair enzymes and remain bound, whereas binding at the wide groove may cause it to protrude. In the latter case, one side will show the displacement phenomenon, leading to repair enzyme recognition and excision.

Schulman questioned whether carcinogenesis by immunosuppressive agents might result from depression of the immune surveillance mechanism rather than through direct interactions with DNA. Brookes replied that while this mechanism would be difficult to completely rule out, indirect evidence argues against it since many immunosuppressive agents do not induce carcinogenesis. Brookes doubted that immunosuppression really allows dormant tumor cells to arise.

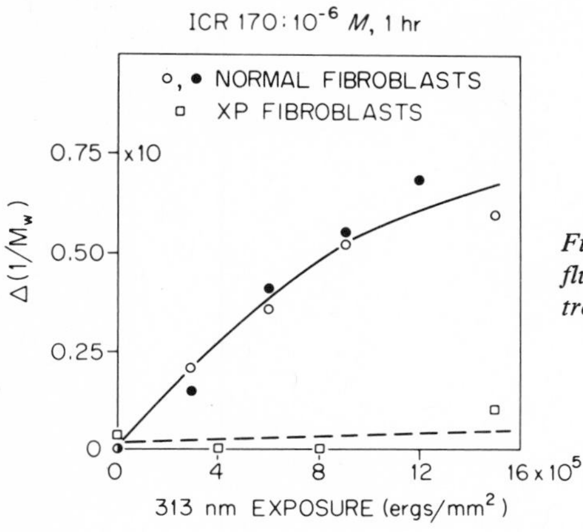

Fig. 4. $\Delta(1/M_w)$ *as a function of increasing fluence of 313-nm radiation after ICR-170 treatment.*

the long-repair inducing agents, as can be seen from Table 1, the amount of base excision differed with the insult. With 200 ergs/mm² of 254-nm UV, about 100 nucleotides were excised and replaced per average repaired region. With *N*-acetoxy-acetoaminofluorene, about 160 nucleotides are excised and with ICR-170 about 40 nucleotides. These are important differences because they indicate some mechanism that controls the amount of base excision during long repair. However, the major distinction between short and long repair is that the former involves excision of only a few nucleotides whereas the latter involves many (40–160) nucleotides per average repaired region. As is clear from Table 1, this is not an exhaustive list of agents and we need to examine many more chemical agents representative of the various classes of chemical carcinogens and mutagens.

We have tested two agents which resulted in alteration of the DNA that was not repair in the usual sense, either short or long. These agents are nitroso-carbaryl (NO-carb) and methylnitrosourethane (MeNU). NO-carb was synthesized by our colleague, William Lijinsky. NO-carb is derived from carbaryl, a widely used pesticide. MeNU is a chemical analogue of NO-carb and is carcinogenic. The structures of these agents are shown in Fig. 5.

In Figure 6 we see the results of sedimentation of profiles of DNA of human cells after 1 h of treatment with either 10^{-4} M carbaryl or 10^{-4} M NO-carb. Also shown are similar profiles from cells 20 h after the treatment was given. These data indicate that the DNA from cells treated with carbaryl and assayed either immediately or 20 h after treatment is essentially the same as numerous control (untreated cultures) human cell DNAs we have examined. By contrast, the DNA of NO-carb-treated cells exhibited a striking reduction in sedimentation rate immediately after or 20 h after treatment (Fig. 6B). These results suggest either the direct induction of many single-strand breaks in the DNA or the formation of numerous alkaline-sensitive bonds, which result in breaks when the cells are

Table 1. Classification of DNA-Damaging Treatments According to the Type of Repair Induced

Treatment	Dose or concentration	Duration	Number of strand breaks/10^8 daltons after BrdUrd incubation and 10^6 ergs/mm^2 of 313-nm radiation	BrdUrd nucleotides inserted/lesion	Type of repair
UV (254 nm)	200 ergs/mm^2		10	25	Long[a]
^{60}Co γ-rays	10 krad		0.6	1	Short[b]
N-Acetoxy-acetoaminofluorene	7×10^{-6} M	60 min	4	~40	Long[a]
4-Nitroquinoline oxide	5×10^{-7} M	90 min	~2		Long and short[c]
Ethyl methane sulfonate	10^{-2} M	120 min	~1.0		Short
Methyl methane sulfonate	5×10^{-5}M	5 min	~0.4		Short
Propane sultone	2×10^{-4} M	2 h	~0.4		Short
ICR-170	10^{-6} M	1 h	~1	~10	Long[a]

[a]Repair in normal cells tenfold greater than in xeroderma cells.
[b]Repair equal in normal and xeroderma cells.
[c]See text for details.

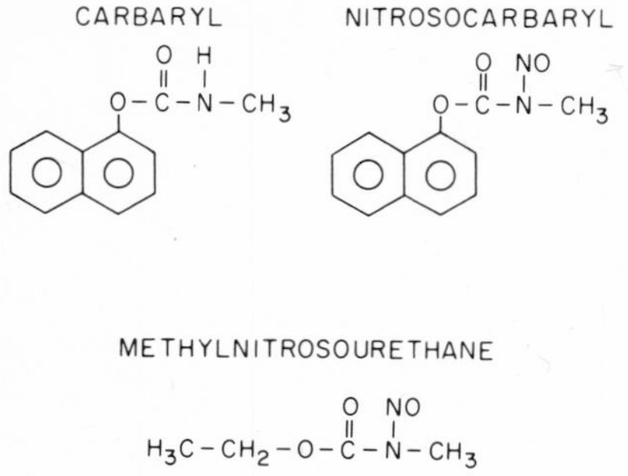

CARBARYL

NITROSOCARBARYL

METHYLNITROSOURETHANE

$H_3C-CH_2-O-C(=O)-N(NO)-CH_3$

Fig. 5. Chemical structure of carbaryl, its nitroso derivative, nitrosocarbaryl, and the methyl analogue of nitrosocarbaryl, methyl nitrosourethane.

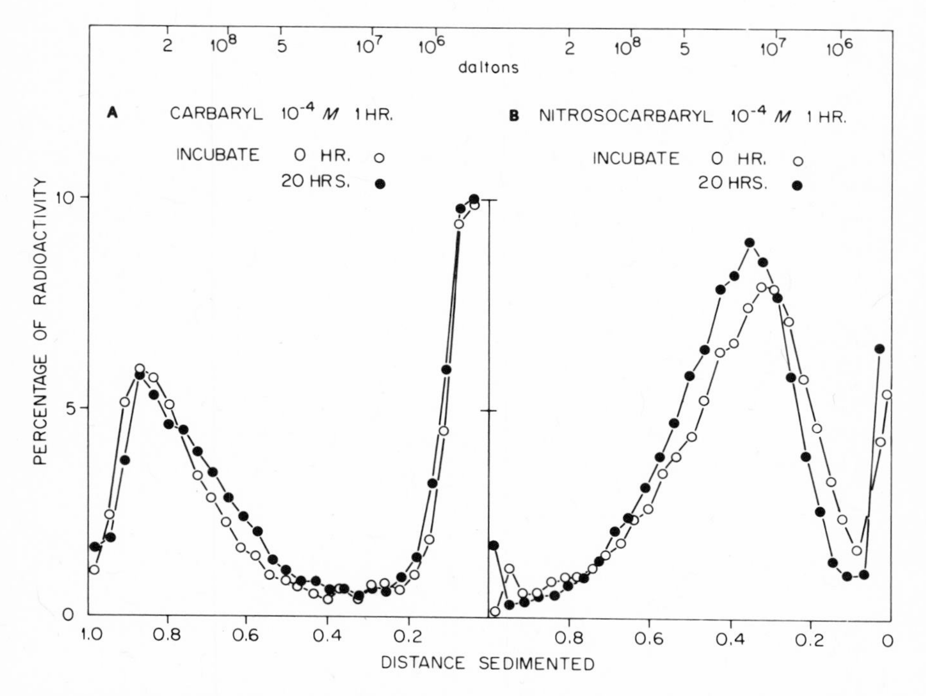

Fig. 6. (A) Effect of 10^{-4} M carbaryl on the sedimentation of DNA from normal human skin cells in alkali immediately after a 1-h treatment (○) and 20 h after treatment (●). The treated cells had been prelabeled with [^{3}H] thymidine. The sedimentation profiles are essentially the same as those from numerous control (untreated) human skin fibroblast cultures we have assayed. (B) Effect of 10^{-4} M NO-carbaryl on the sedimentation in alkali of DNA from ^{3}H-labeled normal human skin cells immediately after a 1-h treatment and after a 20-h incubation without NO-carbaryl.

lysed on top of the alkaline gradients. No evidence of significant cell killing, that is, floating cells or loss of acid insoluble counts, was observed in the cell monolayer either immediately after treatment with the chemical or at 20 h after treatment. MeNu at similar concentrations produced results much the same as observed with NO-carb, whereas urethane, the parent compound of MeNu, showed no effect.

We attempted to investigate the action of these agents on the DNA of human cells by the BrdUrd photolysis assay; however, we could find no evidence for incorporation of BrdUrd during the repair period. This result suggests that no breaks are induced and no bases are inserted. This finding is consistent with an interpretation of many alkaline-sensitive bonds being produced in the DNA rather than many single-strand breaks. If this is the action of NO-carb and MeNU on DNA, it would appear that these alkaline-sensitive regions are essentially irreversible since even though the chemicals are removed for 20 h the same amount of DNA breakage is seen when the cells are lysed on top of alkaline gradients.

We are currently investigating a number of other nitroso derivatives of similar compounds to see if they also induce this effect. We would emphasize that this effect appears not to be long repair or short repair, or any repair for that matter, in terms of DNA strand breakage and nucleotide removal and replacement. Further detailed results of these experiments has been reported elsewhere (17).

ACKNOWLEDGMENT

This research was sponsored by the Energy Research and Development Administration under contract with the Union Carbide Corporation.

REFERENCES

1. J. A. Miller, Carcinogenesis by chemicals: An overview, G. H. A. Clowes Memorial Lecture, *Cancer Res.* **36**, 559–576 (1970).
2. R. B. Setlow and J. K. Setlow, Effects of radiation on polynucleotides, in: *Annual Review of Biophysics and Bioengineering,* Vol. 1 (M. F. Morales, ed.), pp. 293–345, Annual Reviews, Palo Alto, Calif. (1972).
3. J. E. Cleaver, Repair processes for photochemical damage in mammalian cells, in: *Advances in Radiation Biology,* Vol. 4 (J. T. Lett, H. Adler, and M. Zelle, eds.), pp. 1–75, Academic Press, New York (1974).
4. J. D. Regan, R. B. Setlow, and R. D. Ley, Normal and defective repair of damaged DNA in human cells: A sensitivity assay utilizing the photolysis of bromodeoxyuridine, *Proc. Natl. Acad. Sci. USA* **68**, 708–712 (1971).

5. J. D. Regan, R. B. Setlow, M. M. Kaback, R. R. Howell, E. Klein, and G. Burgess, Xeroderma pigmentosum: A rapid, sensitive method for prenatal diagnosis, *Science* **174**, 147–150 (1971).
6. J. D. Regan and R. B. Setlow, Two forms of repair in the DNA of human cells damaged by chemical carcinogens and mutagens, *Cancer Res.* **34**, 3318–3325 (1974).
7. M. B. Lion, Search for a mechanism for the increased sensitivity of 5-bromouracil-substituted DNA to ultraviolet radiation. II. Single strand breaks in the DNA of irradiated 5-bromouracil-substituted T_3 coliphage, *Biochim Biophys. Acta* **209**, 24–33 (1970).
8. R. B. Painter and B. R. Young, Repair replication in mammalian cells after X-irradiation, *Mutat. Res.* **14**, 225–235 (1972).
9. J. E. Cleaver and R. B. Painter, Evidence for repair replication of HeLa cell DNA damaged by ultraviolet light, *Biochim. Biophys. Acta* **161**, 552–554 (1968).
10. M. G. Ormerod and U. Stevens, The rejoining of X-ray-induced strand breaks in the DNA of a murine lymphoma cell (L5178Y), *Biochim. Biophys. Acta* **232**, 72–82 (1971).
11. J. D. Regan and R. B. Setlow, DNA repair in human progeroid cells, *Biochem. Biophys. Res. Commun.* **59**, 858–863 (1974).
12. J. E. Cleaver, Defective repair replication of DNA in xeroderma pigmentosum, *Nature (London)* **218**, 652–656 (1968).
13. R. B. Setlow, J. D. Regan, J. German, and W. L. Carrier, Evidence that Xeroderma pigmentosum cells do not perform the first step in the repair of ultraviolet damage to their DNA, *Proc. Natl. Acad. Sci. USA* **64**, 1035–1041 (1969).
14. J. E. Cleaver, DNA repair with purines and pyrimidines in radiation- and carcinogen-damaged normal and xeroderma pigmentosum human cells, *Cancer Res.* **33**, 362–369 (1973).
15. H. F. Stitch, R. H. C. San, and Y. Kawzoe, Increased sensitivity of xeroderma pigmentosum cells to some chemical carcinogens and mutagens, *Mutat. Res.* **17**, 127–137 (1973).
16. H. F. Stitch, R. H. C. San, J. A. Miller, and E. C. Miller, Various levels of DNA repair synthesis in xeroderma pigmentosum cells exposed to the carcinogens *N*-hydroxy and *N*-acetoxy-2-acetyl aminofluorene, *Nature (London)* **238**, 9–10 (1972).
17. J. D. Regan, R. B. Setlow, A. A. Francis, and W. Lijinsky, Nitrosocarbaryl: Its effect on human DNA, *Mut. Res.* **38**, 293–302 (1976).

57

Discussion

Snyder inquired whether there are any data available on the possible mutagenic or carcinogenic effect of errors in the repair processes.

Regan agreed that this is an important question, on which, however, there is no direct information available because of lack of suitable experimental methods. He noted that in addition to the prereplicational repair systems discussed in his chapter, other mechanisms are known, including postreplication repair which functions on DNA replicated on damaged template. There is some evidence which suggests that this is an error-prone mechanism. One might speculate on this basis that, of the two, postreplication repair might be more important in carcinogenesis. However, it is well known that patients displaying xeroderma pigmentosum, who are defective in prereplication repair, easily develop multiple skin cancers.

58

The Origin, Present Status, and Trends of Toxicology: A Review of the Symposium in Turku, 1975

Herbert Remmer
Institute of Toxicology
University of Tübingen
D-7400 Tübingen, West Germany

In this discussion I will not attempt to summarize all of the papers presented during this symposium because it would be impossible to recount all of the new observations that were presented as well as to evaluate the current status of the several areas that were reviewed. It would seem more appropriate to place these new achievements into perspective within the constantly flowing stream of toxicological knowledge that has been handed down to us from our pharmacological and toxicological predecessors. The work that each of us does, after all, is a link in a chain connecting us in our attempts to gain insights into toxicological phenomena which occur in the cells of man and animals.

There are few better places than Turku for recalling the history of pharmacology and toxicology. Our discipline was conceived in this area of the world in the middle of the nineteenth century by Rudolph Bucheim, who taught at the Baltic-German University of Dorpat which is located in the neighboring country of Estonia just over the Baltic Sea.

Pharmacology as a new science began 130 years ago as unnoticed in the Western countries as a natural mutation. It was not until Otto Schmiedeberg, the famous Baltic-German student of Bucheim, was called to Strasbourg in 1872 that pharmacology became known and acknowledged as a discipline, and, as a result, Strasbourg became the Mecca of pharmacology for many decades. It was the many young pharmacologists from Strasbourg who planted the seed of the new science in countries all over the world.

Bucheim, the ingenious pioneer, translated a book entitled *Materia Medica,* originally written in English by Pereira (1), into German, in which a distinction was made between the effects of drugs on the organisms (i.e., pharmacodynamics) and the action of organisms upon drugs (i.e., pharmacokinetics). Thus this concept which we consider modern today was actually expressed over 120 years ago. Pharmacodynamics was pursued with greater vigor during the succeeding century and it was only with the reawakening of interest in pharmacokinetics by the school of Brodie and Axelrod that the concept of the actions of organisms on drugs was studied in depth.

A new species never develops by single mutations. There were a great number of research workers whose efforts led to an increased emphasis on pharmacokinetics, but one should be mentioned here because of his impact on our discussions in Turku. The most famous German student of Schmiedeberg in Strasbourg may not be very well known in the United States. His name was Wolfgang Heubner (2,3), who taught in Göttingen and later in Berlin. He was the first in Germany to realize that pharmacology could not develop and achieve its goals without investigating the reaction of drugs at the molecular level. Hemoglobin, one of the few proteins available 50 years ago in a purified crystalline state, was his tool for investigating the chemical-induced conversion of hemoglobin to methemoglobin. He considered such modern notions as the involvement of radicals in the autocatalytic oxidation of hemoglobin by phenylhydroxylamine in red blood cells. His students and their successors established that irreversible binding of the metabolites to hemoglobin is one of the critical steps in the conversion of the oxygen-transporting hemeprotein to a less active or inactive substance, thereby causing hypoxia, which is the recognized toxic effect of phenylhydroxylamine. Thus the concept of a role for covalent binding in producing toxicity is a legacy from our scientific past.

We are happy that one of the early pioneers in this field, Dr. J. Miller, is a participant in this symposium. He and his wife were the first to present evidence for the interaction and covalent binding of carcinogenic compounds or their metabolites to nucleic acids (4), and his presentation was one of the highlights of our symposium. In this context I should mention another name, Hecker, whose studies at the end of the 1950s concerning covalent binding of estrogens to proteins by radicals formed in liver microsomes are not so well known. Marks and Hecker (5) demonstrated that covalent binding to proteins is not a toxic event *per se* since it normally occurs in the course of metabolism of estrogenic hormones both *in vivo* and *in vitro* (6). Fortunately, the semiquinone radicals formed in the endoplasmic reticulum do not reach the nucleus since covalent binding to nucleic acids was not observed. Although covalent binding to proteins did not apparently exert a direct toxic action in this case, we face a problem for future elucidation by research workers concerning whether reactive metabolites formed by the microsomal mixed-function oxidase might react as haptens and might be the hidden cause of several different forms of allergic hepatitis. Another related line of work followed by the school of Brodie was presented by

Dr. Gillette, who reported that a correlation exists between the degree of covalent binding to protein molecules and toxic changes which occur in liver and kidney. These observations helped to open up a new field in toxicology.

Considerable discussion, which we believe should be evaluated with caution, has suggested that the origin of tumors produced by carcinogens can now be explained by the reaction between alkylating agents and reactive sites on DNA. Covalent binding is no doubt an essential requirement for many toxic reactions but does not give an insight into the alteration going on in the cell that leads to its death. The results that we heard yesterday and today are of course important steps for the further clarification of carcinogenesis and cell necrosis, but the real problems we have to face in toxicology and cancer research cannot be solved solely by using chemical analytical tests or pure biochemical methods. These have been highly successful and have exerted a profound impact on many investigations in toxicology, but many more approaches with completely different procedures, e.g., morphological and immunological, have to be applied before we can gain a getter understanding of the complicated processes leading to cancer or cell necrosis. An alteration in the nucleus might not have any consequence at all or might not affect an essential feature of cell life. Furthermore, we learned from the report of Dr. Regan that damaged nucleic acids can usually be repaired. Thus the reason for uninhibited cell growth in cancer is an unsolved problem. Obviously there are many parameters involved and it may well be that an increase in cell division is not the only factor that defines cell growth. Toxicologists will have to develop new approaches to the study of dangerous agents at the molecular level such as those relating to morphology, biology, and chemistry if we are to completely understand the deleterious processes occurring in the cell.

We should congratulate Drs. Conney and Gelboin for their very important and stimulating discoveries which have had such an enormous impact in many fields of molecular pharmacology. We must also commend them for the efficiency that they have achieved by applying the concept of team research to problems of great difficulty. In all, there are ten scientists collaborating in the group headed by Dr. Conney and seven listed in Dr. Gelboin's group. Both research teams successfully elucidated the metabolites responsible for the carcinogenic action of benzo[*a*]pyrene. However, it is significant that while two such large groups developed their work independently and competed with each other, they did not neglect to consult and exchange ideas and concepts. These developments would not be possible in countries which organize their investigations centrally. These highly profitable collaborations are welcome to the future of science and we should use them as models for future approaches.

It is important, however, that despite the excellent contribution made by such splendid teams we should carefully avoid suppressing the creativity of the single ingenious brain. We need not mention all of the excellent results we have heard during the past few days which were achieved by our bright young friends and colleagues who have succeeded without having at their disposal a large group of collaborators and costly facilities.

In the name of all participants I would like to express our gratitude to the organizers of the symposium, particularly to Dr. Vainio and Dr. Snyder and our host Professor Hartiala and his associates, for their generous and overwhelming hospitality. Before finishing, let me summarize my impressions by viewing our symposium like an ancient priest looking into the future. Priests of olden times had functions to fulfill which were similar to those of modern scientists. These included interpretation of new findings for the public, providing good counsel, and attempting to project future events.

Haruspex, the Roman priest, performed his function by sacrificing animals and inspecting their liver to determine whether he could divine signs predictive of coming events based on its shape, size, and surface characteristics. Thus he was the first liver morphologist and the predecessor of all hepatologists. The following poem will bring Haruspex into the twentieth century as well as help to review the results of this symposium.

A MODERN HARUSPEX REVIEWS THE SYMPOSIUM IN TURKU, 1975, IN A HEXAMETER

Excising the victim's liver, inspecting its shape and its layers,
Haruspex, the Roman priest, stretched out his arms to the skies,
And called the almighty ruler of heaven and earth with his prayers:
Explain to us, why from this organ such harmful diseases arise!

Intelligent men who pretend to be wise and knowing creature
Complain about life-saving enzymes in liver established by nature,
Producing those dangerous toxic effects and man-killing tumor.
A thunderstorm rose, the strongly built temple started to tremor.

A threatening voice rumored: Precious gifts to benefit mortals
Have to be paid with substantial prices, demanding great hazards.
The matter of all living beings is held by covalent bindings
That cost mutations and death as testify scientists' findings.

Men would destroy themselves with all the ingested pollutants,
Arising from alcohol, smoke, tobacco, and synthesized drugs,
If I had not granted their body mixed-function hydroxylases,
Preventing so dangerous compounds to sweep out from globe all their races.

Nothing is faultless which nature has born and extensively shown.
If all living beings developed on earth had been perfectly grown,
I'm sure that scientists would have discovered the secret stone
To throw against me and pushing me down from my heavenly throne!

REFERENCES

1. J. Pereira, *The Elements of Materia Medica and Therapeutics,* Vol. 1, 3rd American ed. (J. Carson, ed.), pp. 138–140, Blanchard and Lea, Philadelphia (1852).
2. W. L. Lipschitz, Giftung und Entgiftung aromatischer Nitroverbindungen, *Naunyn-Schmiedeberg's Arch. Pharmacol.* **205**, 305–309 (1948).
3. W. Heubner, Giftung aromatischer Nitroverbindungen, *Naunyn Schmiedeberg's Arch. Pharmacol.* **205**, 310 (1948).
4. J. A. Miller, Carcinogenesis by chemicals: An overview, Clowes Memorial Lecture, *Cancer Res.* **30**, 559–576 (1970).
5. F. Marks and E. Hecker, Metabolism and mechanism of action of oestrogens. XII. Structure and mechanism of formation of water-soluble and protein bound metabolites of oestrone in rat-liver microsomes *in vitro* and *in vivo, Biochim. Biophys. Acta* **187**, 250–265 (1969).
6. H. Kappus, H. M. Bolt, and H. Remmer, Irreversible protein binding of metabolites of ethynylestradiol *in vivo* and *in vitro, Steroids* **22**, 203–225 (1973).

List of Participants

Ahokas, Jorma, *Department of Pharmacology, University of Oulu, SF-90220 Oulu 22, Finland*

Aitio, Antero, *Department of Physiology, University of Turku, 20520 Turku 52, Finland*

Anders, M. W., *Institute of Toxicology, University of Tübingen, D-74 Tübingen, Germany*

Arrhenius, Erik, *Environmental Toxicology Unit, Wallenberg Laboratory, S-104 05 Stockholm, Sweden*

Arvela, Pentii, *Department of Pharmacology, University of Oulu, SF-90220 Oulu 22, Finland*

Baron, Jeffrey, *The Toxicology Center, Department of Pharmacology, University of Iowa, Iowa City, Iowa 52242, U.S.A.*

Batt, Anne-Marie, *Laboratoire de Biochimie, Faculté des Sciences Pharmaceutiques et Biologiques, 5400 Nancy, France*

Belvedere, Giorgio, *Istituto di Ricerche Farmacologiche "Mario Negri," Milan, Italy*

Benakis, Achille, *Laboratoire du Métabolisme des Médicaments, Ecole de Médecine, 1211 Genève 4, Switzerland*

Berggren, Margareta, *Department of Forensic Medicine, Karolinska Institutet, S-104 01 Stockholm 60, Sweden*

Bergman, Kerstin, *Department of Toxicology, University of Uppsala, S-751 23 Uppsala, Sweden*

Bodin, Nils-Olov, *Toxicology Laboratory, Astra Pharmaceuticals AB, S-151 85 Södertälje, Sweden*

Borg, Karl Olof, *Analytical Chemistry and Biochemistry, AB Hässle, S-431 20 Mölndal, Sweden*

Breckenridge, Bruce, *Department of Pharmacology, College of Medicine and Dentistry of New Jersey, Rutgers Medical School, Piscataway, N.J. 08854, U.S.A.*

Breimer, Douwe D., *Department of Pharmacology, Subfaculty of Pharmacy, Leiden, The Netherlands*

Brookes, Peter, *Institute of Cancer Research, Pollards Wood Research Station, Chalfont St. Giles, Buckinghamshire, HP8 4SP, England*

Burke, Danny M., *Department of Forensic Medicine, Karolinska Institutet, S-104 01 Stockholm 60, Sweden*

Chasseaud, L. F., *Huntingdon Research Centre, Huntingdon PE18 6ES, England*

Conney, Allan H., *Department of Biochemistry and Drug Metabolism, Hoffman–La Roche Inc., Nutley, N.J. 07110, U.S.A.*

Dagirmanjian, Rose, *Department of Pharmacology, University of Louisville, School of Medicine, Louisville, Ky. 40201, U.S.A.*

D'Argy, Roland, *Department of Toxicology, University of Uppsala, S-751 23 Uppsala, Sweden*

De Matteis, F., *Biochemical Mechanisms Section, MRC Toxicology Unit, Medical Research Council Laboratories, Carshalton, Surrey, England*

De Pierre, Joseph William, *Stureparken 13, S-114 26 Stockholm, Sweden*

Dybing, Erik, *Laboratory of Chemical Pharmacology, National Heart and Lung Institute, National Institutes of Health, Bethesda, Md. 20014, U.S.A.*

Eberts, Floyd S., *The Upjohn Company, Kalamazoo, Mich. 49001, U.S.A.*

El-Hawari, A. Monaem, *Department of Pharmacology, University of Montreal, Montreal, Quebec, Canada*

Ellis, Sydney, *Department of Pharmacology and Toxicology, University of Texas, Medical Branch at Galveston, Galveston, Tex. 77550, U.S.A.*

Elsom, L. E., *Huntingdon Research Centre, Huntingdon PE18 6ES, England*

Fellenius, Erik, *Department of Analytical Chemistry and Biochemistry, AB Hässle, S-431 20 Mölndal, Sweden*

Fiserova-Bergerova Thomas, Vera, *Department of Anesthesiology, University of Miami, School of Medicine, Miami, Fla. 33152, U.S.A.*

Fromson, John Michael, *Hoechst Pharmaceutical Research Laboratory, Milton Keynes MK7 7AJ, Buckinghamshire, England*

Garbe, Andreas, *E. Merck, Darmstadt, Institute of Grafing, D-8018 Grafing, Germany*

Gelboin, Harry V., *Chemistry and Biology Branches, National Cancer Institute, National Institutes of Health, Bethesda, Md. 20014, U.S.A.*

Gielen, J., *Laboratoire de Recherches, Université de Liège, B-4000 Liège, Belgium*

Gillette, James R., *Laboratory of Chemical Pharmacology, National Heart and Lung Institute, National Institutes of Health, Bethesda, Md. 20014, U.S.A.*

Greim, Helmut, *Institute of Toxicology, University of Tübingen, D-74 Tübingen, Germany*

Hagberg, Curt-Eric, *Toxicology Laboratory, Astra Pharmaceuticals AB, S-151 85 Södertälje, Sweden*

Hänninen, Osmo, *Department of Physiology, University of Kuopio, 70101 Kuopio 10, Finland*

Harper, Curtis, *National Institute of Environmental Health Sciences, Pharmacology Branch, Research Triangle Park, N.C. 27709, U.S.A.*

Hartiala, Kaarlo, *Department of Physiology, University of Turku, 20520 Turku 52, Finland*

Hatch, Frederick T., *Lawrence Livermore Laboratory, University of California, Livermore, Calif. 94550, U.S.A.*

Hesse, Sigrun, *Abt. Toxikologie, Ges. für Strahlen- und Umweltforschung, D-8042 Neuherberg, Germany*

Hildebrandt, Alfred G., *Institut für Klinische Pharmakologie, Der Freie Universität Berlin, D1000 Berlin 45, Germany*

Högberg, Johan, *Department of Forensic Medicine, Karolinska Institutet, S-104 01 Stockholm 60, Sweden*

Höjeberg, Bo, *Avd. Biokemi, Arrheniuslaboratoriet, University of Stockholm, S-104 05 Stockholm, Sweden*

Holmstedt, Bo R., *Department of Toxicology, Swedish Medical Research Council, Karolinska Institutet, S-104 01 Stockholm 60, Sweden*

Horning, Evan C., *Institute of Lipid Research, Baylor College of Medicine, Texas Medical Center, Houston, Tex. 77025, U.S.A.*

Horning, Marjorie G., *Institute of Lipid Research, Baylor College of Medicine, Texas Medical Center, Houston, Tex. 77025, U.S.A.*

Housley, J. R., *The Boots Company Ltd., Nottingham NG2 3AA, England*

Hulbert, Peter B., *Department of Pharmaceutical Chemistry, University of Bradford, Bradford, Yorkshire BD7 1DP, England*

Hultmark, Dan, *Wallenberg Laboratory, University of Stockholm, S-104 05 Stockholm 50, Sweden*

Husain, Syed, *Department of Biomedical Research, Stanford Research Institute, Menlo Park, Calif. 94025, U.S.A.*

Järvisalo, Jorma O., *Institute of Occupational Health, Haartmaninkatu 1, 00290 Helsinki 29, Finland*

Jakobsson, Sten, *Department of Forensic Medicine, Karolinska Institutet, S-104 01 Stockholm, Sweden*

Jenssen, Dag, *Wallenberg Laboratory, University of Stockholm, S-104 05 Stockholm 50, Sweden*

Jerina, Donald M., *Laboratory of Chemistry, National Institute of Arthritis, Metabolism, and Digestive Diseases, National Institutes of Health, Bethesda, Md. 20014, U.S.A.*

Jernström, Bengt, *Department of Forensic Medicine, Karolinska Institute, S-104 01 Stockholm 60, Sweden*

Johansson, Eva Britt, *Department of Toxicology, University of Uppsala, S-751 23 Uppsala, Sweden*

Jollow, David, *Department of Pharmacology, Medical University of South Carolina, Charleston, S.C. 29401, U.S.A.*

Jonen, H. G., *Department of Pharmacology, University of Mainz, 6500 Mainz, Germany*

Jonsson, John, *Psychiatric Research Center, University of Uppsala, Ulleråker Hospital, S-750 17 Uppsala, Sweden*

Kahl, Georg F., *Department of Pharmacology, University of Mainz, 6500 Mainz, Germany*

Kangas, Lauri, *Department of Pharmacology, University of Turku, 20520 Turku 52, Finland*

Kato, Ryuichi, *Research Laboratories, Fujisawa Pharmaceutical Co., Ltd., Kashima, Yodogawa-ku, Osaka 532, Japan*

Kocsis, James J., *Department of Pharmacology, Thomas Jefferson University, Philadelphia, Pa. 19107, U.S.A.*

Kristoffersson, Annika, *Department of Forensic Medicine, Karolinska Institutet, S-104 01 Stockholm 60, Sweden*

Kupfer, David, *Worcester Foundation, Shrewsbury, Mass. 01545, U.S.A.*

Laitinen, Matti, *Department of Physiology, University of Kuopio, SF-70101 Kuopio 10, Finland*

Lang, Matti, *Department of Physiology, University of Kuopio, SF-70101 Kuopio 10, Finland*

Larsen, John Chr.., *Institute of Toxicology, Department of Biochemistry, National Food Institute, 2860 Søborg, Denmark*

Leibman, Kenneth C., *Department of Pharmacology, Health Center, University of Florida, Gainesville, Fla. 32610, U.S.A.*

Lenk, Werner, *Pharmakologisches Institut, Universität München, D-8 München, Germany*

Lijinsky, William, *Biology Division, Oak Ridge National Laboratory, Oak Ridge, Tenn. 37830, U.S.A.*

Lind, Christina E., *Avd. Biokemi, Arrheniuslaboratoriet, University of Stockholm, S-104 05 Stockholm, Sweden*

Lindquist, Nils Gunnar, *Research Department, Toxicology, AB Kabi, S-104 25 Stockholm, Sweden*

Lundström, Jan, *Astra Läkemedel, S-151 85 Södertälje, Sweden*

Luoma, Sirpa, *Department of Physiology, University of Turku, 20520 Turku 52, Finland*

Malmfors, Torbjörn, *Toxicology Laboratory, Astra Pharmaceuticals AB, S-151 85 Södertälje, Sweden*

Malnoë, Armand, *Battelle, Geneva Research Centre, 1227-Carouge/Genève, Switzerland*

Marniemi, Jukka, *Department of Physiology, University of Turku, 20520 Turku 52, Finland*

McEwen, John, *Hoechst Pharmaceutical Research Laboratory, Milton Keynes MK7 7AJ, Buckinghamshire, England*

Miller, James A., *McArdle Laboratory for Cancer Research, University of Wisconsin Medical Center, Madison, Wisc. 53706, U.S.A.*

Mitchard, M., *Department of Clinical Pharmacology, The Medical School, University of Birmingham, Birmingham B15 2TJ, England*

Mitchell, Jerry R., *Laboratory of Chemical Pharmacology, National Heart and Lung Institute, National Institutes of Health, Bethesda, Md. 20014, U.S.A.*

Moldéus, Peter, *Department of Forensic Medicine, Karolinska Institutet, S-104 01 Stockholm 60, Sweden*

Mulder, Gerald Johan, *Department of Pharmacology, State University of Groningen, Groningen, The Netherlands*

Muller-Eberhard, Ursula, *Department of Biochemistry, Scripps Clinic and Research Foundation, La Jolla, Calif. 92037, U.S.A.*

Neal, Robert A., *Center in Toxicology, Department of Biochemistry, Vanderbilt University School of Medicine, Nashville, Tenn. 37205, U.S.A.*

Nebert, Daniel W., *National Institute of Child Health and Human Development, National Institutes of Health, Bethesda, Md. 20014, U.S.A.*

Nesnow, Stephen, *Clinical Cancer Center, University of Wisconsin, Madison, Wisc. 53706, U.S.A.*

Netter, K. J., *Department of Pharmacology, University of Mainz, 6500 Mainz, Germany*

Nichol, Charles A., *Burroughs Wellcome Co., Research Triangle Park, N.C. 27709, U.S.A.*

Nienstedt, Walter, *Department of Physiology, University of Turku, 20520 Turku 52, Finland*

Norling, Anja, *Department of Physiology, University of Kuopio, 70101 Kuopio 10, Finland*

Oesch, Franz, *Section of Biochemical Pharmacology, University of Mainz, 6500 Mainz, Germany*

Ohlsson, Agneta, *Institute of Pharmacognosy-BMC, University of Uppsala, S-751 23 Uppsala, Sweden*

Olsson, Sten, *Department of Toxicology, University of Uppsala, S-751 23 Uppsala, Sweden*

Olsson, Ulf, *Wallenberg Laboratory, S-104 05 Stockholm 50, Sweden*

Orrenius, Sten, *Department of Forensic Medicine, Karolinska Institutet, S-104 01 Stockholm 60, Sweden*

Oskarsson, Agneta, *Department of Toxicology, University of Uppsala, S-751 23 Uppsala, Sweden*

Parkki, Max, *Department of Physiology, University of Turku, 20520 Turku 52, Finland*

Pekkarinen, Aimo, *Department of Pharmacology, University of Turku, 20520 Turku 52, Finland*

Pelkonen, Olavi, *Department of Pharmacology, University of Oulu, SF-90220 Oulu 22, Finland*

Penning, Willem, *Biology Group, C.C.R. Euratom, I-21020 Ispra (Va), Italy*

Peterson, Curt, *Department of Pharmacology, Karolinska Institutet, S-104 01 Stockholm 60, Sweden*

Pilotti, Åke, *Svenska Tobaks AB, 10462 Stockholm 17, Sweden*

Plaa, Gabriel L., *Department of Pharmacology, University of Montreal, Montreal 101, Canada*

Prior, Michael, *Animal Pathology Laboratory, Saskatoon, Sask. S7N 0X1, Canada*

Puhakainen, Eino, *Department of Physiology, University of Kuopio, 70101 Kuopio 10, Finland*

Rannug, Ulf, *Wallenberg Laboratory, S-104 05 Stockholm 50, Sweden*

Recknagel, Richard O., *Department of Physiology, School of Medicine, Case Western Reserve University, Cleveland, Ohio 44106, U.S.A.*

Regan, James D., *Biology Division, Oak Ridge National Laboratory, Oak Ridge, Tenn. 37830, U.S.A.*

Remmer, Herbert, *Institute of Toxicology, University of Tübingen, D-7400 Tübingen, Germany*

Rentsch, Gunter A., *Department of Toxicology, Sandoz Ltd., CH 4002 Basel, Switzerland*

Riihimäki, Vesa, *Institute of Occupational Health, 00290 Helsinki 29, Finland*

Roots, Ivar, *Institut für Klinische Pharmakologie, Der Freie Universität Berlin, D-1000 Berlin 45, Germany*

Rosenthal, Otto, *Hospital of the University of Pennsylvania, Philadelphia, Pa. 19104, U.S.A.*

Rubin, Robert J., *Department of Environmental Medicine, Johns Hopkins University, Baltimore, Md. 21205, U.S.A.*

Salmona, Mario, *Istituto di Ricerche Farmacologiche "Mario Negri," Milan, Italy*

Sandberg, Eva, *Department of Toxicology, University of Uppsala, S-751 23 Uppsala, Sweden*

Sato, Ryo, *Institute for Protein Research, Osaka University, Osaka, Japan*

Saukkonen, Jussi, *Department of Microbiology, Thomas Jefferson University, Philadelphia, Pa. 19107, U.S.A.*

Schenkman, John B., *Department of Pharmacology, Yale University School of Medicine, New Haven, Conn. 06510, U.S.A.*

Schleyer, Heinz, *Hospital of the University of Pennsylvania, Philadelphia, Pa. 19104, U.S.A.*

Schulman, M. P., *Department of Pharmacology, University of Illinois, College of Medicine, Chicago, Ill. 60612 U.S.A.*

Schwartz, Morton A., *Department of Biochemistry and Drug Metabolism, Hoffman–La Roche Inc., Nutley, N.J. 07100, U.S.A.*

Selander, Hans, *Apotekarsocieteten, S-111 81 Stockholm, Sweden*

Sims, Peter, *Chester Beatty Research Institute, London SW3, England*

Smith, J. Crispin, *Department of Pharmacology and Toxicology, School of Medicine, University of Rochester, Rochester, N.Y. 14642, U.S.A.*

Snyder, Robert, *Department of Pharmacology, Thomas Jefferson University, Philadelphia, Pa. 19107, U.S.A.*

Somogyi, Arpad, *The Eppley Institute for Research in Cancer, University of Nebraska Medical Center, Omaha, Nebr. 68105, U.S.A.*

Stohs, S. J., *College of Pharmacy, University of Nebraska Medical Center, Lincoln, Nebr. 68508, U.S.A.*

Strolin Benedetti, Margherita, *Battelle, Geneva Research Centre, 1227-Carouge/Genève, Switzerland*

Sundby, Gun-Britt, *Department of Forensic Medicine, Karolinska Institutet, S-104 01 Stockholm 60, Sweden*

Sundwall, Anders, *Research Department of Pharmacology and Toxicology, AB Kabi, S-104 25 Stockholm, Sweden*

Thor, Hjördis, *Department of Forensic Medicine, Karolinska Institutet, S-104 01 Stockholm 60, Sweden*

Tjälve, Hans, *Department of Toxicology, University of Uppsala, S-751 23 Uppsala, Sweden*

Uehleke, Hartmut, *Abt. Toxikologie, Bundesgesundheitsamt, 1 Berlin 33, Germany*

Ullberg, Sven, *Department of Toxicology, University of Uppsala, S-751 23 Uppsala, Sweden*

Ullrich, V., *Department of Physiological Chemistry, D-665 Homburg-Saar, Germany*

Uotila, Pekka, *Department of Physiology, University of Turku, 20520 Turku 52, Finland*

Vadi, Helena, *Department of Forensic Medicine, Karolinska Institutet, S-104 01 Stockholm 60, Sweden*

Vainio, Harri, *Department of Physiology, University of Turku, 20520 Turku 52, Finland*

Valkama, Eeva-Liisa, *Department of Physiology, University of Kuopio, 70101 Kuopio 10, Finland*

Walle, Thomas, *Department of Pharmacology, Medical University of South Carolina, Charleston, S.C. 29401, U.S.A.*

Williams, R. T., *Department of Biochemistry, St. Mary's Hospital Medical School, London W21PG, England*

Witmer, Charlotte, *Department of Pharmacology, Thomas Jefferson University, Philadelphia, Pa. 19107, U.S.A.*

Wolf, Frank J., *Merck Sharp and Dohme Research Laboratories, Rahway, N.J. 07065, U.S.A.*

Wolff, Thomas, *Abt. Toxikologie, Ges. für Strahlen- und Umweltforschung, D-8042 Neuherberg, Germany*

Wooder, Michael, *Department of Chemical Toxicology, Shell Research Ltd., Sittingbourne, Kent, England*

Yesair, David W., *Arthur D. Little, Inc., Cambridge, Mass. 02140, U.S.A.*

Zaratzian, Virginia L., *National Institute of Mental Health, Rockville, Md. 20852, U.S.A.*

Zimmerman, Hyman J., *Veterans Administration Hospital, Washington, D.C. 20422, U.S.A.*

Index